Cryptography

Theory and Practice

Fourth Edition

Textbooks in Mathematics

Series editors:
Al Boggess and Ken Rosen

MATHEMATICAL MODELING FOR BUSINESS ANALYTICS
William P. Fox

ELEMENTARY LINEAR ALGEBRA
James R. Kirkwood and Bessie H. Kirkwood

APPLIED FUNCTIONAL ANALYSIS, THIRD EDITION
J. Tinsley Oden and Leszek Demkowicz

AN INTRODUCTION TO NUMBER THEORY WITH CRYPTOGRAPHY, SECOND EDITION
James R. Kraft and Lawrence Washington

MATHEMATICAL MODELING: BRANCHING BEYOND CALCULUS
Crista Arangala, Nicolas S. Luke and Karen A. Yokley

ELEMENTARY DIFFERENTIAL EQUATIONS, SECOND EDITION
Charles Roberts

ELEMENTARY INTRODUCTION TO THE LEBESGUE INTEGRAL
Steven G. Krantz

LINEAR METHODS FOR THE LIBERAL ARTS
David Hecker and Stephen Andrilli

CRYPTOGRAPHY: THEORY AND PRACTICE, FOURTH EDITION
Douglas R. Stinson and Maura B. Paterson

DISCRETE MATHEMATICS WITH DUCKS, SECOND EDITION
Sarah-Marie Belcastro

BUSINESS PROCESS MODELING, SIMULATION AND DESIGN, THIRD EDITION
Manual Laguna and Johan Marklund

GRAPH THEORY AND ITS APPLICATIONS, THIRD EDITION
Jonathan L. Gross, Jay Yellen and Mark Anderson

Cryptography

Theory and Practice

Fourth Edition

Douglas R. Stinson
Maura B. Paterson

CRC Press
Taylor & Francis Group
Boca Raton London New York

CRC Press is an imprint of the
Taylor & Francis Group, an **informa** business

CRC Press
Taylor & Francis Group
6000 Broken Sound Parkway NW, Suite 300
Boca Raton, FL 33487-2742

Printed on acid-free paper
Version Date: 20180724

International Standard Book Number-13: 978-1-1381-9701-5 (Hardback)

Library of Congress Cataloging-in-Publication Data

Names: Stinson, Douglas R. (Douglas Robert), 1956- author. | Paterson, Maura B., author.
Title: Cryptography : theory and practice / Douglas R. Stinson and Maura B. Paterson.
Description: Fourth edition. | Boca Raton : CRC Press, Taylor & Francis Group, 2018.
Identifiers: LCCN 2018018724 | ISBN 9781138197015
Subjects: LCSH: Coding theory. | Cryptography.
Classification: LCC QA268 .S75 2018 | DDC 005.8/2--dc23
LC record available at https://lccn.loc.gov/2018018724

Visit the Taylor & Francis Web site at
http://www.taylorandfrancis.com

and the CRC Press Web site at
http://www.crcpress.com

To my children, Michela and Aiden
DRS

To my father, Hamish
MBP

Contents

Preface

The first edition of this book was published in 1995. The objective at that time was to produce a general textbook that treated all the essential core areas of cryptography, as well as a selection of more advanced topics. More recently, a second edition was published in 2002 and the third edition appeared in 2006.

There have been many exciting advances in cryptography since the publication of the first edition of this book 23 years ago. At the same time, many of the "core" areas of cryptography that were important then are still relevant now—providing a strong grounding in the fundamentals remains a primary goal of this book. Many decisions had to be made in terms of which older topics to retain and which new subjects should be incorporated into the book. Our choices were guided by criteria such as the relevance to practical applications of cryptography as well as the influence of new approaches and techniques to the design and analysis of cryptographic protocols. In many cases, this involved studying cutting-edge research and attempting to present it in an accessible manner suitable for presentation in the classroom.

In light of the above, the basic core material of secret-key and public-key cryptography is treated in a similar fashion as in previous editions. However, there are many topics that have been added to this edition, the most important being the following:

- There is a brand new chapter on the exciting, emerging area of post-quantum cryptography, which covers the most important cryptosystems that are designed to provide security against attacks by quantum computers (Chapter 9).
- A new high-level, nontechnical overview of the goals and tools of cryptography has been added (Chapter 1).
- A new mathematical appendix is included, which summarizes definitions and main results on number theory and algebra that are used throughout the book. This provides a quick way to reference any mathematical terms or theorems that a reader might wish to find (Appendix A).
- An expanded treatment of stream ciphers is provided, including common design techniques along with a description of the popular stream cipher known as *Trivium*.
- The book now presents additional interesting attacks on cryptosystems, including:

  - padding oracle attack
  - correlation attacks and algebraic attacks on stream ciphers
  - attack on the *DUAL-EC* random bit generator that makes use of a trapdoor.

- A treatment of the sponge construction for hash functions and its use in the new *SHA-3* hash standard is provided. This is a significant new approach to the design of hash functions.

- Methods of key distribution in sensor networks are described.

- There is a section on the basics of visual cryptography. This allows a secure method to split a secret visual message into pieces (shares) that can later be combined to reconstruct the secret.

- The fundamental techniques of cryptocurrencies, as used in BITCOIN and blockchain, are described.

- We explain the basics of the new cryptographic methods employed in messaging protocols such as *Signal*. This includes topics such as deniability and Diffie-Hellman key ratcheting.

We hope that this book can be used in a variety of courses. An introductory undergraduate level course could be based on a selection of material from the first eight chapters. We should point out that, in several chapters, the later sections can be considered to be more advanced than earlier sections. These sections could provide material for graduate courses or for self-study. Material in later chapters can also be included in an introductory or follow-up course, depending on the interests of the instructor.

Cryptography is a broad subject, and it requires knowledge of several areas of mathematics, including number theory, groups, rings and fields, linear algebra, probability and information theory. As well, some familiarity with computational complexity, algorithms, and NP-completeness theory is useful. In our opinion, it is the breadth of mathematical background required that often creates difficulty for students studying cryptography for the first time. With this in mind, we have maintained the mathematical presentation from previous editions. One basic guiding principle is that understanding relevant mathematics is essential to the comprehension of the various cryptographic schemes and topics. At the same time, we try to avoid unnecessarily advanced mathematical techniques—we provide the essentials, but we do not overload the reader with superfluous mathematical concepts.

The following features are common to all editions of this book:

- Mathematical background is provided where it is needed, in a "just-in-time" fashion.

- Informal descriptions of the cryptosystems are given along with more precise pseudo-code descriptions.

- Numerical examples are presented to illustrate the workings of most of the algorithms described in the book.
- The mathematical underpinnings of the algorithms and cryptosystems are explained carefully and rigorously.
- Numerous exercises are included, some of them quite challenging.

We have received useful feedback from various people on the content of this book as we prepared this new edition. In particular, we would like to thank Colleen Swanson for many helpful comments and suggestions. Several anonymous reviewers provided useful suggestions, and we also appreciate comments from Steven Galbraith and Jalaj Upadhyay. Finally, we thank Roberto De Prisco, who prepared the examples of shares in a visual threshold scheme that are included in Chapter 11.

Douglas R. Stinson
Maura B. Paterson

Chapter 1

Introduction to Cryptography

In this chapter, we present a brief overview of the kinds of problems studied in cryptography and the techniques used to solve them. These problems and the cryptographic tools that are employed in their solution are discussed in more detail and rigor in the rest of this book. This introduction may serve to provide an informal, non-technical, non-mathematical summary of the topics to be addressed. As such, it can be considered to be optional reading.

1.1 Cryptosystems and Basic Cryptographic Tools

In this section, we discuss basic notions relating to encryption. This includes secret-key and public-key cryptography, block and stream ciphers, and hybrid cryptography.

1.1.1 Secret-key Cryptosystems

Cryptography has been used for thousands of years to help to provide confidential communications between mutually trusted parties. In its most basic form, two people, often denoted as ***Alice*** and ***Bob***, have agreed on a particular ***secret key***. At some later time, Alice may wish to send a secret message to Bob (or Bob might want to send a message to Alice). The key is used to transform the original message (which is usually termed the ***plaintext***) into a scrambled form that is unintelligible to anyone who does not possess the key. This process is called ***encryption*** and the scrambled message is called the ***ciphertext***. When Bob receives the ciphertext, he can use the key to transform the ciphertext back into the original plaintext; this is the ***decryption*** process. A ***cryptosystem*** constitutes a complete specification of the keys and how they are used to encrypt and decrypt information.

Various types of cryptosystems of increasing sophistication have been used for many purposes throughout history. Important applications have included sensitive communications between political leaders and/or royalty, military maneuvers, etc. However, with the development of the internet and applications such as electronic commerce, many new diverse applications have emerged. These include scenarios such as encryption of passwords, credit card numbers, email, documents, files, and digital media.

It should also be mentioned that cryptographic techniques are also widely used to protect stored data in addition to data that is transmitted from one party to another. For example, users may wish to encrypt data stored on laptops, on external hard disks, in the cloud, in databases, etc. Additionally, it might be useful to be able to perform computations on encrypted data (without first decrypting the data).

The development and deployment of a cryptosystem must address the issue of security. Traditionally, the threat that cryptography addressed was that of an eavesdropping adversary who might intercept the ciphertext and attempt to decrypt it. If the adversary happens to possess the key, then there is nothing that can be done. Thus the main security consideration involves an adversary who does not possess the key, who is still trying to decrypt the ciphertext. The techniques used by the adversary to attempt to "break" the cryptosystem are termed ***cryptanalysis***. The most obvious type of cryptanalysis is to try to guess the key. An attack wherein the adversary tries to decrypt the ciphertext with every possible key in turn is termed an ***exhaustive key search***. When the adversary tries the correct key, the plaintext will be found, but when any other key is used, the "decrypted" ciphertext will likely be random gibberish. So an obvious first step in designing a secure cryptosystem is to specify a very large number of possible keys, so many that the adversary will not be able to test them all in any reasonable amount of time.

The model of cryptography described above is usually called ***secret-key cryptography***. This indicates that there is one secret key, which is known to both Alice and Bob. That is, the key is a "secret" that is known to two parties. This key is employed both to encrypt plaintexts and to decrypt ciphertexts. The actual encryption and decryption functions are thus inverses of each other. Some basic secret-key cryptosystems are introduced and analyzed with respect to different security notions in Chapters 2 and 3.

The drawback of secret-key cryptography is that Alice and Bob must somehow be able to agree on the secret key ahead of time (before they want to send any messages to each other). This might be straightforward if Alice and Bob are in the same place when they choose their secret key. But what if Alice and Bob are far apart, say on different continents? One possible solution is for Alice and Bob to use a public-key cryptosystem.

1.1.2 Public-key Cryptosystems

The revolutionary idea of ***public-key cryptography*** was introduced in the 1970s by Diffie and Hellman. Their idea was that it might be possible to devise a cryptosystem in which there are two distinct keys. A ***public key*** would be used to encrypt the plaintext and a ***private key*** would enable the ciphertext to be decrypted. Note that a public key can be known to "everyone," whereas a private key is known to only one person (namely, the recipient of the encrypted message). So a public-key cryptosystem would enable anyone to encrypt a message to be transmitted to Bob, say, and only Bob could decrypt the message. The first and best-known example of a public-key cryptosystem is the *RSA Cyptosystem* that was invented by Rivest, Shamir and Adleman. Various types of public-key cryptosystems are presented in Chapters 6, 7, and 9.

Public-key cryptography obviates the need for two parties to agree on a prior shared secret key. However, it is still necessary to devise a method to distribute public keys securely. But this is not necessarily a trivial goal to accomplish, the main issue being the correctness or authenticity of purported public keys. Certificates, which we will discuss a bit later, are one common method to deal with this problem.

1.1.3 Block and Stream Ciphers

Cryptosystems are usually categorized as ***block ciphers*** or ***stream ciphers***. In a block cipher, the plaintext is divided into fixed-sized chunks called ***blocks***. A block is specified to be a bitstring (i.e., a string of 0's and 1's) of some fixed length (e.g., 64 or 128 bits). A block cipher will encrypt (or decrypt) one block at a time. In contrast, a stream cipher first uses the key to construct a ***keystream***, which is a bitstring that has exactly the same length as the plaintext (the plaintext is a bitstring of arbitrary length). The encryption operation constructs the ciphertext as the exclusive-or of the plaintext and the keystream. Decryption is accomplished by computing the exclusive-or of the ciphertext and the keystream. Public-key cryptosystems are invariably block ciphers, while secret-key cryptosystems can be block ciphers or stream ciphers. Block ciphers are studied in detail in Chapter 4.

1.1.4 Hybrid Cryptography

One of the drawbacks of public-key cryptosystems is that they are much slower than secret-key cryptosystems. As a consequence, public-key cryptosystems are mainly used to encrypt small amounts of data, e.g., a credit card number. However, there is a nice way to combine secret- and public-key cryptography to achieve the benefits of both. This technique is called ***hybrid cryptography***. Suppose that Alice wants to encrypt a "long" message and send it to Bob. Assume that Alice and Bob do not have a prior shared secret key. Alice can choose a random secret key and encrypt the plaintext, using a (fast) secret-key cryptosystem. Alice then encrypts this secret key using Bob's public key. Alice sends the ciphertext and the encrypted key to Bob. Bob first uses his private decryption key to decrypt the secret key, and then he uses this secret key to decrypt the ciphertext.

Notice that the "slow" public-key cryptosystem is only used to encrypt a short secret key. The much faster secret-key cryptosystem is used to encrypt the longer plaintext. Thus, hybrid cryptography (almost) achieves the efficiency of secret-key cryptography, but it can be used in a situation where Alice and Bob do not have a previously determined secret key.

1.2 Message Integrity

This section discusses various tools that help to achieve integrity of data, including message authentication codes (MACs), signature schemes, and hash functions.

Cryptosystems provide ***secrecy*** (equivalently, ***confidentiality***) against an eavesdropping adversary, which is often called a ***passive adversary***. A passive adversary is assumed to be able to access whatever information is being sent from Alice to Bob; see Figure 1.1. However, there are many other threats that we might want to protect against, particularly when an ***active adversary*** is present. An active adversary is one who can alter information that is transmitted from Alice to Bob.

Figure 1.2 depicts some of the possible actions of an active adversary. An active adversary might

- alter the information that is sent from Alice to Bob,
- send information to Bob in such a way that Bob thinks the information originated from Alice, or
- divert information sent from Alice to Bob in such a way that a third party (Charlie) receives this information instead of Bob.

Possible objectives of an active adversary could include fooling Bob (say) into accepting "bogus" information, or misleading Bob as to who sent the information to him in the first place.

We should note that encryption, by itself, cannot protect against these kinds of active attacks. For example, a stream cipher is susceptible to a ***bit-flipping attack***. If some ciphertext bits are "flipped" (i.e., 0's are replaced by 1's and vice versa), then the effect is to flip the corresponding plaintext bits. Thus, an adversary can modify the plaintext in a predictable way, even though the adversary does not know what the plaintext bits are.

There are various types of "integrity" guarantees that we might seek to provide, in order to protect against the possible actions of an active adversary. Such an adversary might change the information that is being transmitted from Alice to Bob (and note that this information may or may not be encrypted). Alternatively, the adversary might try to "forge" a message and send it to Bob, hoping that he will think that it originated from Alice. Cryptographic tools that protect against these and related types of threats can be constructed in both the secret-key and public-key settings. In the secret-key setting, we will briefly discuss the notion of a ***message authentication code*** (or ***MAC***). In the public-key setting, the tool that serves a roughly similar purpose is a ***signature scheme***.

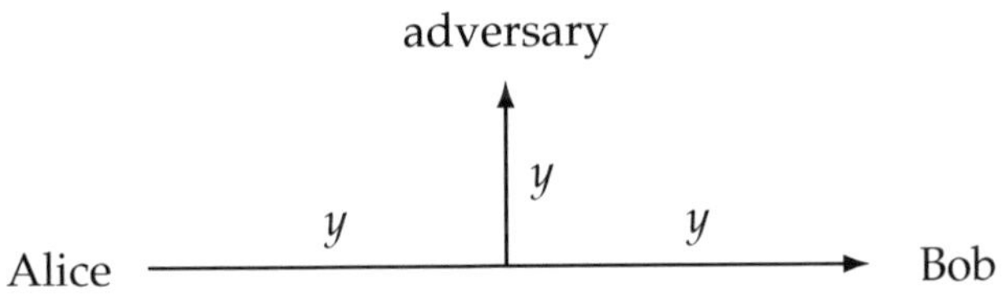

FIGURE 1.1: A passive adversary

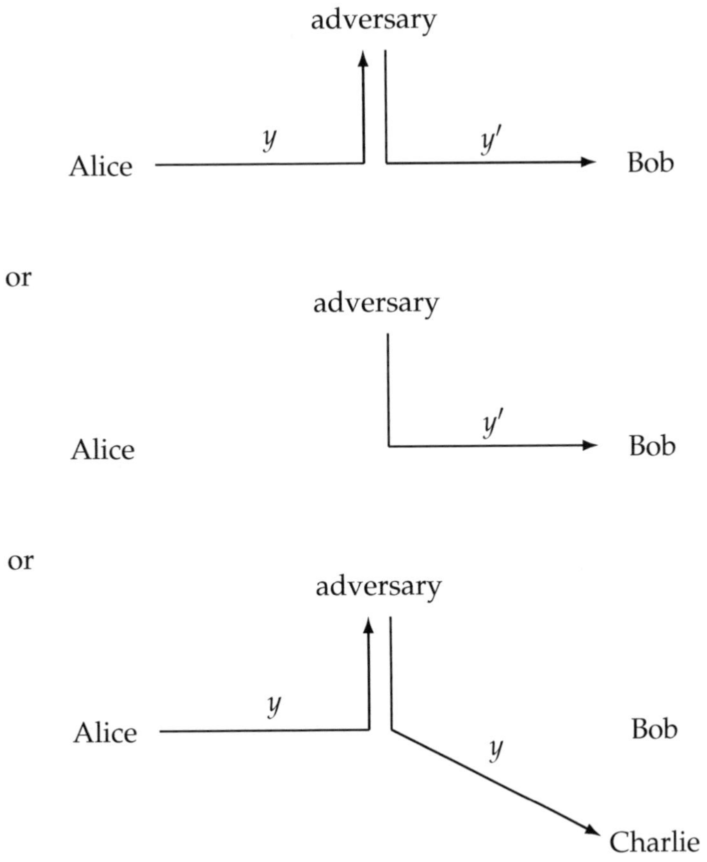

FIGURE 1.2: Active adversaries

1.2.1 Message Authentication Codes

A message authentication code requires Alice and Bob to share a secret key. When Alice wants to send a message to Bob, she uses the secret key to create a ***tag*** that she appends to the message (the tag depends on both the key and the message). When Bob receives the message and tag, he uses the key to re-compute the tag and checks to see if it is the same as the tag that he received. If so, Bob accepts the message as an authentic message from Alice; if not, then Bob rejects the message as being invalid. We note that the message may or may not be encrypted. MACs are discussed in Chapter 5.

If there is no need for confidentiality, then the message can be sent as plaintext. However, if confidentiality is desired, then the plaintext would be encrypted, and then the tag would be computed on the ciphertext. Bob would first verify the correctness of the tag. If the tag is correct, Bob would then decrypt the ciphertext. This process is often called ***encrypt-then-MAC*** (see Section 5.5.3 for a more detailed discussion of this topic).

For a MAC to be considered secure, it should be infeasible for the adversary to compute a correct tag for any message for which they have not already seen a valid tag. Suppose we assume that a secure MAC is being employed by Alice and Bob (and suppose that the adversary does not know the secret key that they are using). Then, if Bob receives a message and a valid tag, he can be confident that Alice created the tag on the given message (provided that Bob did not create it himself) and that neither the message nor the tag was altered by an adversary. A similar conclusion can be reached by Bob when he receives a message from Alice, along with a correct tag.

1.2.2 Signature Schemes

In the public-key setting, a signature scheme provides assurance similar to that provided by a MAC. In a signature scheme, the private key specifies a ***signing algorithm*** that Alice can use to sign messages. Similar to a MAC, the signing algorithm produces an output, which in this case is called a ***signature***, that depends on the message being signed as well as the key. The signature is then appended to the message. Notice that the signing algorithm is known only to Alice. On the other hand, there is a ***verification algorithm*** that is a public key (known to everyone). The verification algorithm takes as input a message and a signature, and outputs *true* or *false* to indicate whether the signature should be accepted as valid. One nice feature of a signature scheme is that anyone can verify Alice's signatures on messages, provided that they have an authentic copy of Alice's verification key. In contrast, in the MAC setting, only Bob can verify tags created by Alice (when Alice and Bob share a secret key). Signature schemes are studied in Chapter 8.

Security requirements for signature schemes are similar to MACs. It should be infeasible for an adversary to create a valid signature on any message not previously signed by Alice. Therefore, if Bob (or anyone else) receives a message and a valid tag (i.e., one that can be verified using Alice's public verification algorithm), then the recipient can be confident that the signature was created by Alice and neither the message nor the signature was modified by an adversary.

One common application of signatures is to facilitate secure software updates. When a user purchases software from an online website, it typically includes a verification algorithm for a signature scheme. Later, when an updated version of the software is downloaded, it includes a signature (on the updated software). This signature can be verified using the verification algorithm that was downloaded when the original version of the software was purchased. This enables the user's computer to verify that the update comes from the same source as the original version of the software.

Signature schemes can be combined with public-key encryption schemes to provide confidentiality along with the integrity guarantees of a signature scheme. Assume that Alice wants to send a signed, encrypted (short) message to Bob. In this situation, the most commonly used technique is for Alice to first create a signature on the plaintext using her private signing algorithm, and then encrypt the plaintext and signature using Bob's public encryption key. When Bob receives the message, he first decrypts it, and then he checks the validity of the signature. This process is called ***sign-then-encrypt***; note that this is in some sense the reverse of the "encrypt-then-MAC" procedure that is used in the secret-key setting.

1.2.3 Nonrepudiation

There is one somewhat subtle difference between MACs and signature schemes. In a signature scheme, the verification algorithm is public. This means that the signature can be verified by anyone. So, if Bob receives a message from Alice containing her valid signature on the message, he can show the message and the signature to anyone else and be confident that the third party will also accept the signature as being valid. Consequently, Alice cannot sign a message and later try to claim that she did not sign the message, a property that is termed ***nonrepudiation***. This is useful in the setting of contracts, where we do not want someone to be able to renege on a signed contract by claiming (falsely) that their signature has been "forged," for example.

However, for a MAC, there is no third-party verifiability because the secret key is required to verify the correctness of the tag, and the key is known only to Alice and Bob. Even if the secret key is revealed to a third party (e.g., as a result of a court order), there is no way to determine if the tag was created by Alice or by Bob, because anything Bob can do, Alice can do as well, and vice versa. So a MAC does not provide nonrepudiation, and for this reason, a MAC is sometimes termed "deniable." It is interesting to note, however, that there are situations where deniability is desirable. This could be the case in real-time communications, where Alice and Bob want to be assured of the authenticity of their communications as they take place, but they do not want a permanent, verifiable record of this communication to exist. Such communication is analogous to an "off-the-record" conversation, e.g., between a journalist and an anonymous source. A MAC is useful in the context of conversations of this type, especially if care is taken, after the conversation is over, to delete the secret keys that are used during the communication.

1.2.4 Certificates

We mentioned that verifying the authenticity of public keys, before they are used, is important. A certificate is a common tool to help achieve this objective. A ***certificate*** will contain information about a particular user or, more commonly, a website, including the website's public keys. These public keys will be signed by a trusted authority. It is assumed that everyone has possession of the trusted authority's public verification key, so anyone can verify the trusted authority's signature on a certificate. See Section 8.6 for more information about certificates.

This technique is used on the internet in *Transport Layer Security* (which is commonly called *TLS*). When a user connects to a secure website, say one belonging to a business engaged in electronic commerce, the website of the company will send a certificate to the user so the user can verify the authenticity of the website's public keys. These public keys will subsequently be used to set up a secure channel, between the user and the website, in which all information is encrypted. Note that the public key of the trusted authority, which is used to verify the public key of the website, is typically hard-coded into the web browser.

1.2.5 Hash Functions

Signature schemes tend to be much less efficient than MACs. So it is not advisable to use a signature scheme to sign "long" messages. (Actually, most signature schemes are designed to only sign messages of a short, fixed length.) In practice, messages are "hashed" before they are signed. A ***cryptographic hash function*** is used to compress a message of arbitrary length to a short, random-looking, fixed-length ***message digest***. Note that a hash function is a public function that is assumed to be known to everyone. Further, a hash function has no key. Hash functions are discussed in Chapter 5.

After Alice hashes the message, she signs the message digest, using her private signing algorithm. The original message, along with the signature on the message, is then transmitted to Bob, say. This process is called ***hash-then-sign***. To verify the signature, Bob will compute the message digest by hashing the message. Then he will use the public verification algorithm to check the validity of the signature on the message digest. When a signature is used along with public-key encryption, the process would actually be ***hash-then-sign-then-encrypt***. That is, the message is hashed, the message digest is then signed, and finally, the message and signature are encrypted.

A cryptographic hash function is very different from a hash function that is used to construct a hash table, for instance. In the context of hash tables, a hash function is generally required only to yield collisions[1] with a sufficiently small probability. On the other hand, if a cryptographic hash function is used, it should

[1] A ***collision*** for a function h occurs when $h(x) = h(y)$ for some $x \neq y$.

be computationally infeasible to find collisions, even though they must exist. Cryptographic hash functions are usually required to satisfy additional security properties, as discussed in Section 5.2.

Cryptographic hash functions also have other uses, such as for ***key derivation***. When used for key derivation, a hash function would be applied to a long random string in order to create a short random key.

Finally, it should be emphasized that hash functions cannot be used for encryption, for two fundamental reasons. First is the fact that hash functions do not have a key. The second is that hash functions cannot be inverted (they are not injective functions) so a message digest cannot be "decrypted" to yield a unique plaintext value.

1.3 Cryptographic Protocols

Cryptographic tools such as cryptosystems, signature schemes, hash functions, etc., can be used on their own to achieve specific security objectives. However, these tools are also used as components in more complicated protocols. (Of course, protocols can also be designed "from scratch," without making use of prior primitives.)

In general, a ***protocol*** (or ***interactive protocol***) refers to a specified sequence of messages exchanged between two (or possibly more) parties. A ***session*** of a protocol between Alice and Bob, say, will consist of one or more ***flows***, where each flow consists of a message sent from Alice to Bob or vice versa. At the end of the session, the parties involved may have established some common shared information, or confirmed possession of some previously shared information.

One important type protocol is an ***identification scheme***, in which one party "proves" their identity to another by demonstrating possession of a password, for example. More sophisticated identification protocols will instead consist of two (or more) flows, for example a challenge followed by a response, where the response is computed from the challenge using a certain secret or private key. Identification schemes are the topic of Chapter 10.

There are many kinds of protocols associated with various aspects of choosing keys or communicating keys from one party to another. In a ***key distribution scheme***, keys might be chosen by a trusted authority and communicated to one or more members of a certain network. Another approach, which does not require the participation of an active trusted authority, is called ***key agreement***. In a key agreement scheme, Alice and Bob (say) are able to end up with a common shared secret key, which should not become known to an adversary. These and related topics are discussed in Chapters 11 and 12.

A ***secret sharing scheme*** involves a trusted authority distributing "pieces" of information (called "shares") in such a way that certain subsets of shares can be suitably combined to reconstruct a certain predefined secret. One common type of secret sharing scheme is a ***threshold scheme***. In a (k, n)-threshold scheme, there are n shares, and any k shares permit the reconstruction of the secret. On the other hand, $k - 1$ or fewer shares provide no information about the value of the secret. Secret sharing schemes are studied in Chapter 11.

1.4 Security

A fundamental goal for a cryptosystem, signature scheme, etc., is for it to be "secure." But what does it mean to be secure and how can we gain confidence that something is indeed secure? Roughly speaking, we would want to say that an adversary cannot succeed in "breaking" a cryptosystem, for example, but we have to make this notion precise. Security in cryptography involves consideration of three different aspects: an ***attack model***, an ***adversarial goal***, and a ***security level***. We will discuss each of these in turn.

The attack model specifies the information that is available to the adversary. We will always assume that the adversary knows the scheme or protocol being used (this is called ***Kerckhoffs' Principle***). The adversary is also assumed to know the public key (if the system is a public-key system). On the other hand, the adversary is assumed not to know any secret or private keys being used. Possible additional information provided to the adversary should be specified in the attack model.

The adversarial goal specifies exactly what it means to "break" the cryptosystem. What is the adversary attempting to do and what information are they trying to determine? Thus, the adversarial goal defines a "successful attack."

The security level attempts to quantify the effort required to break the cryptosystem. Equivalently, what computational resources does the adversary have access to and how much time would it take to carry out an attack using those resources?

A statement of security for a cryptographic scheme will assert that a particular adversarial goal cannot be achieved in a specified attack model, given specified computational resources.

We now illustrate some of the above concepts in relation to a cryptosystem. There are four commonly considered attack models. In a ***known ciphertext attack***, the adversary has access to some amount of ciphertext that is all encrypted with the same unknown key. In a ***known plaintext attack***, the adversary gains access to some plaintext as well as the corresponding ciphertext (all of which is encrypted with the same key). In a ***chosen plaintext attack***, the adversary is allowed to choose plaintext, and then they are given the corresponding ciphertext. Finally, in a ***chosen ciphertext attack***, the adversary chooses some ciphertext and they are then given the corresponding plaintext.

Clearly a chosen plaintext or chosen ciphertext attack provides the adversary with more information than a known ciphertext attack. So they would be considered to be stronger attack models than a known ciphertext attack, since they potentially make the adversary's job easier.

The next aspect to study is the adversarial goal. In a ***complete break*** of a cryptosystem, the adversary determines the private (or secret) key. However, there are other, weaker goals that the adversary could potentially achieve, even if a complete break is not possible. For example, the adversary might be able to decrypt a previously unseen ciphertext with some specified non-zero probability, even though they have not been able to determine the key. Or, the adversary might be able to determine some partial information about the plaintext, given a previously unseen ciphertext, with some specified non-zero probability. "Partial information" could include the values of certain plaintext bits. Finally, as an example of a weak goal, the adversary might be able distinguish between encryptions of two given plaintexts.[2]

Other cryptographic primitives will have different attack models and adversarial goals. In a signature scheme, the attack model would specify what kind of (valid) signatures the adversary has access to. Perhaps the adversary just sees some previously signed messages, or maybe the adversary can request the signer to sign some specific messages of the adversary's choosing. The adversarial goal is typically to sign some "new" message (i.e., one for which the adversary does not already know a valid signature). Perhaps the adversary can find a valid signature for some specific message that the adversary chooses, or perhaps they can find a valid signature for any message. These would represent weak and strong adversarial goals, respectively.

Three levels of security are often studied, which are known as ***computational security***, ***provable security***, and ***unconditional security***.

Computational security means that a specific algorithm to break the system is computationally infeasible, i.e., it cannot be accomplished in a reasonable amount of time using currently available computational resources. Of course, a system that is computationally secure today may not be computationally secure indefinitely. For example, new algorithms might be discovered, computers may get faster, or fundamental new computing paradigms such as quantum computing might become practical. Quantum computing, if it becomes practical, could have an enormous impact on the security of many kinds of public-key cryptography; this is addressed in more detail in Section 9.1.

It is in fact very difficult to predict how long something that is considered secure today will remain secure. There are many examples where many cryptographic schemes have not survived as long as originally expected due to the reasons mentioned above. This has led to rather frequent occurrences of replacing standards with improved standards. For example, in the case of hash functions, there have been a succession of proposed and/or approved standards, denoted as *SHA-0*, *SHA-1*, *SHA-2* and *SHA-3*, as new attacks have been found and old standards have become insecure.

---

[2]Whether or not this kind of limited information can be exploited by the adversary in a malicious way is another question, of course.

An interesting example relating to broken predictions is provided by the public-key *RSA Cryptosystem*. In the August 1977 issue of *Scientific American*, the eminent mathematical expositor Martin Gardner wrote a column on the newly developed *RSA* public-key cryptosystem entitled "A new kind of cipher that would take millions of years to break." Included in the article was a challenge ciphertext, encrypted using a 512-bit key. However, the challenge was solved 17 years later, on April 26, 1994, by factoring the given public key (the plaintext was "the magic words are squeamish ossifrage"). The statement that the cipher would take millions of years to break probably referred to how long it would take to run the best factoring algorithm known in 1977 on the fastest computer available in 1977. However, between 1977 and 1994, there were several developments, including the following:

- computers became much faster,
- improved factoring algorithms were found, and
- the development of the internet facilitated large-scale distributed computations.

Of course, it is basically impossible to predict when new algorithms will be discovered. Also, the third item listed above can be regarded as a "paradigm shift" that was probably not on anyone's radar in 1977.

The next "level" of security we address is provable security (also known as ***reductionist security***), which refers to a situation where breaking the cryptosystem (i.e., achieving the adversarial goal) can be reduced in a complexity-theoretic sense to solving some underlying (assumed difficult) mathematical problem. This would show that breaking the cryptosystem is at least as difficult as solving the given hard problem. Provable security often involves reductions to the factoring problem or the discrete logarithm problem (these problems are studied in Sections 6.6 and 7.2, respectively).

Finally, unconditional security means that the cryptosystem cannot be broken (i.e., the adversarial goal is not achievable), even with unlimited computational resources, because there is not enough information available to the adversary (as specified in the attack model) for them to be able to do this. The most famous example of an unconditionally secure cryptosystem is the *One-time Pad*. In this cryptosystem, the key is a random bitstring having the same length as the plaintext. The ciphertext is formed as the exclusive-or of the plaintext and the key. For the *One-time Pad*, it can be proven mathematically that the adversary can obtain no partial information whatsoever about the plaintext (other than its length), given the ciphertext, provided the key is used to encrypt only one string of plaintext and the key has the same length as the plaintext. The *One-time Pad* is discussed in Chapter 3.

When we analyze a cryptographic scheme, our goal would be to show that the adversary cannot achieve a *weak* adversarial goal in a *strong* attack model, given *significant* computational resources.

The preceding discussion of security has dealt mostly with the situation of a cryptographic primitive such as a cryptosystem. However, cryptographic primitives are generally combined in complicated ways when protocols are defined and ultimately implemented. Even seemingly simple implementation decisions can lead to unexpected vulnerabilities. For example, when data is encrypted using a block cipher, it first needs to be split into fixed length chunks, e.g., 128-bit blocks. If the data does not exactly fill up an integral number of blocks, then some padding has to be introduced. It turns out that a standard padding technique, when used with the common CBC mode of operation, is susceptible to an attack known as a ***padding oracle attack***, which was discovered by Vaudenay in 2002 (see Section 4.7.1 for a description of this attack).

There are also various kinds of attacks against physical implementations of cryptography that are known as ***side channel attacks***. Examples of these include ***timing attacks***, ***fault analysis attacks***, ***power analysis attacks***, and ***cache attacks***. The idea is that information about a secret or private key might be leaked by observing or physically manipulating a device (such as a smart card) on which a particular cryptographic scheme is implemented. One example would be observing the time taken by the device to perform certain computations (a so-called "timing attack"). This leakage of information can take place even though the scheme is "secure."

1.5 Notes and References

There are many monographs and textbooks on the subject of cryptography. We will mention here a few general treatments that may be useful to readers.

For an accessible, non-mathematical treatment, we recommend

- *Everyday Cryptography: Fundamental Principles and Applications, Second Edition* by Keith Martin [127].

For a more mathematical point of view, the following recent texts are helpful:

- *An Introduction to Mathematical Cryptography* by J. Hoffstein, J. Pipher, and J. Silverman [96]
- *Introduction to Modern Cryptography, Second Edition* by J. Katz and Y. Lindell [104]
- *Understanding Cryptography: A Textbook for Students and Practitioners* by C. Paar and J. Pelzl [157]
- *Cryptography Made Simple* by Nigel Smart [185]
- *A Classical Introduction to Cryptography: Applications for Communications Security* by Serge Vaudenay [196].

For mathematical background, especially for public-key cryptography, we recommend

- *Mathematics of Public Key Cryptography* by Stephen Galbraith [84].

Finally, the following is a valuable reference, even though it is quite out of date:

- *Handbook of Applied Cryptography* by A.J. Menezes, P.C. Van Oorschot, and S.A. Vanstone [134].

Chapter 2

Classical Cryptography

In this chapter, we provide a gentle introduction to cryptography and cryptanalysis. We present several simple systems, and describe how they can be "broken." Along the way, we discuss various mathematical techniques that will be used throughout the book.

2.1 Introduction: Some Simple Cryptosystems

The fundamental objective of cryptography is to enable two people, usually referred to as ***Alice*** and ***Bob***, to communicate over an insecure ***channel*** in such a way that an opponent, ***Oscar***, cannot understand what is being said. This channel could be a telephone line or computer network, for example. The information that Alice wants to send to Bob, which we call "plaintext," can be English text, numerical data, or anything at all—its structure is completely arbitrary. Alice encrypts the plaintext, using a predetermined key, and sends the resulting ciphertext over the channel. Oscar, upon seeing the ciphertext in the channel by eavesdropping, cannot determine what the plaintext was; but Bob, who knows the encryption key, can decrypt the ciphertext and reconstruct the plaintext.

These ideas are described formally using the following mathematical notation.

Definition 2.1: A ***cryptosystem*** is a five-tuple $(\mathcal{P}, \mathcal{C}, \mathcal{K}, \mathcal{E}, \mathcal{D})$, where the following conditions are satisfied:

1. $\mathcal{P}$ is a finite set of possible ***plaintexts***;
2. $\mathcal{C}$ is a finite set of possible ***ciphertexts***;
3. $\mathcal{K}$, the ***keyspace***, is a finite set of possible ***keys***;
4. For each $K \in \mathcal{K}$, there is an ***encryption rule*** $e_K \in \mathcal{E}$ and a corresponding ***decryption rule*** $d_K \in \mathcal{D}$. Each $e_K : \mathcal{P} \rightarrow \mathcal{C}$ and $d_K : \mathcal{C} \rightarrow \mathcal{P}$ are functions such that $d_K(e_K(x)) = x$ for every plaintext element $x \in \mathcal{P}$.

The main property is property 4. It says that if a plaintext x is encrypted using e_K, and the resulting ciphertext is subsequently decrypted using d_K, then the original plaintext x results.

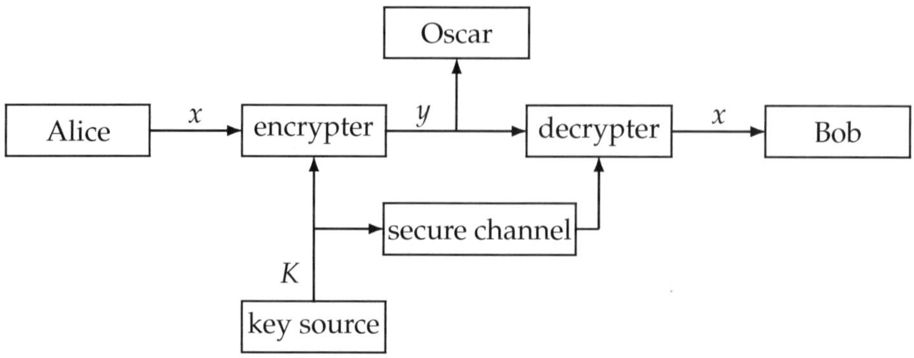

FIGURE 2.1: The communication channel

Alice and Bob will employ the following protocol to use a specific cryptosystem. First, they choose a random key $K \in \mathcal{K}$. This is done when they are in the same place and are not being observed by Oscar, or, alternatively, when they do have access to a secure channel, in which case they can be in different places. At a later time, suppose Alice wants to communicate a message to Bob over an insecure channel. We suppose that this message is a ***string***

$$\mathbf{x} = x_1 x_2 \cdots x_n$$

for some integer $n \geq 1$, where each plaintext symbol $x_i \in \mathcal{P}$, $1 \leq i \leq n$. Each x_i is encrypted using the encryption rule e_K specified by the predetermined key K. Hence, Alice computes $y_i = e_K(x_i)$, $1 \leq i \leq n$, and the resulting ciphertext string

$$\mathbf{y} = y_1 y_2 \cdots y_n$$

is sent over the channel. When Bob receives $y_1 y_2 \cdots y_n$, he decrypts it using the decryption function d_K, obtaining the original plaintext string, $x_1 x_2 \cdots x_n$. See Figure 2.1 for an illustration of the communication channel.

Clearly, it must be the case that each encryption function e_K is an ***injective function*** (i.e., one-to-one); otherwise, decryption could not be accomplished in an unambiguous manner. For example, if

$$y = e_K(x_1) = e_K(x_2)$$

where $x_1 \neq x_2$, then Bob has no way of knowing whether y should decrypt to x_1 or x_2. Note that if $\mathcal{P} = \mathcal{C}$, it follows that each encryption function is a permutation. That is, if the set of plaintexts and ciphertexts are identical, then each encryption function just rearranges (or permutes) the elements of this set.

2.1.1 The Shift Cipher

In this section, we will describe the *Shift Cipher*, which is based on modular arithmetic. But first we review some basic definitions of ***modular arithmetic***.

> **Definition 2.2:** Suppose a and b are integers, and m is a positive integer. Then we write $a \equiv b \pmod{m}$ if m divides $b - a$. The phrase $a \equiv b \pmod{m}$ is called a ***congruence***, and it is read as "a is ***congruent*** to b modulo m." The integer m is called the ***modulus***.

Suppose we divide a and b by m, obtaining integer quotients and remainders, where the remainders are between 0 and $m-1$. That is, $a = q_1m + r_1$ and $b = q_2m + r_2$, where $0 \leq r_1 \leq m-1$ and $0 \leq r_2 \leq m-1$. Then it is not difficult to see that $a \equiv b \pmod{m}$ if and only if $r_1 = r_2$. We will use the notation $a \bmod m$ (without parentheses) to denote the remainder when a is divided by m, i.e., the value r_1 above. Thus $a \equiv b \pmod{m}$ if and only if $a \bmod m = b \bmod m$. If we replace a by $a \bmod m$, we say that a is ***reduced modulo*** m.

We give a couple of examples. To compute $101 \bmod 7$, we write $101 = 7 \times 14 + 3$. Since $0 \leq 3 \leq 6$, it follows that $101 \bmod 7 = 3$. As another example, suppose we want to compute $(-101) \bmod 7$. In this case, we write $-101 = 7 \times (-15) + 4$. Since $0 \leq 4 \leq 6$, it follows that $(-101) \bmod 7 = 4$.

REMARK Many computer programming languages define $a \bmod m$ to be the remainder in the range $-m+1, \ldots, m-1$ having the same sign as a. For example, $(-101) \bmod 7$ would be -3, rather than 4 as we defined it above. But for our purposes, it is much more convenient to define $a \bmod m$ always to be non-negative. ∎

We now define arithmetic modulo m: $\mathbb{Z}_m$ is the set $\{0, \ldots, m-1\}$, equipped with two operations, $+$ and $\times$. Addition and multiplication in $\mathbb{Z}_m$ work exactly like real addition and multiplication, except that the results are reduced modulo m.

For example, suppose we want to compute 11×13 in $\mathbb{Z}_{16}$. As integers, we have $11 \times 13 = 143$. Then we reduce 143 modulo 16 as described above: $143 = 8 \times 16 + 15$, so $143 \bmod 16 = 15$, and hence $11 \times 13 = 15$ in $\mathbb{Z}_{16}$.

These definitions of addition and multiplication in $\mathbb{Z}_m$ satisfy most of the familiar rules of arithmetic. We will list these properties now, without proof:

1. addition is ***closed***, i.e., for any $a, b \in \mathbb{Z}_m$, $a + b \in \mathbb{Z}_m$
2. addition is ***commutative***, i.e., for any $a, b \in \mathbb{Z}_m$, $a + b = b + a$
3. addition is ***associative***, i.e., for any $a, b, c \in \mathbb{Z}_m$, $(a + b) + c = a + (b + c)$
4. 0 is an ***additive identity***, i.e., for any $a \in \mathbb{Z}_m$, $a + 0 = 0 + a = a$
5. the ***additive inverse*** of any $a \in \mathbb{Z}_m$ is $m - a$, i.e., $a + (m - a) = (m - a) + a = 0$ for any $a \in \mathbb{Z}_m$

Cryptosystem 2.1: *Shift Cipher*

Let $\mathcal{P} = \mathcal{C} = \mathcal{K} = \mathbb{Z}_{26}$. For $0 \leq K \leq 25$, define

$$e_K(x) = (x + K) \bmod 26$$

and

$$d_K(y) = (y - K) \bmod 26$$

$(x, y \in \mathbb{Z}_{26})$.

6. multiplication is ***closed***, i.e., for any $a, b \in \mathbb{Z}_m$, $ab \in \mathbb{Z}_m$
7. multiplication is ***commutative***, i.e., for any $a, b \in \mathbb{Z}_m$, $ab = ba$
8. multiplication is ***associative***, i.e., for any $a, b, c \in \mathbb{Z}_m$, $(ab)c = a(bc)$
9. 1 is a ***multiplicative identity***, i.e., for any $a \in \mathbb{Z}_m$, $a \times 1 = 1 \times a = a$
10. the ***distributive property*** is satisfied, i.e., for any $a, b, c \in \mathbb{Z}_m$, $(a + b)c = (ac) + (bc)$ and $a(b + c) = (ab) + (ac)$.

Properties 1, 3–5 say that $\mathbb{Z}_m$ forms an algebraic structure called a ***group*** with respect to the addition operation. Since property 2 also holds, the group is said to be an ***abelian group***.

Properties 1–10 establish that $\mathbb{Z}_m$ is, in fact, a ***ring***. We will see many other examples of groups and rings in this book. Some familiar examples of rings include the integers, $\mathbb{Z}$; the real numbers, $\mathbb{R}$; and the complex numbers, $\mathbb{C}$. However, these are all infinite rings, and our attention will be confined almost exclusively to finite rings.

Since additive inverses exist in $\mathbb{Z}_m$, we can also subtract elements in $\mathbb{Z}_m$. We define $a - b$ in $\mathbb{Z}_m$ to be $(a - b) \bmod m$. That is, we compute the integer $a - b$ and then reduce it modulo m. For example, to compute $11 - 18$ in $\mathbb{Z}_{31}$, we first subtract 18 from 11, obtaining -7, and then compute $(-7) \bmod 31 = 24$.

We present the *Shift Cipher* as Cryptosystem 2.1. It is defined over $\mathbb{Z}_{26}$ since there are 26 letters in the English alphabet, though it could be defined over $\mathbb{Z}_m$ for any modulus m. It is easy to see that the *Shift Cipher* forms a cryptosystem as defined above, i.e., $d_K(e_K(x)) = x$ for every $x \in \mathbb{Z}_{26}$.

REMARK For the particular key $K = 3$, the cryptosystem is often called the *Caesar Cipher*, which was purportedly used by Julius Caesar. ∎

We would use the *Shift Cipher* (with a modulus of 26) to encrypt ordinary English text by setting up a correspondence between alphabetic characters and residues modulo 26 as follows: $A \leftrightarrow 0, B \leftrightarrow 1, \ldots, Z \leftrightarrow 25$. Since we will be using this correspondence in several examples, let's record it for future use:

A	B	C	D	E	F	G	H	I	J	K	L	M
0	1	2	3	4	5	6	7	8	9	10	11	12

N	O	P	Q	R	S	T	U	V	W	X	Y	Z
13	14	15	16	17	18	19	20	21	22	23	24	25

A small example will illustrate.

Example 2.1 Suppose the key for a *Shift Cipher* is $K = 11$, and the plaintext is

`wewillmeetatmidnight.`

We first convert the plaintext to a sequence of integers using the specified correspondence, obtaining the following:

22	4	22	8	11	11	12	4	4	19
0	19	12	8	3	13	8	6	7	19

Next, we add 11 to each value, reducing each sum modulo 26:

7	15	7	19	22	22	23	15	15	4
11	4	23	19	14	24	19	17	18	4

Finally, we convert the sequence of integers to alphabetic characters, obtaining the ciphertext:

`HPHTWWXPPELEXTOYTRSE.`

To decrypt the ciphertext, Bob will first convert the ciphertext to a sequence of integers, then subtract 11 from each value (reducing modulo 26), and finally convert the sequence of integers to alphabetic characters. □

REMARK In the above example we are using upper case letters for ciphertext and lower case letters for plaintext, in order to improve readability. We will do this elsewhere as well. ∎

If a cryptosystem is to be of practical use, it should satisfy certain properties. We informally enumerate two of these properties now.

1. Each encryption function e_K and each decryption function d_K should be efficiently computable.
2. An opponent, upon seeing a ciphertext string $\mathbf{y}$, should be unable to determine the key K that was used, or the plaintext string $\mathbf{x}$.

The second property is defining, in a very vague way, the idea of "security." The process of attempting to compute the key K, given a string of ciphertext $\mathbf{y}$, is called ***cryptanalysis***. (We will make these concepts more precise as we proceed.) Note that, if Oscar can determine K, then he can decrypt $\mathbf{y}$ just as Bob would, using d_K. Hence, determining K is at least as difficult as determining the plaintext string $\mathbf{x}$, given the ciphertext string $\mathbf{y}$.

We observe that the *Shift Cipher* (modulo 26) is not secure, since it can be cryptanalyzed by the obvious method of ***exhaustive key search***. Since there are only 26 possible keys, it is easy to try every possible decryption rule d_K until a "meaningful" plaintext string is obtained. This is illustrated in the following example.

Example 2.2 Given the ciphertext string

```
JBCRCLQRWCRVNBJENBWRWN,
```

we successively try the decryption keys d_0, d_1, etc. The following is obtained:

```
jbcrclqrwcrvnbjenbwrwn
iabqbkpqvbqumaidmavqvm
hzapajopuaptlzhclzupul
gyzozinotzoskygbkytotk
fxynyhmnsynrjxfajxsnsj
ewxmxglmrxmqiweziwrmri
dvwlwfklqwlphvdyhvqlqh
cuvkvejkpvkogucxgupkpg
btujudijoujnftbwftojof
astitchintimesavesnine
```

At this point, we have determined the plaintext to be the phrase "a stitch in time saves nine," and we can stop. The key is $K = 9$. □

On average, a plaintext will be computed using this method after trying $26/2 = 13$ decryption rules.

As the above example indicates, a necessary condition for a cryptosystem to be secure is that an exhaustive key search should be infeasible; i.e., the keyspace should be very large. As might be expected, however, a large keyspace is not sufficient to guarantee security.

2.1.2 The Substitution Cipher

Another well-known cryptosystem is the *Substitution Cipher*, which we define now. This cryptosystem has been used for hundreds of years. Puzzle "cryptograms" in newspapers are examples of *Substitution Ciphers*. This cipher is defined as Cryptosystem 2.2.

Actually, in the case of the *Substitution Cipher*, we might as well take $\mathcal{P}$ and $\mathcal{C}$ both to be the 26-letter English alphabet. We used $\mathbb{Z}_{26}$ in the *Shift Cipher* because

Cryptosystem 2.2: *Substitution Cipher*

Let $\mathcal{P} = \mathcal{C} = \mathbb{Z}_{26}$. $\mathcal{K}$ consists of all possible permutations of the 26 symbols $0, 1, \ldots, 25$. For each permutation $\pi \in \mathcal{K}$, define

$$e_\pi(x) = \pi(x),$$

and define

$$d_\pi(y) = \pi^{-1}(y),$$

where π^{-1} is the inverse permutation to π.

encryption and decryption were algebraic operations. But in the *Substitution Cipher,* it is more convenient to think of encryption and decryption as permutations of alphabetic characters.

Here is an example of a "random" permutation, π, which could comprise an encryption function. (As before, plaintext characters are written in lower case and ciphertext characters are written in upper case.)

a	*b*	*c*	*d*	*e*	*f*	*g*	*h*	*i*	*j*	*k*	*l*	*m*
X	N	Y	A	H	P	O	G	Z	Q	W	B	T

n	*o*	*p*	*q*	*r*	*s*	*t*	*u*	*v*	*w*	*x*	*y*	*z*
S	F	L	R	C	V	M	U	E	K	J	D	I

Thus, $e_\pi(a) = X$, $e_\pi(b) = N$, etc. The decryption function is the inverse permutation. This is formed by writing the second lines first, and then sorting in alphabetical order. The following is obtained:

A	B	C	D	E	F	G	H	I	J	K	L	M
d	*l*	*r*	*y*	*v*	*o*	*h*	*e*	*z*	*x*	*w*	*p*	*t*

N	O	P	Q	R	S	T	U	V	W	X	Y	Z
b	*g*	*f*	*j*	*q*	*n*	*m*	*u*	*s*	*k*	*a*	*c*	*i*

Hence, $d_\pi(A) = d$, $d_\pi(B) = l$, etc.

As an exercise, the reader might decrypt the following ciphertext using this decryption function:

```
MGZVYZLGHCMHJMYXSSFMNHAHYCDLMHA.
```

A key for the *Substitution Cipher* just consists of a permutation of the 26 alphabetic characters. The number of possible permutations is 26!, which is more than 4.0×10^{26}, a very large number. Thus, an exhaustive key search is infeasible, even for a computer. However, we shall see later that a *Substitution Cipher* can easily be cryptanalyzed by other methods.

2.1.3 The Affine Cipher

The *Shift Cipher* is a special case of the *Substitution Cipher*, which includes only 26 of the 26! possible permutations of 26 elements. Another special case of the *Substitution Cipher* is the *Affine Cipher*, which we describe now. In the *Affine Cipher*, we restrict the encryption functions to functions of the form

$$e(x) = (ax + b) \bmod 26,$$

$a, b \in \mathbb{Z}_{26}$. Such a function is called an ***affine function***; hence the name *Affine Cipher*. (Observe that when $a = 1$, we have a *Shift Cipher*.)

In order that decryption is possible, it is necessary to ask when an affine function is injective. In other words, for any $y \in \mathbb{Z}_{26}$, we want the congruence

$$ax + b \equiv y \pmod{26}$$

to have a unique solution for x. This congruence is equivalent to

$$ax \equiv y - b \pmod{26}.$$

Now, as y varies over $\mathbb{Z}_{26}$, so, too, does $y - b$ vary over $\mathbb{Z}_{26}$. Hence, it suffices to study the congruence $ax \equiv y \pmod{26}$ ($y \in \mathbb{Z}_{26}$).

We claim that this congruence has a unique solution for every y if and only if $\gcd(a, 26) = 1$ (where the gcd function denotes the greatest common divisor of its arguments). First, suppose that $\gcd(a, 26) = d > 1$. Then the congruence $ax \equiv 0 \pmod{26}$ has (at least) two distinct solutions in $\mathbb{Z}_{26}$, namely $x = 0$ and $x = 26/d$. In this case $e(x) = (ax + b) \bmod 26$ is not an injective function and hence not a valid encryption function.

For example, since $\gcd(4, 26) = 2$, it follows that $4x + 7$ is not a valid encryption function: x and $x + 13$ will encrypt to the same value, for any $x \in \mathbb{Z}_{26}$.

Let's next suppose that $\gcd(a, 26) = 1$. Suppose for some x_1 and x_2 that

$$ax_1 \equiv ax_2 \pmod{26}.$$

Then

$$a(x_1 - x_2) \equiv 0 \pmod{26},$$

and thus

$$26 \mid a(x_1 - x_2).$$

We now make use of a fundamental property of integer division: if $\gcd(a, b) = 1$ and $a \mid bc$, then $a \mid c$. Since $26 \mid a(x_1 - x_2)$ and $\gcd(a, 26) = 1$, we must therefore have that

$$26 \mid (x_1 - x_2),$$

i.e., $x_1 \equiv x_2 \pmod{26}$.

At this point we have shown that, if $\gcd(a, 26) = 1$, then a congruence of the form $ax \equiv y \pmod{26}$ has, at most, one solution in $\mathbb{Z}_{26}$. Hence, if we let x vary over $\mathbb{Z}_{26}$, then $ax \bmod 26$ takes on 26 distinct values modulo 26. That is, it takes on every value exactly once. It follows that, for any $y \in \mathbb{Z}_{26}$, the congruence $ax \equiv y \pmod{26}$ has a unique solution for x.

There is nothing special about the number 26 in this argument. The following result can be proved in an analogous fashion.

THEOREM 2.1 *The congruence $ax \equiv b \pmod{m}$ has a unique solution $x \in \mathbb{Z}_m$ for every $b \in \mathbb{Z}_m$ if and only if $\gcd(a, m) = 1$.*

Since $26 = 2 \times 13$, the values of $a \in \mathbb{Z}_{26}$ such that $\gcd(a, 26) = 1$ are $a = 1, 3, 5, 7, 9, 11, 15, 17, 19, 21, 23$, and 25. The parameter b can be any element in $\mathbb{Z}_{26}$. Hence the *Affine Cipher* has $12 \times 26 = 312$ possible keys. (Of course, this is much too small to be secure.)

Let's now consider the general setting where the modulus is m. We need another definition from number theory.

> **Definition 2.3:** Suppose $a \geq 1$ and $m \geq 2$ are integers. If $\gcd(a, m) = 1$, then we say that a and m are ***relatively prime***. The number of integers in $\mathbb{Z}_m$ that are relatively prime to m is often denoted by $\phi(m)$ (this function is called the ***Euler phi-function***).

A well-known result from number theory gives the value of $\phi(m)$ in terms of the prime power factorization of m. (An integer $p > 1$ is ***prime*** if it has no positive divisors other than 1 and p. Every integer $m > 1$ can be factored as a product of powers of primes in a unique way. For example, $60 = 2^2 \times 3 \times 5$ and $98 = 2 \times 7^2$.)

We record the formula for $\phi(m)$ in the following theorem.

THEOREM 2.2 *Suppose*

$$m = \prod_{i=1}^{n} p_i^{e_i},$$

where the p_i's are distinct primes and $e_i > 0, 1 \leq i \leq n$. Then

$$\phi(m) = \prod_{i=1}^{n} (p_i^{e_i} - p_i^{e_i - 1}).$$

It follows that the number of keys in the *Affine Cipher* over $\mathbb{Z}_m$ is $m\phi(m)$, where $\phi(m)$ is given by the formula above. (The number of choices for b is m, and the number of choices for a is $\phi(m)$, where the encryption function is $e(x) = ax + b$.) For example, suppose $m = 60$. We have

$$60 = 2^2 \times 3^1 \times 5^1$$

and hence

$$\phi(60) = (4 - 2) \times (3 - 1) \times (5 - 1) = 2 \times 2 \times 4 = 16.$$

The number of keys in the *Affine Cipher* is $60 \times 16 = 960$.

Let's now consider the decryption operation in the *Affine Cipher* with modulus $m = 26$. Suppose that $\gcd(a, 26) = 1$. To decrypt, we need to solve the congruence $y \equiv ax + b \pmod{26}$ for x. The discussion above establishes that the congruence will have a unique solution in $\mathbb{Z}_{26}$, but it does not give us an efficient method of finding the solution. What we require is an efficient algorithm to do this. Fortunately, some further results on modular arithmetic will provide us with the efficient decryption algorithm we seek.

We require the idea of a multiplicative inverse.

Definition 2.4: Suppose $a \in \mathbb{Z}_m$. The ***multiplicative inverse*** of a modulo m, denoted $a^{-1} \bmod m$, is an element $a' \in \mathbb{Z}_m$ such that $aa' \equiv a'a \equiv 1 \pmod{m}$. If m is fixed, we sometimes write a^{-1} for $a^{-1} \bmod m$.

By similar arguments to those used above, it can be shown that a has a multiplicative inverse modulo m if and only if $\gcd(a, m) = 1$; and if a multiplicative inverse exists, it is unique modulo m. Also, observe that if $b = a^{-1}$, then $a = b^{-1}$. If p is prime, then every non-zero element of $\mathbb{Z}_p$ has a multiplicative inverse. A ring in which this is true is called a ***field***.

In Section 6.2.1, we will describe an efficient algorithm for computing multiplicative inverses in $\mathbb{Z}_m$ for any m. However, in $\mathbb{Z}_{26}$, trial and error suffices to find the multiplicative inverses of the elements relatively prime to 26:

$$\begin{aligned}
1^{-1} &= 1,\\
3^{-1} &= 9,\\
5^{-1} &= 21,\\
7^{-1} &= 15,\\
11^{-1} &= 19,\\
17^{-1} &= 23, \text{and}\\
25^{-1} &= 25.
\end{aligned}$$

(All of these can be verified easily. For example, $7 \times 15 = 105 \equiv 1 \pmod{26}$, so $7^{-1} = 15$ and $15^{-1} = 7$.)

Consider our congruence $y \equiv ax + b \pmod{26}$. This is equivalent to

$$ax \equiv y - b \pmod{26}.$$

Since $\gcd(a, 26) = 1$, a has a multiplicative inverse modulo 26. Multiplying both sides of the congruence by a^{-1}, we obtain

$$a^{-1}(ax) \equiv a^{-1}(y - b) \pmod{26}.$$

By associativity of multiplication modulo 26, we have that

$$a^{-1}(ax) \equiv (a^{-1}a)x \equiv 1x \equiv x \pmod{26}.$$

Cryptosystem 2.3: *Affine Cipher*

Let $\mathcal{P} = \mathcal{C} = \mathbb{Z}_{26}$ and let

$$\mathcal{K} = \{(a,b) \in \mathbb{Z}_{26} \times \mathbb{Z}_{26} : \gcd(a,26) = 1\}.$$

For $K = (a,b) \in \mathcal{K}$, define

$$e_K(x) = (ax + b) \bmod 26$$

and

$$d_K(y) = a^{-1}(y - b) \bmod 26$$

$(x, y \in \mathbb{Z}_{26})$.

Consequently, $x = a^{-1}(y - b) \bmod 26$. This is an explicit formula for x, that is, the decryption function is

$$d_K(y) = a^{-1}(y - b) \bmod 26.$$

So, finally, the complete description of the *Affine Cipher* is given as Cryptosystem 2.3.

Let's do a small example.

Example 2.3 Suppose that $K = (7,3)$. As noted above, $7^{-1} \bmod 26 = 15$. The encryption function is

$$e_K(x) = 7x + 3,$$

and the corresponding decryption function is

$$d_K(y) = 15(y - 3) = 15y - 19,$$

where all operations are performed in $\mathbb{Z}_{26}$. It is a good check to verify that $d_K(e_K(x)) = x$ for all $x \in \mathbb{Z}_{26}$. Computing in $\mathbb{Z}_{26}$, we get

$$\begin{aligned} d_K(e_K(x)) &= d_K(7x + 3) \\ &= 15(7x + 3) - 19 \\ &= x + 45 - 19 \\ &= x. \end{aligned}$$

To illustrate, let's encrypt the plaintext *hot*. We first convert the letters *h*, *o*, *t* to residues modulo 26. These are respectively 7, 14, and 19. Now, we encrypt:

$$\begin{aligned} (7 \times 7 + 3) \bmod 26 &= 52 \bmod 26 &&= 0 \\ (7 \times 14 + 3) \bmod 26 &= 101 \bmod 26 &&= 23 \\ (7 \times 19 + 3) \bmod 26 &= 136 \bmod 26 &&= 6. \end{aligned}$$

So the three ciphertext characters are 0, 23, and 6, which corresponds to the alphabetic string *AXG*. We leave the decryption as an exercise for the reader. □

Cryptosystem 2.4: *Vigenère Cipher*

Let m be a positive integer. Define $\mathcal{P} = \mathcal{C} = \mathcal{K} = (\mathbb{Z}_{26})^m$. For a key $K = (k_1, k_2, \ldots, k_m)$, we define

$$e_K(x_1, x_2, \ldots, x_m) = (x_1 + k_1, x_2 + k_2, \ldots, x_m + k_m)$$

and

$$d_K(y_1, y_2, \ldots, y_m) = (y_1 - k_1, y_2 - k_2, \ldots, y_m - k_m),$$

where all operations are performed in $\mathbb{Z}_{26}$.

2.1.4 The Vigenère Cipher

In both the *Shift Cipher* and the *Substitution Cipher*, once a key is chosen, each alphabetic character is mapped to a unique alphabetic character. For this reason, these cryptosystems are called ***monoalphabetic cryptosystems***. We now present a cryptosystem that is not monoalphabetic, the well-known *Vigenère Cipher*, as Cryptosystem 2.4. This cipher is named after Blaise de Vigenère, who lived in the sixteenth century.

Using the correspondence $A \leftrightarrow 0, B \leftrightarrow 1, \ldots, Z \leftrightarrow 25$ described earlier, we can associate each key K with an alphabetic string of length m, called a ***keyword***. The *Vigenère Cipher* encrypts m alphabetic characters at a time: each plaintext element is equivalent to m alphabetic characters.

Let's do a small example.

Example 2.4 Suppose $m = 6$ and the keyword is $CIPHER$. This corresponds to the numerical equivalent $K = (2, 8, 15, 7, 4, 17)$. Suppose the plaintext is the string

`thiscryptosystemisnotsecure.`

We convert the plaintext elements to residues modulo 26, write them in groups of six, and then "add" the keyword modulo 26, as follows:

19	7	8	18	2	17	24	15	19	14	18	24
2	8	15	7	4	17	2	8	15	7	4	17
21	15	23	25	6	8	0	23	8	21	22	15

18	19	4	12	8	18	13	14	19	18	4	2
2	8	15	7	4	17	2	8	15	7	4	17
20	1	19	19	12	9	15	22	8	25	8	19

20	17	4
2	8	15
22	25	19

The alphabetic equivalent of the ciphertext string would thus be:

```
VPXZGIAXIVWPUBTTMJPWIZITWZT.
```

To decrypt, we can use the same keyword, but we would subtract it modulo 26 from the ciphertext, instead of adding it. □

Observe that the number of possible keywords of length m in a *Vigenère Cipher* is 26^m, so even for relatively small values of m, an exhaustive key search would require a long time. For example, if we take $m = 5$, then the keyspace has size exceeding 1.1×10^7. This is already large enough to preclude exhaustive key search by hand (but not by computer).

In a *Vigenère Cipher* having keyword length m, an alphabetic character can be mapped to one of m possible alphabetic characters (assuming that the keyword contains m distinct characters). Such a cryptosystem is called a ***polyalphabetic cryptosystem***. In general, cryptanalysis is more difficult for polyalphabetic than for monoalphabetic cryptosystems.

2.1.5 The Hill Cipher

In this section, we describe another polyalphabetic cryptosystem called the *Hill Cipher*. This cipher was invented in 1929 by Lester S. Hill. Let m be a positive integer, and define $\mathcal{P} = \mathcal{C} = (\mathbb{Z}_{26})^m$. The idea is to take m linear combinations of the m alphabetic characters in one plaintext element, thus producing the m alphabetic characters in one ciphertext element.

For example, if $m = 2$, we could write a plaintext element as $x = (x_1, x_2)$ and a ciphertext element as $y = (y_1, y_2)$. Here, y_1 would be a linear combination of x_1 and x_2, as would y_2. We might take

$$\begin{aligned} y_1 &= (11x_1 + 3x_2) \bmod 26 \\ y_2 &= (8x_1 + 7x_2) \bmod 26. \end{aligned}$$

Of course, this can be written more succinctly in matrix notation as follows:

$$(y_1, y_2) = (x_1, x_2) \begin{pmatrix} 11 & 8 \\ 3 & 7 \end{pmatrix},$$

where all operations are performed in $\mathbb{Z}_{26}$. In general, we will take an $m \times m$ matrix K as our key. If the entry in row i and column j of K is $k_{i,j}$, then we write $K = (k_{i,j})$. For $x = (x_1, \dots, x_m) \in \mathcal{P}$ and $K \in \mathcal{K}$, we compute $y = e_K(x) = (y_1, \dots, y_m)$ as follows:

$$(y_1, y_2, \dots, y_m) = (x_1, x_2, \dots, x_m) \begin{pmatrix} k_{1,1} & k_{1,2} & \dots & k_{1,m} \\ k_{2,1} & k_{2,2} & \dots & k_{2,m} \\ \vdots & \vdots & & \vdots \\ k_{m,1} & k_{m,2} & \dots & k_{m,m} \end{pmatrix}.$$

In other words, using matrix notation, $y = xK$.

We say that the ciphertext is obtained from the plaintext by means of a ***linear transformation***. We have to consider how decryption will work, that is, how x can be computed from y. Readers familiar with linear algebra will realize that we will use the inverse matrix K^{-1} to decrypt. The ciphertext is decrypted using the matrix equation $x = yK^{-1}$.

Here are the definitions of necessary concepts from linear algebra. If $A = (a_{i,j})$ is an $\ell \times m$ matrix and $B = (b_{j,k})$ is an $m \times n$ matrix, then we define the ***matrix product*** $AB = (c_{i,k})$ by the formula

$$c_{i,k} = \sum_{j=1}^{m} a_{i,j} b_{j,k}$$

for $1 \leq i \leq \ell$ and $1 \leq k \leq n$. That is, the entry in row i and column k of AB is formed by taking the ith row of A and the kth column of B, multiplying corresponding entries together, and summing. Note that AB is an $\ell \times n$ matrix.

Matrix multiplication is associative (that is, $(AB)C = A(BC)$) but not, in general, commutative (it is not always the case that $AB = BA$, even for square matrices A and B).

The $m \times m$ ***identity matrix***, denoted by I_m, is the $m \times m$ matrix with 1's on the main diagonal and 0's elsewhere. Thus, the 2×2 identity matrix is

$$I_2 = \begin{pmatrix} 1 & 0 \\ 0 & 1 \end{pmatrix}.$$

I_m is termed an identity matrix since $AI_m = A$ for any $\ell \times m$ matrix A and $I_m B = B$ for any $m \times n$ matrix B. Now, the ***inverse matrix*** of an $m \times m$ matrix A (if it exists) is the matrix A^{-1} such that $AA^{-1} = A^{-1}A = I_m$. Not all matrices have inverses, but if an inverse exists, it is unique.

With these facts at hand, it is easy to derive the decryption formula given above, assuming that K has an inverse matrix K^{-1}. Since $y = xK$, we can multiply both sides of the formula by K^{-1}, obtaining

$$yK^{-1} = (xK)K^{-1} = x(KK^{-1}) = xI_m = x.$$

(Note the use of the associativity property.)

We can verify that the example encryption matrix defined above has an inverse in $\mathbb{Z}_{26}$:

$$\begin{pmatrix} 11 & 8 \\ 3 & 7 \end{pmatrix}^{-1} = \begin{pmatrix} 7 & 18 \\ 23 & 11 \end{pmatrix}$$

since

$$\begin{aligned} \begin{pmatrix} 11 & 8 \\ 3 & 7 \end{pmatrix} \begin{pmatrix} 7 & 18 \\ 23 & 11 \end{pmatrix} &= \begin{pmatrix} 11 \times 7 + 8 \times 23 & 11 \times 18 + 8 \times 11 \\ 3 \times 7 + 7 \times 23 & 3 \times 18 + 7 \times 11 \end{pmatrix} \\ &= \begin{pmatrix} 261 & 286 \\ 182 & 131 \end{pmatrix} \\ &= \begin{pmatrix} 1 & 0 \\ 0 & 1 \end{pmatrix}. \end{aligned}$$

(Remember that all arithmetic operations are done modulo 26.)

Let's now do an example to illustrate encryption and decryption in the *Hill Cipher*.

Example 2.5 Suppose the key is

$$K = \begin{pmatrix} 11 & 8 \\ 3 & 7 \end{pmatrix}.$$

From the computations above, we have that

$$K^{-1} = \begin{pmatrix} 7 & 18 \\ 23 & 11 \end{pmatrix}.$$

Suppose we want to encrypt the plaintext *july*. We have two elements of plaintext to encrypt: $(9, 20)$ (corresponding to *ju*) and $(11, 24)$ (corresponding to *ly*). We compute as follows:

$$(9, 20) \begin{pmatrix} 11 & 8 \\ 3 & 7 \end{pmatrix} = (99 + 60, 72 + 140) = (3, 4)$$

and

$$(11, 24) \begin{pmatrix} 11 & 8 \\ 3 & 7 \end{pmatrix} = (121 + 72, 88 + 168) = (11, 22).$$

Hence, the encryption of *july* is *DELW*. To decrypt, Bob would compute:

$$(3, 4) \begin{pmatrix} 7 & 18 \\ 23 & 11 \end{pmatrix} = (9, 20)$$

and

$$(11, 22) \begin{pmatrix} 7 & 18 \\ 23 & 11 \end{pmatrix} = (11, 24).$$

Hence, the correct plaintext is obtained. □

At this point, we have shown that decryption is possible if K has an inverse. In fact, for decryption to be possible, it is necessary that K has an inverse. (This follows fairly easily from elementary linear algebra, but we will not give a proof here.) So we are interested precisely in those matrices K that are invertible.

The invertibility of a (square) matrix depends on the value of its determinant, which we define now.

Definition 2.5: Suppose that $A = (a_{i,j})$ is an $m \times m$ matrix. For $1 \leq i \leq m$, $1 \leq j \leq m$, define A_{ij} to be the matrix obtained from A by deleting the ith row and the jth column. The ***determinant*** of A, denoted $\det A$, is the value $a_{1,1}$ if $m = 1$. If $m > 1$, then $\det A$ is computed recursively from the formula

$$\det A = \sum_{j=1}^{m} (-1)^{i+j} a_{i,j} \det A_{ij},$$

where i is any fixed integer between 1 and m.

It is not at all obvious that the value of $\det A$ is independent of the choice of i in the formula given above, but it can be proved that this is indeed the case. It will be useful to write out the formulas for determinants of 2×2 and 3×3 matrices. If $A = (a_{i,j})$ is a 2×2 matrix, then

$$\det A = a_{1,1}a_{2,2} - a_{1,2}a_{2,1}.$$

If $A = (a_{i,j})$ is a 3×3 matrix, then

$$\begin{aligned}\det A &= a_{1,1}a_{2,2}a_{3,3} + a_{1,2}a_{2,3}a_{3,1} + a_{1,3}a_{2,1}a_{3,2}\\ &\quad -(a_{1,1}a_{2,3}a_{3,2} + a_{1,2}a_{2,1}a_{3,3} + a_{1,3}a_{2,2}a_{3,1}).\end{aligned}$$

For large m, the recursive formula given in the definition above is not usually a very efficient method of computing the determinant of an $m \times m$ square matrix. A preferred method is to compute the determinant using so-called "elementary row operations"; see any text on linear algebra.

Two important properties of determinants that we will use are $\det I_m = 1$ and the multiplication rule $\det(AB) = \det A \times \det B$.

A real matrix K has an inverse if and only if its determinant is non-zero. However, it is important to remember that we are working over $\mathbb{Z}_{26}$. The relevant result for our purposes is that a matrix K has an inverse modulo 26 if and only if $\gcd(\det K, 26) = 1$. To see that this condition is necessary, suppose K has an inverse, denoted K^{-1}. By the multiplication rule for determinants, we have

$$1 = \det I = \det(KK^{-1}) = \det K \det K^{-1}.$$

Hence, $\det K$ is invertible in $\mathbb{Z}_{26}$, which is true if and only if $\gcd(\det K, 26) = 1$.

Sufficiency of this condition can be established in several ways. We will give an explicit formula for the inverse of the matrix K. Define a matrix K^* to have as its (i,j)-entry the value $(-1)^{i+j} \det K_{ji}$. (Recall that K_{ji} is obtained from K by deleting the jth row and the ith column.) K^* is called the ***adjoint matrix*** of K. We state the following theorem, concerning inverses of matrices over $\mathbb{Z}_n$, without proof.

THEOREM 2.3 *Suppose* $K = (k_{i,j})$ *is an* $m \times m$ *matrix over* $\mathbb{Z}_n$ *such that* $\det K$ *is invertible in* $\mathbb{Z}_n$. *Then* $K^{-1} = (\det K)^{-1} K^*$, *where* K^* *is the adjoint matrix of* K.

REMARK The above formula for K^{-1} is not very efficient computationally, except for small values of m (e.g., $m = 2, 3$). For larger m, the preferred method of computing inverse matrices would involve performing elementary row operations on the matrix K. ∎

In the 2×2 case, we have the following formula, which is an immediate corollary of Theorem 2.3.

COROLLARY 2.4 *Suppose*

$$K = \begin{pmatrix} k_{1,1} & k_{1,2} \\ k_{2,1} & k_{2,2} \end{pmatrix}$$

is a matrix having entries in $\mathbb{Z}_n$*, and* $\det K = k_{1,1}k_{2,2} - k_{1,2}k_{2,1}$ *is invertible in* $\mathbb{Z}_n$*. Then*

$$K^{-1} = (\det K)^{-1} \begin{pmatrix} k_{2,2} & -k_{1,2} \\ -k_{2,1} & k_{1,1} \end{pmatrix}.$$

Let's look again at the example considered earlier. First, we have

$$\begin{aligned} \det \begin{pmatrix} 11 & 8 \\ 3 & 7 \end{pmatrix} &= (11 \times 7 - 8 \times 3) \bmod 26 \\ &= (77 - 24) \bmod 26 \\ &= 53 \bmod 26 \\ &= 1. \end{aligned}$$

Now, $1^{-1} \bmod 26 = 1$, so the inverse matrix is

$$\begin{pmatrix} 11 & 8 \\ 3 & 7 \end{pmatrix}^{-1} = \begin{pmatrix} 7 & 18 \\ 23 & 11 \end{pmatrix},$$

as we verified earlier.

Here is another example, using a 3×3 matrix.

Example 2.6 Suppose that

$$K = \begin{pmatrix} 10 & 5 & 12 \\ 3 & 14 & 21 \\ 8 & 9 & 11 \end{pmatrix},$$

where all entries are in $\mathbb{Z}_{26}$. The reader can verify that $\det K = 7$. In $\mathbb{Z}_{26}$, we have that $7^{-1} \bmod 26 = 15$. The adjoint matrix is

$$K^* = \begin{pmatrix} 17 & 1 & 15 \\ 5 & 14 & 8 \\ 19 & 2 & 21 \end{pmatrix}.$$

Finally, the inverse matrix is

$$K^{-1} = 15K^* = \begin{pmatrix} 21 & 15 & 17 \\ 23 & 2 & 16 \\ 25 & 4 & 3 \end{pmatrix}.$$

□

As mentioned above, encryption in the *Hill Cipher* is done by multiplying the plaintext by the matrix K, while decryption multiplies the ciphertext by the inverse matrix K^{-1}. We now give a precise mathematical description of the *Hill Cipher* over $\mathbb{Z}_{26}$; see Cryptosystem 2.5.

Cryptosystem 2.5: *Hill Cipher*

Let $m \geq 2$ be an integer. Let $\mathcal{P} = \mathcal{C} = (\mathbb{Z}_{26})^m$ and let

$$\mathcal{K} = \{m \times m \text{ invertible matrices over } \mathbb{Z}_{26}\}.$$

For a key K, we define

$$e_K(x) = xK$$

and

$$d_K(y) = yK^{-1},$$

where all operations are performed in $\mathbb{Z}_{26}$.

2.1.6 The Permutation Cipher

All of the cryptosystems we have discussed so far involve substitution: plaintext characters are replaced by different ciphertext characters. The idea of a permutation cipher is to keep the plaintext characters unchanged, but to alter their positions by rearranging them using a permutation.

A ***permutation*** of a finite set X is a bijective function $\pi : X \rightarrow X$. In other words, the function π is one-to-one (***injective***) and onto (***surjective***). It follows that, for every $x \in X$, there is a unique element $x' \in X$ such that $\pi(x') = x$. This allows us to define the ***inverse permutation***, $\pi^{-1} : X \rightarrow X$ by the rule

$$\pi^{-1}(x) = x' \quad \text{if and only if} \quad \pi(x') = x.$$

Then π^{-1} is also a permutation of X.

The *Permutation Cipher* (also known as the *Transposition Cipher*) is defined formally as Cryptosystem 2.6. This cryptosystem has been in use for hundreds of years. In fact, the distinction between the *Permutation Cipher* and the *Substitution Cipher* was pointed out as early as 1563 by Giovanni Porta.

As with the *Substitution Cipher*, it is more convenient to use alphabetic characters as opposed to residues modulo 26, since there are no algebraic operations being performed in encryption or decryption.

Here is an example to illustrate:

Example 2.7 Suppose $m = 6$ and the key is the following permutation π:

x	1	2	3	4	5	6
$\pi(x)$	3	5	1	6	4	2

Note that the first row of the above diagram lists the values of x, $1 \leq x \leq 6$, and the second row lists the corresponding values of $\pi(x)$. Then the inverse permutation π^{-1} can be constructed by interchanging the two rows, and rearranging the

Cryptosystem 2.6: *Permutation Cipher*

Let m be a positive integer. Let $\mathcal{P} = \mathcal{C} = (\mathbb{Z}_{26})^m$ and let $\mathcal{K}$ consist of all permutations of $\{1, \ldots, m\}$. For a key (i.e., a permutation) π, we define

$$e_\pi(x_1, \ldots, x_m) = (x_{\pi(1)}, \ldots, x_{\pi(m)})$$

and

$$d_\pi(y_1, \ldots, y_m) = (y_{\pi^{-1}(1)}, \ldots, y_{\pi^{-1}(m)}),$$

where π^{-1} is the inverse permutation to π.

columns so that the first row is in increasing order. Carrying out these operations, we see that the permutation π^{-1} is the following:

x	1	2	3	4	5	6
$\pi^{-1}(x)$	3	6	1	5	2	4

Now, suppose we are given the plaintext

`shesellsseashellsbytheseashore.`

We first partition the plaintext into groups of six letters:

`shesel | lsseas | hellsb | ythese | ashore`

Now each group of six letters is rearranged according to the permutation π, yielding the following:

`EESLSH | SALSES | LSHBLE | HSYEET | HRAEOS`

So, the ciphertext is:

`EESLSHSALSESLSHBLEHSYEETHRAEOS.`

The ciphertext can be decrypted in a similar fashion, using the inverse permutation π^{-1}. □

We now show that the *Permutation Cipher* is a special case of the *Hill Cipher*. Given a permutation π of the set $\{1, \ldots, m\}$, we can define an associated $m \times m$ permutation matrix $K_\pi = (k_{i,j})$ according to the formula

$$k_{i,j} = \begin{cases} 1 & \text{if } i = \pi(j) \\ 0 & \text{otherwise.} \end{cases}$$

(A ***permutation matrix*** is a matrix in which every row and column contains exactly one "1," and all other values are "0." A permutation matrix can be obtained from an identity matrix by permuting rows or columns.)

It is not difficult to see that Hill encryption using the matrix K_π is, in fact, equivalent to permutation encryption using the permutation π. Moreover, ${K_\pi}^{-1} = K_{\pi^{-1}}$, i.e., the inverse matrix to K_π is the permutation matrix defined by the permutation π^{-1}. Thus, Hill decryption is equivalent to permutation decryption.

For the permutation π used in the example above, the associated permutation matrices are

$$K_\pi = \begin{pmatrix} 0 & 0 & 1 & 0 & 0 & 0 \\ 0 & 0 & 0 & 0 & 0 & 1 \\ 1 & 0 & 0 & 0 & 0 & 0 \\ 0 & 0 & 0 & 0 & 1 & 0 \\ 0 & 1 & 0 & 0 & 0 & 0 \\ 0 & 0 & 0 & 1 & 0 & 0 \end{pmatrix}$$

and

$${K_\pi}^{-1} = \begin{pmatrix} 0 & 0 & 1 & 0 & 0 & 0 \\ 0 & 0 & 0 & 0 & 1 & 0 \\ 1 & 0 & 0 & 0 & 0 & 0 \\ 0 & 0 & 0 & 0 & 0 & 1 \\ 0 & 0 & 0 & 1 & 0 & 0 \\ 0 & 1 & 0 & 0 & 0 & 0 \end{pmatrix}.$$

The reader can verify that the product of these two matrices is the identity matrix.

2.1.7 Stream Ciphers

In the cryptosystems we have studied so far, successive plaintext elements are encrypted using the same key, K. That is, the ciphertext string $\mathbf{y}$ is obtained as follows:

$$\mathbf{y} = y_1 y_2 \cdots = e_K(x_1) e_K(x_2) \cdots .$$

Cryptosystems of this type are often called ***block ciphers***.

An alternative approach is to use what are called stream ciphers. The basic idea is to generate a keystream $\mathbf{z} = z_1 z_2 \cdots$, and use it to encrypt a plaintext string $\mathbf{x} = x_1 x_2 \cdots$ according to the rule

$$\mathbf{y} = y_1 y_2 \cdots = e_{z_1}(x_1) e_{z_2}(x_2) \cdots .$$

The simplest type of stream cipher is one in which the keystream is constructed from the key, independent of the plaintext string, using some specified algorithm. This type of stream cipher is called "synchronous" and can be defined formally as follows:

Definition 2.6: A ***synchronous stream cipher*** is a tuple $(\mathcal{P}, \mathcal{C}, \mathcal{K}, \mathcal{L}, \mathcal{E}, \mathcal{D})$, together with a function g, such that the following conditions are satisfied:

1. $\mathcal{P}$ is a finite set of possible plaintexts
2. $\mathcal{C}$ is a finite set of possible ciphertexts
3. $\mathcal{K}$, the keyspace, is a finite set of possible keys
4. $\mathcal{L}$ is a finite set called the ***keystream alphabet***
5. g is the ***keystream generator***. g takes a key K as input, and generates an infinite string $z_1z_2\cdots$ called the ***keystream***, where $z_i \in \mathcal{L}$ for all $i \geq 1$.
6. For each $z \in \mathcal{L}$, there is an encryption rule $e_z \in \mathcal{E}$ and a corresponding decryption rule $d_z \in \mathcal{D}$. $e_z : \mathcal{P} \rightarrow \mathcal{C}$ and $d_z : \mathcal{C} \rightarrow \mathcal{P}$ are functions such that $d_z(e_z(x)) = x$ for every plaintext element $x \in \mathcal{P}$.

To illustrate this definition, we show how the *Vigenère Cipher* can be defined as a synchronous stream cipher. Suppose that m is the keyword length of a *Vigenère Cipher*. Define $\mathcal{K} = (\mathbb{Z}_{26})^m$ and $\mathcal{P} = \mathcal{C} = \mathcal{L} = \mathbb{Z}_{26}$; and define $e_z(x) = (x+z) \bmod 26$ and $d_z(y) = (y-z) \bmod 26$. Finally, define the keystream $z_1z_2\cdots$ as follows:

$$z_i = \begin{cases} k_i & \text{if } 1 \leq i \leq m \\ z_{i-m} & \text{if } i \geq m+1, \end{cases}$$

where $K = (k_1, \ldots, k_m)$. This generates the keystream

$$k_1k_2\cdots k_mk_1k_2\cdots k_mk_1k_2\cdots$$

from the key $K = (k_1, k_2, \ldots, k_m)$.

REMARK We can think of a block cipher as a special case of a stream cipher where the keystream is constant: $z_i = K$ for all $i \geq 1$. ∎

A stream cipher is a ***periodic stream cipher*** with period d if $z_{i+d} = z_i$ for all integers $i \geq 1$. The *Vigenère Cipher* with keyword length m, as described above, can be thought of as a periodic stream cipher with period m.

Stream ciphers are often described in terms of binary alphabets, i.e., $\mathcal{P} = \mathcal{C} = \mathcal{L} = \mathbb{Z}_2$. In this situation, the encryption and decryption operations are just addition modulo 2:

$$e_z(x) = (x+z) \bmod 2$$

and

$$d_z(y) = (y+z) \bmod 2.$$

If we think of "0" as representing the boolean value "false" and "1" as representing "true," then addition modulo 2 corresponds to the ***exclusive-or*** operation. Hence, encryption (and decryption) can be implemented very efficiently in hardware.

Let's look at another method of generating a (synchronous) keystream. We will work over binary alphabets. Suppose we start with a binary m-tuple $(k_1, \ldots, k_m)$ and let $z_i = k_i$, $1 \leq i \leq m$ (as before). Now we generate the keystream using a ***linear recurrence*** of degree m:

$$z_{i+m} = \sum_{j=0}^{m-1} c_j z_{i+j} \bmod 2,$$

for all $i \geq 1$, where $c_0, \ldots, c_{m-1} \in \mathbb{Z}_2$ are specified constants.

REMARK This recurrence is said to have ***degree m*** since each term depends on the previous m terms. It is a ***linear recurrence*** because z_{i+m} is a linear function of previous terms. Note that we can take $c_0 = 1$ without loss of generality, for otherwise the recurrence will be of degree (at most) $m - 1$. ▮

Here, the key K consists of the $2m$ values $k_1, \ldots, k_m, c_0, \ldots, c_{m-1}$. If

$$(k_1, \ldots, k_m) = (0, \ldots, 0),$$

then the keystream consists entirely of 0's. Of course, this should be avoided, as the ciphertext will then be identical to the plaintext. However, if the constants $c_0, \ldots, c_{m-1}$ are chosen in a suitable way, then any other initialization vector $(k_1, \ldots, k_m)$ will give rise to a periodic keystream having period $2^m - 1$. So a "short" key can give rise to a keystream having a very long period. This is certainly a desirable property: we will see in a later section how the *Vigenère Cipher* can be cryptanalyzed by exploiting the fact that the keystream has a short period.

Here is an example to illustrate.

Example 2.8 Suppose $m = 4$ and the keystream is generated using the linear recurrence

$$z_{i+4} = (z_i + z_{i+1}) \bmod 2,$$

$i \geq 1$. If the keystream is initialized with any vector other than $(0,0,0,0)$, then we obtain a keystream of period 15. For example, starting with $(1,0,0,0)$, the keystream is

$$1\,0\,0\,0\,1\,0\,0\,1\,1\,0\,1\,0\,1\,1\,1 \cdots .$$

Any other non-zero initialization vector will give rise to a cyclic permutation of the same keystream. □

Another appealing aspect of this method of keystream generation is that the keystream can be produced efficiently in hardware using a ***linear feedback shift register***, or ***LFSR***. We would use a shift register with m ***stages***. The vector $(k_1, \ldots, k_m)$ would be used to initialize the shift register. At each time unit, the following operations would be performed concurrently:

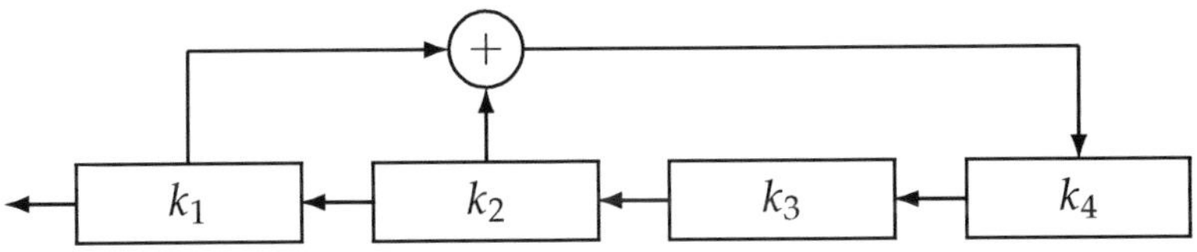

FIGURE 2.2: A linear feedback shift register

1. k_1 would be tapped as the next keystream bit
2. $k_2, \ldots, k_m$ would each be shifted one stage to the left
3. the "new" value of k_m would be computed to be

$$\sum_{j=0}^{m-1} c_j k_{j+1}$$

(this is the "linear feedback").

At any given point in time, the shift register contains m consecutive keystream elements, say $z_i, \ldots, z_{i+m-1}$. After one time unit, the shift register contains $z_{i+1}, \ldots, z_{i+m}$.

Observe that the linear feedback is carried out by tapping certain stages of the register (as specified by the constants c_j having the value "1") and computing a sum modulo 2 (which is an exclusive-or). This is illustrated in Figure 2.2, where we depict the LFSR that will generate the keystream of Example 2.8.

A ***non-synchronous stream cipher*** is a stream cipher in which each keystream element z_i depends on previous plaintext or ciphertext elements ($x_1, \ldots, x_{i-1}$ and/or $y_1, \ldots, y_{i-1}$) as well as the key K. A simple type of non-synchronous stream cipher, known as the *Autokey Cipher*, is presented as Cryptosystem 2.7. It is apparently due to Vigenère. The reason for the terminology "autokey" is that the plaintext is used to construct the keystream (aside from the initial "priming key" K). Of course, the *Autokey Cipher* is insecure since there are only 26 possible keys.

Here is an example to illustrate:

Example 2.9 Suppose the key is $K = 8$, and the plaintext is

`rendezvous.`

We first convert the plaintext to a sequence of integers:

17 4 13 3 4 25 21 14 20 18

The keystream is as follows:

8 17 4 13 3 4 25 21 14 20

Cryptosystem 2.7: *Autokey Cipher*

Let $\mathcal{P} = \mathcal{C} = \mathcal{K} = \mathcal{L} = \mathbb{Z}_{26}$. Let $z_1 = K$, and define $z_i = x_{i-1}$ for all $i \geq 2$. For $0 \leq z \leq 25$, define

$$e_z(x) = (x + z) \bmod 26$$

and

$$d_z(y) = (y - z) \bmod 26$$

$(x, y \in \mathbb{Z}_{26})$.

Now we add corresponding elements, reducing modulo 26:

$$25 \quad 21 \quad 17 \quad 16 \quad 7 \quad 3 \quad 20 \quad 9 \quad 8 \quad 12$$

In alphabetic form, the ciphertext is:

`ZVRQHDUJIM.`

Now let's look at how the ciphertext would be decrypted. First, we convert the alphabetic string to the numeric string

$$25 \quad 21 \quad 17 \quad 16 \quad 7 \quad 3 \quad 20 \quad 9 \quad 8 \quad 12$$

Then we compute

$$x_1 = d_8(25) = (25 - 8) \bmod 26 = 17.$$

Next,

$$x_2 = d_{17}(21) = (21 - 17) \bmod 26 = 4,$$

and so on. Each time we obtain another plaintext character, we also use it as the next keystream element. □

In the next section, we discuss methods that can be used to cryptanalyze the various cryptosystems we have presented.

2.2 Cryptanalysis

In this section, we discuss some techniques of cryptanalysis. The general assumption that is usually made is that the opponent, Oscar, knows the cryptosystem being used. This is usually referred to as ***Kerckhoffs' Principle***. Of course, if Oscar does not know the cryptosystem being used, that will make his task more difficult. But we do not want to base the security of a cryptosystem on the (possibly shaky) premise that Oscar does not know what system is being employed. Hence, our goal in designing a cryptosystem will be to obtain security while assuming that Kerckhoffs' principle holds.

First, we want to differentiate between different attack models on cryptosystems. The ***attack model*** specifies the information available to the adversary when he mounts his attack. The most common types of attack models are enumerated as follows.

ciphertext-only attack
The opponent possesses a string of ciphertext, $\mathbf{y}$.

known plaintext attack
The opponent possesses a string of plaintext, $\mathbf{x}$, and the corresponding ciphertext, $\mathbf{y}$.

chosen plaintext attack
The opponent has obtained temporary access to the encryption machinery. Hence he can choose a plaintext string, $\mathbf{x}$, and construct the corresponding ciphertext string, $\mathbf{y}$.

chosen ciphertext attack
The opponent has obtained temporary access to the decryption machinery. Hence he can choose a ciphertext string, $\mathbf{y}$, and construct the corresponding plaintext string, $\mathbf{x}$.

In each case, the objective of the adversary is to determine the key that was used. This would allow the opponent to decrypt a specific "target" ciphertext string, and further, to decrypt any additional ciphertext strings that are encrypted using the same key.

At first glance, a chosen ciphertext attack may seem to be a bit artificial. For, if there is only one ciphertext string of interest to the opponent, then the opponent can obviously decrypt that ciphertext string if a chosen ciphertext attack is permitted. However, we are suggesting that the opponent's objective normally includes determining the key that is used by Alice and Bob, so that other ciphertext strings can be decrypted (at a later time, perhaps). A chosen ciphertext attack makes sense in this context.

We first consider the weakest type of attack, namely a ciphertext-only attack (this is sometimes called a ***known ciphertext attack***). We also assume that the plaintext string is ordinary English text, without punctuation or "spaces." (This makes cryptanalysis more difficult than if punctuation and spaces were encrypted.)

Many techniques of cryptanalysis use statistical properties of the English language. Various people have estimated the relative frequencies of the 26 letters by compiling statistics from numerous novels, magazines, and newspapers. The estimates in Table 2.1 were obtained by Beker and Piper. On the basis of these probabilities, Beker and Piper partition the 26 letters into five groups as follows:

TABLE 2.1: Probabilities of occurrence of the 26 letters

letter	probability	letter	probability
A	.082	N	.067
B	.015	O	.075
C	.028	P	.019
D	.043	Q	.001
E	.127	R	.060
F	.022	S	.063
G	.020	T	.091
H	.061	U	.028
I	.070	V	.010
J	.002	W	.023
K	.008	X	.001
L	.040	Y	.020
M	.024	Z	.001

1. E, having probability about 0.120
2. T, A, O, I, N, S, H, R, each having probability between 0.06 and 0.09
3. D, L, each having probability around 0.04
4. $C, U, M, W, F, G, Y, P, B$, each having probability between 0.015 and 0.028
5. V, K, J, X, Q, Z, each having probability less than 0.01.

It is also useful to consider sequences of two or three consecutive letters, called ***digrams*** and ***trigrams***, respectively. The 30 most common digrams are (in decreasing order):

$$TH, HE, IN, ER, AN, RE, ED, ON, ES, ST,$$
$$EN, AT, TO, NT, HA, ND, OU, EA, NG, AS,$$
$$OR, TI, IS, ET, IT, AR, TE, SE, HI, OF.$$

The twelve most common trigrams are:

$$THE, ING, AND, HER, ERE, ENT,$$
$$THA, NTH, WAS, ETH, FOR, DTH.$$

2.2.1 Cryptanalysis of the Affine Cipher

As a simple illustration of how cryptanalysis can be performed using statistical data, let's look first at the *Affine Cipher*. Suppose Oscar has intercepted the ciphertext shown in the following example:

TABLE 2.2: Frequency of occurrence of the 26 ciphertext letters

letter	frequency	letter	frequency
A	2	N	1
B	1	O	1
C	0	P	2
D	7	Q	0
E	5	R	8
F	4	S	3
G	0	T	0
H	5	U	2
I	0	V	4
J	0	W	0
K	5	X	2
L	2	Y	1
M	2	Z	0

Example 2.10 Ciphertext obtained from an *Affine Cipher*

```
FMXVEDKAPHFERBNDKRXRSREFMORUDSDKDVSHVUFEDK
APRKDLYEVLRHHRH
```

The frequency analysis of this ciphertext is given in Table 2.2.

There are only 57 characters of ciphertext, but this is usually sufficient to cryptanalyze an *Affine Cipher*. The most frequent ciphertext characters are: R (8 occurrences), D (7 occurrences), E, H, K (5 occurrences each), and F, S, V (4 occurrences each). As a first guess, we might hypothesize that R is the encryption of e and D is the encryption of t, since e and t are (respectively) the two most common letters. Expressed numerically, we have $e_K(4) = 17$ and $e_K(19) = 3$. Recall that $e_K(x) = ax + b$, where a and b are unknowns. So we get two linear equations in two unknowns:

$$\begin{aligned} 4a + b &= 17 \\ 19a + b &= 3. \end{aligned}$$

This system has the unique solution $a = 6$, $b = 19$ (in $\mathbb{Z}_{26}$). But this is an illegal key, since $\gcd(a, 26) = 2 > 1$. So our hypothesis must be incorrect.

Our next guess might be that R is the encryption of e and E is the encryption of t. Proceeding as above, we obtain $a = 13$, which is again illegal. So we try the next possibility, that R is the encryption of e and H is the encryption of t. This yields $a = 8$, again impossible. Continuing, we suppose that R is the encryption of e and K is the encryption of t. This produces $a = 3$, $b = 5$, which is at least a legal key. It remains to compute the decryption function corresponding to $K = (3, 5)$, and then to decrypt the ciphertext to see if we get a meaningful string of English, or nonsense. This will confirm the validity of $(3, 5)$.

TABLE 2.3: Frequency of occurrence of the **26** ciphertext letters

letter	frequency	letter	frequency
A	0	N	9
B	1	O	0
C	15	P	1
D	13	Q	4
E	7	R	10
F	11	S	3
G	1	T	2
H	4	U	5
I	5	V	5
J	11	W	8
K	1	X	6
L	0	Y	10
M	16	Z	20

If we perform these operations, we obtain $d_K(y) = 9y - 19$ and the given ciphertext decrypts to yield:

```
algorithmsarequitegeneraldefinitionsofarit
hmeticprocesses
```

We conclude that we have determined the correct key. □

2.2.2 Cryptanalysis of the Substitution Cipher

Here, we look at the more complicated situation, the *Substitution Cipher*. Consider the ciphertext in the following example:

Example 2.11 Ciphertext obtained from a *Substitution Cipher*

```
YIFQFMZRWQFYVECFMDZPCVMRZWNMDZVEJBTXCDDUMJ
NDIFEFMDZCDMQZKCEYFCJMYRNCWJCSZREXCHZUNMXZ
NZUCDRJXYYSMRTMEYIFZWDYVZVYFZUMRZCRWNZDZJJ
XZWGCHSMRNMDHNCMFQCHZJMXJZWIEJYUCFWDJNZDIR
```

The frequency analysis of this ciphertext is given in Table 2.3.

Since Z occurs significantly more often than any other ciphertext character, we might conjecture that $d_K(Z) = e$. The remaining ciphertext characters that occur at least ten times (each) are C, D, F, J, M, R, Y. We might expect that these letters are encryptions of (a subset of) t, a, o, i, n, s, h, r, but the frequencies really do not vary enough to tell us what the correspondence might be.

At this stage we might look at digrams, especially those of the form $-Z$ or $Z-$, since we conjecture that Z decrypts to e. We find that the most common digrams of this type are DZ and ZW (four times each); NZ and ZU (three times each); and

RZ, HZ, XZ, FZ, ZR, ZV, ZC, ZD, and ZJ (twice each). Since ZW occurs four times and WZ not at all, and W occurs less often than many other characters, we might guess that $d_K(W) = d$. Since DZ occurs four times and ZD occurs twice, we would think that $d_K(D) \in \{r, s, t\}$, but it is not clear which of the three possibilities is the correct one.

If we proceed on the assumption that $d_K(Z) = e$ and $d_K(W) = d$, we might look back at the ciphertext and notice that we have ZRW occurring near the beginning of the ciphertext, and RW occurs again later on. Since R occurs frequently in the ciphertext and *nd* is a common digram, we might try $d_K(R) = n$ as the most likely possibility.

At this point, we have the following:

```
------end---------e----ned---e------------
YIFQFMZRWQFYVECFMDZPCVMRZWNMDZVEJBTXCDDUMJ

--------e----e---------n--d---en----e----e
NDIFEFMDZCDMQZKCEYFCJMYRNCWJCSZREXCHZUNMXZ

-e---n------n------ed---e---e--ne-nd-e-e--
NZUCDRJXYYSMRTMEYIFZWDYVZVYFZUMRZCRWNZDZJJ

-ed-----n-----------e----ed-------d---e--n
XZWGCHSMRNMDHNCMFQCHZJMXJZWIEJYUCFWDJNZDIR
```

Our next step might be to try $d_K(N) = h$, since NZ is a common digram and ZN is not. If this is correct, then the segment of plaintext *ne* $-$ *ndhe* suggests that $d_K(C) = a$. Incorporating these guesses, we have:

```
------end-----a---e-a--nedh--e------a-----
YIFQFMZRWQFYVECFMDZPCVMRZWNMDZVEJBTXCDDUMJ

h-------ea---e-a---a---nhad-a-en--a-e-h--e
NDIFEFMDZCDMQZKCEYFCJMYRNCWJCSZREXCHZUNMXZ

he-a-n------n------ed---e---e--neandhe-e--
NZUCDRJXYYSMRTMEYIFZWDYVZVYFZUMRZCRWNZDZJJ

-ed-a---nh---ha---a-e----ed-----a-d--he--n
XZWGCHSMRNMDHNCMFQCHZJMXJZWIEJYUCFWDJNZDIR
```

Now, we might consider M, the second most common ciphertext character. The ciphertext segment RNM, which we believe decrypts to $nh-$, suggests that $h-$ begins a word, so M probably represents a vowel. We have already accounted for *a* and *e*, so we expect that $d_K(M) = i$ or o. Since *ai* is a much more likely digram than *ao*, the ciphertext digram CM suggests that we try $d_K(M) = i$ first. Then we have:

```
-----iend-----a-i-e-a-inedhi-e------a---i-
YIFQFMZRWQFYVECFMDZPCVMRZWNMDZVEJBTXCDDUMJ

h-----i-ea-i-e-a---a-i-nhad-a-en--a-e-hi-e
NDIFEFMDZCDMQZKCEYFCJMYRNCWJCSZREXCHZUNMXZ

he-a-n-----in-i----ed---e---e-ineandhe-e--
NZUCDRJXYYSMRTMEYIFZWDYVZVYFZUMRZCRWNZDZJJ

-ed-a--inhi--hai--a-e-i--ed-----a-d--he--n
XZWGCHSMRNMDHNCMFQCHZJMXJZWIEJYUCFWDJNZDIR
```

Next, we might try to determine which letter is the encryption of o. Since o is a common plaintext character, we guess that the corresponding ciphertext character is one of D, F, J, Y. Y seems to be the most likely possibility; otherwise, we would get long strings of vowels, namely *aoi* from CFM or CJM. Hence, let's suppose $d_K(Y) = o$.

The three most frequent remaining ciphertext letters are D, F, J, which we conjecture could decrypt to r, s, t in some order. Two occurrences of the trigram NMD suggest that $d_K(D) = s$, giving the trigram *his* in the plaintext (this is consistent with our earlier hypothesis that $d_K(D) \in \{r, s, t\}$). The segment $HNCMF$ could be an encryption of *chair*, which would give $d_K(F) = r$ (and $d_K(H) = c$) and so we would then have $d_K(J) = t$ by process of elimination. Now, we have:

```
o-r-riend-ro--arise-a-inedhise--t---ass-it
YIFQFMZRWQFYVECFMDZPCVMRZWNMDZVEJBTXCDDUMJ

hs-r-riseasi-e-a-orationhadta-en--ace-hi-e
NDIFEFMDZCDMQZKCEYFCJMYRNCWJCSZREXCHZUNMXZ

he-asnt-oo-in-i-o-redso-e-ore-ineandhesett
NZUCDRJXYYSMRTMEYIFZWDYVZVYFZUMRZCRWNZDZJJ

-ed-ac-inhischair-aceti-ted--to-ardsthes-n
XZWGCHSMRNMDHNCMFQCHZJMXJZWIEJYUCFWDJNZDIR
```

It is now very easy to determine the plaintext and the key for Example 2.11. The complete decryption is the following:

> Our friend from Paris examined his empty glass with surprise, as if evaporation had taken place while he wasn't looking. I poured some more wine and he settled back in his chair, face tilted up towards the sun.[1]

□

[1] P. Mayle, *A Year in Provence*, A. Knopf, Inc., 1989.

2.2.3 Cryptanalysis of the Vigenère Cipher

In this section we describe some methods for cryptanalyzing the *Vigenère Cipher*. The first step is to determine the keyword length, which we denote by m. There are a couple of techniques that can be employed. The first of these is the so-called Kasiski test and the second uses the index of coincidence.

The ***Kasiski test*** was described by Friedrich Kasiski in 1863; however, it was apparently discovered earlier, around 1854, by Charles Babbage. It is based on the observation that two identical segments of plaintext will be encrypted to the same ciphertext whenever their occurrence in the plaintext is δ positions apart, where $\delta \equiv 0 \pmod{m}$. Conversely, if we observe two identical segments of ciphertext, each of length at least three, say, then there is a good chance that they correspond to identical segments of plaintext.

The Kasiski test works as follows. We search the ciphertext for pairs of identical segments of length at least three, and record the distance between the starting positions of the two segments. If we obtain several such distances, say $\delta_1, \delta_2, \ldots$, then we would conjecture that m divides all of the δ_i's, and hence m divides the greatest common divisor of the δ_i's.

Further evidence for the value of m can be obtained by the index of coincidence. This concept was defined by William Friedman in 1920, as follows.

Definition 2.7: Suppose $\mathbf{x} = x_1 x_2 \cdots x_n$ is a string of n alphabetic characters. The ***index of coincidence*** of $\mathbf{x}$, denoted $I_c(\mathbf{x})$, is defined to be the probability that two random elements of $\mathbf{x}$ are identical.

Suppose we denote the frequencies of $A, B, C, \ldots, Z$ in $\mathbf{x}$ by $f_0, f_1, \ldots, f_{25}$ (respectively). We can choose two elements of $\mathbf{x}$ in $\binom{n}{2}$ ways.[2] For each i, $0 \leq i \leq 25$, there are $\binom{f_i}{2}$ ways of choosing both elements to be i. Hence, we have the formula

$$I_c(\mathbf{x}) = \frac{\sum_{i=0}^{25} \binom{f_i}{2}}{\binom{n}{2}} = \frac{\sum_{i=0}^{25} f_i(f_i - 1)}{n(n-1)}.$$

Suppose $\mathbf{x}$ is a string of English language text. Denote the expected probabilities of occurrence of the letters $A, B, \ldots, Z$ in Table 2.1 by $p_0, \ldots, p_{25}$, respectively. Then, we would expect that

$$I_c(\mathbf{x}) \approx \sum_{i=0}^{25} {p_i}^2 = 0.065,$$

since the probability that two random elements both are A is ${p_0}^2$, the probability that both are B is ${p_1}^2$, etc. The same reasoning applies if $\mathbf{x}$ is a ciphertext string obtained using any monoalphabetic cipher. In this case, the individual probabilities will be permuted, but the quantity $\sum {p_i}^2$ will be unchanged.

[2]The ***binomial coefficient*** $\binom{n}{k} = n!/(k!(n-k)!)$ denotes the number of ways of choosing a subset of k objects from a set of n objects.

Now, suppose we start with a ciphertext string $\mathbf{y} = y_1 y_2 \cdots y_n$ that has been constructed by using a *Vigenère Cipher*. Define m substrings of $\mathbf{y}$, denoted $\mathbf{y}_1, \mathbf{y}_2, \dots, \mathbf{y}_m$, by writing out the ciphertext, in columns, in a rectangular array of dimensions $m \times (n/m)$. The rows of this matrix are the substrings $\mathbf{y}_i$, $1 \leq i \leq m$. In other words, we have that

$$\begin{aligned} \mathbf{y}_1 &= y_1 y_{m+1} y_{2m+1} \cdots, \\ \mathbf{y}_2 &= y_2 y_{m+2} y_{2m+2} \cdots, \\ \vdots \; &\;\; \vdots \\ \mathbf{y}_m &= y_m y_{2m} y_{3m} \cdots. \end{aligned}$$

If $\mathbf{y}_1, \mathbf{y}_2, \dots, \mathbf{y}_m$ are constructed in this way, and m is indeed the keyword length, then each value $I_c(\mathbf{y}_i)$ should be roughly equal to 0.065. On the other hand, if m is not the keyword length, then the substrings $\mathbf{y}_i$ will look much more random, since they will have been obtained by shift encryption with different keys. Observe that a completely random string will have

$$I_c \approx 26 \left(\frac{1}{26} \right)^2 = \frac{1}{26} = 0.038.$$

The two values 0.065 and 0.038 are sufficiently far apart that we will often be able to determine the correct keyword length by this method (or confirm a guess that has already been made using the Kasiski test).

Let us illustrate these two techniques with an example.

Example 2.12 Ciphertext obtained from a *Vigenère Cipher*

```
CHREEVOAHMAERATBIAXXWTNXBEEOPHBSBQMQEQERBW
RVXUOAKXAOSXXWEAHBWGJMMQMNKGRFVGXWTRZXWIAK
LXFPSKAUTEMNDCMGTSXMXBTUIADNGMGPSRELXNJELX
VRVPRTULHDNQWTWDTYGBPHXTFALJHASVBFXNGLLCHR
ZBWELEKMSJIKNBHWRJGNMGJSGLXFEYPHAGNRBIEQJT
AMRVLCRREMNDGLXRRIMGNSNRWCHRQHAEYEVTAQEBBI
PEEWEVKAKOEWADREMXMTBHHCHRTKDNVRZCHRCLQOHP
WQAIIWXNRMGWOIIFKEE
```

First, let's try the Kasiski test. The ciphertext string *CHR* occurs in five places in the ciphertext, beginning at positions 1, 166, 236, 276, and 286. The distances from the first occurrence to the other four occurrences are (respectively) 165, 235, 275, and 285. The greatest common divisor of these four integers is 5, so that is very likely the keyword length.

Let's see if computation of indices of coincidence gives the same conclusion. With $m = 1$, the index of coincidence is 0.045. With $m = 2$, the two indices are 0.046 and 0.041. With $m = 3$, we get 0.043, 0.050, 0.047. With $m = 4$, we have indices 0.042, 0.039, 0.045, 0.040. Then, trying $m = 5$, we obtain the values 0.063, 0.068, 0.069, 0.061, and 0.072. This also provides strong evidence that the keyword length is five. □

Assuming that we have determined the correct value of m, how do we determine the actual key, $K = (k_1, k_2, \ldots, k_m)$? We describe a simple and effective method now. Let $1 \leq i \leq m$, and let $f_0, \ldots, f_{25}$ denote the frequencies of $A, B, \ldots, Z$, respectively, in the string $\mathbf{y}_i$. Also, let $n' = n/m$ denote the length of the string $\mathbf{y}_i$. Then the probability distribution of the 26 letters in $\mathbf{y}_i$ is

$$\frac{f_0}{n'}, \ldots, \frac{f_{25}}{n'}.$$

Now, recall that the substring $\mathbf{y}_i$ is obtained by shift encryption of a subset of the plaintext elements using a shift k_i. Therefore, we would hope that the shifted probability distribution

$$\frac{f_{k_i}}{n'}, \ldots, \frac{f_{25+k_i}}{n'}$$

would be "close to" the ideal probability distribution $p_0, \ldots, p_{25}$ tabulated in Table 2.1, where subscripts in the above formula are evaluated modulo 26.

Suppose that $0 \leq g \leq 25$, and define the quantity

$$M_g = \sum_{i=0}^{25} \frac{p_i f_{i+g}}{n'}. \tag{2.1}$$

If $g = k_i$, then we would expect that

$$M_g \approx \sum_{i=0}^{25} {p_i}^2 = 0.065,$$

as in the consideration of the index of coincidence. If $g \neq k_i$, then M_g will usually be significantly smaller than 0.065 (see the Exercises for a justification of this statement). Hopefully this technique will allow us to determine the correct value of k_i for each value of i, $1 \leq i \leq m$.

Let us illustrate by returning to Example 2.12.

Example 2.12 (Cont.) We have hypothesized that the keyword length is 5. We now compute the values M_g as described above, for $1 \leq i \leq 5$. These values are tabulated in Table 2.4. For each i, we look for a value of M_g that is close to 0.065. These g's determine the shifts $k_1, \ldots, k_5$.

From the data in Table 2.4, we see that the key is likely to be $K = (9, 0, 13, 4, 19)$, and hence the keyword likely is *JANET*. This is correct, and the complete decryption of the ciphertext is the following:

> The almond tree was in tentative blossom. The days were longer, often ending with magnificent evenings of corrugated pink skies. The hunting season was over, with hounds and guns put away for six months. The vineyards were busy again as the well-organized farmers treated their vines and the more lackadaisical neighbors hurried to do the pruning they should have done in November.[3]

□

[3]P. Mayle, *A Year in Provence*, A. Knopf, Inc., 1989.

TABLE 2.4: Values of M_g

i	value of $M_g(\mathbf{y}_i)$								
1	.035	.031	.036	.037	.035	.039	.028	.028	.048
	.061	.039	.032	.040	.038	.038	.045	.036	.030
	.042	.043	.036	.033	.049	.043	.042	.036	
2	.069	.044	.032	.035	.044	.034	.036	.033	.029
	.031	.042	.045	.040	.045	.046	.042	.037	.032
	.034	.037	.032	.034	.043	.032	.026	.047	
3	.048	.029	.042	.043	.044	.034	.038	.035	.032
	.049	.035	.031	.035	.066	.035	.038	.036	.045
	.027	.035	.034	.034	.036	.035	.046	.040	
4	.045	.032	.033	.038	.060	.034	.034	.034	.050
	.033	.033	.043	.040	.033	.029	.036	.040	.044
	.037	.050	.034	.034	.039	.044	.038	.035	
5	.034	.031	.035	.044	.047	.037	.043	.038	.042
	.037	.033	.032	.036	.037	.036	.045	.032	.029
	.044	.072	.037	.027	.031	.048	.036	.037	

2.2.4 Cryptanalysis of the Hill Cipher

The *Hill Cipher* can be difficult to break with a ciphertext-only attack, but it succumbs easily to a known plaintext attack. Let us first assume that the opponent has determined the value of m being used. Suppose they have at least m distinct plaintext-ciphertext pairs, say

$$x_j = (x_{1,j}, x_{2,j}, \ldots, x_{m,j})$$

and

$$y_j = (y_{1,j}, y_{2,j}, \ldots, y_{m,j}),$$

for $1 \leq j \leq m$, such that $y_j = e_K(x_j)$, $1 \leq j \leq m$. If we define two $m \times m$ matrices $X = (x_{i,j})$ and $Y = (y_{i,j})$, then we have the matrix equation $Y = XK$, where the $m \times m$ matrix K is the unknown key. Provided that the matrix X is invertible, Oscar can compute $K = X^{-1}Y$ and thereby break the system. (If X is not invertible, then it will be necessary to try other sets of m plaintext-ciphertext pairs.)

Let's look at a simple example.

Example 2.13 Suppose the plaintext *friday* is encrypted using a *Hill Cipher* with $m = 2$, to give the ciphertext *PQCFKU*.

We have that $e_K(5,17) = (15,16)$, $e_K(8,3) = (2,5)$ and $e_K(0,24) = (10,20)$. From the first two plaintext-ciphertext pairs, we get the matrix equation

$$\begin{pmatrix} 15 & 16 \\ 2 & 5 \end{pmatrix} = \begin{pmatrix} 5 & 17 \\ 8 & 3 \end{pmatrix} K.$$

Using Corollary 2.4, it is easy to compute

$$\begin{pmatrix} 5 & 17 \\ 8 & 3 \end{pmatrix}^{-1} = \begin{pmatrix} 9 & 1 \\ 2 & 15 \end{pmatrix},$$

so

$$K = \begin{pmatrix} 9 & 1 \\ 2 & 15 \end{pmatrix} \begin{pmatrix} 15 & 16 \\ 2 & 5 \end{pmatrix} = \begin{pmatrix} 7 & 19 \\ 8 & 3 \end{pmatrix}.$$

This can be verified by using the third plaintext-ciphertext pair. □

What would the opponent do if they do not know m? Assuming that m is not too big, they could simply try $m = 2, 3, \ldots,$ until the key is found. If a guessed value of m is incorrect, then an $m \times m$ matrix found by using the algorithm described above will not agree with further plaintext-ciphertext pairs. In this way, the value of m can be determined if it is not known ahead of time.

2.2.5 Cryptanalysis of the LFSR Stream Cipher

Recall that the ciphertext is the sum modulo 2 of the plaintext and the keystream, i.e., $y_i = (x_i + z_i) \bmod 2$. The keystream is produced from an initial m-tuple, $(z_1, \ldots, z_m) = (k_1, \ldots, k_m)$, using the linear recurrence

$$z_{m+i} = \sum_{j=0}^{m-1} c_j z_{i+j} \bmod 2,$$

$i \geq 1$, where $c_0, \ldots, c_{m-1} \in \mathbb{Z}_2$.

Since all operations in this cryptosystem are linear, we might suspect that the cryptosystem is vulnerable to a known-plaintext attack, as is the case with the *Hill Cipher*. Suppose Oscar has a plaintext string $x_1 x_2 \cdots x_n$ and the corresponding ciphertext string $y_1 y_2 \cdots y_n$. Then he can compute the keystream bits $z_i = (x_i + y_i) \bmod 2$, $1 \leq i \leq n$. Let us also suppose that Oscar knows the value of m. Then Oscar needs only to compute $c_0, \ldots, c_{m-1}$ in order to be able to reconstruct the entire keystream. In other words, he needs to be able to determine the values of m unknowns.

Now, for any $i \geq 1$, we have

$$z_{m+i} = \sum_{j=0}^{m-1} c_j z_{i+j} \bmod 2,$$

which is a linear equation in the m unknowns. If $n \geq 2m$, then there are m linear equations in m unknowns, which can subsequently be solved.

The system of m linear equations can be written in matrix form as follows:

$$(z_{m+1}, z_{m+2}, \ldots, z_{2m}) = (c_0, c_1, \ldots, c_{m-1}) \begin{pmatrix} z_1 & z_2 & \ldots & z_m \\ z_2 & z_3 & \ldots & z_{m+1} \\ \vdots & \vdots & & \vdots \\ z_m & z_{m+1} & \ldots & z_{2m-1} \end{pmatrix}.$$

If the coefficient matrix has an inverse (modulo 2), we obtain the solution

$$(c_0, c_1, \ldots, c_{m-1}) = (z_{m+1}, z_{m+2}, \ldots, z_{2m}) \begin{pmatrix} z_1 & z_2 & \cdots & z_m \\ z_2 & z_3 & \cdots & z_{m+1} \\ \vdots & \vdots & & \vdots \\ z_m & z_{m+1} & \cdots & z_{2m-1} \end{pmatrix}^{-1}.$$

In fact, the matrix will have an inverse if m is the degree of the recurrence used to generate the keystream (see the Exercises for a proof).

Let's illustrate with an example.

Example 2.14 Suppose Oscar obtains the ciphertext string

101101011110010

corresponding to the plaintext string

011001111111000.

Then he can compute the keystream bits:

110100100001010.

Suppose also that Oscar knows that the keystream was generated using a 5-stage LFSR. Then he would solve the following matrix equation, which is obtained from the first 10 keystream bits:

$$(0,1,0,0,0) = (c_0, c_1, c_2, c_3, c_4) \begin{pmatrix} 1 & 1 & 0 & 1 & 0 \\ 1 & 0 & 1 & 0 & 0 \\ 0 & 1 & 0 & 0 & 1 \\ 1 & 0 & 0 & 1 & 0 \\ 0 & 0 & 1 & 0 & 0 \end{pmatrix}.$$

It can be verified that

$$\begin{pmatrix} 1 & 1 & 0 & 1 & 0 \\ 1 & 0 & 1 & 0 & 0 \\ 0 & 1 & 0 & 0 & 1 \\ 1 & 0 & 0 & 1 & 0 \\ 0 & 0 & 1 & 0 & 0 \end{pmatrix}^{-1} = \begin{pmatrix} 0 & 1 & 0 & 0 & 1 \\ 1 & 0 & 0 & 1 & 0 \\ 0 & 0 & 0 & 0 & 1 \\ 0 & 1 & 0 & 1 & 1 \\ 1 & 0 & 1 & 1 & 0 \end{pmatrix},$$

by checking that the product of the two matrices, computed modulo 2, is the identity matrix. This yields

$$\begin{aligned} (c_0, c_1, c_2, c_3, c_4) &= (0,1,0,0,0) \begin{pmatrix} 0 & 1 & 0 & 0 & 1 \\ 1 & 0 & 0 & 1 & 0 \\ 0 & 0 & 0 & 0 & 1 \\ 0 & 1 & 0 & 1 & 1 \\ 1 & 0 & 1 & 1 & 0 \end{pmatrix} \\ &= (1,0,0,1,0). \end{aligned}$$

Thus the recurrence used to generate the keystream is

$$z_{i+5} = (z_i + z_{i+3}) \bmod 2.$$

□

2.3 Notes and References

Material on classical cryptography is covered in various textbooks and monographs, such as

- *Decrypted Secrets: Methods and Maxims of Cryptology* by Friedrich Bauer [10]
- *Cryptology* by Albrecht Beutelspacher [28]
- *Code Breaking: A History and Exploration* by Rudolf Kippenhahn [106]
- *Basic Methods of Cryptography* by Jan van der Lubbe [123].

We have used the statistical data on frequency of English letters that is reported in Beker and Piper [13].

A good reference for elementary number theory is

- *Elementary Number Theory, 7th Edition* by David Burton [53].

Background in linear algebra can be found in

- *Linear Algebra and Its Applications, 5th Edition* by David Lay, Steven Lay, and Judi McDonald [118].

Two very enjoyable and readable books that provide interesting histories of cryptography are

- *The Codebreakers: The Comprehensive History of Secret Communication from Ancient Times to the Internet* by David Kahn [103]
- *The Code Book: The Science of Secrecy from Ancient Egypt to Quantum Cryptography* by Simon Singh [183].

Exercises

2.1 Evaluate the following:

(a) 7503 mod 81

(b) $(-7503) \bmod 81$

(c) $81 \bmod 7503$

(d) $(-81) \bmod 7503$.

2.2 Suppose that $a, m > 0$, and $a \not\equiv 0 \pmod{m}$. Prove that

$$(-a) \bmod m = m - (a \bmod m).$$

2.3 Prove that $a \bmod m = b \bmod m$ if and only if $a \equiv b \pmod{m}$.

2.4 Prove that $a \bmod m = a - \lfloor \frac{a}{m} \rfloor m$, where $\lfloor x \rfloor = \max\{y \in \mathbb{Z} : y \leq x\}$.

2.5 Use exhaustive key search to decrypt the following ciphertext, which was encrypted using a *Shift Cipher*:

```
BEEAKFYDJXUQYHYJIQRYHTYJIQFBQDUYJIIKFUHCQD.
```

2.6 If an encryption function e_K is identical to the decryption function d_K, then the key K is said to be an ***involutory key***. Find all the involutory keys in the *Shift Cipher* over $\mathbb{Z}_{26}$.

2.7 Determine the number of keys in an *Affine Cipher* over $\mathbb{Z}_m$ for $m = 30, 100$ and 1225.

2.8 List all the invertible elements in $\mathbb{Z}_m$ for $m = 28, 33$, and 35.

2.9 For $1 \leq a \leq 28$, determine $a^{-1} \bmod 29$ by trial and error.

2.10 Suppose that $K = (5, 21)$ is a key in an *Affine Cipher* over $\mathbb{Z}_{29}$.

(a) Express the decryption function $d_K(y)$ in the form $d_K(y) = a'y + b'$, where $a', b' \in \mathbb{Z}_{29}$.

(b) Prove that $d_K(e_K(x)) = x$ for all $x \in \mathbb{Z}_{29}$.

2.11 (a) Suppose that $K = (a, b)$ is a key in an *Affine Cipher* over $\mathbb{Z}_n$. Prove that K is an involutory key if and only if $a^{-1} \bmod n = a$ and $b(a + 1) \equiv 0 \pmod{n}$.

(b) Determine all the involutory keys in the *Affine Cipher* over $\mathbb{Z}_{15}$.

(c) Suppose that $n = pq$, where p and q are distinct odd primes. Prove that the number of involutory keys in the *Affine Cipher* over $\mathbb{Z}_n$ is $n + p + q + 1$.

2.12 (a) Let p be prime. Prove that the number of 2×2 matrices that are invertible over $\mathbb{Z}_p$ is $(p^2 - 1)(p^2 - p)$.

HINT Since p is prime, $\mathbb{Z}_p$ is a field. Use the fact that a matrix over a field is invertible if and only if its rows are linearly independent vectors (i.e., there does not exist a non-zero linear combination of the rows whose sum is the vector of all 0's).

(b) For p prime and $m \geq 2$ an integer, find a formula for the number of $m \times m$ matrices that are invertible over $\mathbb{Z}_p$.

2.13 For $n = 6, 9$, and 26, how many 2×2 matrices are there that are invertible over $\mathbb{Z}_n$?

2.14 (a) Prove that $\det A \equiv \pm 1 \pmod{26}$ if A is a matrix over $\mathbb{Z}_{26}$ such that $A = A^{-1}$.

(b) Use the formula given in Corollary 2.4 to determine the number of involutory keys in the *Hill Cipher* (over $\mathbb{Z}_{26}$) in the case $m = 2$.

2.15 Determine the inverses of the following matrices over $\mathbb{Z}_{26}$:

(a) $\begin{pmatrix} 2 & 5 \\ 9 & 5 \end{pmatrix}$

(b) $\begin{pmatrix} 1 & 11 & 12 \\ 4 & 23 & 2 \\ 17 & 15 & 9 \end{pmatrix}$

2.16 (a) Suppose that π is the following permutation of $\{1, \ldots, 8\}$:

x	1	2	3	4	5	6	7	8
$\pi(x)$	4	1	6	2	7	3	8	5

Compute the permutation π^{-1}.

(b) Decrypt the following ciphertext, for a *Permutation Cipher* with $m = 8$, which was encrypted using the key π:

TGEEMNELNNTDROEOAAHDOETCSHAEIRLM.

2.17 (a) Prove that a permutation π in the *Permutation Cipher* is an involutory key if and only if $\pi(i) = j$ implies $\pi(j) = i$, for all $i, j \in \{1, \ldots, m\}$.

(b) Determine the number of involutory keys in the *Permutation Cipher* for $m = 2, 3, 4, 5$, and 6.

2.18 Consider the following linear recurrence over $\mathbb{Z}_2$ of degree four:

$$z_{i+4} = (z_i + z_{i+1} + z_{i+2} + z_{i+3}) \bmod 2,$$

$i \geq 0$. For each of the 16 possible initialization vectors $(z_0, z_1, z_2, z_3) \in (\mathbb{Z}_2)^4$, determine the period of the resulting keystream.

2.19 Redo the preceding question, using the recurrence

$$z_{i+4} = (z_i + z_{i+3}) \bmod 2,$$

$i \geq 0$.

2.20 Suppose we construct a keystream in a synchronous stream cipher using the following method. Let $K \in \mathcal{K}$ be the key, let $\mathcal{L}$ be the keystream alphabet, and let Σ be a finite set of states. First, an initial state $\sigma_0 \in \Sigma$ is determined from K by some method. For all $i \geq 1$, the state σ_i is computed from the previous state σ_{i-1} according to the following rule:

$$\sigma_i = f(\sigma_{i-1}, K),$$

where $f : \Sigma \times \mathcal{K} \rightarrow \Sigma$. Also, for all $i \geq 1$, the keystream element z_i is computed using the following rule:

$$z_i = g(\sigma_i, K),$$

where $g : \Sigma \times \mathcal{K} \rightarrow \mathcal{L}$. Prove that any keystream produced by this method has period at most $|\Sigma|$.

2.21 Below are given four examples of ciphertext, one obtained from a *Substitution Cipher*, one from a *Vigenère Cipher*, one from an *Affine Cipher*, and one unspecified. In each case, the task is to determine the plaintext.

Give a clearly written description of the steps you followed to decrypt each ciphertext. This should include all statistical analysis and computations you performed.

The first two plaintexts were taken from *The Diary of Samuel Marchbanks*, by Robertson Davies, Clarke Irwin, 1947; the fourth was taken from *Lake Wobegon Days*, by Garrison Keillor, Viking Penguin, Inc., 1985.

(a) *Substitution Cipher*:

```
EMGLOSUDCGDNCUSWYSFHNSFCYKDPUMLWGYICOXYSIPJCK
QPKUGKMGOLICGINCGACKSNISACYKZSCKXECJCKSHYSXCG
OIDPKZCNKSHICGIWYGKKGKGOLDSILKGOIUSIGLEDSPWZU
GFZCCNDGYYSFUSZCNXEOJNCGYEOWEUPXEZGACGNFGLKNS
ACIGOIYCKXCJUCIUZCFZCCNDGYYSFEUEKUZCSOCFZCCNC
IACZEJNCSHFZEJZEGMXCYHCJUMGKUCY
```

HINT F decrypts to w.

(b) *Vigenère Cipher*:

```
KCCPKBGUFDPHQTYAVINRRTMVGRKDNBVFDETDGILTXRGUD
DKOTFMBPVGEGLTGCKQRACQCWDNAWCRXIZAKFTLEWRPTYC
QKYVXCHKFTPONCQQRHJVAJUWETMCMSPKQDYHJVDAHCTRL
SVSKCGCZQQDZXGSFRLSWCWSJTBHAFSIASPRJAHKJRJUMV
GKMITZHFPDISPZLVLGWTFPLKKEBDPGCEBSHCTJRWXBAFS
PEZQNRWXCVYCGAONWDDKACKAWBBIKFTIOVKCGGHJVLNHI
FFSQESVYCLACNVRWBBIREPBBVFEXOSCDYGZWPFDTKFQIY
CWHJVLNHIQIBTKHJVNPIST
```

(c) *Affine Cipher*:

```
KQEREJEBCPPCJCRKIEACUZBKRVPKRBCIBQCARBJCVFCUP
KRIOFKPACUZQEPBKRXPEIIEABDKPBCPFCDCCAFIEABDKP
BCPFEQPKAZBKRHAIBKAPCCIBURCCDKDCCJCIDFUIXPAFF
ERBICZDFKABICBBENEFCUPJCVKABPCYDCCDPKBCOCPERK
IVKSCPICBRKIJPKABI
```

(d) unspecified cipher:

```
BNVSNSIHQCEELSSKKYERIFJKXUMBGYKAMQLJTYAVFBKVT
DVBPVVRJYYLAOKYMPQSCGDLFSRLLPROYGESEBUUALRWXM
MASAZLGLEDFJBZAVVPXWICGJXASCBYEHOSNMULKCEAHTQ
OKMFLEBKFXLRRFDTZXCIWBJSICBGAWDVYDHAVFJXZIBKC
GJIWEAHTTOEWTUHKRQVVRGZBXYIREMMASCSPBNLHJMBLR
FFJELHWEYLWISTFVVYFJCMHYUYRUFSFMGESIGRLWALSWM
NUHSIMYYITCCQPZSICEHBCCMZFEGVJYOCDEMMPGHVAAUM
ELCMOEHVLTIPSUYILVGFLMVWDVYDBTHFRAYISYSGKVSUU
HYHGGCKTMBLRX
```

2.22 (a) Suppose that $p_1, \ldots, p_n$ and $q_1, \ldots, q_n$ are both probability distributions, and $p_1 \geq \cdots \geq p_n$. Let $q'_1, \ldots, q'_n$ be any permutation of $q_1, \ldots, q_n$. Prove that the quantity

$$\sum_{i=1}^{n} p_i q'_i$$

is maximized when $q'_1 \geq \cdots \geq q'_n$.

(b) Explain why the expression in Equation (2.1) is likely to be maximized when $g = k_i$.

2.23 Suppose we are told that the plaintext

`breathtaking`

yields the ciphertext

`RUPOTENTOIFV`

where the *Hill Cipher* is used (but m is not specified). Determine the encryption matrix.

2.24 An *Affine-Hill Cipher* is the following modification of a *Hill Cipher*: Let m be a positive integer, and define $\mathcal{P} = \mathcal{C} = (\mathbb{Z}_{26})^m$. In this cryptosystem, a key K consists of a pair (L, b), where L is an $m \times m$ invertible matrix over $\mathbb{Z}_{26}$, and $b \in (\mathbb{Z}_{26})^m$. For $x = (x_1, \ldots, x_m) \in \mathcal{P}$ and $K = (L, b) \in \mathcal{K}$, we compute $y = e_K(x) = (y_1, \ldots, y_m)$ by means of the formula $y = xL + b$. Hence, if

$L = (\ell_{i,j})$ and $b = (b_1, \ldots, b_m)$, then

$$(y_1, \ldots, y_m) = (x_1, \ldots, x_m) \begin{pmatrix} \ell_{1,1} & \ell_{1,2} & \ldots & \ell_{1,m} \\ \ell_{2,1} & \ell_{2,2} & \ldots & \ell_{2,m} \\ \vdots & \vdots & & \vdots \\ \ell_{m,1} & \ell_{m,2} & \ldots & \ell_{m,m} \end{pmatrix} + (b_1, \ldots, b_m).$$

Suppose Oscar has learned that the plaintext

`adisplayedequation`

is encrypted to give the ciphertext

`DSRMSIOPLXLJBZULLM`

and Oscar also knows that $m = 3$. Determine the key, showing all computations.

2.25 Here is how we might cryptanalyze the *Hill Cipher* using a ciphertext-only attack. Suppose that we know that $m = 2$. Break the ciphertext into blocks of length two letters (digrams). Each such digram is the encryption of a plaintext digram using the unknown encryption matrix. Pick out the most frequent ciphertext digram and assume it is the encryption of a common digram in the list following Table 2.1 (for example, *TH* or *ST*). For each such guess, proceed as in the known-plaintext attack, until the correct encryption matrix is found.

Here is a sample of ciphertext for you to decrypt using this method:

```
LMQETXYEAGTXCTUIEWNCTXLZEWUAISPZYVAPEWLMGQWYA
XFTCJMSQCADAGTXLMDXNXSNPJQSYVAPRIQSMHNOCVAXFV
```

2.26 We describe a special case of a *Permutation Cipher*. Let m, n be positive integers. Write out the plaintext, by rows, in $m \times n$ rectangles. Then form the ciphertext by taking the columns of these rectangles. For example, if $m = 3$, $n = 4$, then we would encrypt the plaintext "*cryptography*" by forming the following rectangle:

```
cryp
togr
aphy
```

The ciphertext would be "*CTAROPYGHPRY*."

(a) Describe how Bob would decrypt a ciphertext string (given values for m and n).

(b) Decrypt the following ciphertext, which was obtained by using this method of encryption:

MYAMRARUYIQTENCTORAHROYWDSOYEOUARRGDERNOGW

2.27 The purpose of this exercise is to prove the statement made in Section 2.2.5 that the $m \times m$ coefficient matrix is invertible. This is equivalent to saying that the rows of this matrix are linearly independent vectors over $\mathbb{Z}_2$.

Suppose that the recurrence has the form

$$z_{m+i} = \sum_{j=0}^{m-1} c_j z_{i+j} \bmod 2,$$

where $(z_1, \dots, z_m)$ comprises the initialization vector. For $i \geq 1$, define

$$v_i = (z_i, \dots, z_{i+m-1}).$$

Note that the coefficient matrix has the vectors $v_1, \dots, v_m$ as its rows, so our objective is to prove that these m vectors are linearly independent.

Prove the following assertions:

(a) For any $i \geq 1$,

$$v_{m+i} = \sum_{j=0}^{m-1} c_j v_{i+j} \bmod 2.$$

(b) Choose h to be the minimum integer such that there exists a non-trivial linear combination of the vectors $v_1, \dots, v_h$ which sums to the vector $(0, \dots, 0)$ modulo 2. Then

$$v_h = \sum_{j=0}^{h-2} \alpha_j v_{j+1} \bmod 2,$$

and not all the α_j's are zero. Observe that $h \leq m+1$, since any $m+1$ vectors in an m-dimensional vector space are dependent.

(c) Prove that the keystream must satisfy the recurrence

$$z_{h-1+i} = \sum_{j=0}^{h-2} \alpha_j z_{j+i} \bmod 2$$

for any $i \geq 1$.

(d) If $h \leq m$, then the keystream satisfies a linear recurrence of degree less than m. Show that this is impossible, by considering the initialization vector $(0, \dots, 0, 1)$. Hence, conclude that $h = m + 1$, and therefore the matrix must be invertible.

2.28 Decrypt the following ciphertext, obtained from the *Autokey Cipher*, by using exhaustive key search:

MALVVMAFBHBUQPTSOXALTGVWWRG

2.29 We describe a stream cipher that is a modification of the *Vigenère Cipher*. Given a keyword $(K_1, \ldots, K_m)$ of length m, construct a keystream by the rule $z_i = K_i$ $(1 \leq i \leq m)$, $z_{i+m} = (z_i + 1) \bmod 26$ $(i \geq 1)$. In other words, each time we use the keyword, we replace each letter by its successor modulo 26. For example, if *SUMMER* is the keyword, we use *SUMMER* to encrypt the first six letters, we use *TVNNFS* for the next six letters, and so on.

(a) Describe how you can use the concept of index of coincidence to first determine the length of the keyword, and then actually find the keyword.

(b) Test your method by cryptanalyzing the following ciphertext:

```
IYMYSILONRFNCQXQJEDSHBUIBCJUZBOLFQYSCHATPEQGQ
JEJNGNXZWHHGWFSUKULJQACZKKJOAAHGKEMTAFGMKVRDO
PXNEHEKZNKFSKIFRQVHHOVXINPHMRTJPYWQGJWPUUVKFP
OAWPMRKKQZWLQDYAZDRMLPBJKJOBWIWPSEPVVQMBCRYVC
RUZAAOUMBCHDAGDIEMSZFZHALIGKEMJJFPCIWKRMLMPIN
AYOFIREAOLDTHITDVRMSE
```

The plaintext was taken from *The Codebreakers*, by D. Kahn, Scribner, 1996.

2.30 We describe another stream cipher, which incorporates one of the ideas from the *Enigma* machime used by Germany in World War II. Suppose that π is a fixed permutation of $\mathbb{Z}_{26}$. The key is an element $K \in \mathbb{Z}_{26}$. For all integers $i \geq 1$, the keystream element $z_i \in \mathbb{Z}_{26}$ is defined according to the rule $z_i = (K + i - 1) \bmod 26$. Encryption and decryption are performed using the permutations π and π^{-1}, respectively, as follows:

$$e_z(x) = \pi(x) + z \bmod 26$$

and

$$d_z(y) = \pi^{-1}(y - z \bmod 26),$$

where $z \in \mathbb{Z}_{26}$.

Suppose that π is the following permutation of $\mathbb{Z}_{26}$:

x	0	1	2	3	4	5	6	7	8	9	10	11	12
$\pi(x)$	23	13	24	0	7	15	14	6	25	16	22	1	19

x	13	14	15	16	17	18	19	20	21	22	23	24	25
$\pi(x)$	18	5	11	17	2	21	12	20	4	10	9	3	8

The following ciphertext has been encrypted using this stream cipher; use exhaustive key search to decrypt it:

```
WRTCNRLDSAFARWKXFTXCZRNHNYPDTZUUKMPLUSOXNEUDO
KLXRMCBKGRCCURR
```

Chapter 3

Shannon's Theory, Perfect Secrecy, and the One-Time Pad

This chapter introduces notions of cryptographic security, concentrating on the concept of unconditional security. The *One-time Pad* is presented and concepts such as information theory, entropy, and perfect secrecy are discussed.

3.1 Introduction

In 1949, Claude Shannon published a paper entitled *Communication Theory of Secrecy Systems* in the *Bell Systems Technical Journal*. This paper had a great influence on the scientific study of cryptography. In this chapter, we discuss several of Shannon's ideas. First, however, we consider some of the various approaches to evaluating the security of a cryptosystem. We define some of the most useful criteria now.

computational security

This measure concerns the computational effort required to break a cryptosystem. We might define a cryptosystem to be ***computationally secure*** if the best algorithm for breaking it requires at least N operations, where N is some specified, very large number. The problem is that no known practical cryptosystem can be proved to be secure under this definition. In practice, people often study the computational security of a cryptosystem with respect to certain specific types of attacks (e.g., an exhaustive key search). Of course, security against one specific type of attack does not guarantee security against some other type of attack.

provable security

Another approach is to provide evidence of security by means of a reduction. In other words, we show that if the cryptosystem can be "broken" in some specific way, then it would be possible to efficiently solve some well-studied problem that is thought to be difficult. For example, it may be possible to prove a statement of the type "a given cryptosystem is secure if a given integer n cannot be factored." Cryptosystems of this type are sometimes termed ***provably secure***, but it must be understood that this approach only provides a proof of security relative to some other problem, not an absolute proof of security. This is a similar situation to proving that a problem is NP-complete: it proves that the given problem is at least as difficult as any other NP-complete problem, but it does not provide an absolute proof of the computational difficulty of the problem.

unconditional security

This measure concerns the security of cryptosystems when there is no bound placed on the amount of computation that Oscar is allowed to do. A cryptosystem is defined to be ***unconditionally secure*** if it cannot be broken, even with infinite computational resources.

When we discuss the security of a cryptosystem, we should also specify the type of attack that is being considered. For example, in Chapter 2, we saw that neither the *Shift Cipher*, the *Substitution Cipher*, nor the *Vigenère Cipher* is computationally secure against a ciphertext-only attack (given a sufficient amount of ciphertext).

After introducing some basics of probability theory in Section 3.2, we will develop a theory of cryptosystems that are unconditionally secure against a ciphertext-only attack in Section 3.3. This theory allows us to prove mathematically that certain cryptosystems are secure if the amount of ciphertext is sufficiently small. For example, it turns out that the *Shift Cipher* and the *Substitution Cipher* are both unconditionally secure if a single element of plaintext is encrypted with a given key. Similarly, the *Vigenère Cipher* with keyword length m is unconditionally secure if the key is used to encrypt only one element of plaintext (which consists of m alphabetic characters).

Section 3.4 presents the concept of entropy, which is used in Section 3.5 to analyze the unicity distance of a cryptosystem.

3.2 Elementary Probability Theory

The unconditional security of a cryptosystem obviously cannot be studied from the point of view of computational complexity because we allow computation time to be infinite. The appropriate framework in which to study unconditional security is probability theory. We need only elementary facts concerning probability; the main definitions are reviewed now. First, we define the idea of a random variable.

Definition 3.1: A ***discrete random variable***, say $\mathbf{X}$, consists of a finite set X and a ***probability distribution*** defined on X. The probability that the random variable $\mathbf{X}$ takes on the value x is denoted $\mathbf{Pr}[\mathbf{X} = x]$; sometimes we will abbreviate this to $\mathbf{Pr}[x]$ if the random variable $\mathbf{X}$ is fixed. It must be the case that $0 \leq \mathbf{Pr}[x]$ for all $x \in X$, and

$$\sum_{x \in X} \mathbf{Pr}[x] = 1.$$

As an example, we could consider a coin toss to be a random variable defined on the set $\{heads, tails\}$. The associated probability distribution would be $\mathbf{Pr}[heads] = \mathbf{Pr}[tails] = 1/2$.

Suppose we have random variable $\mathbf{X}$ defined on X, and $E \subseteq X$. The probability that $\mathbf{X}$ takes on a value in the subset E is computed to be

$$\mathbf{Pr}[x \in E] = \sum_{x \in E} \mathbf{Pr}[x]. \tag{3.1}$$

The subset E is often called an ***event***.

Example 3.1 Suppose we consider a random throw of a pair of dice. This can be modeled by a random variable $\mathbf{Z}$ defined on the set

$$Z = \{1,2,3,4,5,6\} \times \{1,2,3,4,5,6\},$$

where $\mathbf{Pr}[(i,j)] = 1/36$ for all $(i,j) \in Z$. Let's consider the sum of the two dice. Each possible sum defines an event, and the probabilities of these events can be computed using equation (3.1). For example, suppose that we want to compute the probability that the sum is 4. This corresponds to the event

$$S_4 = \{(1,3),(2,2),(3,1)\},$$

and therefore $\mathbf{Pr}[S_4] = 3/36 = 1/12$.

The probabilities of all the sums can be computed in a similar fashion. If we denote by S_j the event that the sum is j, then we obtain the following: $\mathbf{Pr}[S_2] = \mathbf{Pr}[S_{12}] = 1/36$, $\mathbf{Pr}[S_3] = \mathbf{Pr}[S_{11}] = 1/18$, $\mathbf{Pr}[S_4] = \mathbf{Pr}[S_{10}] = 1/12$, $\mathbf{Pr}[S_5] = \mathbf{Pr}[S_9] = 1/9$, $\mathbf{Pr}[S_6] = \mathbf{Pr}[S_8] = 5/36$, and $\mathbf{Pr}[S_7] = 1/6$.

Since the events $S_2, \dots, S_{12}$ partition the set S, it follows that we can consider the value of the sum of a pair of dice to be a random variable in its own right, which has the probability distribution computed above. □

We next consider the concepts of joint and conditional probabilities.

Definition 3.2: Suppose $\mathbf{X}$ and $\mathbf{Y}$ are random variables defined on finite sets X and Y, respectively. The ***joint probability*** $\mathbf{Pr}[x,y]$ is the probability that $\mathbf{X}$ takes on the value x and $\mathbf{Y}$ takes on the value y. The ***conditional probability*** $\mathbf{Pr}[x|y]$ denotes the probability that $\mathbf{X}$ takes on the value x given that $\mathbf{Y}$ takes on the value y. The random variables $\mathbf{X}$ and $\mathbf{Y}$ are said to be ***independent random variables*** if $\mathbf{Pr}[x,y] = \mathbf{Pr}[x]\mathbf{Pr}[y]$ for all $x \in X$ and $y \in Y$.

Joint probability can be related to conditional probability by the formula

$$\mathbf{Pr}[x,y] = \mathbf{Pr}[x|y]\mathbf{Pr}[y].$$

Interchanging x and y, we have that

$$\mathbf{Pr}[x,y] = \mathbf{Pr}[y|x]\mathbf{Pr}[x].$$

From these two expressions, we immediately obtain the following result, which is known as Bayes' theorem.

THEOREM 3.1 (Bayes' theorem) *If* $\mathbf{Pr}[y] > 0$, *then*

$$\mathbf{Pr}[x|y] = \frac{\mathbf{Pr}[x]\mathbf{Pr}[y|x]}{\mathbf{Pr}[y]}.$$

COROLLARY 3.2 $\mathbf{X}$ *and* $\mathbf{Y}$ *are independent random variables if and only if* $\mathbf{Pr}[x|y] = \mathbf{Pr}[x]$ *for all* $x \in X$ *and* $y \in Y$.

Example 3.2 Suppose we consider a random throw of a pair of dice. Let $\mathbf{X}$ be the random variable defined on the set $X = \{2, \ldots, 12\}$, obtained by considering the sum of two dice, as in Example 3.1. Further, suppose that $\mathbf{Y}$ is a random variable which takes on the value D if the two dice are the same (i.e., if we throw "doubles"), and the value N, otherwise. Then we have that $\mathbf{Pr}[D] = 1/6$, $\mathbf{Pr}[N] = 5/6$.

It is straightforward to compute joint and conditional probabilities for these random variables. For example, the reader can check that $\mathbf{Pr}[D|4] = 1/3$ and $\mathbf{Pr}[4|D] = 1/6$, so

$$\mathbf{Pr}[D|4]\mathbf{Pr}[4] = \mathbf{Pr}[D]\mathbf{Pr}[4|D],$$

as stated by Bayes' theorem. □

3.3 Perfect Secrecy

Throughout this section, we assume that a cryptosystem $(\mathcal{P}, \mathcal{C}, \mathcal{K}, \mathcal{E}, \mathcal{D})$ is specified, and a particular key $K \in \mathcal{K}$ is used for only one encryption. Let us suppose that there is a probability distribution on the plaintext space, $\mathcal{P}$. Thus the plaintext element defines a random variable, denoted $\mathbf{x}$. We denote the *a priori* probability that plaintext x occurs by $\mathbf{Pr}[\mathbf{x} = x]$. We also assume that the key K is chosen (by Alice and Bob) using some fixed probability distribution (often a key is chosen at random, so all keys will be equiprobable, but this need not be the case). So the key also defines a random variable, which we denote by $\mathbf{K}$. Denote the probability that key K is chosen by $\mathbf{Pr}[\mathbf{K} = K]$. Recall that the key is chosen before Alice knows what the plaintext will be. Hence, we make the reasonable assumption that the key and the plaintext are independent random variables.

The two probability distributions on $\mathcal{P}$ and $\mathcal{K}$ induce a probability distribution on $\mathcal{C}$. Thus, we can also consider the ciphertext element to be a random variable, say $\mathbf{y}$. It is not hard to compute the probability $\mathbf{Pr}[\mathbf{y} = y]$ that y is the ciphertext that is transmitted. For a key $K \in \mathcal{K}$, define

$$C(K) = \{e_K(x) : x \in \mathcal{P}\}.$$

That is, $C(K)$ represents the set of possible ciphertexts if K is the key. Then, for every $y \in \mathcal{C}$, we have that

$$\mathbf{Pr}[\mathbf{y} = y] = \sum_{\{K: y \in C(K)\}} \mathbf{Pr}[\mathbf{K} = K]\mathbf{Pr}[\mathbf{x} = d_K(y)].$$

We also observe that, for any $y \in \mathcal{C}$ and $x \in \mathcal{P}$, we can compute the conditional probability $\mathbf{Pr}[\mathbf{y} = y | \mathbf{x} = x]$ (i.e., the probability that y is the ciphertext, given that x is the plaintext) to be

$$\mathbf{Pr}[\mathbf{y} = y | \mathbf{x} = x] = \sum_{\{K: x = d_K(y)\}} \mathbf{Pr}[\mathbf{K} = K].$$

It is now possible to compute the conditional probability $\mathbf{Pr}[\mathbf{x} = x | \mathbf{y} = y]$ (i.e., the probability that x is the plaintext, given that y is the ciphertext) using Bayes' theorem. The following formula is obtained:

$$\mathbf{Pr}[\mathbf{x} = x | \mathbf{y} = y] = \frac{\mathbf{Pr}[\mathbf{x} = x] \times \displaystyle\sum_{\{K: x = d_K(y)\}} \mathbf{Pr}[\mathbf{K} = K]}{\displaystyle\sum_{\{K: y \in C(K)\}} \mathbf{Pr}[\mathbf{K} = K]\mathbf{Pr}[\mathbf{x} = d_K(y)]}.$$

Observe that all these calculations can be performed by anyone who knows the probability distributions.

We present a toy example to illustrate the computation of these probability distributions.

Example 3.3 Let $\mathcal{P} = \{a, b\}$ with $\mathbf{Pr}[a] = 1/4, \mathbf{Pr}[b] = 3/4$. Let $\mathcal{K} = \{K_1, K_2, K_3\}$ with $\mathbf{Pr}[K_1] = 1/2, \mathbf{Pr}[K_2] = \mathbf{Pr}[K_3] = 1/4$. Let $\mathcal{C} = \{1, 2, 3, 4\}$, and suppose the encryption functions are defined to be $e_{K_1}(a) = 1, e_{K_1}(b) = 2$; $e_{K_2}(a) = 2, e_{K_2}(b) = 3$; and $e_{K_3}(a) = 3, e_{K_3}(b) = 4$. This cryptosystem can be represented by the following ***encryption matrix***:

	a	b
K_1	1	2
K_2	2	3
K_3	3	4

We now compute the probability distribution on $\mathcal{C}$. We obtain the following:

$$\begin{aligned}
\mathbf{Pr}[1] &= \frac{1}{8} \\
\mathbf{Pr}[2] &= \frac{3}{8} + \frac{1}{16} = \frac{7}{16} \\
\mathbf{Pr}[3] &= \frac{3}{16} + \frac{1}{16} = \frac{1}{4} \\
\mathbf{Pr}[4] &= \frac{3}{16}.
\end{aligned}$$

Now we can compute the conditional probability distributions on the plaintext, given that a certain ciphertext has been observed. We have:

$$\begin{aligned}
\mathbf{Pr}[a|1] &= 1 & \mathbf{Pr}[b|1] &= 0 \\
\mathbf{Pr}[a|2] &= \frac{1}{7} & \mathbf{Pr}[b|2] &= \frac{6}{7} \\
\mathbf{Pr}[a|3] &= \frac{1}{4} & \mathbf{Pr}[b|3] &= \frac{3}{4} \\
\mathbf{Pr}[a|4] &= 0 & \mathbf{Pr}[b|4] &= 1.
\end{aligned}$$

□

We are now ready to define the concept of perfect secrecy. Informally, perfect secrecy means that Oscar can obtain no information about the plaintext by observing the ciphertext. This idea is made precise by formulating it in terms of the probability distributions we have defined, as follows.

Definition 3.3: A cryptosystem has ***perfect secrecy*** if $\mathbf{Pr}[x|y] = \mathbf{Pr}[x]$ for all $x \in \mathcal{P}$, $y \in \mathcal{C}$. That is, the *a posteriori* probability that the plaintext is x, given that the ciphertext y is observed, is identical to the *a priori* probability that the plaintext is x.

In Example 3.3, the perfect secrecy property is satisfied for the ciphertext $y = 3$, but not for the other three ciphertexts.

We now prove that the *Shift Cipher* provides perfect secrecy. This seems quite obvious intuitively. For, if we are given any ciphertext element $y \in \mathbb{Z}_{26}$, then any plaintext element $x \in \mathbb{Z}_{26}$ is a possible decryption of y, depending on the value of the key. The following theorem gives the formal statement and proof using probability distributions.

THEOREM 3.3 *Suppose the 26 keys in the Shift Cipher are used with equal probability 1/26. Then for any plaintext probability distribution, the Shift Cipher has perfect secrecy.*

PROOF Recall that $\mathcal{P} = \mathcal{C} = \mathcal{K} = \mathbb{Z}_{26}$, and for $0 \leq K \leq 25$, the encryption rule e_K is defined as $e_K(x) = (x + K) \bmod 26$ $(x \in \mathbb{Z}_{26})$. First, we compute the probability distribution on $\mathcal{C}$. Let $y \in \mathbb{Z}_{26}$; then

$$\begin{aligned}
\mathbf{Pr}[\mathbf{y} = y] &= \sum_{K \in \mathbb{Z}_{26}} \mathbf{Pr}[\mathbf{K} = K]\mathbf{Pr}[\mathbf{x} = d_K(y)] \\
&= \sum_{K \in \mathbb{Z}_{26}} \frac{1}{26}\mathbf{Pr}[\mathbf{x} = y - K] \\
&= \frac{1}{26} \sum_{K \in \mathbb{Z}_{26}} \mathbf{Pr}[\mathbf{x} = y - K].
\end{aligned}$$

Now, for fixed y, the values $(y - K) \bmod 26$ comprise a permutation of $\mathbb{Z}_{26}$. Hence we have that

$$\begin{aligned}
\sum_{K \in \mathbb{Z}_{26}} \mathbf{Pr}[\mathbf{x} = y - K] &= \sum_{x \in \mathbb{Z}_{26}} \mathbf{Pr}[\mathbf{x} = x] \\
&= 1.
\end{aligned}$$

Consequently,

$$\mathbf{Pr}[y] = \frac{1}{26}$$

for any $y \in \mathbb{Z}_{26}$.

Next, we have that

$$\begin{aligned}
\mathbf{Pr}[y|x] &= \mathbf{Pr}[\mathbf{K} = (y - x) \bmod 26] \\
&= \frac{1}{26}
\end{aligned}$$

for every x, y. (This is true because, for every x, y, the unique key K such that $e_K(x) = y$ is $K = (y - x) \bmod 26$.) Now, using Bayes' theorem, it is trivial to compute

$$\begin{aligned}
\mathbf{Pr}[x|y] &= \frac{\mathbf{Pr}[x]\mathbf{Pr}[y|x]}{\mathbf{Pr}[y]} \\
&= \frac{\mathbf{Pr}[x]\frac{1}{26}}{\frac{1}{26}} \\
&= \mathbf{Pr}[x],
\end{aligned}$$

so we have perfect secrecy. ∎

Hence, the *Shift Cipher* is "unbreakable" provided that a new random key is used to encrypt every plaintext character.

It might be worthwhile to pause and consider why an exhaustive key search will not succeed in breaking a cryptosystem that achieves perfect secrecy. We will discuss this using the preceding example of the *Shift Cipher*, but a similar analysis applies to any cryptosystem that satisfies the "perfect secrecy" property. Remember that it is only allowed to encrypt one plaintext character using an unknown secret key K. When a ciphertext y is observed, an exhaustive key search would consider all the possible keys, $K = 0, 1, \ldots, 25$. For purposes of illustration, consider $y = 10$. We could certainly make a list of the decryptions of this ciphertext under all 26 possible keys. We would then see that $K = 0 \leftrightarrow x = 10$, $K = 1 \leftrightarrow x = 9$, $K = 2 \leftrightarrow x = 8, \ldots, K = 25 \leftrightarrow x = 11$. As we consider all 26 possible keys, we get a corresponding list of all 26 possible plaintexts. So no plaintexts can be ruled out by this process!

Let us next investigate perfect secrecy in general. If $\mathbf{Pr}[x_0] = 0$ for some $x_0 \in \mathcal{P}$, then it is trivially the case that $\mathbf{Pr}[x_0|y] = \mathbf{Pr}[x_0]$ for all $y \in \mathcal{C}$. So we need only consider those plaintext elements $x \in \mathcal{P}$ such that $\mathbf{Pr}[x] > 0$. For such plaintexts, we observe that, using Bayes' theorem, the condition that $\mathbf{Pr}[x|y] = \mathbf{Pr}[x]$ for all $y \in \mathcal{C}$ is equivalent to $\mathbf{Pr}[y|x] = \mathbf{Pr}[y]$ for all $y \in \mathcal{C}$. Now, let us make the reasonable assumption that $\mathbf{Pr}[y] > 0$ for all $y \in \mathcal{C}$ (if $\mathbf{Pr}[y] = 0$, then ciphertext y is never used and can be omitted from $\mathcal{C}$).

Fix any $x \in \mathcal{P}$. For each $y \in \mathcal{C}$, we have $\mathbf{Pr}[y|x] = \mathbf{Pr}[y] > 0$. Hence, for each $y \in \mathcal{C}$, there must be at least one key K such that $e_K(x) = y$. It follows that $|\mathcal{K}| \geq |\mathcal{C}|$. In any cryptosystem, we must have $|\mathcal{C}| \geq |\mathcal{P}|$ since each encoding rule is injective. In the case of equality, where $|\mathcal{K}| = |\mathcal{C}| = |\mathcal{P}|$, we can give a nice characterization of when perfect secrecy can be obtained. This characterization is originally due to Shannon.

THEOREM 3.4 *Suppose $(\mathcal{P}, \mathcal{C}, \mathcal{K}, \mathcal{E}, \mathcal{D})$ is a cryptosystem where $|\mathcal{K}| = |\mathcal{C}| = |\mathcal{P}|$. Then the cryptosystem provides perfect secrecy if and only if every key is used with equal probability $1/|\mathcal{K}|$, and for every $x \in \mathcal{P}$ and every $y \in \mathcal{C}$, there is a unique key K such that $e_K(x) = y$.*

PROOF Suppose the given cryptosystem provides perfect secrecy. As observed above, for each $x \in \mathcal{P}$ and $y \in \mathcal{C}$, there must be at least one key K such that $e_K(x) = y$. So we have the inequalities:

$$\begin{aligned} |\mathcal{C}| &= |\{e_K(x) : K \in \mathcal{K}\}| \\ &\leq |\mathcal{K}|. \end{aligned}$$

But we are assuming that $|\mathcal{C}| = |\mathcal{K}|$. Hence, it must be the case that

$$|\{e_K(x) : K \in \mathcal{K}\}| = |\mathcal{K}|.$$

That is, there do not exist two distinct keys K_1 and K_2 such that $e_{K_1}(x) = e_{K_2}(x) = y$. Hence, we have shown that for any $x \in \mathcal{P}$ and $y \in \mathcal{C}$, there is exactly one key K such that $e_K(x) = y$.

Denote $n = |\mathcal{K}|$. Let $\mathcal{P} = \{x_i : 1 \leq i \leq n\}$ and fix a ciphertext element $y \in \mathcal{C}$. We can name the keys $K_1, K_2, \ldots, K_n$, in such a way that $e_{K_i}(x_i) = y$, $1 \leq i \leq n$. Using

Cryptosystem 3.1: *One-time Pad*

Let $n \geq 1$ be an integer, and take $\mathcal{P} = \mathcal{C} = \mathcal{K} = (\mathbb{Z}_2)^n$. For $K \in (\mathbb{Z}_2)^n$, define $e_K(x)$ to be the vector sum modulo 2 of K and x (or, equivalently, the exclusive-or of the two associated bitstrings). So, if $x = (x_1, \ldots, x_n)$ and $K = (K_1, \ldots, K_n)$, then

$$e_K(x) = (x_1 + K_1, \ldots, x_n + K_n) \bmod 2.$$

Decryption is identical to encryption. If $y = (y_1, \ldots, y_n)$, then

$$d_K(y) = (y_1 + K_1, \ldots, y_n + K_n) \bmod 2.$$

Bayes' theorem, we have

$$\begin{aligned}\mathbf{Pr}[x_i|y] &= \frac{\mathbf{Pr}[y|x_i]\mathbf{Pr}[x_i]}{\mathbf{Pr}[y]} \\ &= \frac{\mathbf{Pr}[\mathbf{K} = K_i]\mathbf{Pr}[x_i]}{\mathbf{Pr}[y]}.\end{aligned}$$

Consider the perfect secrecy condition $\mathbf{Pr}[x_i|y] = \mathbf{Pr}[x_i]$. From this, it follows that $\mathbf{Pr}[K_i] = \mathbf{Pr}[y]$, for $1 \leq i \leq n$. This says that all the keys are used with equal probability (namely, $\mathbf{Pr}[y]$). But since the number of keys is $|\mathcal{K}|$, we must have that $\mathbf{Pr}[K] = 1/|\mathcal{K}|$ for every $K \in \mathcal{K}$.

Conversely, suppose the two hypothesized conditions are satisfied. Then the cryptosystem is easily seen to provide perfect secrecy for any plaintext probability distribution, in a manner similar to the proof of Theorem 3.3. We leave the details for the reader. ∎

One well-known realization of perfect secrecy is the *One-time Pad*, which was first described by Gilbert Vernam in 1917 for use in automatic encryption and decryption of telegraph messages. It is interesting that the *One-time Pad* was thought for many years to be an "unbreakable" cryptosystem, but there was no mathematical proof of this until Shannon developed the concept of perfect secrecy over 30 years later. The *One-time Pad* is presented as Cryptosystem 3.1.

Using Theorem 3.4, it is easily seen that the *One-time Pad* provides perfect secrecy. The system is also attractive because of the ease of encryption and decryption. Vernam patented his idea in the hope that it would have widespread commercial use. Unfortunately, there are major disadvantages to unconditionally secure cryptosystems such as the *One-time Pad*. The fact that $|\mathcal{K}| \geq |\mathcal{P}|$ means that the amount of key that must be communicated securely is at least as big as the amount of plaintext. For example, in the case of the *One-time Pad*, we require n bits of key to encrypt n bits of plaintext. This would not be a major problem if the same key could be used to encrypt different messages; however, the security of unconditionally secure cryptosystems depends on the fact that each key is used for only one encryption. (This is the reason for the adjective "one-time" in the *One-time Pad*.)

For example, the *One-time Pad* is vulnerable to a known-plaintext attack, since K can be computed as the exclusive-or of the bitstrings x and $e_K(x)$. Hence, a new key needs to be generated and communicated over a secure channel for every message that is going to be sent. This creates severe key management problems, which has limited the use of the *One-time Pad* in commercial applications. However, the *One-time Pad* has been employed in military and diplomatic contexts, where unconditional security may be of great importance.

The historical development of cryptography has been to try to design cryptosystems where one key can be used to encrypt a relatively long string of plaintext (i.e., one key can be used to encrypt many messages) and still maintain some measure of computational security. Cryptosystems of this type include the *Data Encryption Standard* and the *Advanced Encryption Standard*, which we will discuss in the next chapter.

3.4 Entropy

In the previous section, we discussed the concept of perfect secrecy. We restricted our attention to the special situation where a key is used for only one encryption. We now want to look at what happens as more and more plaintexts are encrypted using the same key, and how likely a cryptanalyst will be able to carry out a successful ciphertext-only attack, given sufficient time.

The basic tool in studying this question is the idea of entropy, a concept from information theory introduced by Shannon in 1948. Entropy can be thought of as a mathematical measure of information or uncertainty, and is computed as a function of a probability distribution.

Suppose we have a discrete random variable $\mathbf{X}$ which takes values from a finite set X according to a specified probability distribution. What is the information gained by the outcome of an experiment which takes place according to this probability distribution? Equivalently, if the experiment has not (yet) taken place, what is the uncertainty about the outcome? This quantity is called the entropy of $\mathbf{X}$ and is denoted by $H(\mathbf{X})$.

These ideas may seem rather abstract, so let's look at a more concrete example. Suppose our random variable $\mathbf{X}$ represents the toss of a coin. As mentioned earlier, the associated probability distribution is $\mathbf{Pr}[heads] = \mathbf{Pr}[tails] = 1/2$. It would seem reasonable to say that the information, or entropy, of a coin toss is one bit, since we could encode *heads* by 1 and *tails* by 0, for example. In a similar fashion, the entropy of n independent coin tosses is n, since the n coin tosses can be encoded by a bitstring of length n.

As a slightly more complicated example, suppose we have a random variable $\mathbf{X}$ that takes on three possible values x_1, x_2, x_3 with probabilities $1/2, 1/4, 1/4$ respectively. Suppose we encode the three possible outcomes as follows: x_1 is encoded as 0, x_2 is encoded as 10, and x_3 is encoded as 11. Then the (weighted) average number of bits in this encoding of $\mathbf{X}$ is

$$\frac{1}{2} \times 1 + \frac{1}{4} \times 2 + \frac{1}{4} \times 2 = \frac{3}{2}.$$

The above examples suggest that an event which occurs with probability 2^{-n} could perhaps be encoded as a bitstring of length n. More generally, we could plausibly imagine that an outcome occurring with probability p might be encoded by a bitstring of length approximately $-\log_2 p$. Given an arbitrary probability distribution, taking on the values $p_1, p_2, \dots, p_n$ for a random variable $\mathbf{X}$, we take the weighted average of the quantities $-\log_2 p_i$ to be our measure of information. This motivates the following formal definition.

Definition 3.4: Suppose $\mathbf{X}$ is a discrete random variable that takes on values from a finite set X. Then, the ***entropy*** of the random variable $\mathbf{X}$ is defined to be the quantity

$$H(\mathbf{X}) = -\sum_{x \in X} \mathbf{Pr}[x] \log_2 \mathbf{Pr}[x].$$

REMARK Observe that $\log_2 y$ is undefined if $y = 0$. Hence, entropy is sometimes defined to be the relevant sum over all the non-zero probabilities. However, since $\lim_{y \to 0} y \log_2 y = 0$, there is no real difficulty with allowing $\mathbf{Pr}[x] = 0$ for some x's. Also, we note that the choice of two as the base of the logarithms is arbitrary: another base would only change the value of the entropy by a constant factor. ∎

Note that if $|X| = n$ and $\mathbf{Pr}[x] = 1/n$ for all $x \in X$, then $H(\mathbf{X}) = \log_2 n$. Also, it is easy to see that $H(\mathbf{X}) \geq 0$ for any random variable $\mathbf{X}$, and $H(\mathbf{X}) = 0$ if and only if $\mathbf{Pr}[x_0] = 1$ for some $x_0 \in X$ and $\mathbf{Pr}[x] = 0$ for all $x \neq x_0$.

Let us look at the entropy of the various components of a cryptosystem. We can think of the key as being a random variable $\mathbf{K}$ that takes on values in $\mathcal{K}$, and hence we can compute the entropy $H(\mathbf{K})$. Similarly, we can compute entropies $H(\mathbf{P})$ and $H(\mathbf{C})$ of random variables associated with the plaintext and ciphertext, respectively.

To illustrate, we compute the entropies of the cryptosystem of Example 3.3.

Example 3.3 (Cont.) We compute as follows:

$$\begin{aligned} H(\mathbf{P}) &= -\frac{1}{4}\log_2 \frac{1}{4} - \frac{3}{4}\log_2 \frac{3}{4} \\ &= -\frac{1}{4}(-2) - \frac{3}{4}(\log_2 3 - 2) \\ &= 2 - \frac{3}{4}(\log_2 3) \\ &\approx 0.81. \end{aligned}$$

Similar calculations yield $H(\mathbf{K}) = 1.5$ and $H(\mathbf{C}) \approx 1.85$. □

3.4.1 Properties of Entropy

In this section, we prove some fundamental results concerning entropy. First, we state a fundamental result, known as Jensen's inequality, that will be very useful to us. Jensen's inequality involves concave functions, which we now define.

Definition 3.5: A real-valued function f is a ***concave function*** on an interval I if

$$f\left(\frac{x+y}{2}\right) \geq \frac{f(x)+f(y)}{2}$$

for all $x, y \in I$. f is a ***strictly concave function*** on an interval I if

$$f\left(\frac{x+y}{2}\right) > \frac{f(x)+f(y)}{2}$$

for all $x, y \in I$, $x \neq y$.

Here is ***Jensen's inequality***, which we state without proof.

THEOREM 3.5 (Jensen's inequality) *Suppose f is a continuous strictly concave function on the interval I. Suppose further that*

$$\sum_{i=1}^{n} a_i = 1$$

and $a_i > 0$, $1 \leq i \leq n$. Then

$$\sum_{i=1}^{n} a_i f(x_i) \leq f\left(\sum_{i=1}^{n} a_i x_i\right),$$

where $x_i \in I$, $1 \leq i \leq n$. Further, equality occurs if and only if $x_1 = \cdots = x_n$.

We now proceed to derive several results on entropy. In the next theorem, we make use of the fact that the function $\log_2 x$ is strictly concave on the interval $(0, \infty)$. (In fact, this follows easily from elementary calculus since the second derivative of the logarithm function is negative on the interval $(0, \infty)$.)

THEOREM 3.6 *Suppose $\mathbf{X}$ is a random variable having a probability distribution that takes on the values $p_1, p_2, \ldots, p_n$, where $p_i > 0$, $1 \leq i \leq n$. Then $H(\mathbf{X}) \leq \log_2 n$, with equality if and only if $p_i = 1/n$, $1 \leq i \leq n$.*

PROOF Applying Jensen's inequality, we have the following:

$$\begin{aligned} H(\mathbf{X}) &= -\sum_{i=1}^{n} p_i \log_2 p_i \\ &= \sum_{i=1}^{n} p_i \log_2 \frac{1}{p_i} \\ &\leq \log_2 \sum_{i=1}^{n} \left(p_i \times \frac{1}{p_i}\right) \\ &= \log_2 n. \end{aligned}$$

Further, equality occurs if and only if $p_i = 1/n, 1 \leq i \leq n$. ∎

THEOREM 3.7 $H(\mathbf{X}, \mathbf{Y}) \leq H(\mathbf{X}) + H(\mathbf{Y})$, *with equality if and only if* **X** *and* **Y** *are independent random variables.*

PROOF Suppose **X** takes on values x_i, $1 \leq i \leq m$, and **Y** takes on values y_j, $1 \leq j \leq n$. Denote $p_i = \mathbf{Pr}[\mathbf{X} = x_i]$, $1 \leq i \leq m$, and $q_j = \mathbf{Pr}[\mathbf{Y} = y_j]$, $1 \leq j \leq n$. Then define $r_{ij} = \mathbf{Pr}[\mathbf{X} = x_i, \mathbf{Y} = y_j]$, $1 \leq i \leq m$, $1 \leq j \leq n$ (this is the joint probability distribution).

Observe that

$$p_i = \sum_{j=1}^{n} r_{ij}$$

($1 \leq i \leq m$), and

$$q_j = \sum_{i=1}^{m} r_{ij}$$

($1 \leq j \leq n$). We compute as follows:

$$\begin{aligned} H(\mathbf{X}) + H(\mathbf{Y}) &= -\left(\sum_{i=1}^{m} p_i \log_2 p_i + \sum_{j=1}^{n} q_j \log_2 q_j\right) \\ &= -\left(\sum_{i=1}^{m}\sum_{j=1}^{n} r_{ij} \log_2 p_i + \sum_{j=1}^{n}\sum_{i=1}^{m} r_{ij} \log_2 q_j\right) \\ &= -\sum_{i=1}^{m}\sum_{j=1}^{n} r_{ij} \log_2 p_i q_j. \end{aligned}$$

On the other hand,

$$H(\mathbf{X}, \mathbf{Y}) = -\sum_{i=1}^{m}\sum_{j=1}^{n} r_{ij} \log_2 r_{ij}.$$

Combining, we obtain the following:

$$\begin{aligned}
H(\mathbf{X},\mathbf{Y}) - H(\mathbf{X}) - H(\mathbf{Y}) &= \sum_{i=1}^{m}\sum_{j=1}^{n} r_{ij}\log_2 \frac{1}{r_{ij}} + \sum_{i=1}^{m}\sum_{j=1}^{n} r_{ij}\log_2 p_i q_j \\
&= \sum_{i=1}^{m}\sum_{j=1}^{n} r_{ij}\log_2 \frac{p_i q_j}{r_{ij}} \\
&\leq \log_2 \sum_{i=1}^{m}\sum_{j=1}^{n} p_i q_j \\
&= \log_2 1 \\
&= 0.
\end{aligned}$$

(In the above computations, we apply Jensen's inequality, using the fact that the r_{ij}'s are positive real numbers that sum to 1.)

We can also say when equality occurs: it must be the case that there is a constant c such that $p_i q_j / r_{ij} = c$ for all i, j. Using the fact that

$$\sum_{j=1}^{n}\sum_{i=1}^{m} r_{ij} = \sum_{j=1}^{n}\sum_{i=1}^{m} p_i q_j = 1,$$

it follows that $c = 1$. Hence, equality occurs if and only if $r_{ij} = p_i q_j$, i.e., if and only if

$$\mathbf{Pr}[\mathbf{X} = x_i, \mathbf{Y} = y_j] = \mathbf{Pr}[\mathbf{X} = x_i]\mathbf{Pr}[\mathbf{Y} = y_j],$$

$1 \leq i \leq m, 1 \leq j \leq n$. But this says that **X** and **Y** are independent. ∎

We next define the idea of conditional entropy.

Definition 3.6: Suppose **X** and **Y** are two random variables. Then for any fixed value y of **Y**, we get a (conditional) probability distribution on X; we denote the associated random variable by $\mathbf{X}|y$. Clearly,

$$H(\mathbf{X}|y) = -\sum_{x} \mathbf{Pr}[x|y] \log_2 \mathbf{Pr}[x|y].$$

We define the ***conditional entropy***, denoted $H(\mathbf{X}|\mathbf{Y})$, to be the weighted average (with respect to the probabilities $\mathbf{Pr}[y]$) of the entropies $H(\mathbf{X}|y)$ over all possible values y. It is computed to be

$$H(\mathbf{X}|\mathbf{Y}) = -\sum_{y}\sum_{x} \mathbf{Pr}[y]\mathbf{Pr}[x|y] \log_2 \mathbf{Pr}[x|y].$$

The conditional entropy measures the average amount of information about **X** that is not revealed by **Y**.

The next two results are straightforward; we leave the proofs as exercises.

THEOREM 3.8 $H(\mathbf{X},\mathbf{Y}) = H(\mathbf{Y}) + H(\mathbf{X}|\mathbf{Y})$.

COROLLARY 3.9 *$H(\mathbf{X}|\mathbf{Y}) \leq H(\mathbf{X})$, with equality if and only if $\mathbf{X}$ and $\mathbf{Y}$ are independent.*

3.5 Spurious Keys and Unicity Distance

In this section, we apply the entropy results we have proved to cryptosystems. First, we show a fundamental relationship exists among the entropies of the components of a cryptosystem. The conditional entropy $H(\mathbf{K}|\mathbf{C})$ is called the ***key equivocation***; it is a measure of the amount of uncertainty of the key remaining when the ciphertext is known.

THEOREM 3.10 *Let $(\mathcal{P},\mathcal{C},\mathcal{K},\mathcal{E},\mathcal{D})$ be a cryptosystem. Then*

$$H(\mathbf{K}|\mathbf{C}) = H(\mathbf{K}) + H(\mathbf{P}) - H(\mathbf{C}).$$

PROOF First, observe that $H(\mathbf{K},\mathbf{P},\mathbf{C}) = H(\mathbf{C}|\mathbf{K},\mathbf{P}) + H(\mathbf{K},\mathbf{P})$. Now, the key and plaintext determine the ciphertext uniquely, since $y = e_K(x)$. This implies that $H(\mathbf{C}|\mathbf{K},\mathbf{P}) = 0$. Hence, $H(\mathbf{K},\mathbf{P},\mathbf{C}) = H(\mathbf{K},\mathbf{P})$. But $\mathbf{K}$ and $\mathbf{P}$ are independent, so $H(\mathbf{K},\mathbf{P}) = H(\mathbf{K}) + H(\mathbf{P})$. Hence,

$$H(\mathbf{K},\mathbf{P},\mathbf{C}) = H(\mathbf{K},\mathbf{P}) = H(\mathbf{K}) + H(\mathbf{P}).$$

In a similar fashion, since the key and ciphertext determine the plaintext uniquely (i.e., $x = d_K(y)$), we have that $H(\mathbf{P}|\mathbf{K},\mathbf{C}) = 0$ and hence $H(\mathbf{K},\mathbf{P},\mathbf{C}) = H(\mathbf{K},\mathbf{C})$.

Now, we compute as follows:

$$\begin{aligned} H(\mathbf{K}|\mathbf{C}) &= H(\mathbf{K},\mathbf{C}) - H(\mathbf{C}) \\ &= H(\mathbf{K},\mathbf{P},\mathbf{C}) - H(\mathbf{C}) \\ &= H(\mathbf{K}) + H(\mathbf{P}) - H(\mathbf{C}), \end{aligned}$$

giving the desired formula. ∎

Let us return to Example 3.3 to illustrate this result.

Example 3.1 (Cont.) We have already computed $H(\mathbf{P}) \approx 0.81$, $H(\mathbf{K}) = 1.5$, and $H(\mathbf{C}) \approx 1.85$. Theorem 3.10 tells us that $H(\mathbf{K}|\mathbf{C}) \approx 1.5 + 0.81 - 1.85 \approx 0.46$. This can be verified directly by applying the definition of conditional entropy, as follows. First, we need to compute the probabilities $\mathbf{Pr}[\mathbf{K} = K_i|\mathbf{y} = j]$, $1 \leq i \leq 3$,

$1 \leq j \leq 4$. This can be done using Bayes' theorem, and the following values result:

$$\begin{array}{lll} \mathbf{Pr}[K_1|1] = 1 & \mathbf{Pr}[K_2|1] = 0 & \mathbf{Pr}[K_3|1] = 0 \\ \mathbf{Pr}[K_1|2] = \frac{6}{7} & \mathbf{Pr}[K_2|2] = \frac{1}{7} & \mathbf{Pr}[K_3|2] = 0 \\ \mathbf{Pr}[K_1|3] = 0 & \mathbf{Pr}[K_2|3] = \frac{3}{4} & \mathbf{Pr}[K_3|3] = \frac{1}{4} \\ \mathbf{Pr}[K_1|4] = 0 & \mathbf{Pr}[K_2|4] = 0 & \mathbf{Pr}[K_3|4] = 1. \end{array}$$

Now we compute

$$H(\mathbf{K}|\mathbf{C}) = \frac{1}{8} \times 0 + \frac{7}{16} \times 0.59 + \frac{1}{4} \times 0.81 + \frac{3}{16} \times 0 = 0.46,$$

agreeing with the value predicted by Theorem 3.10. □

Suppose $(\mathcal{P}, \mathcal{C}, \mathcal{K}, \mathcal{E}, \mathcal{D})$ is the cryptosystem being used, and a string of plaintext

$$x_1 x_2 \cdots x_n$$

is encrypted with one key, producing a string of ciphertext

$$y_1 y_2 \cdots y_n.$$

Recall that the basic goal of the cryptanalyst is to determine the key. We are looking at ciphertext-only attacks, and we assume that Oscar has infinite computational resources. We also assume that Oscar knows that the plaintext is a "natural" language, such as English. In general, Oscar will be able to rule out certain keys, but many "possible" keys may remain, only one of which is the correct key. The remaining possible, but incorrect, keys are called ***spurious keys***.

For example, suppose Oscar obtains the ciphertext string *WNAJW*, which has been obtained by encryption using a shift cipher. It is easy to see that there are two "meaningful" plaintext strings, namely *river* and *arena*, corresponding respectively to the possible encryption keys F $(= 5)$ and W $(= 22)$. Of these two keys, one will be the correct key and the other will be spurious. (It is rather difficult to find a ciphertext of length exceeding 5 for the *Shift Cipher* that has two meaningful decryptions; see the Exercises.)

Our goal is to prove a bound on the expected number of spurious keys. First, we have to define what we mean by the entropy (per letter) of a natural language L, which we denote H_L. H_L should be a measure of the average information per letter in a "meaningful" string of plaintext. (Note that a random string of alphabetic characters would have entropy (per letter) equal to $\log_2 26 \approx 4.70$.) As a "first-order" approximation to H_L, we could take $H(\mathbf{P})$. In the case where L is the English language, we get $H(\mathbf{P}) \approx 4.19$ by using the probability distribution given in Table 2.1.

Of course, successive letters in a language are not independent, and correlations among successive letters reduce the entropy. For example, in English, the letter "Q" is almost always followed by the letter "U." For a "second-order" approximation, we would compute the entropy of the probability distribution of all digrams and then divide by 2. In general, define $\mathbf{P}^n$ to be the random variable that has as its probability distribution that of all n-grams of plaintext. We make use of the following definitions.

Definition 3.7: Suppose L is a natural language. The ***entropy*** of L is defined to be the quantity

$$H_L = \lim_{n \to \infty} \frac{H(\mathbf{P}^n)}{n}$$

and the ***redundancy*** of L is defined to be

$$R_L = 1 - \frac{H_L}{\log_2 |\mathcal{P}|}.$$

REMARK H_L measures the entropy per letter of the language L. A random language would have entropy $\log_2 |\mathcal{P}|$. So the quantity R_L measures the fraction of "excess characters," which we think of as redundancy. ∎

In the case of the English language, a tabulation of a large number of digrams and their frequencies would produce an estimate for $H(\mathbf{P}^2)$. $H(\mathbf{P}^2)/2 \approx 3.90$ is one estimate obtained in this way. One could continue, tabulating trigrams, etc. and thus obtain an estimate for H_L. In fact, various experiments have yielded the empirical result that $1.0 \leq H_L \leq 1.5$. That is, the average information content in English is something like one to one-and-a-half bits per letter!

Using 1.25 as our estimate of H_L gives a redundancy of about 0.75. This means that the English language is 75% redundant! (This is not to say that one can arbitrarily remove three out of every four letters from English text and hope to still be able to read it. What it does mean is that it is possible to find a certain "encoding" of n-grams, for a large enough value of n, which will compress English text to about one quarter of its original length.)

Given probability distributions on $\mathcal{K}$ and $\mathcal{P}^n$, we can define the induced probability distribution on $\mathcal{C}^n$, the set of n-grams of ciphertext (we already did this in the case $n = 1$). We have defined $\mathbf{P}^n$ to be a random variable representing an n-gram of plaintext. Similarly, define $\mathbf{C}^n$ to be a random variable representing an n-gram of ciphertext.

Given $\mathbf{y} \in \mathbf{C}^n$, define

$$K(\mathbf{y}) = \{K \in \mathcal{K} : \exists \mathbf{x} \in \mathcal{P}^n \text{ such that } \mathbf{Pr}[\mathbf{x}] > 0 \text{ and } e_K(\mathbf{x}) = \mathbf{y}\}.$$

That is, $K(\mathbf{y})$ is the set of keys K for which $\mathbf{y}$ is the encryption of a meaningful string of plaintext of length n, i.e., the set of "possible" keys, given that $\mathbf{y}$ is the ciphertext. If $\mathbf{y}$ is the observed string of ciphertext, then the number of spurious keys is $|K(\mathbf{y})| - 1$, since only one of the "possible" keys is the correct key. The average number of spurious keys (over all possible ciphertext strings of length n) is denoted by $\overline{s}_n$. Its value is computed to be

$$\begin{aligned}
\overline{s}_n &= \sum_{\mathbf{y} \in \mathcal{C}^n} \mathbf{Pr}[\mathbf{y}](|K(\mathbf{y})| - 1) \\
&= \sum_{\mathbf{y} \in \mathcal{C}^n} \mathbf{Pr}[\mathbf{y}]|K(\mathbf{y})| - \sum_{\mathbf{y} \in \mathcal{C}^n} \mathbf{Pr}[\mathbf{y}] \\
&= \sum_{\mathbf{y} \in \mathcal{C}^n} \mathbf{Pr}[\mathbf{y}]|K(\mathbf{y})| - 1.
\end{aligned}$$

From Theorem 3.10, we have that

$$H(\mathbf{K}|\mathbf{C}^n) = H(\mathbf{K}) + H(\mathbf{P}^n) - H(\mathbf{C}^n).$$

Also, we can use the estimate

$$H(\mathbf{P}^n) \approx nH_L = n(1 - R_L)\log_2 |\mathcal{P}|,$$

provided n is reasonably large. Certainly,

$$H(\mathbf{C}^n) \leq n \log_2 |\mathcal{C}|.$$

Then, if $|\mathcal{C}| = |\mathcal{P}|$, it follows that

$$H(\mathbf{K}|\mathbf{C}^n) \geq H(\mathbf{K}) - nR_L \log_2 |\mathcal{P}|. \tag{3.2}$$

Next, we relate the quantity $H(\mathbf{K}|\mathbf{C}^n)$ to the number of spurious keys, $\overline{s}_n$. We compute as follows:

$$\begin{aligned}
H(\mathbf{K}|\mathbf{C}^n) &= \sum_{\mathbf{y} \in \mathcal{C}^n} \mathbf{Pr}[\mathbf{y}] H(\mathbf{K}|\mathbf{y}) \\
&\leq \sum_{\mathbf{y} \in \mathcal{C}^n} \mathbf{Pr}[\mathbf{y}] \log_2 |K(\mathbf{y})| \\
&\leq \log_2 \sum_{\mathbf{y} \in \mathcal{C}^n} \mathbf{Pr}[\mathbf{y}]|K(\mathbf{y})| \\
&= \log_2(\overline{s}_n + 1),
\end{aligned}$$

where we apply Jensen's inequality (Theorem 3.5) with $f(x) = \log_2 x$. Thus we obtain the inequality

$$H(\mathbf{K}|\mathbf{C}^n) \leq \log_2(\overline{s}_n + 1). \tag{3.3}$$

Combining the two inequalities (3.2) and (3.3), we get that

$$\log_2(\overline{s}_n + 1) \geq H(\mathbf{K}) - nR_L \log_2 |\mathcal{P}|.$$

In the case where keys are chosen equiprobably (which maximizes $H(\mathbf{K})$), we have the following result.

THEOREM 3.11 *Suppose $(\mathcal{P}, \mathcal{C}, \mathcal{K}, \mathcal{E}, \mathcal{D})$ is a cryptosystem where $|\mathcal{C}| = |\mathcal{P}|$ and keys are chosen equiprobably. Let R_L denote the redundancy of the underlying language. Then given a string of ciphertext of length n, where n is sufficiently large, the expected number of spurious keys, $\overline{s}_n$, satisfies*

$$\overline{s}_n \geq \frac{|\mathcal{K}|}{|\mathcal{P}|^{nR_L}} - 1.$$

The quantity $|\mathcal{K}|/|\mathcal{P}|^{nR_L} - 1$ approaches 0 exponentially quickly as n increases. Also, note that the estimate may not be accurate for small values of n, especially since $H(\mathbf{P}^n)/n$ may not be a good estimate for H_L if n is small.

We have one more concept to define.

Definition 3.8: The ***unicity distance*** of a cryptosystem is defined to be the value of n, denoted by n_0, at which the expected number of spurious keys becomes zero; i.e., the average amount of ciphertext required for an opponent to be able to uniquely compute the key, given enough computing time.

If we set $\overline{s}_n = 0$ in Theorem 3.11 and solve for n, we get an estimate for the unicity distance, namely

$$n_0 \approx \frac{\log_2 |\mathcal{K}|}{R_L \log_2 |\mathcal{P}|}.$$

As an example, consider the *Substitution Cipher*. In this cryptosystem, $|\mathcal{P}| = 26$ and $|\mathcal{K}| = 26!$. If we take $R_L = 0.75$, then we get an estimate for the unicity distance of

$$n_0 \approx \frac{88.4}{0.75 \times 4.7} \approx 25.$$

This suggests that, given a ciphertext string of length at least 25, (usually) a unique decryption is possible.

3.6 Notes and References

The idea of perfect secrecy and the use of entropy techniques in cryptography was pioneered by Claude Shannon [178]. The concept of entropy was also defined by Shannon, in [177]. Good introductions to entropy and related topics can be found in the following books:

- *Codes and Cryptography* by Dominic Welsh [200]
- *Communication Theory* by Charles Goldie and Richard Pinch [88].

The results of Section 3.5 are due to Beauchemin and Brassard [11], who generalized earlier results of Shannon.

Exercises

3.1 Referring to Example 3.1, suppose we define the event

$$T_d\{(i,j) \in Z : |i-j| = d\},$$

for $0 \leq d \leq 5$. (That is, the event T_d corresponds to the situation where the difference of a pair of dice is equal to d.) Compute the probabilities $\mathbf{Pr}[T_d]$, $0 \leq d \leq 5$

3.2 Referring to Example 3.2, determine all the joint and conditional probabilities, $\mathbf{Pr}[x,y]$, $\mathbf{Pr}[x|y]$, and $\mathbf{Pr}[y|x]$, where $x \in \{2,\ldots,12\}$ and $y \in \{D,N\}$.

3.3 Let n be a positive integer. A ***Latin square*** of order n is an $n \times n$ array L of the integers $1,\ldots,n$ such that every one of the n integers occurs exactly once in each row and each column of L. An example of a Latin square of order 3 is as follows:

1	2	3
3	1	2
2	3	1

Given any Latin square L of order n, we can define a related *Latin Square Cryptosystem*. Take $\mathcal{P} = \mathcal{C} = \mathcal{K} = \{1,\ldots,n\}$. For $1 \leq i \leq n$, the encryption rule e_i is defined to be $e_i(j) = L(i,j)$. (Hence each row of L gives rise to one encryption rule.)

Give a complete proof that this *Latin Square Cryptosystem* achieves perfect secrecy provided that every key is used with equal probability.

3.4 Let $\mathcal{P} = \{a,b\}$ and let $\mathcal{K} = \{K_1,K_2,K_3,K_4,K_5\}$. Let $\mathcal{C} = \{1,2,3,4,5\}$, and suppose the encryption functions are represented by the following encryption matrix:

	a	b
K_1	1	2
K_2	2	3
K_3	3	1
K_4	4	5
K_5	5	4

Now choose two positive real numbers α and β such that $\alpha + \beta = 1$, and define $\mathbf{Pr}[K_1] = \mathbf{Pr}[K_2] = \mathbf{Pr}[K_3] = \alpha/3$ and $\mathbf{Pr}[K_4] = \mathbf{Pr}[K_5] = \beta/2$.

Prove that this cryptosystem achieves perfect secrecy.

3.5 (a) Prove that the *Affine Cipher* achieves perfect secrecy if every key is used with equal probability $1/312$.

(b) More generally, suppose we are given a probability distribution on the set

$$\{a \in \mathbb{Z}_{26} : \gcd(a, 26) = 1\}.$$

Suppose that every key (a, b) for the *Affine Cipher* is used with probability $\mathbf{Pr}[a]/26$. Prove that the *Affine Cipher* achieves perfect secrecy when this probability distribution is defined on the keyspace.

3.6 Suppose a cryptosystem achieves perfect secrecy for a particular plaintext probability distribution. Prove that perfect secrecy is maintained for any plaintext probability distribution.

3.7 Prove that if a cryptosystem has perfect secrecy and $|\mathcal{K}| = |\mathcal{C}| = |\mathcal{P}|$, then every ciphertext is equally probable.

3.8 Suppose that y and y' are two ciphertext elements (i.e., binary n-tuples) in the *One-time Pad* that were obtained by encrypting plaintext elements x and x', respectively, using the same key, K. Prove that $x + x' \equiv y + y' \pmod 2$.

3.9 (a) Construct the encryption matrix (as defined in Example 3.3) for the *One-time Pad* with $n = 3$.

(b) For any positive integer n, give a direct proof that the encryption matrix of a *One-time Pad* defined over $(\mathbb{Z}_2)^n$ is a Latin square of order 2^n, in which the symbols are the elements of $(\mathbb{Z}_2)^n$.

3.10 Suppose that $\mathbf{S}$ is a random variable representing the sum of a pair of dice (see Example 3.1). Compute $H(\mathbf{S})$.

3.11 Prove from first principles (i.e., using the definition) that the function $f(x) = x^2$ is concave over the interval $(-\infty, \infty)$.

3.12 Prove that $H(\mathbf{X}, \mathbf{Y}) = H(\mathbf{Y}) + H(\mathbf{X}|\mathbf{Y})$. Then show as a corollary that $H(\mathbf{X}|\mathbf{Y}) \leq H(\mathbf{X})$, with equality if and only if $\mathbf{X}$ and $\mathbf{Y}$ are independent.

3.13 Prove that a cryptosystem has perfect secrecy if and only if $H(\mathbf{P}|\mathbf{C}) = H(\mathbf{P})$.

3.14 Prove that, in any cryptosystem, $H(\mathbf{K}|\mathbf{C}) \geq H(\mathbf{P}|\mathbf{C})$. (Intuitively, this result says that, given a ciphertext, the opponent's uncertainty about the key is at least as great as his uncertainty about the plaintext.)

3.15 Consider a cryptosystem in which $\mathcal{P} = \{a, b, c\}$, $\mathcal{K} = \{K_1, K_2, K_3\}$ and $\mathcal{C} = \{1, 2, 3, 4\}$. Suppose the encryption matrix is as follows:

	a	b	c
K_1	1	2	3
K_2	2	3	4
K_3	3	4	1

Given that keys are chosen equiprobably, and the plaintext probability distribution is $\mathbf{Pr}[a] = 1/2$, $\mathbf{Pr}[b] = 1/3$, $\mathbf{Pr}[c] = 1/6$, compute $H(\mathbf{P})$, $H(\mathbf{C})$, $H(\mathbf{K})$, $H(\mathbf{K}|\mathbf{C})$, and $H(\mathbf{P}|\mathbf{C})$.

3.16 Compute $H(\mathbf{K}|\mathbf{C})$ and $H(\mathbf{K}|\mathbf{P},\mathbf{C})$ for the *Affine Cipher*, assuming that keys are used equiprobably and the plaintexts are equiprobable.

3.17 Suppose that *APNDJI* or *XYGROBO* are ciphertexts that are obtained from encryption using the *Shift Cipher*. Show in each case that there are two "meaningful" plaintexts that could encrypt to the given ciphertext. (Thanks to John van Rees for these examples.)

3.18 Consider a *Vigenère Cipher* with keyword length m. Show that the unicity distance is $1/R_L$, where R_L is the redundancy of the underlying language. (This result is interpreted as follows. If n_0 denotes the number of alphabetic characters being encrypted, then the "length" of the plaintext is n_0/m, since each plaintext element consists of m alphabetic characters. So, a unicity distance of $1/R_L$ corresponds to a plaintext consisting of m/R_L alphabetic characters.)

3.19 Show that the unicity distance of the *Hill Cipher* (with an $m \times m$ encryption matrix) is less than m/R_L. (Note that the number of alphabetic characters in a plaintext of this length is m^2/R_L.)

3.20 A *Substitution Cipher* over a plaintext space of size n has $|\mathcal{K}| = n!$ ***Stirling's formula*** gives the following estimate for $n!$:

$$n! \approx \sqrt{2\pi n}\left(\frac{n}{e}\right)^n.$$

(a) Using Stirling's formula, derive an estimate of the unicity distance of the *Substitution Cipher*.

(b) Let $m \geq 1$ be an integer. The *m-gram Substitution Cipher* is the *Substitution Cipher* where the plaintext (and ciphertext) spaces consist of all 26^m m-grams. Estimate the unicity distance of the *m-gram Substitution Cipher* if $R_L = 0.75$.

Chapter 4

Block Ciphers and Stream Ciphers

This chapter discusses various aspects of block and stream ciphers. We introduce the substitution-permutation network as a design technique for block ciphers and we discuss some standard attacks. We look at standards such as the *Data Encryption Standard* and *Advanced Encryption Standard*. Modes of operation are discussed and we also provide a brief treatment of stream ciphers.

4.1 Introduction

Most modern-day block ciphers incorporate a sequence of permutation and substitution operations. A commonly used design is that of an ***iterated cipher***. An iterated cipher requires the specification of a ***round function*** and a ***key schedule***, and the encryption of a plaintext will proceed through $\mathcal{N}$ similar ***rounds***.

Let K be a random binary key of some specified length. K is used to construct $\mathcal{N}$ ***round keys*** (also called ***subkeys***), which are denoted $K^1, \ldots, K^{\mathcal{N}}$. The list of round keys, $(K^1, \ldots, K^{\mathcal{N}})$, is the key schedule. The key schedule is constructed from K using a fixed, public algorithm.

The round function, say g, takes two inputs: a round key (K^r) and a current ***state*** (which we denote w^{r-1}). The next state is defined as $w^r = g(w^{r-1}, K^r)$. The initial state, w^0, is defined to be the plaintext, x. The ciphertext, y, is defined to be the state after all $\mathcal{N}$ rounds have been performed. Therefore, the encryption operation is carried out as follows:

$$\begin{array}{rcl} w^0 & \leftarrow & x \\ w^1 & \leftarrow & g(w^0, K^1) \\ w^2 & \leftarrow & g(w^1, K^2) \\ \vdots & \vdots & \vdots \\ w^{\mathcal{N}-1} & \leftarrow & g(w^{\mathcal{N}-2}, K^{\mathcal{N}-1}) \\ w^{\mathcal{N}} & \leftarrow & g(w^{\mathcal{N}-1}, K^{\mathcal{N}}) \\ y & \leftarrow & w^{\mathcal{N}}. \end{array}$$

In order for decryption to be possible, the function g must have the property that it is injective (i.e., one-to-one) if its second argument is fixed. This is equivalent to saying that there exists a function g^{-1} with the property that

$$g^{-1}(g(w,y),y) = w$$

for all w and y. Then decryption can be accomplished as follows:

$$\begin{array}{rcl} w^{\mathcal{N}} & \leftarrow & y \\ w^{\mathcal{N}-1} & \leftarrow & g^{-1}(w^{\mathcal{N}}, K^{\mathcal{N}}) \\ \vdots & \vdots & \vdots \\ w^1 & \leftarrow & g^{-1}(w^2, K^2) \\ w^0 & \leftarrow & g^{-1}(w^1, K^1) \\ x & \leftarrow & w^0. \end{array}$$

In Section 4.2, we describe a simple type of iterated cipher, the substitution-permutation network, which illustrates many of the main principles used in the design of practical block ciphers. Linear and differential attacks on substitution-permutation networks are described in Sections 4.3 and 4.4, respectively. In Section 4.5, we discuss Feistel-type ciphers and the *Data Encryption Standard*. In Section 4.6, we present the *Advanced Encryption Standard*. Finally, modes of operation of block ciphers are the topic of Section 4.7 and stream ciphers are discussed in Section 4.8.

4.2 Substitution-Permutation Networks

We begin by defining a ***substitution-permutation network***, or ***SPN***. (An SPN is a special type of iterated cipher with a couple of small changes that we will indicate.) Suppose that ℓ and m are positive integers. A plaintext and ciphertext will both be binary vectors of length ℓm (i.e., ℓm is the ***block length*** of the cipher). An SPN is built from two components, which are denoted π_S and π_P.

$$\pi_S : \{0,1\}^\ell \rightarrow \{0,1\}^\ell$$

is a permutation of the 2^ℓ bitstrings of length ℓ, and

$$\pi_P : \{1,\dots,\ell m\} \rightarrow \{1,\dots,\ell m\}$$

is also a permutation, of the integers $1,\dots,\ell m$. The permutation π_S is called an ***S-box*** (the letter "S" denotes "substitution"). It is used to replace ℓ bits with a different set of ℓ bits. π_P, on the other hand, is used to permute ℓm bits by changing their order.

Given an ℓm-bit binary string, say $x = (x_1,\dots,x_{\ell m})$, we can regard x as the concatenation of m ℓ-bit substrings, which we denote $x_{<1>},\dots,x_{<m>}$. Thus

$$x = x_{<1>} \parallel \dots \parallel x_{<m>}$$

Cryptosystem 4.1: *Substitution-Permutation Network*

Let ℓ, m, and $\mathcal{N}$ be positive integers, let $\pi_S : \{0,1\}^\ell \rightarrow \{0,1\}^\ell$ be a permutation, and let $\pi_P : \{1,\ldots,\ell m\} \rightarrow \{1,\ldots,\ell m\}$ be a permutation. Let $\mathcal{P} = \mathcal{C} = \{0,1\}^{\ell m}$, and let $\mathcal{K} \subseteq (\{0,1\}^{\ell m})^{\mathcal{N}+1}$ consist of all possible key schedules that could be derived from an initial key K using the key scheduling algorithm. For a key schedule $(K^1,\ldots,K^{\mathcal{N}+1})$, we encrypt the plaintext x using Algorithm 4.1.

and for $1 \leq i \leq m$, we have that

$$x_{<i>} = \left(x_{(i-1)\ell+1},\ldots,x_{i\ell}\right).$$

The SPN will consist of $\mathcal{N}$ rounds. In each round (except for the last round, which is slightly different), we will perform m substitutions using π_S, followed by a permutation using π_P. Prior to each substitution operation, we will incorporate round key bits via a simple exclusive-or operation. We now present an SPN, based on π_S and π_P, as Cryptosystem 4.1.

In Algorithm 4.1, u^r is the input to the S-boxes in round r, and v^r is the output of the S-boxes in round r. w^r is obtained from v^r by applying the permutation π_P, and then u^{r+1} is constructed from w^r by x-or-ing with the round key K^{r+1} (this is called ***round key mixing***). In the last round, the permutation π_P is not applied. As a consequence, the encryption algorithm can also be used for decryption, if appropriate modifications are made to the key schedule and the S-boxes are replaced by their inverses (see the Exercises).

Notice that the very first and last operations performed in this SPN are x-ors with subkeys. This is called ***whitening***, and it is regarded as a useful way to prevent an attacker from even beginning to carry out an encryption or decryption operation if the key is not known.

We illustrate the above general description with a particular SPN.

Example 4.1 Suppose that $\ell = m = \mathcal{N} = 4$. Let π_S be defined as follows, where the input (i.e., z) and the output (i.e., $\pi_S(z)$) are written in hexadecimal notation, $(0 \leftrightarrow (0,0,0,0),\ 1 \leftrightarrow (0,0,0,1),\ \ldots,\ 9 \leftrightarrow (1,0,0,1),\ A \leftrightarrow (1,0,1,0),\ \ldots,\ F \leftrightarrow (1,1,1,1))$:

z	0	1	2	3	4	5	6	7	8	9	A	B	C	D	E	F
$\pi_S(z)$	E	4	D	1	2	F	B	8	3	A	6	C	5	9	0	7

Further, let π_P be defined as follows:

z	1	2	3	4	5	6	7	8	9	10	11	12	13	14	15	16
$\pi_P(z)$	1	5	9	13	2	6	10	14	3	7	11	15	4	8	12	16

See Figure 4.1 for a pictorial representation of this particular SPN. (In this diagram,

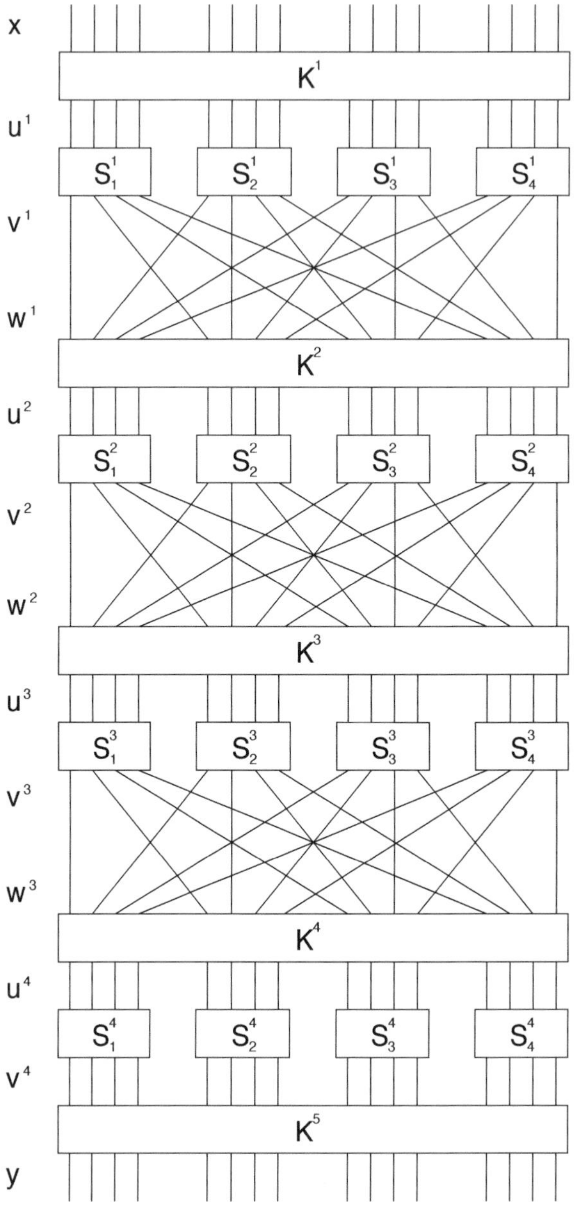

FIGURE 4.1: A substitution-permutation network

Algorithm 4.1: SPN$(x, \pi_S, \pi_P, (K^1, \dots, K^{\mathcal{N}+1}))$

$w^0 \leftarrow x$
for $r \leftarrow 1$ **to** $\mathcal{N} - 1$
do $\begin{cases} u^r \leftarrow w^{r-1} \oplus K^r \\ \textbf{for } i \leftarrow 1 \textbf{ to } m \\ \quad \textbf{do } v^r_{<i>} \leftarrow \pi_S(u^r_{<i>}) \\ w^r \leftarrow (v^r_{\pi_P(1)}, \dots, v^r_{\pi_P(\ell m)}) \end{cases}$
$u^{\mathcal{N}} \leftarrow w^{\mathcal{N}-1} \oplus K^{\mathcal{N}}$
for $i \leftarrow 1$ **to** m
  do $v^{\mathcal{N}}_{<i>} \leftarrow \pi_S(u^{\mathcal{N}}_{<i>})$
$y \leftarrow v^{\mathcal{N}} \oplus K^{\mathcal{N}+1}$
output (y)

we have named the S-boxes S_i^r ($1 \leq i \leq 4, 1 \leq r \leq 4$) for ease of later reference. All 16 S-boxes incorporate the same substitution function based on π_S.)

In order to complete the description of the SPN, we need to specify a key scheduling algorithm. Here is a simple possibility: suppose that we begin with a 32-bit key $K = (k_1, \dots, k_{32}) \in \{0,1\}^{32}$. For $1 \leq r \leq 5$, define K^r to consist of 16 consecutive bits of K, beginning with k_{4r-3}. (This is not a very secure way to define a key schedule; we have just chosen something easy for purposes of illustration.)

Now let's work out a sample encryption using this SPN. We represent all data in binary notation. Suppose the key is

$$K = \texttt{0011 1010 1001 0100 1101 0110 0011 1111}.$$

Then the round keys are as follows:

$$\begin{aligned} K^1 &= \texttt{0011 1010 1001 0100} \\ K^2 &= \texttt{1010 1001 0100 1101} \\ K^3 &= \texttt{1001 0100 1101 0110} \\ K^4 &= \texttt{0100 1101 0110 0011} \\ K^5 &= \texttt{1101 0110 0011 1111}. \end{aligned}$$

Suppose that the plaintext is

$$x = \texttt{0010 0110 1011 0111}.$$

Then the encryption of x proceeds as shown in Figure 4.2, yielding the ciphertext y.

□

$$
\begin{aligned}
w^0 &= \texttt{0010 0110 1011 0111}\\
K^1 &= \texttt{0011 1010 1001 0100}\\
u^1 &= \texttt{0001 1100 0010 0011}\\
v^1 &= \texttt{0100 0101 1101 0001}\\
w^1 &= \texttt{0010 1110 0000 0111}\\
K^2 &= \texttt{1010 1001 0100 1101}\\
u^2 &= \texttt{1000 0111 0100 1010}\\
v^2 &= \texttt{0011 1000 0010 0110}\\
w^2 &= \texttt{0100 0001 1011 1000}\\
K^3 &= \texttt{1001 0100 1101 0110}\\
u^3 &= \texttt{1101 0101 0110 1110}\\
v^3 &= \texttt{1001 1111 1011 0000}\\
w^3 &= \texttt{1110 0100 0110 1110}\\
K^4 &= \texttt{0100 1101 0110 0011}\\
u^4 &= \texttt{1010 1001 0000 1101}\\
v^4 &= \texttt{0110 1010 1110 1001}\\
K^5 &= \texttt{1101 0110 0011 1111}\\
y &= \texttt{1011 1100 1101 0110}
\end{aligned}
$$

FIGURE 4.2: Encryption using a substitution-permutation network

SPNs have several attractive features. First, the design is simple and very efficient, in both hardware and software. In software, an S-box is usually implemented in the form of a look-up table. Observe that the memory requirement of the S-box $\pi_S : \{0,1\}^\ell \rightarrow \{0,1\}^\ell$ is $\ell\, 2^\ell$ bits, since we have to store 2^ℓ values, each of which needs ℓ bits of storage. Hardware implementations, in particular, necessitate the use of relatively small S-boxes.

In Example 4.1, we used four identical S-boxes in each round. The memory requirement of the S-box is $2^4 \times 4 = 2^6$ bits. If we instead used one S-box that mapped 16 bits to 16 bits, the memory requirement would be increased to $2^{16} \times 16 = 2^{20}$ bits, which would be prohibitively high for some applications. The S-box used in the *Advanced Encryption Standard* (to be discussed in Section 4.6) maps eight bits to eight bits.

The SPN in Example 4.1 is not secure, if for no other reason than the key length (32 bits) is small enough that an exhaustive key search is feasible. However, "larger" SPNs can be designed that are secure against all known attacks. A practical, secure SPN would have a larger key size and block length than Example 4.1, would most likely use larger S-boxes, and would have more rounds. *Rijndael,* which was chosen to be the *Advanced Encryption Standard,* is an example of an

SPN that is similar to Example 4.1 in many respects. *Rijndael* has a minimum key size of 128 bits, a block length of 128, a minimum of 10 rounds; and its S-box maps eight bits to eight bits (see Section 4.6 for a complete description).

Many variations of SPNs are possible. One common modification would be to use more than one S-box. In Example, 4.1, we could use four different S-boxes in each round if we so desired, instead of using the same S-box four times. This feature can be found in the *Data Encryption Standard*, which employs eight different S-boxes in each round (see Section 4.5.1). Another popular design strategy is to include an invertible linear transformation in each round, either as a replacement for, or in addition to, the permutation operation. This is done in the *Advanced Encryption Standard* (see Section 4.6.1).

4.3 Linear Cryptanalysis

We begin by informally describing the strategy behind linear cryptanalysis. The idea can be applied, in principle, to any iterated cipher. Suppose that it is possible to find a probabilistic linear relationship between a subset of plaintext bits and a subset of state bits immediately preceding the substitutions performed in the last round. In other words, there exists a subset of bits whose exclusive-or behaves in a non-random fashion (it takes on the value 0, say, with probability bounded away from $1/2$). Now assume that an attacker has a large number of plaintext-ciphertext pairs, all of which are encrypted using the same unknown key K (i.e., we consider a known-plaintext attack). For each of the plaintext-ciphertext pairs, we will begin to decrypt the ciphertext, using all possible candidate keys for the last round of the cipher. For each candidate key, we compute the values of the relevant state bits involved in the linear relationship, and determine if the above-mentioned linear relationship holds. Whenever it does, we increment a counter corresponding to the particular candidate key. At the end of this process, we hope that the candidate key that has a frequency count furthest from $1/2$ times the number of plaintext-ciphertext pairs contains the correct values for these key bits.

We will illustrate the above description with a detailed example later in this section. First, we need to establish some results from probability theory to provide a (non-rigorous) justification for the techniques involved in the attack.

4.3.1 The Piling-up Lemma

We use terminology and concepts introduced in Section 3.2. Suppose that $\mathbf{X_1}, \mathbf{X_2}, \ldots$ are independent random variables taking on values from the set $\{0, 1\}$. Let $p_1, p_2, \ldots$ be real numbers such that $0 \leq p_i \leq 1$ for all i, and suppose that

$$\mathbf{Pr}[\mathbf{X_i} = 0] = p_i,$$

$i = 1, 2, \ldots$. Hence,

$$\mathbf{Pr}[\mathbf{X_i} = 1] = 1 - p_i,$$

$i = 1, 2, \ldots$.

Suppose that $i \neq j$. The independence of $\mathbf{X_i}$ and $\mathbf{X_j}$ implies that

$$\begin{aligned}
\mathbf{Pr}[\mathbf{X_i} = 0, \mathbf{X_j} = 0] &= p_i p_j \\
\mathbf{Pr}[\mathbf{X_i} = 0, \mathbf{X_j} = 1] &= p_i(1 - p_j) \\
\mathbf{Pr}[\mathbf{X_i} = 1, \mathbf{X_j} = 0] &= (1 - p_i)p_j, \quad \text{and} \\
\mathbf{Pr}[\mathbf{X_i} = 1, \mathbf{X_j} = 1] &= (1 - p_i)(1 - p_j).
\end{aligned}$$

Now consider the discrete random variable $\mathbf{X_i} \oplus \mathbf{X_j}$ (this is the same thing as $\mathbf{X_i} + \mathbf{X_j} \bmod 2$). It is easy to see that $\mathbf{X_i} \oplus \mathbf{X_j}$ has the following probability distribution:

$$\begin{aligned}
\mathbf{Pr}[\mathbf{X_i} \oplus \mathbf{X_j} = 0] &= p_i p_j + (1 - p_i)(1 - p_j) \\
\mathbf{Pr}[\mathbf{X_i} \oplus \mathbf{X_j} = 1] &= p_i(1 - p_j) + (1 - p_i)p_j.
\end{aligned}$$

It is often convenient to express a probability distribution of a random variable taking on the values 0 and 1 in terms of a quantity called the bias of the distribution. The ***bias*** of $\mathbf{X_i}$ is defined to be the quantity

$$\epsilon_i = p_i - \frac{1}{2}.$$

Observe the following facts:

$$\begin{aligned}
&-\frac{1}{2} \leq \epsilon_i \leq \frac{1}{2}, \\
&\mathbf{Pr}[\mathbf{X_i} = 0] = \frac{1}{2} + \epsilon_i, \text{and} \\
&\mathbf{Pr}[\mathbf{X_i} = 1] = \frac{1}{2} - \epsilon_i,
\end{aligned}$$

for $i = 1, 2, \ldots$.

The following result, which gives a formula for the bias of the random variable $\mathbf{X_{i_1}} \oplus \cdots \oplus \mathbf{X_{i_k}}$, is known as the ***piling-up lemma***.

LEMMA 4.1 (Piling-up lemma) *Let $\epsilon_{i_1, i_2, \ldots, i_k}$ denote the bias of the random variable $\mathbf{X_{i_1}} \oplus \cdots \oplus \mathbf{X_{i_k}}$. Then*

$$\epsilon_{i_1, i_2, \ldots, i_k} = 2^{k-1} \prod_{j=1}^{k} \epsilon_{i_j}.$$

PROOF The proof is by induction on k. Clearly the result is true when $k = 1$. We next prove the result for $k = 2$, where we want to determine the bias of $\mathbf{X_{i_1}} \oplus \mathbf{X_{i_2}}$. Using the equations presented above, we have that

$$\begin{aligned}
\mathbf{Pr}[\mathbf{X_{i_1}} \oplus \mathbf{X_{i_2}} = 0] &= \left(\frac{1}{2} + \epsilon_{i_1}\right)\left(\frac{1}{2} + \epsilon_{i_2}\right) + \left(\frac{1}{2} - \epsilon_{i_1}\right)\left(\frac{1}{2} - \epsilon_{i_2}\right) \\
&= \frac{1}{2} + 2\epsilon_{i_1}\epsilon_{i_2}.
\end{aligned}$$

Hence, the bias of $\mathbf{X_{i_1}} \oplus \mathbf{X_{i_2}}$ is $2\epsilon_{i_1}\epsilon_{i_2}$, as claimed.

Now, as an induction hypothesis, assume that the result is true for $k = \ell$, for some positive integer $\ell \geq 2$. We will prove that the formula is true for $k = \ell + 1$.

We want to determine the bias of $\mathbf{X_{i_1}} \oplus \cdots \oplus \mathbf{X_{i_{\ell+1}}}$. We split this random variable into two parts, as follows:

$$\mathbf{X_{i_1}} \oplus \cdots \oplus \mathbf{X_{i_{\ell+1}}} = (\mathbf{X_{i_1}} \oplus \cdots \oplus \mathbf{X_{i_\ell}}) \oplus \mathbf{X_{i_{\ell+1}}}.$$

The bias of $\mathbf{X_{i_1}} \oplus \cdots \oplus \mathbf{X_{i_\ell}}$ is $2^{\ell-1}\prod_{j=1}^{\ell}\epsilon_{i_j}$ (by induction) and the bias of $\mathbf{X_{i_{\ell+1}}}$ is $\epsilon_{i_{\ell+1}}$. Then, by induction (more specifically, using the formula for $k = 2$), the bias of $\mathbf{X_{i_1}} \oplus \cdots \oplus \mathbf{X_{i_{\ell+1}}}$ is

$$2 \times \left(2^{\ell-1}\prod_{j=1}^{\ell}\epsilon_{i_j}\right) \times \epsilon_{i_{\ell+1}} = 2^{\ell}\prod_{j=1}^{\ell+1}\epsilon_{i_j},$$

as desired. By induction, the proof is complete. ∎

COROLLARY 4.2 *Let $\epsilon_{i_1,i_2,\ldots,i_k}$ denote the bias of the random variable $\mathbf{X_{i_1}} \oplus \cdots \oplus \mathbf{X_{i_k}}$. Suppose that $\epsilon_{i_j} = 0$ for some j. Then $\epsilon_{i_1,i_2,\ldots,i_k} = 0$.*

It is important to realize that Lemma 4.1 holds, in general, only when the relevant random variables are independent. We illustrate this by considering an example. Suppose that $\epsilon_1 = \epsilon_2 = \epsilon_3 = 1/4$. Applying Lemma 4.1, we see that $\epsilon_{1,2} = \epsilon_{2,3} = \epsilon_{1,3} = 1/8$. Now, consider the random variable $\mathbf{X_1} \oplus \mathbf{X_3}$. It is clear that

$$\mathbf{X_1} \oplus \mathbf{X_3} = (\mathbf{X_1} \oplus \mathbf{X_2}) \oplus (\mathbf{X_2} \oplus \mathbf{X_3}).$$

If the two random variables $\mathbf{X_1} \oplus \mathbf{X_2}$ and $\mathbf{X_2} \oplus \mathbf{X_3}$ were independent, then Lemma 4.1 would say that $\epsilon_{1,3} = 2(1/8)^2 = 1/32$. However, we already know that this is not the case: $\epsilon_{1,3} = 1/8$. Lemma 4.1 does not yield the correct value of $\epsilon_{1,3}$ because $\mathbf{X_1} \oplus \mathbf{X_2}$ and $\mathbf{X_2} \oplus \mathbf{X_3}$ are not independent.

4.3.2 Linear Approximations of S-boxes

Consider an S-box $\pi_S : \{0,1\}^m \rightarrow \{0,1\}^n$. (We do not assume that π_S is a permutation, or even that $m = n$.) Let us write an input m-tuple as $X = (x_1, \ldots, x_m)$. This m-tuple is chosen uniformly at random from $\{0,1\}^m$, which means that each co-ordinate x_i defines a random variable $\mathbf{X_i}$ taking on values 0 and 1, having bias $\epsilon_i = 0$. Further, these m random variables are independent.

Now write an output n-tuple as $Y = (y_1, \ldots, y_n)$. Each co-ordinate y_j defines a random variable $\mathbf{Y_j}$ taking on values 0 and 1. These n random variables are, in general, not independent from each other or from the $\mathbf{X_i}$'s. In fact, it is not hard to see that the following formula holds:

$$\mathbf{Pr}[\mathbf{X_1} = x_1, \ldots, \mathbf{X_m} = x_m, \mathbf{Y_1} = y_1, \ldots, \mathbf{Y_n} = y_n] = 0$$

TABLE 4.1: Random variables defined by an S-box

X_1	X_2	X_3	X_4	Y_1	Y_2	Y_3	Y_4
0	0	0	0	1	1	1	0
0	0	0	1	0	1	0	0
0	0	1	0	1	1	0	1
0	0	1	1	0	0	0	1
0	1	0	0	0	0	1	0
0	1	0	1	1	1	1	1
0	1	1	0	1	0	1	1
0	1	1	1	1	0	0	0
1	0	0	0	0	0	1	1
1	0	0	1	1	0	1	0
1	0	1	0	0	1	1	0
1	0	1	1	1	1	0	0
1	1	0	0	0	1	0	1
1	1	0	1	1	0	0	1
1	1	1	0	0	0	0	0
1	1	1	1	0	1	1	1

if $(y_1, \ldots, y_n) \neq \pi_S(x_1, \ldots, x_m)$; and

$$\mathbf{Pr}[\mathbf{X_1} = x_1, \ldots, \mathbf{X_m} = x_m, \mathbf{Y_1} = y_1, \ldots, \mathbf{Y_n} = y_n] = 2^{-m}$$

if $(y_1, \ldots, y_n) = \pi_S(x_1, \ldots, x_m)$. (The last formula holds because

$$\mathbf{Pr}[\mathbf{X_1} = x_1, \ldots, \mathbf{X_m} = x_m] = 2^{-m}$$

and

$$\mathbf{Pr}[\mathbf{Y_1} = y_1, \ldots, \mathbf{Y_n} = y_n | \mathbf{X_1} = x_1, \ldots, \mathbf{X_m} = x_m] = 1$$

if $(y_1, \ldots, y_n) = \pi_S(x_1, \ldots, x_m)$.)

It is now relatively straightforward to compute the bias of a random variable of the form

$$\mathbf{X_{i_1}} \oplus \cdots \oplus \mathbf{X_{i_k}} \oplus \mathbf{Y_{j_1}} \oplus \cdots \oplus \mathbf{Y_{j_\ell}}$$

using the formulas stated above. (A linear cryptanalytic attack can potentially be mounted when a random variable of this form has a bias that is bounded away from zero.)

Let's consider a small example.

Example 4.2 We use the S-box from Example 4.1, which is defined by a permutation $\pi_S : \{0,1\}^4 \to \{0,1\}^4$. We record the possible values taken on by the eight random variables $\mathbf{X_1}, \ldots, \mathbf{X_4}, \mathbf{Y_1}, \ldots, \mathbf{Y_4}$ in the rows of Table 4.1.

Now, consider the random variable $\mathbf{X_1} \oplus \mathbf{X_4} \oplus \mathbf{Y_2}$. The probability that this random variable takes on the value 0 can be determined by counting the number of rows in the above table in which $\mathbf{X_1} \oplus \mathbf{X_4} \oplus \mathbf{Y_2} = 0$, and then dividing by 16 ($16 = 2^4$ is the total number of rows in the table). It is seen that

$$\mathbf{Pr}[\mathbf{X_1} \oplus \mathbf{X_4} \oplus \mathbf{Y_2} = 0] = \frac{1}{2}$$

(and therefore

$$\mathbf{Pr}[\mathbf{X_1} \oplus \mathbf{X_4} \oplus \mathbf{Y_2} = 1] = \frac{1}{2},$$

as well.) Hence, the bias of this random variable is 0. □

If we instead analyzed the random variable $\mathbf{X_3} \oplus \mathbf{X_4} \oplus \mathbf{Y_1} \oplus \mathbf{Y_4}$, we would find that the bias is $-3/8$. (We suggest that the reader verify this computation.) Indeed, it is not difficult to compute the biases of all $2^8 = 256$ possible random variables of this form.

We record this information using the following notation. We represent each of the relevant random variables in the form

$$\left(\bigoplus_{i=1}^{4} a_i \mathbf{X_i} \right) \oplus \left(\bigoplus_{i=1}^{4} b_i \mathbf{Y_i} \right),$$

where $a_i \in \{0,1\}$, $b_i \in \{0,1\}$, $i = 1,2,3,4$. Then, in order to have a compact notation, we treat each of the binary vectors (a_1, a_2, a_3, a_4) and (b_1, b_2, b_3, b_4) as a hexadecimal digit (these are called the ***input sum*** and ***output sum***, respectively). In this way, each of the 256 random variables is named by a (unique) pair of hexadecimal digits, representing the input and output sum.

As an example, consider the random variable $\mathbf{X_1} \oplus \mathbf{X_4} \oplus \mathbf{Y_2}$. The input sum is $(1,0,0,1)$, which is 9 in hexadecimal; the output sum is $(0,1,0,0)$, which is 4 in hexadecimal.

Definition 4.1: For a random variable having (hexadecimal) input sum a and output sum b (where $a = (a_1, a_2, a_3, a_4)$ and $b = (b_1, b_2, b_3, b_4)$, in binary), let $N_L(a,b)$ denote the number of binary eight-tuples $(x_1, x_2, x_3, x_4, y_1, y_2, y_3, y_4)$ such that

$$(y_1, y_2, y_3, y_4) = \pi_S(x_1, x_2, x_3, x_4)$$

and

$$\left(\bigoplus_{i=1}^{4} a_i x_i \right) \oplus \left(\bigoplus_{i=1}^{4} b_i y_i \right) = 0.$$

Observe that the bias of the random variable having input sum a and output sum b is computed as $\epsilon(a,b) = (N_L(a,b) - 8)/16$.

We computed $N_L(9,4) = 8$, and hence $\epsilon(9,4) = 0$, in Example 4.2. The table of all values N_L is called the ***linear approximation table***; see Table 4.2.

TABLE 4.2: Linear approximation table: values of $N_L(a,b)$

a \ b	0	1	2	3	4	5	6	7	8	9	A	B	C	D	E	F
0	16	8	8	8	8	8	8	8	8	8	8	8	8	8	8	8
1	8	8	6	6	8	8	6	14	10	10	8	8	10	10	8	8
2	8	8	6	6	8	8	6	6	8	8	10	10	8	8	2	10
3	8	8	8	8	8	8	8	8	10	2	6	6	10	10	6	6
4	8	10	8	6	6	4	6	8	8	6	8	10	10	4	10	8
5	8	6	6	8	6	8	12	10	6	8	4	10	8	6	6	8
6	8	10	6	12	10	8	8	10	8	6	10	12	6	8	8	6
7	8	6	8	10	10	4	10	8	6	8	10	8	12	10	8	10
8	8	8	8	8	8	8	8	8	6	10	10	6	10	6	6	2
9	8	8	6	6	8	8	6	6	4	8	6	10	8	12	10	6
A	8	12	6	10	4	8	10	6	10	10	8	8	10	10	8	8
B	8	12	8	4	12	8	12	8	8	8	8	8	8	8	8	8
C	8	6	12	6	6	8	10	8	10	8	10	12	8	10	8	6
D	8	10	10	8	6	12	8	10	4	6	10	8	10	8	8	10
E	8	10	10	8	6	4	8	10	6	8	8	6	4	10	6	8
F	8	6	4	6	6	8	10	8	8	6	12	6	6	8	10	8

4.3.3 A Linear Attack on an SPN

Linear cryptanalysis requires finding a set of linear approximations of S-boxes that can be used to derive a linear approximation of the entire SPN (excluding the last round). We will illustrate the procedure using the SPN from Example 4.1. The diagram in Figure 4.3 illustrates the structure of the approximation we will use. This diagram can be interpreted as follows: Lines with arrows correspond to random variables that will be involved in linear approximations. The labeled S-boxes are the ones used in these approximations (they are called the ***active S-boxes*** in the approximation).

This approximation incorporates four active S-boxes:

- In S_2^1, the random variable $\mathbf{T_1} = \mathbf{U_5^1} \oplus \mathbf{U_7^1} \oplus \mathbf{U_8^1} \oplus \mathbf{V_6^1}$ has bias $1/4$
- In S_2^2, the random variable $\mathbf{T_2} = \mathbf{U_6^2} \oplus \mathbf{V_6^2} \oplus \mathbf{V_8^2}$ has bias $-1/4$
- In S_2^3, the random variable $\mathbf{T_3} = \mathbf{U_6^3} \oplus \mathbf{V_6^3} \oplus \mathbf{V_8^3}$ has bias $-1/4$
- In S_4^3, the random variable $\mathbf{T_4} = \mathbf{U_{14}^3} \oplus \mathbf{V_{14}^3} \oplus \mathbf{V_{16}^3}$ has bias $-1/4$

Note that $\mathbf{U_i^r}$ ($1 \leq i \leq 4$) are random variables corresponding to the inputs to the S-boxes in round r, and $\mathbf{V_i^r}$ are random variables corresponding to the outputs of the same S-boxes. The four random variables $\mathbf{T_1}, \mathbf{T_2}, \mathbf{T_3}, \mathbf{T_4}$ have biases that are high in absolute value.

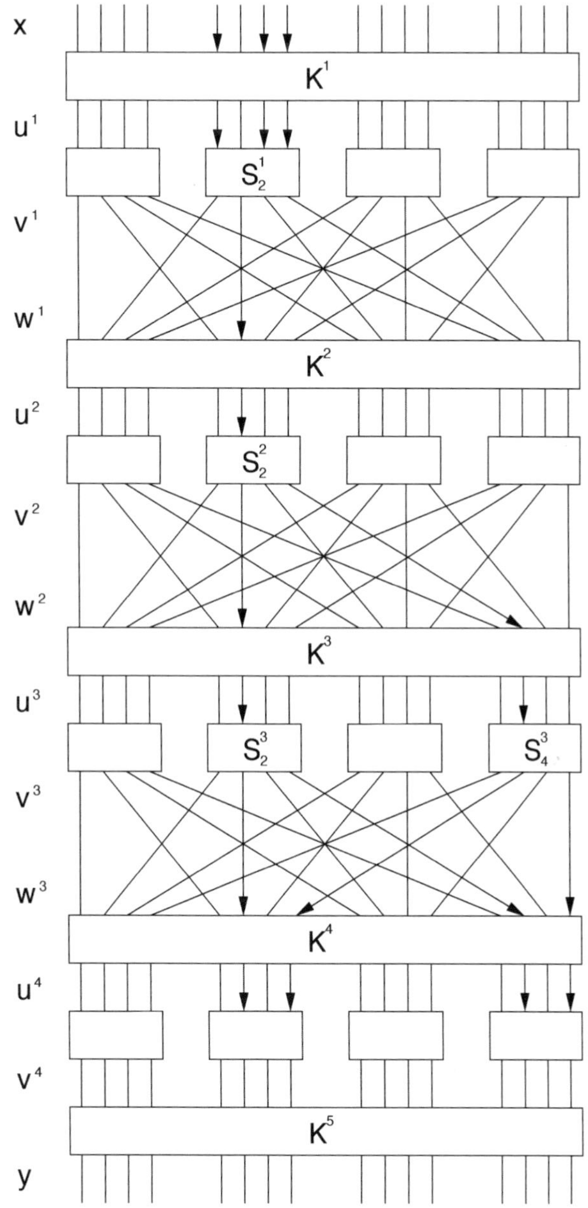

FIGURE 4.3: A linear approximation of a substitution-permutation network

If we make the assumption that these four random variables are independent, then we can compute the bias of their x-or using the piling-up lemma (Lemma 4.1). (The random variables are in fact not independent, which means that we cannot provide a mathematical justification of this approximation. Nevertheless, the approximation seems to work in practice, as we shall demonstrate.) We therefore hypothesize that the random variable

$$\mathbf{T_1} \oplus \mathbf{T_2} \oplus \mathbf{T_3} \oplus \mathbf{T_4}$$

has bias equal to $2^3(1/4)(-1/4)^3 = -1/32$.

The random variables $\mathbf{T_1}$, $\mathbf{T_2}$, $\mathbf{T_3}$, and $\mathbf{T_4}$ have been carefully constructed so that the exclusive-or $\mathbf{T_1} \oplus \mathbf{T_2} \oplus \mathbf{T_3} \oplus \mathbf{T_4}$ will lead to cancellations of "intermediate" random variables. This happens because the "output" random variables in $\mathbf{T_r}$ correspond to the "input" random variables in $\mathbf{T_{r+1}}$. For example, the term $\mathbf{U^2_6}$ in $\mathbf{T_2}$ can be expressed as $\mathbf{V^1_6} \oplus \mathbf{K^2_6}$. The random variable $\mathbf{T_1}$ contains a term $\mathbf{V^1_6}$. Thus, if we compute $\mathbf{T_1} \oplus \mathbf{T_2}$, the two occurrences of the term $\mathbf{V^1_6}$ cancel each other out.

Thus, the random variables $\mathbf{T_1}, \mathbf{T_2}, \mathbf{T_3}$, and $\mathbf{T_4}$ have the property that their x-or can be expressed in terms of plaintext bits, bits of u^4 (the input to the last round of S-boxes), and key bits. This can be done as follows: First, we have the following relations, which can be easily verified by inspecting Figure 4.3:

$$\begin{aligned}
\mathbf{T_1} &= \mathbf{U^1_5} \oplus \mathbf{U^1_7} \oplus \mathbf{U^1_8} \oplus \mathbf{V^1_6} &&= \mathbf{X_5} \oplus \mathbf{K^1_5} \oplus \mathbf{X_7} \oplus \mathbf{K^1_7} \oplus \mathbf{X_8} \oplus \mathbf{K^1_8} \oplus \mathbf{V^1_6}\\
\mathbf{T_2} &= \mathbf{U^2_6} \oplus \mathbf{V^2_6} \oplus \mathbf{V^2_8} &&= \mathbf{V^1_6} \oplus \mathbf{K^2_6} \oplus \mathbf{V^2_6} \oplus \mathbf{V^2_8}\\
\mathbf{T_3} &= \mathbf{U^3_6} \oplus \mathbf{V^3_6} \oplus \mathbf{V^3_8} &&= \mathbf{V^2_6} \oplus \mathbf{K^3_6} \oplus \mathbf{V^3_6} \oplus \mathbf{V^3_8}\\
\mathbf{T_4} &= \mathbf{U^3_{14}} \oplus \mathbf{V^3_{14}} \oplus \mathbf{V^3_{16}} &&= \mathbf{V^2_8} \oplus \mathbf{K^3_{14}} \oplus \mathbf{V^3_{14}} \oplus \mathbf{V^3_{16}}.
\end{aligned}$$

If we compute the x-or of the random variables on the right sides of the above equations, we see that the random variable

$$\mathbf{X_5} \oplus \mathbf{X_7} \oplus \mathbf{X_8} \oplus \mathbf{V^3_6} \oplus \mathbf{V^3_8} \oplus \mathbf{V^3_{14}} \oplus \mathbf{V^3_{16}} \oplus \mathbf{K^1_5} \oplus \mathbf{K^1_7} \oplus \mathbf{K^1_8} \oplus \mathbf{K^2_6} \oplus \mathbf{K^3_6} \oplus \mathbf{K^3_{14}} \tag{4.1}$$

has bias equal to $-1/32$. The next step is to replace the terms $\mathbf{V^3_i}$ in the above formula by expressions involving $\mathbf{U^4_i}$ and further key bits:

$$\begin{aligned}
\mathbf{V^3_6} &= \mathbf{U^4_6} \oplus \mathbf{K^4_6}\\
\mathbf{V^3_8} &= \mathbf{U^4_{14}} \oplus \mathbf{K^4_{14}}\\
\mathbf{V^3_{14}} &= \mathbf{U^4_8} \oplus \mathbf{K^4_8}\\
\mathbf{V^3_{16}} &= \mathbf{U^4_{16}} \oplus \mathbf{K^4_{16}}
\end{aligned}$$

Now we substitute these four expressions into (4.1), to get the following:

$$\begin{aligned}
&\mathbf{X_5} \oplus \mathbf{X_7} \oplus \mathbf{X_8} \oplus \mathbf{U^4_6} \oplus \mathbf{U^4_8} \oplus \mathbf{U^4_{14}} \oplus \mathbf{U^4_{16}}\\
&\qquad \oplus \mathbf{K^1_5} \oplus \mathbf{K^1_7} \oplus \mathbf{K^1_8} \oplus \mathbf{K^2_6} \oplus \mathbf{K^3_6} \oplus \mathbf{K^3_{14}} \oplus \mathbf{K^4_6} \oplus \mathbf{K^4_8} \oplus \mathbf{K^4_{14}} \oplus \mathbf{K^4_{16}}
\end{aligned} \tag{4.2}$$

This expression only involves plaintext bits, bits of u^4, and key bits. Suppose that the key bits in (4.2) are fixed. Then the random variable

$$\mathbf{K}_5^1 \oplus \mathbf{K}_7^1 \oplus \mathbf{K}_8^1 \oplus \mathbf{K}_6^2 \oplus \mathbf{K}_6^3 \oplus \mathbf{K}_{14}^3 \oplus \mathbf{K}_6^4 \oplus \mathbf{K}_8^4 \oplus \mathbf{K}_{14}^4 \oplus \mathbf{K}_{16}^4$$

has the (fixed) value 0 or 1. It follows that the random variable

$$\mathbf{X}_5 \oplus \mathbf{X}_7 \oplus \mathbf{X}_8 \oplus \mathbf{U}_6^4 \oplus \mathbf{U}_8^4 \oplus \mathbf{U}_{14}^4 \oplus \mathbf{U}_{16}^4 \tag{4.3}$$

has bias equal to $\pm 1/32$, where the sign of this bias depends on the values of unknown key bits. Note that the random variable (4.3) involves only plaintext bits and bits of u^4. The fact that (4.3) has bias bounded away from 0 allows us to carry out the linear attack mentioned at the beginning of Section 4.3.

Suppose that we have T plaintext-ciphertext pairs, all of which use the same unknown key, K. (It will turn out that we need $T \approx 8000$ in order for the attack to succeed.) Denote this set of T pairs by $\mathcal{T}$. The attack will allow us to obtain the eight key bits in $K^5_{<2>}$ and $K^5_{<4>}$, namely,

$$K_5^5, K_6^5, K_7^5, K_8^5, K_{13}^5, K_{14}^5, K_{15}^5, \text{ and } K_{16}^5.$$

These are the eight key bits that are x-ored with the output of the S-boxes S_2^4 and S_4^4. Notice that there are $2^8 = 256$ possibilities for this list of eight key bits. We will refer to a binary 8-tuple (comprising values for these eight key bits) as a ***candidate subkey***.

For each $(x, y) \in \mathcal{T}$ and for each candidate subkey, it is possible to compute a partial decryption of y and obtain the resulting value for $u^4_{<2>}$ and $u^4_{<4>}$. Then we compute the value

$$x_5 \oplus x_7 \oplus x_8 \oplus u_6^4 \oplus u_8^4 \oplus u_{14}^4 \oplus u_{16}^4 \tag{4.4}$$

taken on by the random variable (4.3). We maintain an array of counters indexed by the 256 possible candidate subkeys, and increment the counter corresponding to a particular subkey whenever (4.4) has the value 0. (This array is initialized to have all values equal to 0.)

At the end of this counting process, we expect that most counters will have a value close to $T/2$, but the counter for the correct candidate subkey will have a value that is close to $T/2 \pm T/32$. This will (hopefully) allow us to identify eight subkey bits.

The algorithm for this particular linear attack is presented as Algorithm 4.2. In this algorithm, the variables L_1 and L_2 take on hexadecimal values. The set $\mathcal{T}$ is the set of T plaintext-ciphertext pairs used in the attack. $\pi_S{}^{-1}$ is the permutation corresponding to the inverse of the S-box; this is used to partially decrypt the ciphertexts. The output, *maxkey*, contains the "most likely" eight subkey bits identified in the attack.

Algorithm 4.2 is not very complicated. As mentioned previously, we are just computing (4.4) for every plaintext-ciphertext pair $(x, y) \in \mathcal{T}$ and for every possible candidate subkey (L_1, L_2). In order to do this, we refer to Figure 4.3. First, we compute the exclusive-ors $L_1 \oplus y_{<2>}$ and $L_2 \oplus y_{<4>}$. These yield $v^4_{<2>}$ and

Algorithm 4.2: LINEARATTACK($\mathcal{T}, T, \pi_S{}^{-1}$)

for $(L_1, L_2) \leftarrow (0,0)$ **to** (F,F)
  do $Count[L_1, L_2] \leftarrow 0$
for each $(x,y) \in \mathcal{T}$
  do {
    for $(L_1, L_2) \leftarrow (0,0)$ **to** (F,F)
      do {
        $v^4_{<2>} \leftarrow L_1 \oplus y_{<2>}$
        $v^4_{<4>} \leftarrow L_2 \oplus y_{<4>}$
        $u^4_{<2>} \leftarrow \pi_S{}^{-1}(v^4_{<2>})$
        $u^4_{<4>} \leftarrow \pi_S{}^{-1}(v^4_{<4>})$
        $z \leftarrow x_5 \oplus x_7 \oplus x_8 \oplus u^4_6 \oplus u^4_8 \oplus u^4_{14} \oplus u^4_{16}$
        if $z = 0$
          then $Count[L_1, L_2] \leftarrow Count[L_1, L_2] + 1$
      }
  }
$max \leftarrow -1$
for $(L_1, L_2) \leftarrow (0,0)$ **to** (F,F)
  do {
    $Count[L_1, L_2] \leftarrow |Count[L_1, L_2] - T/2|$
    if $Count[L_1, L_2] > max$
      then { $max \leftarrow Count[L_1, L_2]$
        $maxkey \leftarrow (L_1, L_2)$ }
  }
output $(maxkey)$

$v^4_{<4>}$, respectively, when (L_1, L_2) is the correct subkey. $u^4_{<2>}$ and $u^4_{<4>}$ can then be computed from $v^4_{<2>}$ and $v^4_{<4>}$ by using the inverse S-box $\pi_S{}^{-1}$; again, the values obtained are correct if (L_1, L_2) is the correct subkey. Then we compute (4.4) and we increment the counter for the pair (L_1, L_2) if (4.4) has the value 0. After having computed all the relevant counters, we just find the pair (L_1, L_2) corresponding to the maximum counter; this is the output of Algorithm 4.2.

In general, it is suggested that a linear attack based on a linear approximation having bias equal to ϵ will be successful if the number of plaintext-ciphertext pairs, which we denote by T, is approximately $c\,\epsilon^{-2}$, for some "small" constant c. We implemented the attack described in Algorithm 4.2, and found that the attack was usually successful if we took $T = 8000$. Note that $T = 8000$ corresponds to $c \approx 8$, because $\epsilon^{-2} = 1024$.

4.4 Differential Cryptanalysis

Differential cryptanalysis is similar to linear cryptanalysis in many respects. The main difference from linear cryptanalysis is that differential cryptanalysis involves comparing the x-or of two inputs to the x-or of the corresponding two outputs. In general, we will be looking at inputs x and x^* (which are assumed to be binary strings) having a specified (fixed) x-or value denoted by $x' = x \oplus x^*$. Throughout this section, we will use prime markings ($'$) to indicate the x-or of two bitstrings.

Differential cryptanalysis is a chosen-plaintext attack. We assume that an attacker has a large number of tuples (x, x^*, y, y^*), where the x-or value $x' = x \oplus x^*$ is fixed. The plaintext elements (i.e., x and x^*) are encrypted using the same unknown key, K, yielding the ciphertexts y and y^*, respectively. For each of these tuples, we will begin to decrypt the ciphertexts y and y^*, using all possible candidate keys for the last round of the cipher. For each candidate key, we compute the values of certain state bits, and determine if their x-or has a certain value (namely, the most likely value for the given input x-or). Whenever it does, we increment a counter corresponding to the particular candidate key. At the end of this process, we hope that the candidate key that has the highest frequency count contains the correct values for these key bits. (As we did with linear cryptanalysis, we will illustrate the attack with a particular example.)

Definition 4.2: Let $\pi_S : \{0,1\}^m \rightarrow \{0,1\}^n$ be an S-box. Consider an (ordered) pair of bitstrings of length m, say (x, x^*). We say that the ***input x-or*** of the S-box is $x \oplus x^*$ and the ***output x-or*** is $\pi_S(x) \oplus \pi_S(x^*)$. Note that the output x-or is a bitstring of length n.

For any $x' \in \{0,1\}^m$, define the set $\Delta(x')$ to consist of all the ordered pairs (x, x^*) having input x-or equal to x'.

It is easy to see that any set $\Delta(x')$ contains 2^m pairs, and that

$$\Delta(x') = \{(x, x \oplus x') : x \in \{0,1\}^m\}.$$

For each pair in $\Delta(x')$, we can compute the output x-or of the S-box. Then we can tabulate the resulting distribution of output x-ors. There are 2^m output x-ors, which are distributed among 2^n possible values. A non-uniform output distribution will be the basis for a successful differential attack.

Example 4.3 We again use the S-box from Example 4.1. Suppose we consider input x-or $x' = 1011$. Then

$$\Delta(1011) = \{(0000, 1011), (0001, 1010), \ldots, (1111, 0100)\}.$$

For each ordered pair in the set $\Delta(1011)$, we compute output x-or of π_S in Table 4.3. In each row of this table, we have $x \oplus x^* = 1011$, $y = \pi_S(x)$, $y^* = \pi_S(x^*)$, and $y' = y \oplus y^*$.

Looking at the last column of Table 4.3, we obtain the following distribution of output x-ors:

0000	0001	0010	0011	0100	0101	0110	0111
0	0	8	0	0	2	0	2

TABLE 4.3: Input and output x-ors

x	x^*	y	y^*	y'
0000	1011	1110	1100	0010
0001	1010	0100	0110	0010
0010	1001	1101	1010	0111
0011	1000	0001	0011	0010
0100	1111	0010	0111	0101
0101	1110	1111	0000	1111
0110	1101	1011	1001	0010
0111	1100	1000	0101	1101
1000	0011	0011	0001	0010
1001	0010	1010	1101	0111
1010	0001	0110	0100	0010
1011	0000	1100	1110	0010
1100	0111	0101	1000	1101
1101	0110	1001	1011	0010
1110	0101	0000	1111	1111
1111	0100	0111	0010	0101

1000	1001	1010	1011	1100	1101	1110	1111
0	0	0	0	0	2	0	2

□

In Example 4.3, only five of the 16 possible output x-ors actually occur. This particular example has a very non-uniform distribution.

We can carry out computations, as was done in Example 4.3, for any possible input x-or. It will be convenient to have some notation to describe the distributions of the output x-ors, so we state the following definition.

Definition 4.3: For a bitstring x' of length m and a bitstring y' of length n, define

$$N_D(x', y') = |\{(x, x^*) \in \Delta(x') : \pi_S(x) \oplus \pi_S(x^*) = y'\}|.$$

In other words, $N_D(x', y')$ counts the number of pairs with input x-or equal to x' that also have output x-or equal to y' (for a given S-box). All the values $N_D(a', b')$ for the S-box from Example 4.1 are tabulated in Table 4.4 (a' and b' are the hexadecimal representations of the input and output x-ors, respectively). Observe that the distribution computed in Example 4.3 corresponds to row "B" in the table in Table 4.4.

Recall that the input to the ith S-box in round r of the SPN from Example 4.1 is denoted $u^r_{<i>}$, and

$$u^r_{<i>} = w^{r-1}_{<i>} \oplus K^r_{<i>}.$$

TABLE 4.4: Difference distribution table: values of $N_D(a', b')$

a' \ b'	0	1	2	3	4	5	6	7	8	9	A	B	C	D	E	F
0	16	0	0	0	0	0	0	0	0	0	0	0	0	0	0	0
1	0	0	0	2	0	0	0	2	0	2	4	0	4	2	0	0
2	0	0	0	2	0	6	2	2	0	2	0	0	0	0	2	0
3	0	0	2	0	2	0	0	0	0	4	2	0	2	0	0	4
4	0	0	0	2	0	0	6	0	0	2	0	4	2	0	0	0
5	0	4	0	0	0	2	2	0	0	0	4	0	2	0	0	2
6	0	0	0	4	0	4	0	0	0	0	0	0	2	2	2	2
7	0	0	2	2	2	0	2	0	0	2	2	0	0	0	0	4
8	0	0	0	0	0	0	2	2	0	0	0	4	0	4	2	2
9	0	2	0	0	2	0	0	4	2	0	2	2	2	0	0	0
A	0	2	2	0	0	0	0	0	6	0	0	2	0	0	4	0
B	0	0	8	0	0	2	0	2	0	0	0	0	0	2	0	2
C	0	2	0	0	2	2	2	0	0	0	0	2	0	6	0	0
D	0	4	0	0	0	0	0	4	2	0	2	0	2	0	2	0
E	0	0	2	4	2	0	0	0	6	0	0	0	0	0	2	0
F	0	2	0	0	6	0	0	0	0	4	0	2	0	0	2	0

An input x-or is computed as

$$\begin{aligned} u^r_{<i>} \oplus (u^r_{<i>})^* &= (w^{r-1}_{<i>} \oplus K^r_{<i>}) \oplus ((w^{r-1}_{<i>})^* \oplus K^r_{<i>}) \\ &= w^{r-1}_{<i>} \oplus (w^{r-1}_{<i>})^* \end{aligned}$$

Therefore, this input x-or does not depend on the subkey bits used in round r; it is equal to the (permuted) output x-or of round $r-1$. (However, the output x-or of round r certainly does depend on the subkey bits in round r.)

Let a' denote an input x-or and let b' denote an output x-or. The pair (a', b') is called a ***differential***. Each entry in the difference distribution table gives rise to an ***x-or propagation ratio*** (or more simply, a ***propagation ratio***) for the corresponding differential.

Definition 4.4: The propagation ratio $R_p(a', b')$ for the differential (a', b') is defined as follows:

$$R_p(a', b') = \frac{N_D(a', b')}{2^m}.$$

$R_p(a', b')$ can be interpreted as a conditional probability:

$$\mathbf{Pr}[\text{output x-or } = b' | \text{ input x-or } = a'] = R_p(a', b').$$

Suppose we find propagation ratios for differentials in consecutive rounds of the SPN, such that the input x-or of a differential in any round is the same as the (permuted) output x-ors of the differentials in the previous round. Then these differentials can be combined to form a ***differential trail***. We make the assumption that the various propagation ratios in a differential trail are independent (an assumption that may not be mathematically valid, in fact). This assumption allows us to multiply the propagation ratios of the differentials in order to obtain the propagation ratio of the differential trail.

We illustrate this process by returning to the SPN from Example 4.1. A particular differential trail is shown in Figure 4.4. Arrows are used to highlight the "1" bits in the input and output x-ors of the differentials that are used in the differential trail.

The differential attack arising from Figure 4.4 uses the following propagation ratios of differentials, all of which can be verified from Figure 4.4:

- In S_2^1, $R_p(1011, 0010) = 1/2$
- In S_3^2, $R_p(0100, 0110) = 3/8$
- In S_2^3, $R_p(0010, 0101) = 3/8$
- In S_3^3, $R_p(0010, 0101) = 3/8$

These differentials can be combined to form a differential trail. We therefore obtain a propagation ratio for a differential trail of the first three rounds of the SPN:

$$R_p(\texttt{0000 1011 0000 0000}, \texttt{0000 0101 0101 0000}) = \frac{1}{2} \times \left(\frac{3}{8}\right)^3 = \frac{27}{1024}.$$

In other words,

$$x' = \texttt{0000 1011 0000 0000} \Rightarrow (v^3)' = \texttt{0000 0101 0101 0000}$$

with probability $27/1024$. However,

$$(v^3)' = \texttt{0000 0101 0101 0000} \Leftrightarrow (u^4)' = \texttt{0000 0110 0000 0110}.$$

Hence, it follows that

$$x' = \texttt{0000 1011 0000 0000} \Rightarrow (u^4)' = \texttt{0000 0110 0000 0110}$$

with probability $27/1024$. Note that $(u^4)'$ is the x-or of two inputs to the last round of S-boxes.

Now we can present an algorithm, for this particular example, based on the informal description at the beginning of this section; see Algorithm 4.3. The input and output of this algorithm are similar to linear attack; the main difference is that $\mathcal{T}$ is a set of tuples of the form (x, x^*, y, y^*), where x' is fixed, in the differential attack.

Algorithm 4.3 makes use of a certain ***filtering operation***. Tuples (x, x^*, y, y^*)

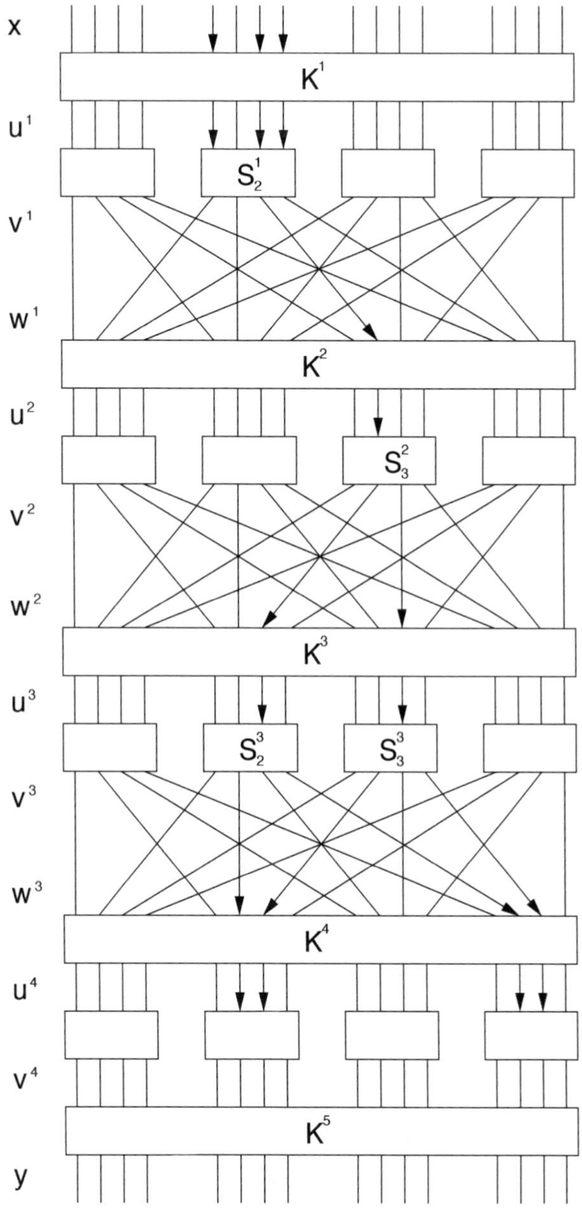

FIGURE 4.4: A differential trail for a substitution-permutation network

Algorithm 4.3: DIFFERENTIALATTACK($\mathcal{T}, T, \pi_S{}^{-1}$)

for $(L_1, L_2) \leftarrow (0,0)$ **to** (F, F)
  do $Count[L_1, L_2] \leftarrow 0$
for each $(x, y, x^*, y^*) \in \mathcal{T}$
  do
    if $(y_{<1>} = (y_{<1>})^*)$ **and** $(y_{<3>} = (y_{<3>})^*)$
      then
        for $(L_1, L_2) \leftarrow (0,0)$ **to** (F, F)
          do
            $v^4_{<2>} \leftarrow L_1 \oplus y_{<2>}$
            $v^4_{<4>} \leftarrow L_2 \oplus y_{<4>}$
            $u^4_{<2>} \leftarrow \pi_S{}^{-1}(v^4_{<2>})$
            $u^4_{<4>} \leftarrow \pi_S{}^{-1}(v^4_{<4>})$
            $(v^4_{<2>})^* \leftarrow L_1 \oplus (y_{<2>})^*$
            $(v^4_{<4>})^* \leftarrow L_2 \oplus (y_{<4>})^*$
            $(u^4_{<2>})^* \leftarrow \pi_S{}^{-1}((v^4_{<2>})^*)$
            $(u^4_{<4>})^* \leftarrow \pi_S{}^{-1}((v^4_{<4>})^*)$
            $(u^4_{<2>})' \leftarrow u^4_{<2>} \oplus (u^4_{<2>})^*$
            $(u^4_{<4>})' \leftarrow u^4_{<4>} \oplus (u^4_{<4>})^*$
            if $((u^4_{<2>})' = 0110)$ **and** $((u^4_{<4>})' = 0110)$
              then $Count[L_1, L_2] \leftarrow Count[L_1, L_2] + 1$
$max \leftarrow -1$
for $(L_1, L_2) \leftarrow (0,0)$ **to** (F, F)
  do
    if $Count[L_1, L_2] > max$
      then
        $max \leftarrow Count[L_1, L_2]$
        $maxkey \leftarrow (L_1, L_2)$
output $(maxkey)$

for which the differential holds are often called ***right pairs***, and it is the right pairs that allow us to determine the relevant key bits. (Tuples that are not right pairs basically constitute "random noise" that provides no useful information.) A right pair has

$$(u^4_{<1>})' = (u^4_{<3>})' = 0000.$$

Hence, it follows that a right pair must have $y_{<1>} = (y_{<1>})^*$ and $y_{<3>} = (y_{<3>})^*$. If a tuple (x, x^*, y, y^*) does not satisfy these conditions, then we know that it is not a right pair, and we can discard it. This filtering process increases the efficiency of the attack.

The workings of Algorithm 4.3 can be summarized as follows. For each tuple $(x, x^*, y, y^*) \in \mathcal{T}$, we first perform the filtering operation. If (x, x^*, y, y^*) is a right pair, then we test each possible candidate subkey (L_1, L_2) and increment an appropriate counter if a certain x-or is observed. The steps include computing an exclusive-or with candidate subkeys and applying the inverse S-box (as was done in Algorithm 4.2), followed by computation of the relevant x-or value.

A differential attack based on a differential trail having propagation ratio equal to ϵ will often be successful if the number of tuples (x, x^*, y, y^*), which we denote by T, is approximately $c\,\epsilon^{-1}$, for a "small" constant c. We implemented the attack described in Algorithm 4.3, and found that the attack was often successful if we took T between 50 and 100. In this example, $\epsilon^{-1} \approx 38$.

4.5 The Data Encryption Standard

On May 15, 1973, the National Bureau of Standards (now the *National Institute of Standards and Technology*, or *NIST*) published a solicitation for cryptosystems in the Federal Register. This led ultimately to the adoption of the *Data Encryption Standard*, or *DES*, which became the most widely used cryptosystem in the world. *DES* was developed at IBM, as a modification of an earlier system known as *Lucifer*. *DES* was first published in the Federal Register of March 17, 1975. After a considerable amount of public discussion, *DES* was adopted as a standard for "unclassified" applications on January 15, 1977. It was initially expected that DES would only be used as a standard for 10–15 years; however, it proved to be much more durable. *DES* was reviewed approximately every five years after its adoption. Its last renewal was in January 1999; by that time, development of a replacement, the *Advanced Encryption Standard*, had already begun (see Section 4.6).

4.5.1 Description of DES

A complete description of the *Data Encryption Standard* is given in the *Federal Information Processing Standards* (*FIPS*) Publication 46, dated January 15, 1977. *DES* is a special type of iterated cipher called a ***Feistel cipher***. We describe the basic form of a Feistel cipher now, using the terminology from Section 4.1. In a Feistel cipher, each state u^i is divided into two halves of equal length, say L^i and R^i. The round function g has the following form: $g(L^{i-1}, R^{i-1}, K^i) = (L^i, R^i)$, where

$$\begin{aligned} L^i &= R^{i-1} \\ R^i &= L^{i-1} \oplus f(R^{i-1}, K^i). \end{aligned}$$

We observe that the function f does not need to satisfy any type of injective property. This is because a Feistel-type round function is always invertible, given the round key:

$$\begin{aligned} L^{i-1} &= R^i \oplus f(L^i, K^i) \\ R^{i-1} &= L^i. \end{aligned}$$

DES is a 16-round Feistel cipher having block length 64: it encrypts a plaintext bitstring x (of length 64) using a 56-bit key, K, obtaining a ciphertext bitstring (of length 64). Prior to the 16 rounds of encryption, there is a fixed ***initial permutation***

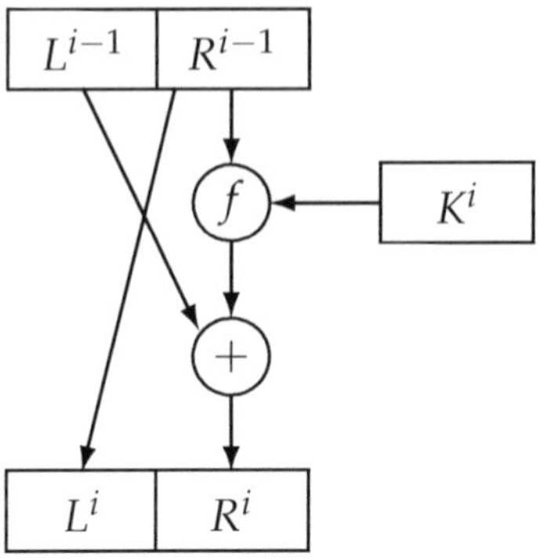

FIGURE 4.5: One round of DES encryption

IP that is applied to the plaintext. We denote

$$\mathbf{IP}(x) = L^0 R^0.$$

After the 16 rounds of encryption, the inverse permutation $\mathbf{IP}^{-1}$ is applied to the bitstring $R^{16}L^{16}$, yielding the ciphertext y. That is,

$$y = \mathbf{IP}^{-1}(R^{16}L^{16})$$

(note that L^{16} and R^{16} are swapped before $\mathbf{IP}^{-1}$ is applied). The application of **IP** and $\mathbf{IP}^{-1}$ has no cryptographic significance, and is often ignored when the security of *DES* is discussed. One round of *DES* encryption is depicted in Figure 4.5.

Each L^i and R^i is 32 bits in length. The function

$$f : \{0,1\}^{32} \times \{0,1\}^{48} \rightarrow \{0,1\}^{32}$$

takes as input a 32-bit string (the right half of the current state) and a round key. The key schedule, $(K^1, K^2, \ldots, K^{16})$, consists of 48-bit round keys that are derived from the 56-bit key, K. Each K^i is a certain permuted selection of bits from K.

The f function is shown in Figure 4.6. Basically, it consists of a substitution (using an S-box) followed by a (fixed) permutation, denoted **P**. Suppose we denote the first argument of f by A, and the second argument by J. Then, in order to compute $f(A, J)$, the following steps are executed.

1. A is "expanded" to a bitstring of length 48 according to a fixed ***expansion function*** **E**. $\mathbf{E}(A)$ consists of the 32 bits from A, permuted in a certain way, with 16 of the bits appearing twice.

2. Compute $\mathbf{E}(A) \oplus J$ and write the result as the concatenation of eight 6-bit strings $B = B_1B_2B_3B_4B_5B_6B_7B_8$.

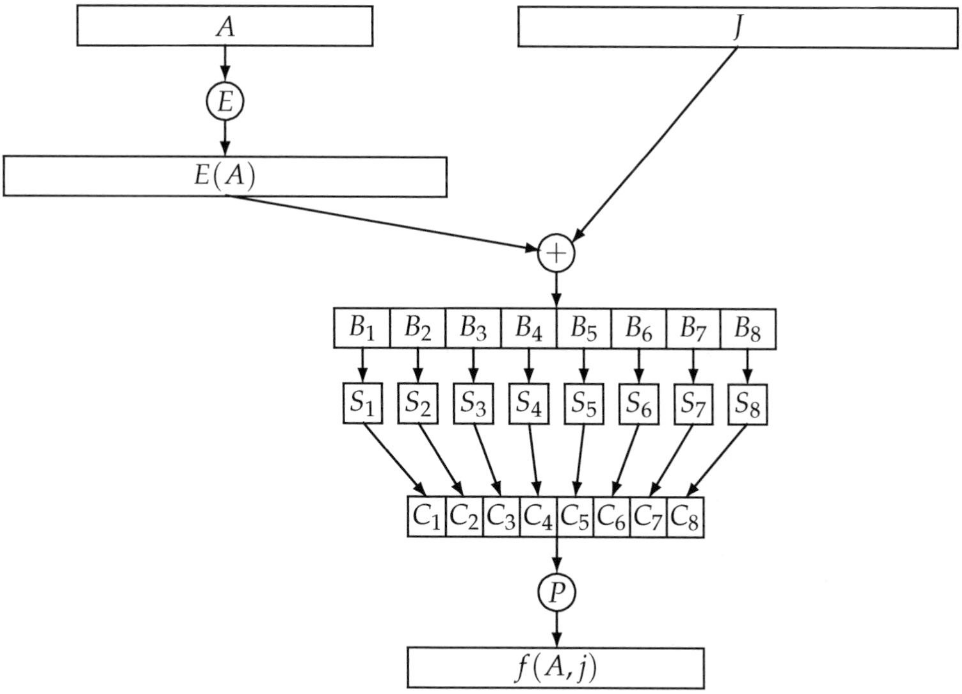

FIGURE 4.6: The DES f function

3. The next step uses eight S-boxes, denoted $S_1, \ldots, S_8$. Each S-box

$$S_i : \{0,1\}^6 \rightarrow \{0,1\}^4$$

maps six bits to four bits. Using these eight S-boxes, we compute $C_j = S_j(B_j)$, $1 \leq j \leq 8$.

4. The bitstring

$$C = C_1C_2C_3C_4C_5C_6C_7C_8$$

of length 32 is permuted according to the permutation $\mathbf{P}$. The resulting bitstring $\mathbf{P}(C)$ is defined to be $f(A, J)$.

4.5.2 Analysis of DES

When *DES* was proposed as a standard, there was considerable criticism. One objection to *DES* concerned the S-boxes. All computations in *DES*, with the sole exception of the S-boxes, are linear, i.e., computing the exclusive-or of two outputs is the same as forming the exclusive-or of two inputs and then computing the output. The S-boxes, being the non-linear components of the cryptosystem, are vital to its security. (We saw in Chapter 2 how linear cryptosystems, such as the *Hill*

Cipher, could easily be cryptanalyzed by a known plaintext attack.) At the time that *DES* was proposed, several people suggested that its S-boxes might contain hidden "trapdoors" which would allow the National Security Agency to easily decrypt messages while claiming falsely that *DES* is "secure." It is, of course, impossible to disprove such a speculation, but no evidence ever came to light that indicated that trapdoors in *DES* do, in fact, exist.

Actually, it was eventually revealed that the *DES* S-boxes were designed to prevent certain types of attacks. When Biham and Shamir invented the technique of differential cryptanalysis (which we discussed in Section 4.4) in the early 1990s, it was acknowledged that the purpose of certain unpublished design criteria of the S-boxes was to make differential cryptanalysis of *DES* infeasible. Differential cryptanalysis was known to IBM researchers at the time that *DES* was being developed, but it was kept secret for almost 20 years, until Biham and Shamir independently discovered the attack.

The most pertinent criticism of *DES* is that the size of the keyspace, 2^{56}, is too small to be really secure. The IBM *Lucifer* cryptosystem, a predecessor of *DES*, had a 128-bit key. The original proposal for *DES* had a 64-bit key, but this was later reduced to a 56-bit key. IBM claimed that the reason for this reduction was that it was necessary to include eight parity-check bits in the key, meaning that 64 bits of storage could only contain a 56-bit key.

Even in the 1970s, it was argued that a special-purpose machine could be built to carry out a known plaintext attack, which would essentially perform an exhaustive search for the key. That is, given a 64-bit plaintext x and corresponding ciphertext y, every possible key would be tested until a key K is found such that $e_K(x) = y$ (note that there may be more than one such key K). As early as 1977, Diffie and Hellman suggested that one could build a VLSI chip which could test 10^6 keys per second. A machine with 10^6 chips could search the entire key space in about a day. They estimated that such a machine could be built, at that time, for about \$20,000,000.

Later, at the *CRYPTO '93* Rump Session, Michael Wiener gave a very detailed design of a *DES* key search machine. The machine is based on a key search chip that is pipelined so that 16 encryptions take place simultaneously. This chip would test 5×10^7 keys per second, and could have been built using 1993 technology for \$10.50 per chip. A frame consisting of 5760 chips could be built for \$100,000. This would allow a *DES* key to be found in about 1.5 days on average. A machine using ten frames would cost \$1,000,000, but would reduce the average search time to about 3.5 hours.

Wiener's machine was never built, but a key search machine costing \$250,000 was built in 1998 by the Electronic Frontier Foundation. This computer, called *DES Cracker*, contained 1536 chips and could search 88 billion keys per second. It won RSA Laboratory's *DES Challenge II-2* by successfully finding a *DES* key in 56 hours in July 1998. In January 1999, RSA Laboratory's *DES Challenge III* was solved by the DES Cracker working in conjunction with a worldwide network (of 100,000 computers) known as `distributed.net`. This co-operative effort found a *DES* key in 22 hours, 15 minutes, testing over 245 billion keys per second.

More recently, `crack.sh` has built a special-purpose key search device consisting of 48 FPGAs that can exhaustively search all 2^{56} possible *DES* keys in 26 hours. In fact, they offer a commercial service to find *DES* keys in a known plaintext attack.

Other than exhaustive key search, the two most important cryptanalytic attacks on *DES* are differential cryptanalysis and linear cryptanalysis. (For SPNs, these attacks were described in Sections 4.4 and 4.3, respectively.) In the case of *DES*, linear cryptanalysis is the more efficient of the two attacks, and an actual implementation of linear cryptanalysis was carried out in 1994 by its inventor, Matsui. This linear cryptanalysis of *DES* is a known-plaintext attack using 2^{43} plaintext-ciphertext pairs, all of which are encrypted using the same (unknown) key. It took 40 days to generate the 2^{43} pairs, and it took 10 days to actually find the key. This cryptanalysis did not have a practical impact on the security of *DES*, however, due to the extremely large number of plaintext-ciphertext pairs that are required to mount the attack: it is unlikely in practice that an adversary will be able to accumulate such a large number of plaintext-ciphertext pairs that are all encrypted using the same key.

4.6 The Advanced Encryption Standard

On January 2, 1997, NIST began the process of choosing a replacement for *DES*. The replacement would be called the *Advanced Encryption Standard*, or *AES*. A formal call for algorithms was made on September 12, 1997. It was required that the *AES* have a block length of 128 bits and support key lengths of 128, 192, and 256 bits. It was also necessary that the *AES* should be available worldwide on a royalty-free basis.

Submissions were due on June 15, 1998. Of the 21 submitted cryptosystems, 15 met all the necessary criteria and were accepted as *AES* candidates. NIST announced the 15 *AES* candidates at the *First AES Candidate Conference* on August 20, 1998. A *Second AES Candidate Conference* was held in March 1999. Then, in August 1999, five of the candidates were chosen by NIST as finalists: *MARS*, *RC6*, *Rijndael*, *Serpent*, and *Twofish*.

The *Third AES Candidate Conference* was held in April 2000. On October 2, 2000, *Rijndael* was selected to be the *Advanced Encryption Standard*. On February 28, 2001, NIST announced that a draft Federal Information Processing Standard for the *AES* was available for public review and comment. *AES* was adopted as a standard on November 26, 2001, and it was published as FIPS 197 in the *Federal Register* on December 4, 2001.

The selection process for the *AES* was notable for its openness and its international flavor. The three candidate conferences, as well as official solicitations for public comments, provided ample opportunity for feedback and public discussion and analysis of the candidates, and the process was viewed very favorably by everyone involved. The "international" aspect of *AES* is demonstrated by the variety of countries represented by the authors of the 15 candidate ciphers: Australia, Belgium, Canada, Costa Rica, France, Germany, Israel, Japan, Korea, Norway, the United Kingdom, and the USA. *Rijndael,* which was ultimately selected as the *AES,* was invented by two Belgian researchers, Daemen and Rijmen. Another interesting departure from past practice was that the *Second AES Candidate Conference* was held outside the U.S., in Rome, Italy.

AES candidates were evaluated for their suitability according to three main criteria:

- security
- cost
- algorithm and implementation characteristics

Security of the proposed algorithm was absolutely essential, and any algorithm found not to be secure would not be considered further. "Cost" refers to the computational efficiency (speed and memory requirements) of various types of implementations, including software, hardware and smart cards. Algorithm and implementation characteristics include flexibility and algorithm simplicity, among other factors.

In the end, the five finalists were all felt to be secure. *Rijndael* was selected because its combination of security, performance, efficiency, implementability, and flexibility was judged to be superior to the other finalists.

4.6.1 Description of AES

As mentioned above, the *AES* has block length 128, and there are three allowable key lengths, namely 128 bits, 192 bits, and 256 bits. *AES* is an iterated cipher; the number of rounds, which we denote by $\mathcal{N}$, depends on the key length. $\mathcal{N} = 10$ if the key length is 128 bits; $\mathcal{N} = 12$ if the key length is 192 bits; and $\mathcal{N} = 14$ if the key length is 256 bits.

We first give a high-level description of *AES*. The algorithm proceeds as follows:

1. Given a plaintext x, initialize **State** to be x and perform an operation ADDROUNDKEY, which x-ors the **RoundKey** with **State**.

2. For each of the first $\mathcal{N} - 1$ rounds, perform a substitution operation called SUBBYTES on **State** using an S-box; perform a permutation SHIFTROWS on **State**; perform an operation MIXCOLUMNS on **State**; and perform ADDROUNDKEY.

3. Perform SUBBYTES; perform SHIFTROWS; and perform ADDROUNDKEY.

4. Define the ciphertext y to be **State**.

Algorithm 4.4: SUBBYTES($a_7a_6a_5a_4a_3a_2a_1a_0$)

external FIELDINV, BINARYTOFIELD, FIELDTOBINARY
$z \leftarrow$ BINARYTOFIELD($a_7a_6a_5a_4a_3a_2a_1a_0$)
if $z \neq 0$
  then $z \leftarrow$ FIELDINV(z)
$(a_7a_6a_5a_4a_3a_2a_1a_0) \leftarrow$ FIELDTOBINARY(z)
$(c_7c_6c_5c_4c_3c_2c_1c_0) \leftarrow (01100011)$
comment: In the following loop, all subscripts are to be reduced modulo 8

for $i \leftarrow 0$ **to** 7
  do $b_i \leftarrow (a_i + a_{i+4} + a_{i+5} + a_{i+6} + a_{i+7} + c_i) \bmod 2$
return $(b_7b_6b_5b_4b_3b_2b_1b_0)$

From this high-level description, we can see that the structure of the *AES* is very similar in many respects to the SPN discussed in Section 4.2. In every round of both these cryptosystems, we have round key mixing, a substitution step, and a permutation step. Both ciphers also include whitening. *AES* is "larger" and it also includes an additional linear transformation (MIXCOLUMNS) in each round.

We now give precise descriptions of all the operations used in the *AES*; describe the structure of **State**; and discuss the construction of the key schedule. All operations in *AES* are byte-oriented operations, and all variables used are considered to be formed from an appropriate number of bytes. The plaintext x consists of 16 bytes, denoted $x_0, \ldots, x_{15}$. **State** is represented as a four by four array of bytes, as follows:

$s_{0,0}$	$s_{0,1}$	$s_{0,2}$	$s_{0,3}$
$s_{1,0}$	$s_{1,1}$	$s_{1,2}$	$s_{1,3}$
$s_{2,0}$	$s_{2,1}$	$s_{2,2}$	$s_{2,3}$
$s_{3,0}$	$s_{3,1}$	$s_{3,2}$	$s_{3,3}$

Initially, **State** is defined to consist of the 16 bytes of the plaintext x, as follows:

$s_{0,0}$	$s_{0,1}$	$s_{0,2}$	$s_{0,3}$		x_0	x_4	x_8	x_{12}
$s_{1,0}$	$s_{1,1}$	$s_{1,2}$	$s_{1,3}$	$\leftarrow$	x_1	x_5	x_9	x_{13}
$s_{2,0}$	$s_{2,1}$	$s_{2,2}$	$s_{2,3}$		x_2	x_6	x_{10}	x_{14}
$s_{3,0}$	$s_{3,1}$	$s_{3,2}$	$s_{3,3}$		x_3	x_7	x_{11}	x_{15}

We will often use hexadecimal notation to represent the contents of a byte. Each byte therefore consists of two hexadecimal digits.

The operation SUBBYTES performs a substitution on each byte of **State** independently, using an S-box, say π_S, which is a permutation of $\{0,1\}^8$. To present this π_S, we represent bytes in hexadecimal notation. π_S is depicted as a 16 by 16 array, where the rows and columns are indexed by hexadecimal digits. The entry

TABLE 4.5: The AES S-box

	Y															
X	0	1	2	3	4	5	6	7	8	9	A	B	C	D	E	F
0	63	7C	77	7B	F2	6B	6F	C5	30	01	67	2B	FE	D7	AB	76
1	CA	82	C9	7D	FA	59	47	F0	AD	D4	A2	AF	9C	A4	72	C0
2	B7	FD	93	26	36	3F	F7	CC	34	A5	E5	F1	71	D8	31	15
3	04	C7	23	C3	18	96	05	9A	07	12	80	E2	EB	27	B2	75
4	09	83	2C	1A	1B	6E	5A	A0	52	3B	D6	B3	29	E3	2F	84
5	53	D1	00	ED	20	FC	B1	5B	6A	CB	BE	39	4A	4C	58	CF
6	D0	EF	AA	FB	43	4D	33	85	45	F9	02	7F	50	3C	9F	A8
7	51	A3	40	8F	92	9D	38	F5	BC	B6	DA	21	10	FF	F3	D2
8	CD	0C	13	EC	5F	97	44	17	C4	A7	7E	3D	64	5D	19	73
9	60	81	4F	DC	22	2A	90	88	46	EE	B8	14	DE	5E	0B	DB
A	E0	32	3A	0A	49	06	24	5C	C2	D3	AC	62	91	95	E4	79
B	E7	C8	37	6D	8D	D5	4E	A9	6C	56	F4	EA	65	7A	AE	08
C	BA	78	25	2E	1C	A6	B4	C6	E8	DD	74	1F	4B	BD	8B	8A
D	70	3E	B5	66	48	03	F6	0E	61	35	57	B9	86	C1	1D	9E
E	E1	F8	98	11	69	D9	8E	94	9B	1E	87	E9	CE	55	28	DF
F	8C	A1	89	0D	BF	E6	42	68	41	99	2D	0F	B0	54	BB	16

in row X and column Y is $\pi_S(XY)$. The array representation of π_S is presented in Table 4.5.

In contrast to the S-boxes in *DES*, which are apparently "random" substitutions, the *AES* S-box can be defined algebraically. The algebraic formulation of the *AES* S-box involves operations in a finite field (finite fields are discussed in detail in Section 7.4). We include the following description for the benefit of readers who are already familiar with finite fields (other readers may want to skip this description, or read Section 7.4 first): The permutation π_S incorporates operations in the finite field

$$\mathbb{F}_{2^8} = \mathbb{Z}_2[x]/(x^8 + x^4 + x^3 + x + 1).$$

Let FIELDINV denote the multiplicative inverse of a field element; let BINARYTOFIELD convert a byte to a field element; and let FIELDTOBINARY perform the inverse conversion. This conversion is done in the obvious way: the field element

$$\sum_{i=0}^{7} a_i x^i$$

corresponds to the byte

$$a_7 a_6 a_5 a_4 a_3 a_2 a_1 a_0,$$

where $a_i \in \mathbb{Z}_2$ for $0 \leq i \leq 7$. Then the permutation π_S is defined according to Algorithm 4.4. In this algorithm, the eight input bits $a_7 a_6 a_5 a_4 a_3 a_2 a_1 a_0$ are replaced by the eight output bits $b_7 b_6 b_5 b_4 b_3 b_2 b_1 b_0$.

Example 4.4 We do a small example to illustrate Algorithm 4.4, where we also include the conversions to hexadecimal. Suppose we begin with (hexadecimal) 53. In binary, this is

$$01010011,$$

Algorithm 4.5: MIXCOLUMN(c)

```
external FIELDMULT, BINARYTOFIELD, FIELDTOBINARY
for i ← 0 to 3
  do t_i ← BINARYTOFIELD(s_{i,c})
u_0 ← FIELDMULT(x, t_0) ⊕ FIELDMULT(x + 1, t_1) ⊕ t_2 ⊕ t_3
u_1 ← FIELDMULT(x, t_1) ⊕ FIELDMULT(x + 1, t_2) ⊕ t_3 ⊕ t_0
u_2 ← FIELDMULT(x, t_2) ⊕ FIELDMULT(x + 1, t_3) ⊕ t_0 ⊕ t_1
u_3 ← FIELDMULT(x, t_3) ⊕ FIELDMULT(x + 1, t_0) ⊕ t_1 ⊕ t_2
for i ← 0 to 3
  do s_{i,c} ← FIELDTOBINARY(u_i)
```

which represents the field element

$$x^6 + x^4 + x + 1.$$

The multiplicative inverse (in the field $\mathbb{F}_{2^8}$) can be shown to be

$$x^7 + x^6 + x^3 + x.$$

Therefore, in binary notation, we have

$$(a_7a_6a_5a_4a_3a_2a_1a_0) = (11001010).$$

Next, we compute

$$\begin{aligned} b_0 &= a_0 + a_4 + a_5 + a_6 + a_7 + c_0 \bmod 2 \\ &= 0 + 0 + 0 + 1 + 1 + 1 \bmod 2 \\ &= 1, \end{aligned}$$

followed by

$$\begin{aligned} b_1 &= a_1 + a_5 + a_6 + a_7 + a_0 + c_1 \bmod 2 \\ &= 1 + 0 + 1 + 1 + 0 + 1 \bmod 2 \\ &= 0, \end{aligned}$$

etc. The result is that

$$(b_7b_6b_5b_4b_3b_2b_1b_0) = (11101101).$$

In hexadecimal notation, 11101101 is *ED*.

This computation can be checked by verifying that the entry in row 5 and column 3 of Table 4.5 is *ED*. □

The operation SHIFTROWS acts on **State** as shown in the following diagram:

$s_{0,0}$	$s_{0,1}$	$s_{0,2}$	$s_{0,3}$		$s_{0,0}$	$s_{0,1}$	$s_{0,2}$	$s_{0,3}$
$s_{1,0}$	$s_{1,1}$	$s_{1,2}$	$s_{1,3}$	$\leftarrow$	$s_{1,1}$	$s_{1,2}$	$s_{1,3}$	$s_{1,0}$
$s_{2,0}$	$s_{2,1}$	$s_{2,2}$	$s_{2,3}$		$s_{2,2}$	$s_{2,3}$	$s_{2,0}$	$s_{2,1}$
$s_{3,0}$	$s_{3,1}$	$s_{3,2}$	$s_{3,3}$		$s_{3,3}$	$s_{3,0}$	$s_{3,1}$	$s_{3,2}$

The operation MIXCOLUMNS is carried out on each of the four columns of **State**; it is presented as Algorithm 4.5. Each column of **State** is replaced by a new column which is formed by multiplying that column by a certain matrix of elements of the field $\mathbb{F}_{2^8}$. Here, "multiplication" means multiplication in the field $\mathbb{F}_{2^8}$. We assume that the external procedure FIELDMULT takes as input two field elements and computes their product in the field. In Algorithm 4.5, we are multiplying by the field elements x and $x + 1$; these correspond to the bitstrings 00000010 and 00000011, respectively.

Field addition is just componentwise addition modulo 2 (i.e., the x-or of the corresponding bitstrings). This operation is denoted by "$\oplus$" in Algorithm 4.5.

It remains to discuss the key schedule for the *AES*. We describe how to construct the key schedule for the 10-round version of *AES*, which uses a 128-bit key (key schedules for 12- and 14-round versions are similar to 10-round *AES*, but there are some minor differences in the key scheduling algorithm). We need 11 round keys, each of which consists of 16 bytes. The key scheduling algorithm is word-oriented (a ***word*** consists of 4 bytes, or, equivalently, 32 bits). Therefore each round key is comprised of four words. The concatenation of the round keys is called the ***expanded key***, which consists of 44 words. It is denoted $w[0], \ldots, w[43]$, where each $w[i]$ is a word. The expanded key is constructed using the operation KEYEXPANSION, which is presented as Algorithm 4.6.

The input to this algorithm is the 128-bit key, *key*, which is treated as an array of bytes, $key[0], \ldots, key[15]$; and the output is the array of words, w, that was introduced above.

KEYEXPANSION incorporates two other operations, which are named ROTWORD and SUBWORD. ROTWORD(B_0, B_1, B_2, B_3) performs a cyclic shift of the four bytes B_0, B_1, B_2, B_3, i.e.,

$$\text{ROTWORD}(B_0, B_1, B_2, B_3) = (B_1, B_2, B_3, B_0).$$

SUBWORD(B_0, B_1, B_2, B_3) applies the *AES* S-box to each of the four bytes B_0, B_1, B_2, B_3, i.e.,

$$\text{SUBWORD}(B_0, B_1, B_2, B_3) = (B'_0, B'_1, B'_2, B'_3)$$

where $B'_i = \text{SUBBYTES}(B_i)$, $i = 0, 1, 2, 3$. *RCon* is an array of 10 words, denoted $RCon[1], \ldots, RCon[10]$. These are constants that are defined in hexadecimal notation at the beginning of Algorithm 4.6.

We have now described all the operations required to perform an encryption operation in the *AES*. In order to decrypt, it is necessary to perform all operations

Algorithm 4.6: KEYEXPANSION(*key*)

```
external RotWord, SubWord
RCon[1] ← 01000000
RCon[2] ← 02000000
RCon[3] ← 04000000
RCon[4] ← 08000000
RCon[5] ← 10000000
RCon[6] ← 20000000
RCon[7] ← 40000000
RCon[8] ← 80000000
RCon[9] ← 1B000000
RCon[10] ← 36000000
for i ← 0 to 3
  do w[i] ← (key[4i], key[4i + 1], key[4i + 2], key[4i + 3])
for i ← 4 to 43
  do { temp ← w[i − 1]
       if i ≡ 0 (mod 4)
         then temp ← SubWord(RotWord(temp)) ⊕ RCon[i/4]
       w[i] ← w[i − 4] ⊕ temp
return (w[0], . . . , w[43])
```

in the reverse order, and use the key schedule in reverse order. Further the operations SHIFTROWS, SUBBYTES, and MIXCOLUMNS must be replaced by their inverse operations (the operation ADDROUNDKEY is its own inverse). It is also possible to construct an "equivalent inverse cipher" that performs *AES* decryption by doing a sequence of (inverse) operations in the same order as is done for *AES* encryption. It is suggested that this can lead to implementation efficiencies.

4.6.2 Analysis of AES

Obviously, the *AES* is secure against all known attacks. Various aspects of its design incorporate specific features that help provide security against specific attacks. For example, the use of the finite field inversion operation in the construction of the S-box yields linear approximation and difference distribution tables in which the entries are close to uniform. This provides security against differential and linear attacks. As well, the linear transformation, MIXCOLUMNS, makes it impossible to find differential and linear attacks that involve "few" active S-boxes (the designers refer to this feature as the ***wide trail strategy***).

There are apparently no known "general" attacks on *AES* that are significantly faster than exhaustive search. The best such attack is called the ***biclique attack***. It is due to Bogdanov, Khovratovich, and Rechberger and was published in 2011.

This attack reduces the complexity of an exhaustive search by a factor of four or five; it applies to all three variants of *AES*.

There are also some attacks against reduced-round variants of *AES*. The strongest results involve so-called ***related-key attacks***, which exploit certain weaknesses in the key schedule. In a related-key attack, the adversary is provided with ciphertexts that have been encrypted using two or more unknown keys that have some specified relation between them (of course, this is quite a powerful attack model, and it is probably not realistic in practice). For example, there are several attacks on *AES-256* published in 2009 by Biryukov, Dunkelman, Keller, Khovratovich, and Shamir. One of their attacks uses two related keys and takes 2^{39} time to recover the key for 9-round *AES-256*, which is quite impressive. However, their attacks do not extend to the "full" 14-round *AES-256*.

4.7 Modes of Operation

Four ***modes of operation*** were developed for *DES*. They were standardized in FIPS Publication 81 in December 1980. These modes of operation can be used (with minor changes) for any block cipher in which the plaintext and ciphertext spaces are identical, i.e., whenever the block cipher is ***endomorphic***). More recently, some additional modes of operation have been proposed for *AES*. The following seven modes of operation are presented as popular examples of modes, many of which are commonly used in practice.

- ***electronic codebook mode*** (ECB mode),
- ***cipher block chaining mode*** (CBC mode),
- ***output feedback mode*** (OFB mode),
- ***cipher feedback mode*** (CFB mode),
- ***counter mode*** (CTR mode),
- ***counter with cipher-block chaining MAC*** (CCM mode), and
- ***Galois/counter mode*** (GCM).

Here are short descriptions of these modes of operation:

ECB mode

This mode corresponds to the naive use of a block cipher: given a sequence $x_1x_2\ldots$ of plaintext blocks, each x_i is encrypted with the same key K, producing a string of ciphertext blocks, $y_1y_2\ldots$.

ECB mode is virtually never used in practice. One obvious weakness of ECB mode is that the encryption of identical plaintext blocks yields identical ciphertext blocks. This is a serious weakness if the underlying message blocks are chosen from a "low entropy" plaintext space. To take an extreme example, if a plaintext block always consists entirely of 0's or entirely of 1's, then ECB mode is essentially useless.

CBC mode

In CBC mode, each ciphertext block y_i is x-ored with the next plaintext block, x_{i+1}, before being encrypted with the key K. More formally, we start with an ***initialization vector***, denoted by IV, and define $y_0 = \text{IV}$. (Note that IV has the same length as a plaintext block.) Then we construct $y_1, y_2, \ldots$, using the rule

$$y_i = e_K(y_{i-1} \oplus x_i),$$

$i \geq 1$.

Encryption and decryption using CBC mode is depicted in Figure 4.7.

Observe that, if a plaintext block x_i is changed in CBC mode, then y_i and all subsequent ciphertext blocks will be affected. This property means that CBC mode is useful for purposes of authentication. More specifically, this mode can be used to produce a message authentication code, or MAC. The MAC is appended to a sequence of plaintext blocks, and is used to convince Bob that the given sequence of plaintext originated with Alice and was not tampered with by Oscar. Thus the MAC guarantees the integrity (or authenticity) of a message (but it does not provide secrecy, of course). We will say much more about MACs in Chapter 5. The use of CBC modes to construct MACs is studied further in Section 5.5.2.

A couple of general comments about initialization vectors (IVs) are in order. An IV is not usually secret; however, in the context of encryption, it is important to never use the same IV more than once with a given key (see the Exercises to examine the consequences of re-using an IV). Thus, an IV is typically chosen using a suitable pseudorandom number generator, and transmitted in unencrypted form along with the ciphertext.

OFB mode

In OFB mode, a keystream is generated, which is then x-ored with the plaintext (i.e., it operates as a stream cipher, cf. Section 2.1.7). OFB mode is actually a synchronous stream cipher: the keystream is produced by repeatedly encrypting an initialization vector, IV. We define $z_0 = \text{IV}$, and then compute the keystream $z_1 z_2 \ldots$ using the rule

$$z_i = e_K(z_{i-1}),$$

for all $i \geq 1$. The plaintext sequence $x_1 x_2 \ldots$ is then encrypted by computing

$$y_i = x_i \oplus z_i,$$

for all $i \geq 1$.

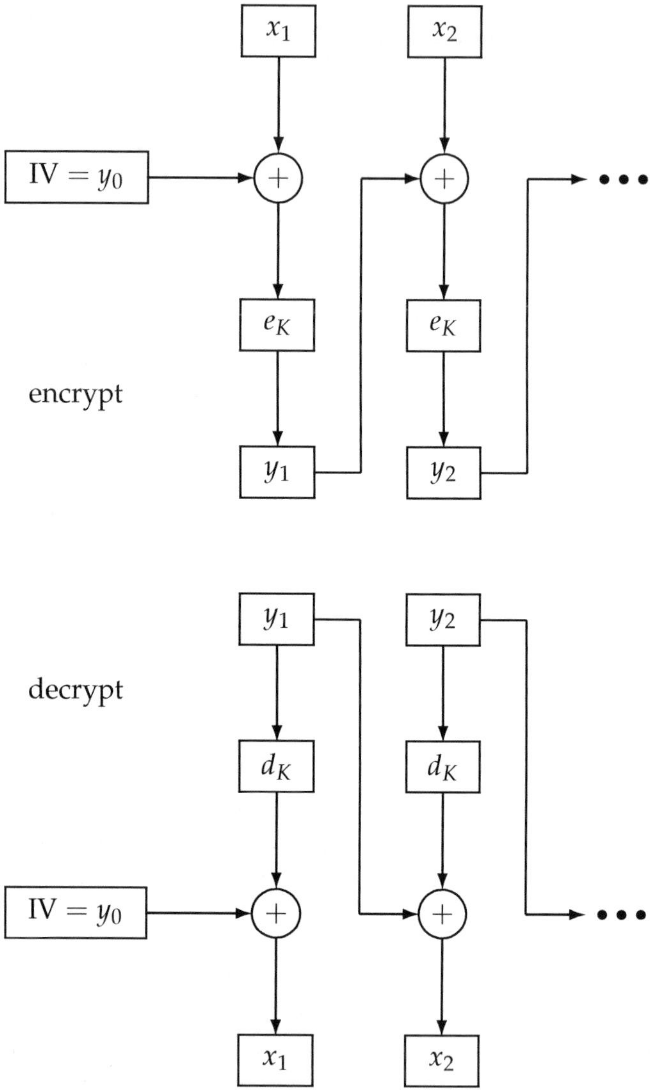

FIGURE 4.7: CBC mode

Decryption is straightforward. First, recompute the keystream $z_1 z_2 \ldots$, and then compute

$$x_i = y_i \oplus z_i,$$

for all $i \geq 1$. Note that the encryption function e_K is used for both encryption and decryption in OFB mode.

CFB mode

CFB mode also generates a keystream for use in a stream cipher, but this time the resulting stream cipher is asynchronous. We start with $y_0 = \text{IV}$ (an initialization vector) and we produce the keystream element z_i by encrypting the previous ciphertext block. That is,

$$z_i = e_K(y_{i-1}),$$

for all $i \geq 1$. As in OFB mode, we encrypt using the formula

$$y_i = x_i \oplus z_i,$$

for all $i \geq 1$. Again, the encryption function e_K is used for both encryption and decryption in CFB mode.

The use of CFB mode is depicted in Figure 4.8.

CTR mode

Counter mode is similar to OFB mode; the only difference is in how the keystream is constructed. Suppose that the length of a plaintext block is denoted by m. In counter mode, we choose a ***counter***, denoted ctr, which is a bitstring of length m. Then we construct a sequence of bitstrings of length m, denoted $T_1, T_2, \ldots$, defined as follows:

$$T_i = ctr + i - 1 \bmod 2^m$$

for all $i \geq 1$. Then we encrypt the plaintext blocks $x_1, x_2, \ldots$ by computing

$$y_i = x_i \oplus e_K(T_i),$$

for all $i \geq 1$. Observe that the keystream in counter mode is obtained by encrypting the sequence of counters using the key K.

As in the case of OFB mode, the keystream in counter mode can be constructed independently of the plaintext. However, in counter mode, there is no need to iteratively compute a sequence of encryptions; each keystream element $e_K(T_i)$ can be computed independently of any other keystream element. (In contrast, OFB mode requires one to compute z_{i-1} prior to computing z_i.) This feature of counter mode permits very efficient implementations in software or hardware by exploiting opportunities for parallelism (see the Exercises).

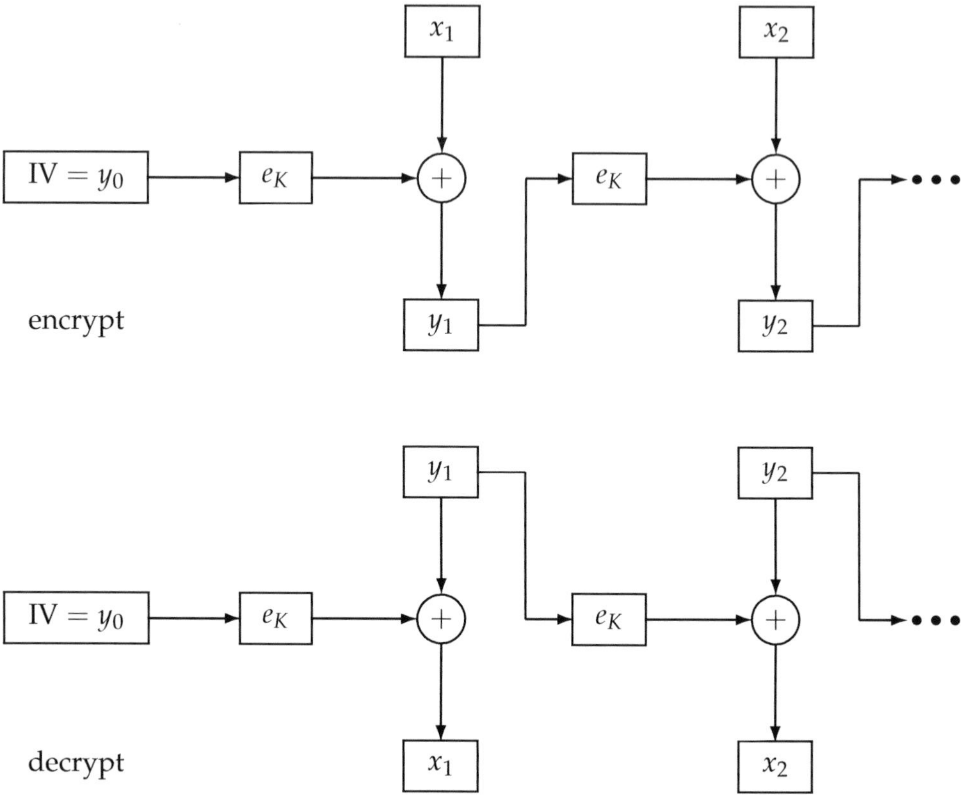

FIGURE 4.8: CFB mode

CCM mode

Basically, CCM mode combines the use of counter mode (for encryption) with CBC-mode (for authentication). This mode, which is discussed further in Section 5.5.3, is used for authenticated encryption.

GCM

GCM is another mode used for authenticated encryption. See Section 5.5.3 for details.

4.7.1 Padding Oracle Attack on CBC Mode

In this section, we describe an unusual and ingenious attack on encryption using CBC mode in conjunction with a certain padding scheme. This attack, which is known as a "padding oracle attack," was first presented by Vaudenay in 2002.

It exploits the requirement for plaintext data to be "padded" so that its length is a multiple of the block size before it is encrypted.

Let's assume that our plaintext data consists of some integral number of bytes, and suppose that we are using a block cipher with block size 128 bits (i.e., 16 bytes). The plaintext would be partitioned into blocks, with a possibly incomplete block at the end. This last block will be padded with extra data so that it fills out the entire 128 bits. The ***padding scheme*** describes how this will be done.

We illustrate using *PKCS #7*, which is a common padding scheme. We use hexadecimal notation. The rule is that 15 bytes of data will be padded with the byte 01 (i.e., the eight bits 00000001). 14 bytes will be padded with the two bytes 02 02; 13 bytes will be padded with 03 03 03, etc., and one byte of data will be padded with 15 copies of $0F$. Finally, if the last block is a complete block, then we concatenate an extra block consisting of 16 repetitions of 00.

Suppose we have a sequence of ciphertext blocks $y_0, y_1 \dots, y_n$ (as usual, y_0 is the IV). After decryption, the last block is checked to see if it is padded correctly. If so, then the padding is discarded. However, if the padding is invalid, then some kind of error would be raised.

A ***padding oracle attack*** refers to an attack model where the adversary is allowed to submit ciphertext blocks to an "oracle" that reports if the resulting plaintext is correctly padded (note that the actual plaintext is not given to the adversary). Mathematically, we can describe the oracle as function $\mathcal{O}(y_0, y_1 \dots, y_n)$ which returns *true* if the plaintext is correctly padded, and *false* otherwise.

Let's consider the first block of actual ciphertext, namely, y_1. The plaintext x_1 is computed as

$$x_1 = d_K(y_1) \oplus y_0,$$

where y_0 is the IV. The adversary is free to choose any values it likes for y_0; we will use y_0' to denote a value chosen by the adversary. Suppose we write y_0' as the concatenation of 16 bytes: $y_0' = r_1 r_2 \cdots r_{16}$. The first 15 bytes are chosen randomly and r_{16} will successively take on all 256 possible values $00, 01, \dots, FF$. Now consider what happens if the adversary invokes the oracle to compute $\mathcal{O}(y_0', y_1)$ for the various possible values of y_0'. There will be exactly one value of r_{16} that will result in the last byte of $d_K(y_1) \oplus y_0'$ having the value 01. When this happens, the oracle will output the value *true*. But this allows the adversary to compute the last byte of x_1: the last byte of x_1 is equal to $r_{16} \oplus 01$. Thus the adversary is able to compute the last byte of x_1 after a maximum of 256 calls to the padding oracle.

There is one small technical detail that we should mention. There is a small probability that $d_K(y_1) \oplus y_0'$ is padded correctly, but the padding is not 01 (it could be one of 02 02, 03 03 03, etc.). But these are much less likely to occur than 01, and we will not worry about how to handle these situations.

Having computed the last byte of x_1, it is now possible to compute the second last byte of x_1. Our starting point is that we have a value r_{16} such that the last byte of $d_K(y_1) \oplus y_0'$ is equal to 01. Suppose we increment r_{16} by 1 and denote the resulting 16 bytes by y_0''. Then the last byte of $d_K(y_1) \oplus y_0''$ would be equal to 02 and the padding would be valid if the second last byte of $d_K(y_1) \oplus y_0''$ is also equal to 02. So the strategy is to consider all 256 possible values for the second last byte of y_0'', which is denoted by r_{15}. When the oracle responds $\mathcal{O}(y_0'', y_1) = \mathit{true}$ for some y_0'', we know that the second last byte of x_1 is equal to $r_{15} \oplus 02$. Thus we have computed the second last byte of x_1 using at most 256 additional calls to the oracle.

This process can be repeated, to successively compute all 16 bytes of x_1 one at a time. The number of calls to the oracle is at most $16 \times 256 = 4096$.

In fact, any plaintext block can computed by this technique. We used y_1 along with suitably manipulated values of y_0 to compute x_1. Analogously, we can use y_2 along with altered values of y_1 to compute x_2, using the equation

$$x_2 = d_K(y_2) \oplus y_1.$$

In general, we employ one ciphertext block, along with appropriate modifications of the previous ciphertext block, to determine a given plaintext block.

Finally, we should point out that this kind of attack has been carried out against various web browser platforms implementing *TLS* (*Transport Layer Security*), so it is not just a "theoretical" attack.

4.8 Stream Ciphers

In this section, we discuss some common approaches to the design of practical stream ciphers. We will restrict our attention to stream ciphers that encrypt and decrypt a binary plaintext using an exclusive-or (i.e., an x-or) with a binary keystream. Virtually all stream ciphers used in practice are of this type.

We introduced stream ciphers in Section 2.1.7, where we mentioned the use of a linear feedback shift register (LFSR) as a possible technique to generate a keystream. However, an LFSR does not yield a secure stream cipher, as we showed in Section 2.2.5. Nevertheless, the idea of using LFSRs to construct stream ciphers is very appealing due to the efficiency of LFSRs and the fact that they can have a large period. So various techniques have been proposed to "combine" LFSRs in such a way that an efficient and secure stream cipher is obtained. That is, instead of taking the output of an LFSR to be the keystream, we produce a keystream from some number (one or greater than one) of LFSRs by using a suitable boolean function or some other mechanism. Three of the most common methods of doing this are the following:

- ***combination generator***,
- ***filter generator***, and
- ***shrinking generator***.

Here are short descriptions of these generators:

combination generator

In a combination generator, we have some number, say r, of LFSRs. Suppose that the jth LFSR generates the keystream $z_1^j, z_2^j, \ldots$. The basic idea is to use a boolean function $f : (\mathbb{Z}_2)^r \rightarrow \mathbb{Z}_2$ to combine the r keystreams into a new keystream $z_1 z_2 \ldots$, via the rule

$$z_i = f(z_i^1, \ldots, z_i^r),$$

$i = 1, 2, \ldots$. The function f is called the ***combining function***. Note that it is desirable that the r LFSRs have periods that are pairwise relatively prime—this will ensure that that the input to the combining function has the longest possible period (namely, the product of the periods of the r LFSRs).

filter generator

In a filter generator, we use a single LFSR, having m stages, say. But instead of taking keystream bits to be the bits that are produced by the LFSR, we apply a boolean function (having m inputs) to the entire m-bit state of the LFSR. The output of the boolean function at any given time is a keystream bit.

shrinking generator

In a shrinking generator, we use two LFSRs. The keystream bits are obtained from the first LFSR. However, some of these bits are discarded, depending on the output of the second LFSR. If the second LFSR outputs a zero, then the output of the first LFSR is discarded; if the second LFSR outputs a one, then the output of the first LFSR is the next keystream bit.

4.8.1 Correlation Attack on a Combination Generator

There has been a considerable amount of research done on these various types of generators, including a variety of possible attacks. In this section, we describe an attack on the combination generator, which is known as a ***correlation attack***. This attack can be carried out when there are correlations between outputs of the LFSRs (which are the inputs to the combining function) and the output of the combining function.

Suppose we have $r = 3$ LFSRs and the combining function is

$$f(z_1, z_2, z_3) = (z_1 \wedge z_2) \oplus (z_1 \wedge z_3) \oplus (z_2 \wedge z_3).$$[1]

This is sometimes called the "majority function" since it outputs the most frequently occurring bit among the three input bits. We tabulate the eight possible

[1]The boolean operation $\wedge$ is used to denote the logical "and" of the two inputs.

inputs to f along with its output:

z_1	z_2	z_3	$f(z_1, z_2, z_3)$
0	0	0	0
0	0	1	0
0	1	0	0
0	1	1	1
1	0	0	0
1	0	1	1
1	1	0	1
1	1	1	1

Let us assume that each input triple (z_1, z_2, z_3) is equally likely. We can associate a random variable $\mathbf{z_j}$ with each input variable z_j, and we associate the random variable $\mathbf{z}$ with the output $f(z_1, z_2, z_3)$. Then it is easy to see from the table above that

$$\mathbf{Pr}[\mathbf{z} = \mathbf{z_j}] = \frac{3}{4},$$

for $j = 1, 2, 3$. It turns out that we can use this correlation to search for the initial key for each of the three LFSRs.

To make the attack precise, suppose that the three LFSRs (i.e., the linear recurrence relations) are known. We also assume that the combining function is known. The key then consists of the initial states of the three component LFSRs. Suppose that the jth LFSR corresponds to a linear recurrence of degree L_j. Then the key for the resulting stream cipher has length

$$L = L_1 + L_2 + L_3.$$

As we did in Section 2.2.5, we will consider a known plaintext attack, which immediately allows us to compute keystream bits. The objective will be to determine the L initial keystream bits. For purposes of comparison, we observe that there is always the option of carrying out a brute force search for the key. For each possible L-bit key, we can generate a keystream using the three LFSRs along with the given combining function. If the generated keystream is identical to the keystream that we have determined from the known plaintext attack, then we can be confident that we have found the correct key. Because this is a brute force search, we may have to consider all 2^L keys until we find the correct one. If L is sufficiently large, this might not be feasible.

However, by making use of the correlations between the inputs to f and its output, we can make the search much more efficient. The correlations allow us to search for the initial state of each LFSR separately (hence this is sometimes referred to as a "divide-and-conquer" attack). Here is how this can be done. Suppose we focus on the first LFSR and we guess an initial keystream consisting of L_1 bits. Then we generate a sequence of bits using this LFSR and compare it to the sequence of actual keystream bits. If the guessed initial key is correct, then we would expect 75% of the bits generated by the LFSR to agree with the keystream.

However, if the guessed initial key is incorrect, then we would expect 50% of the bits generated by the LFSR to agree with the keystream. So the strategy will be to consider all 2^{L_1} possible initial keys for this LFSR, and for each initial key, generate a stream of bits. We keep track of which stream of bits most closely matches the actual keystream bits. If we have a sufficient number of keystream bits, then we can be very confident that this initial key is indeed the correct one.

Thus, by doing an exhaustive search over 2^{L_1} possible initial states, we are hopefully able to determine L_1 bits of the key. We can repeat this attack for the second and third LFSR and thereby obtain the entire L-bit key. The total computation required by this attack can be estimated to be

$$2^{L_1} + 2^{L_2} + 2^{L_3},$$

since the attacks on the three LFSRs are carried out separately. This is much smaller than the brute force attack which requires approximately

$$2^L = 2^{L_1} \times 2^{L_2} \times 2^{L_3}$$

sequences to be tested in the worst case.

Looking at some particular parameters will help to make the comparison more concrete. Suppose $L_1 = 19$, $L_2 = 21$, and $L_3 = 23$. Then the brute force search requires testing 2^{63} possible keys, which is a very large number. However, the correlation attack tests

$$2^{19} + 2^{21} + 2^{23} < 2^{24}$$

keys, which can be done very quickly.

We consider a small example to illustrate how the attack can be carried out in practice.

Example 4.5 Suppose we have three LFSRs, with $L_1 = 5$, $L_2 = 7$, and $L_3 = 9$. These LFSRs (respectively) implement the following linear recurrence relations:

$$\begin{aligned} a_i &= a_{i-3} + a_{i-5} \\ b_i &= b_{i-6} + b_{i-7} \\ c_i &= c_{i-5} + c_{i-9} \end{aligned}$$

where all arithmetic is modulo 2. Suppose we have obtained the following 90 bits of the keystream from a known plaintext attack:

011011001010000111101100000110
001111100101000111110100010111
001100110010100001001111000100

We begin by comparing 30 bits of the keystream to 30 bits generated by the first LFSR using the $2^5 - 1$ possible different nonzero initial keys. For each bit-string generated by the LFSR, we count the number of agreements with the true keystream, obtaining the data presented in Table 4.6.

TABLE 4.6: Possible keystreams

000010010110011111000110111010	15
000100101100111110001101110101	12
000110111010100001001011001111	15
001001011001111100011011101010	12
001011001111100011011101010000	19
001101110101000010010110011111	12
001111100011011101010000100101	15
010000100101100111110001101110	15
010010110011111000110111010100	12
010100001001011001111100011011	15
010110011111000110111010100001	16
011001111100011011101010000100	19
011011101010000100101100111110	24
011101010000100101100111110001	15
011111000110111010100001001011	16
100001001011001111100011011101	16
100011011101010000100101100111	15
100101100111110001101110101000	12
100111110001101110101000010010	15
101000010010110011111000110111	16
101010000100101100111110001101	15
101100111110001101110101000010	16
101110101000010010110011111000	11
110001101110101000010010110011	11
110011111000110111010100001001	16
110101000010010110011111000110	19
110111010100001001011001111100	12
111000110111010100001001011001	11
111010100001001011001111100011	16
111100011011101010000100101100	15
111110001101110101000010010110	16

For each row of Table 4.6, the first five bits comprise the initial key for the LFSR. We observe that the initial key 01101 generates a bitstring that agrees with the keystream in 24 (out of 30) bits, while no other initial key generates a bitstring that agrees with the keystream in more than 19 bits. Hence we would strongly suspect that 01101 is the initial key for the first LFSR.

We can repeat this process for the other two LFSRs. It is probably advisable to carry out these computations with more key bits, because the number of possible initial keys is greater (to be precise, the number of initial keys is $2^7 - 1 = 127$ and

$2^9 - 1 = 511$, respectively, for the last two LFSRs). Suppose we use 60 keystream bits to attack the second LFSR and 90 keystream bits to attack the third LFSR.

For the second LFSR, the initial key 1100110 yields 48 matches (out of 60 bits) whereas no other possible initial key yields more than 39 matches. For the third LFSR, we observe a similar kind of "separation." The initial key 011011001 yields 67 matches (out of 90 bits), but no other possible initial key yields more than 59 matches.

Having identified what we think are the three components of the 21-bit key, it is then a good check to see that we really have the correct key. This is done easily by using the generator to compute keystream bits and then comparing them to the true keystream bits obtained from the known plaintext attack. If we do indeed have the correct 21-bit key, the two keystreams should be identical. In this example, we can confirm that the correct keystream is obtained from the three initial keys that we have identified. □

We note that techniques from probability theory can be used to predict how many keystream bits are required in order for the attack to succeed with high probability. In general, the number of required keystream bits will depend on the correlation (a higher correlation corresponds to fewer required keystream bits) and the degree of the recurrence relations (i.e., the number of stages in the LFSRs). Larger LFSRs will usually require more keystream bits.

It has been suggested that an attack on an LFSR having L_i stages will succeed if the number of keystream bits, N, is at least

$$\frac{L_i}{\left(p - \frac{1}{2}\right)^2},$$

where p is the predicted correlation. Note that Example 4.5 succeeded with a smaller number of keystream bits.

4.8.2 Algebraic Attack on a Filter Generator

In this section, we describe another type of attack called an ***algebraic attack***. Algebraic attacks can be launched against various types of block and stream ciphers. We illustrate the basic idea by presenting an algebraic attack against a filter generator. This can be done provided a "sufficient" number of bits of the keystream are known.

We already saw in Section 2.2.5 that we could attack the *LFSR Stream Cipher* by solving a system of linear equations. However, if a nonlinear filter generator is used to generate a keystream, we instead have to solve a system of polynomial equations in several variables to break the system. We illustrate this attack using a toy example.

Suppose the attacker knows the linear recurrence relation of the underlying LFSR, as well as the filtering function. The initial state of the LFSR is the secret key that they wish to learn.

TABLE 4.7: States and output bits for a filter generator

state	output
(1,0,0,0)	0
(0,0,0,1)	0
(0,0,1,0)	0
(0,1,0,0)	0
(1,0,0,1)	0
(0,0,1,1)	1
(0,1,1,0)	0
(1,1,0,1)	1
(1,0,1,0)	0
(0,1,0,1)	0

Suppose we have a four stage LFSR with initial state $(z_0, z_1, z_2, z_3) = (1,0,0,0)$ that satisfies the recurrence relation $z_{n+4} = z_{n+1} + z_n$ for $n \geq 0$. Suppose further that we use the filtering function $f(z_0, z_1, z_2, z_3) = z_0 z_1 + z_2 z_3$ to generate an output bit from each state. Then the first ten states and corresponding output bits are shown in Table 4.7.

Each output bit can be used to derive an equation in the initial state variables z_0, z_1, z_2, and z_3. The first equation is simply based directly on the filtering function: $f(z_0, z_1, z_2, z_3) = 0$, which gives

$$z_0 z_1 + z_2 z_3 = 0.$$

Now, the next output bit leads to the equation $f(z_1, z_2, z_3, z_4) = 1$. Using the underlying recurrence relation, we know that $z_4 = z_1 + z_0$, so the resulting polynomial equation becomes

$$\begin{aligned} 0 &= f(z_1, z_2, z_3, z_0 + z_1) \\ &= z_1 z_2 + z_3(z_0 + z_1) \\ &= z_1 z_2 + z_0 z_3 + z_1 z_3. \end{aligned}$$

Because the operation of updating the state is linear, we can describe it using matrix multiplication. In this case we have $(z_{i+1}, z_{i+2}, z_{i+3}, z_{i+4}) = (z_i, z_{i+1}, z_{i+2}, z_{i+3})A$, where A is the matrix

$$\begin{pmatrix} 0 & 0 & 0 & 1 \\ 1 & 0 & 0 & 1 \\ 0 & 1 & 0 & 0 \\ 0 & 1 & 0 & 0 \\ 0 & 0 & 1 & 0 \end{pmatrix}.$$

This means that once the LFSR has been clocked n times, its state will be $(z_0, z_1, z_2, z_3)A^n$. Hence the nth output bit, say y_n, gives rise to the polynomial equation $y_n = f((z_0, z_1, z_2, z_3)A^n)$. Determining the resulting expressions for the first 10 output bits gives us the following system of polynomial equations:

$$\begin{aligned}
z_0z_1 + z_2z_3 &= 0\\
z_0z_3 + z_1z_2 + z_1z_3 &= 0\\
z_0z_1 + z_0z_2 + z_1 + z_1z_2 + z_2z_3 &= 0\\
z_0z_3 + z_1z_2 + z_2 + z_2z_3 &= 0\\
z_0z_1 + z_0z_3 + z_1 + z_1z_3 + z_2z_3 + z_3 &= 0\\
z_0 + z_0z_1 + z_0z_2 + z_0z_3 + z_1z_3 + z_2 &= 1\\
z_0z_1 + z_0z_2 + z_1z_3 + z_3 &= 0\\
z_0 + z_0z_2 + z_1 + z_1z_3 &= 1\\
z_0z_2 + z_1 + z_1z_2 + z_1z_3 + z_2 &= 0\\
z_0z_2 + z_1z_2 + z_1z_3 + z_2 + z_2z_3 + z_3 &= 0.
\end{aligned}$$

We note that the values of each of the variables z_0, z_1, z_2, and z_3 are either 0 or 1. The fact that $0^2 = 0$ and $1^2 = 1$ means we can replace any instance of z_i^2 by z_i in these equations. If we can find a solution to this system of equations, it will allow us to recover the initial state of the generator. In this case we have enough equations to allow us to use an approach known as ***linearization***.

The above polynomials are sums of terms that are either a single variable or a product of two different variables. As we are working with the variables z_0, z_1, z_2, and z_3, there are thus $4 + \binom{4}{2} = 10$ distinct possible terms. We can replace each of these ten possible terms with a new variable, for example, by setting $X_0 = z_0$, $X_1 = z_0z_1$, $X_2 = z_0z_2$, $X_3 = z_0z_3$, $X_4 = z_1$, $X_5 = z_1z_2$, $X_6 = z_1z_3$, $X_7 = z_2$, $X_8 = z_2z_3$, and $X_9 = z_3$. Written in terms of these new variables, our polynomial equations become the following linear equations:

$$\begin{aligned}
X_1 + X_8 &= 0\\
X_3 + X_5 + X_6 &= 0\\
X_1 + X_2 + X_4 + X_5 + X_8 &= 0\\
X_3 + X_5 + X_7 + X_8 &= 0\\
X_1 + X_3 + X_4 + X_6 + X_8 + X_9 &= 0\\
X_0 + X_1 + X_2 + X_3 + X_6 + X_7 &= 1\\
X_1 + X_2 + X_6 + X_9 &= 0\\
X_0 + X_2 + X_4 + X_6 &= 1\\
X_2 + X_4 + X_5 + X_6 + X_7 &= 0\\
X_2 + X_5 + X_6 + X_7 + X_8 + X_9 &= 0.
\end{aligned}$$

We now have ten linear equations in ten variables, and it turns out that there is a unique solution $(1,0,0,0,0,0,0,0,0,0)$. Translating back into our original variables, this solution tells us that $z_0 = 1$ and $z_1 = z_2 = z_3 = 0$.

More generally, if the LFSR has m stages and the filter function involves terms of degree at most d, then the polynomial system we derive from it involves $\sum_{i=1}^{d} \binom{m}{i}$ distinct terms. Hence, in order to use linearization to obtain a solution we would need to know $O(m^d)$ keystream values. If we have fewer equations, then we may be able to use a more sophisticated technique for solving the system of polynomial equations, such as a ***Gröbner basis algorithm***. The more keystream values we have, the more likely it is that this computation will be feasible.

4.8.3 Trivium

Trivium is one of the more popular recently proposed stream ciphers. It was designed by De Cannière and Preneel in 2005. *Trivium* is very efficient and is secure against known attacks. It is one of the recommended ciphers resulting from the eSTREAM project.

Trivium has a simple and attractive design. It employs three registers, having states of sizes 93, 84, and 111 (comprising 288 bits in total). The three registers are similar to, but not exactly the same as, LFSRs. They are also "linked" together, in that they feed bits into each other.

Suppose we denote the three registers by A, B, and C. These three registers are used to generate sequences of bits that we denote by a_i, b_i, and c_i, respectively. Three recurrence relations are used to accomplish this:

$$\begin{aligned} a_i &= c_{i-66} \oplus c_{i-111} \oplus (c_{i-110} \wedge c_{i-109}) \oplus a_{i-69} \\ b_i &= a_{i-66} \oplus a_{i-93} \oplus (a_{i-92} \wedge a_{i-91}) \oplus b_{i-78} \\ c_i &= b_{i-69} \oplus b_{i-84} \oplus (b_{i-83} \wedge b_{i-82}) \oplus c_{i-87}. \end{aligned}$$

Notice that A depends on bits from A and C, B depends on bits from B and A, and C depends on bits from C and B.

The keystream bits are computed from the three registers using the following formulas:

$$r_i = c_{i-66} \oplus c_{i-111} \oplus a_{i-66} \oplus a_{i-93} \oplus b_{i-69} \oplus b_{i-84}.$$

The stream cipher has an 80-bit key, which is loaded into the high-order (leftmost) bits of the A register; the remaining bits of the A register are set to 0. An 80-bit non-secret initialization vector (IV) is loaded into the high-order bits of the B register; the remaining bits of the B register are set to 0. Finally, the three low-order bits of the C register are set equal to 1 and the remaining bits of this register are set equal to 0.

After the above-described initialization has taken place, 1152 bits of output are generated and discarded. Following that, all output bits are used as keystream bits.

4.9 Notes and References

The following book is a good introduction to block ciphers, including many of the topics discussed in this chapter:

- *The Block Cipher Companion* by Lars Knudsen and Matthew Robshaw [109].

For a recent book on stream ciphers, see

- *Stream Ciphers* by Andreas Klein [107].

The technique of differential cryptanalysis was developed by Biham and Shamir [29]. Linear cryptanalysis was invented by Matsui [128]. Our treatment of differential and linear cryptanalysis is based closely on the excellent tutorial by Heys [93]; we have also used the differential and linear attacks on SPNs that are described in [93]. General design principles for substitution-permutation networks that are resistant to linear and differential cryptanalysis are presented by Heys and Tavares [94].

A description of *DES* can be found in the 1999 Federal Information Processing Standards (FIPS) publication 46-3 [146]; this was withdrawn in 2005 but this paper is still available on the NIST website. *AES* is presented in the 2001 FIPS publication 197 [149]. Daemen and Rijmen have also written a monograph [63] on *Rijndael* and the design strategies they incorporated into its design.

The related key attacks on *AES-256* have been published in Biryukov, Dunkelman, Keller, Khovratovich, and Shamir [31] and the biclique attack is presented in Bogdanov, Khovratovich, and Rechberger [39]. For a book discussing algebraic aspects of *AES*, see Cid, Murphy, and Robshaw [58].

Standardizations of the ECB, CBC, CFB, OFB, and CTR modes of operation for block ciphers are presented in the NIST special publication 800-38A [76]. Vaudenay's padding oracle attack on CBC mode was published in [195].

Correlation attacks were introduced by Siegenthaler [181]. Improvements have been described by several authors, including Meier and Staffelbach [132]. The other main types of attack against stream ciphers are algebraic attacks; see, for example, Courtois [62] for a thorough treatment.

Trivium was first proposed by De Cannière in [65]. The version by De Cannière and Preneel [66] is the the *eSTREAM* submitted paper describing *Trivium*.

Exercises

4.1 Let y be the output of Algorithm 4.1 on input x, where π_S and π_P are defined as in Example 4.1. In other words,

$$y = \text{SPN}\left(x, \pi_S, \pi_P, (K^1, \ldots, K^{\mathcal{N}+1})\right),$$

where $(K^1, \ldots, K^{\mathcal{N}+1})$ is the key schedule. Find a substitution π_{S^*} and a permutation π_{P^*} such that

$$x = \text{SPN}\left(y, \pi_{S^*}, \pi_{P^*}, (L^{\mathcal{N}+1}, \ldots, L^1)\right),$$

where each L^i is a permutation of K^i.

4.2 Prove that decryption in a Feistel cipher can be done by applying the encryption algorithm to the ciphertext, with the key schedule reversed.

4.3 Let $DES(x, K)$ represent the encryption of plaintext x with key K using the *DES* cryptosystem. Suppose $y = DES(x, K)$ and $y' = DES(c(x), c(K))$, where $c(\cdot)$ denotes the bitwise complement of its argument. Prove that $y' = c(y)$ (i.e., if we complement the plaintext and the key, then the ciphertext is also complemented). Note that this can be proved using only the "high-level" description of *DES*—the actual structure of S-boxes and other components of the system are irrelevant.

4.4 Suppose that we have the following 128-bit *AES* key, given in hexadecimal notation:

`2B7E151628AED2A6ABF7158809CF4F3C`

Construct the complete key schedule arising from this key.

4.5 Compute the encryption of the following plaintext (given in hexadecimal notation) using the 10-round *AES*:

`3243F6A8885A308D313198A2E0370734`

Use the 128-bit key from the previous exercise.

4.6 Prove that decryption in CBC mode or CFB mode can be parallelized efficiently. More precisely, suppose we have n ciphertext blocks and n processors. Show that it is possible to decrypt all n ciphertext blocks in constant time.

4.7 Describe in detail how both encryption and decryption in CTR mode can be parallelized efficiently.

4.8 Suppose that $X = (x_1, \ldots, x_n)$ and $X' = (x'_1, \ldots, x'_n)$ are two sequences of n plaintext blocks. Define

$$\mathbf{same}(X, X') = \max\{j : x_i = x'_i \text{ for all } i \leq j\}.$$

Suppose X and X' are encrypted in CBC or CFB mode using the same key and the same IV. Show that it is easy for an adversary to compute $\mathbf{same}(X, X')$.

4.9 Suppose that $X = (x_1, \ldots, x_n)$ and $X' = (x'_1, \ldots, x'_n)$ are two sequences of n plaintext blocks. Suppose X and X' are encrypted in OFB mode using the same key and the same IV. Show that it is easy for an adversary to compute $X \oplus X'$. Show that a similar result holds for CTR mode if *ctr* is reused.

4.10 Suppose a sequence of plaintext blocks, $x_1 \ldots x_n$, yields the ciphertext sequence $y_1 \ldots y_n$. Suppose that one ciphertext block, say y_i, is transmitted incorrectly (i.e., some 1's are changed to 0's and vice versa). Show that the number of plaintext blocks that will be decrypted incorrectly is equal to one if ECB or OFB modes are used for encryption; and equal to two if CBC or CFB modes are used.

4.11 The purpose of this question is to investigate a time-memory trade-off for a chosen plaintext attack on a certain type of cipher. Suppose we have a cryptosystem in which $\mathcal{P} = \mathcal{C} = \mathcal{K}$, which attains perfect secrecy. Then it must be the case that $e_K(x) = e_{K_1}(x)$ implies $K = K_1$. Denote $\mathcal{P} = Y = \{y_1, \ldots, y_N\}$. Let x be a fixed plaintext. Define the function $g : Y \rightarrow Y$ by the rule $g(y) = e_y(x)$. Define a directed graph G having vertex set Y, in which the edge set consists of all the directed edges of the form $(y_i, g(y_i))$, $1 \leq i \leq N$.

Algorithm 4.7: TIME-MEMORY TRADE-OFF(y)

$y_0 \leftarrow y$
backup $\leftarrow$ **false**
while $g(y) \neq y_0$
do $\begin{cases} \textbf{if } y = z_j \text{ for some } j \textbf{ and not } \textit{backup} \\ \quad \textbf{then } \begin{cases} y \leftarrow g^{-T}(z_j) \\ \textit{backup} \leftarrow \textbf{true} \end{cases} \\ \quad \textbf{else } \begin{cases} y \leftarrow g(y) \\ K \leftarrow y \end{cases} \end{cases}$

(a) Prove that G consists of the union of disjoint directed cycles.

(b) Let T be a desired time parameter. Suppose we have a set of elements $Z = \{z_1, \ldots, z_m\} \subseteq Y$ such that, for every element $y_i \in Y$, either y_i is contained in a cycle of length at most T, or there exists an element $z_j \neq y_i$ such that the distance from y_i to z_j (in G) is at most T. Prove that there exists such a set Z such that

$$|Z| \leq \frac{2N}{T},$$

so $|Z|$ is $O(N/T)$.

(c) For each $z_j \in Z$, define $g^{-T}(z_j)$ to be the element y_i such that $g^T(y_i) = z_j$, where g^T is the function that consists of T iterations of g. Construct a table X consisting of the ordered pairs $(z_j, g^{-T}(z_j))$, sorted with respect to their first coordinates.

A pseudo-code description of an algorithm to find K, given $y = e_K(x)$, is presented. Prove that this algorithm finds K in at most T steps. (Hence the time-memory trade-off is $O(N)$.)

(d) Describe a pseudo-code algorithm to construct the desired set Z in time $O(NT)$ without using an array of size N.

4.12 Suppose that $\mathbf{X_1}, \mathbf{X_2}$, and $\mathbf{X_3}$ are independent discrete random variables defined on the set $\{0,1\}$. Let ϵ_i denote the bias of $\mathbf{X_i}$, for $i = 1,2,3$. Prove that $\mathbf{X_1} \oplus \mathbf{X_2}$ and $\mathbf{X_2} \oplus \mathbf{X_3}$ are independent if and only if $\epsilon_1 = 0$, $\epsilon_3 = 0$, or $\epsilon_2 = \pm 1/2$.

4.13 Suppose that $\pi_S : \{0,1\}^m \rightarrow \{0,1\}^n$ is an S-box. Prove the following facts about the function N_L (as defined in Definition 4.1).

(a) $N_L(0,0) = 2^m$.

(b) $N_L(a,0) = 2^{m-1}$ for all integers a such that $0 < a \leq 2^m - 1$.

(c) For all integers b such that $0 \leq b \leq 2^n - 1$, it holds that

$$\sum_{a=0}^{2^m-1} N_L(a,b) = 2^{2m-1} \pm 2^{m-1}.$$

(d) It holds that

$$\sum_{a=0}^{2^m-1} \sum_{b=0}^{2^n-1} N_L(a,b) \in \{2^{n+2m-1}, 2^{n+2m-1} + 2^{n+m-1}\}.$$

4.14 An S-box $\pi_S : \{0,1\}^m \rightarrow \{0,1\}^n$ is said to be ***balanced*** if

$$|{\pi_S}^{-1}(y)| = 2^{n-m}$$

for all $y \in \{0,1\}^n$. Prove the following facts about the function N_L for a balanced S-box.

(a) $N_L(0,b) = 2^{m-1}$ for all integers b such that $0 < b \leq 2^n - 1$.

(b) For all integers a such that $0 \leq a \leq 2^m - 1$, it holds that

$$\sum_{b=0}^{2^n-1} N_L(a,b) = 2^{m+n-1} - 2^{m-1} + i2^n,$$

where i is an integer such that $0 \leq i \leq 2^{m-n}$.

4.15 Suppose that the S-box of Example 4.1 is replaced by the S-box defined by the following substitution $\pi_{S'}$:

z	0	1	2	3	4	5	6	7	8	9	A	B	C	D	E	F
$\pi_{S'}(z)$	8	4	2	1	C	6	3	D	A	5	E	7	F	B	9	0

(a) Compute the linear approximation table for this S-box.

(b) Find a linear approximation using three active S-boxes, and use the piling-up lemma to estimate the bias of the random variable

$$\mathbf{X_{16}} \oplus \mathbf{U_1^4} \oplus \mathbf{U_9^4}.$$

(c) Describe a linear attack, analogous to Algorithm 4.3, that will find eight subkey bits in the last round.

(d) Implement your attack and test it to see how many plaintexts are required in order for the algorithm to find the correct subkey bits (approximately 1000–1500 plaintexts should suffice; this attack is more efficient than Algorithm 4.3 because the bias is larger by a factor of 2, which means that the number of plaintexts can be reduced by a factor of about 4).

4.16 Suppose that the S-box of Example 4.1 is replaced by the S-box defined by the following substitution $\pi_{S''}$:

z	0	1	2	3	4	5	6	7	8	9	A	B	C	D	E	F
$\pi_{S''}(z)$	E	2	1	3	D	9	0	6	F	4	5	A	8	C	7	B

(a) Compute the table of values N_D (as defined in Definition 4.3) for this S-box.

(b) Find a differential trail using four active S-boxes, namely, S_1^1, S_4^1, S_4^2, and S_4^3, that has propagation ratio $27/2048$.

(c) Describe a differential attack, analogous to Algorithm 4.3, that will find eight subkey bits in the last round.

(d) Implement your attack and test it to see how many plaintexts are required in order for the algorithm to find the correct subkey bits (approximately 100–200 plaintexts should suffice; this attack is not as efficient as Algorithm 4.3 because the propagation ratio is smaller by a factor of 2).

4.17 Suppose that we use the SPN presented in Example 4.1, but the S-box is replaced by a function π_T that is not a permutation. This means, in particular, that π_T is not surjective. Use this fact to derive a ciphertext-only attack that can be used to determine the key bits in the last round, given a sufficient number of ciphertexts that all have been encrypted using the same key.

4.18 The *Geffe Generator* is the combining function $F : (\mathbb{Z}_2)^3 \rightarrow \mathbb{Z}_2$ defined by the following formula:

$$F(z_1, z_2, z_3) = (z_1 \wedge z_2) \oplus (\neg z_1 \wedge z_3).^2$$

Determine the correlations between the inputs and output of this function, as was done in Section 4.8.1 for the majority function.

[2] The notation $\neg z$ denotes the negation of a boolean variable z.

4.19 Describe how the correlations computed in the previous exercise can be used to mount a correlation attack against the *Geffe Generator*. Note that this is a bit more complicated than the attack against the majority function generator because not all three correlations are bounded away from 1/2.

4.20 A function is ***balanced*** if it takes on the values 0 and 1 equally often. Construct a balanced combining function $F : (\mathbb{Z}_2)^3 \rightarrow \mathbb{Z}_2$ such that

$$\mathbf{Pr}[\mathbf{z_j} = \mathbf{z}] = \frac{1}{2},$$

for $j = 1, 2, 3$.

Chapter 5

Hash Functions and Message Authentication

This chapter concerns mechanisms for data integrity, specifically hash functions and message authentication codes. We discuss design techniques for hash functions, including iterated hash functions and the sponge construction. We look at various algorithms that have been approved as standards for hash functions. As far as message authentication codes are concerned, we provide a treatment of design techniques, attacks, and applications to authenticated encryption.

5.1 Hash Functions and Data Integrity

So far, we have mainly been considering methods to achieve confidentiality (or secrecy) by encrypting messages using a suitable cryptosystem. This is sufficient to protect against a passive adversary who is only observing messages that are transmitted between Alice and Bob. However, there are many other threats that we need to address. One natural scenario is when there is an active adversary who is able to change the content of messages. We may not be able to prevent the adversary from modifying messages, but appropriate cryptographic tools will enable us to detect when a modification has occurred.

Encryption by itself is not sufficient to alleviate these kinds of threats. For example, suppose that a message is encrypted using a stream cipher, by computing the exclusive-or of the plaintext and the keystream. Suppose an adversary is able to modify the ciphertext that is transmitted from Alice to Bob. The adversary could just complement arbitrary bits of the ciphertext (i.e., change 1's to 0's and vice versa). This attack, which is known as a ***bit-flipping attack***, has the effect of complementing exactly the same bits of the plaintext. Even though the adversary does not know what the plaintext is, he can modify it in a predictable way.

Thus, our goal is to detect modifications of transmitted messages (encrypted or not). This objective is often referred to as ***data integrity***. A ***cryptographic hash function*** can provide assurance of data integrity in certain settings. A hash function is used to construct a short "fingerprint" of some data; if the data is altered, then the fingerprint will (with high probability) no longer be valid. Suppose that the fingerprint is stored in a secure place. Then, even if the data is stored in an insecure place, its integrity can be checked from time to time by recomputing the fingerprint and verifying that the fingerprint has not changed.

Let h be a hash function and let x be some data. As an illustrative example, x could be a binary string of arbitrary length. The corresponding fingerprint is defined to be $y = h(x)$. This fingerprint is often referred to as a ***message digest***. A message digest would typically be a fairly short binary string; 160 bits or 256 bits are common choices.

As mentioned above, we assume that y is stored in a secure place, but x is not. If x is changed, to x', say, then we hope that the "old" message digest, y, is not also a message digest for x'. If this is indeed the case, then the fact that x has been altered can be detected simply by computing the message digest $y' = h(x')$ and verifying that $y' \neq y$.

A particularly important application of hash functions occurs in the context of digital signature schemes, which will be studied in Chapter 8.

The motivating example discussed above assumes the existence of a single, fixed hash function. It is also useful to study a ***hash family***, which is just a family of ***keyed hash functions***. There is a different hash function for each possible key. A keyed hash function is often used as a ***message authentication code***, or ***MAC***. Suppose that Alice and Bob share a secret key, K, which determines a hash function, say h_K. For a message, say x, the corresponding ***authentication tag*** (or more simply, ***tag***), is $y = h_K(x)$. This tag can be computed by either Alice or Bob. The pair (x, y) can be transmitted over an insecure channel from Alice to Bob (or from Bob to Alice). Suppose Bob receives the pair (x, y) from Alice. Then he can check if $y = h_K(x)$ by recomputing the tag. If this condition holds, then Bob is confident that neither x nor y was altered by an adversary, provided that the hash family is "secure." In particular, Bob is assured that the message x originates from Alice (assuming that Bob did not transmit the message himself).

Notice the distinction between the assurance of data integrity provided by an unkeyed, as opposed to a keyed, hash function. In the case of an unkeyed hash function, the message digest must be securely stored so it cannot be altered by an adversary. On the other hand, if Alice and Bob use a secret key K to specify the hash function they are using, then they can transmit both the data and the authentication tag over an insecure channel.

In the remainder of this chapter, we will study hash functions, as well as keyed hash families. We begin by giving definitions for a keyed hash family.

Definition 5.1: A ***hash family*** is a four-tuple $(\mathcal{X}, \mathcal{Y}, \mathcal{K}, \mathcal{H})$, where the following conditions are satisfied:

1. $\mathcal{X}$ is a set of possible ***messages***
2. $\mathcal{Y}$ is a finite set of possible ***message digests*** or ***authentication tags*** (or just ***tags***)
3. $\mathcal{K}$, the ***keyspace***, is a finite set of possible ***keys***
4. For each $K \in \mathcal{K}$, there is a ***hash function*** $h_K \in \mathcal{H}$. Each $h_K : \mathcal{X} \rightarrow \mathcal{Y}$.

In the above definition, $\mathcal{X}$ could be a finite or infinite set; $\mathcal{Y}$ is always a finite set. If $\mathcal{X}$ is a finite set and $\mathcal{X} > \mathcal{Y}$, the function is sometimes called a ***compression function***. In this situation, we will often assume the stronger condition that $|\mathcal{X}| \geq 2|\mathcal{Y}|$.

An ***unkeyed hash function*** is a function $h : \mathcal{X} \rightarrow \mathcal{Y}$, where $\mathcal{X}$ and $\mathcal{Y}$ are the same as in Definition 5.1. We could think of an unkeyed hash function simply as a hash family in which there is only one possible key, i.e., one in which $|\mathcal{K}| = 1$.

We typically use the terminology "message digest" for the output of an unkeyed hash function, whereas the term "tag" refers to the output of a keyed hash function.

A pair $(x, y) \in \mathcal{X} \times \mathcal{Y}$ is said to be a ***valid pair*** under a hash function h if $h(x) = y$. Here h could be a keyed or unkeyed hash function. Much of what we discuss in this chapter concerns methods to prevent the construction of certain types of valid pairs by an adversary.

Let $\mathcal{F}^{\mathcal{X},\mathcal{Y}}$ denote the set of all functions from $\mathcal{X}$ to $\mathcal{Y}$. Suppose that $|\mathcal{X}| = N$ and $|\mathcal{Y}| = M$. Then it is clear that $|\mathcal{F}^{\mathcal{X},\mathcal{Y}}| = M^N$. (This follows because, for each of the N possible inputs $x \in \mathcal{X}$, there are M possible values for the corresponding output $h(x) \in \mathcal{Y}$.) Any hash family $\mathcal{F}$ consisting of functions with domain $\mathcal{X}$ and range $\mathcal{Y}$ can be considered to be a subset of $\mathcal{F}^{\mathcal{X},\mathcal{Y}}$, i.e., $\mathcal{F} \subseteq \mathcal{F}^{\mathcal{X},\mathcal{Y}}$. Such a hash family is termed an ***(N, M)-hash family***.

The remaining sections of this chapter are organized as follows. In Section 5.2, we introduce concepts of security for hash functions, in particular, the idea of collision resistance. We also study the exact security of "ideal" hash functions using the "random oracle model" in this section; and we discuss the birthday paradox, which provides an estimate of the difficulty of finding collisions for an arbitrary hash function. In Section 5.3, we introduce the important design technique of iterated hash functions. We discuss how this method is used in the design of practical hash functions, as well as in the construction of a provably secure hash function from a secure compression function. Section 5.4 concerns another, newer, design technique called the "sponge construction" and its application to the most recent hash function standard, *SHA-3*. Section 5.5 provides a treatment of message authentication codes, where we again present some general constructions and security proofs. Unconditionally secure MACs, and their construction using strongly universal hash families, are considered in Section 5.6.

5.2 Security of Hash Functions

Suppose that $h : \mathcal{X} \rightarrow \mathcal{Y}$ is an unkeyed hash function. Let $x \in \mathcal{X}$, and define $y = h(x)$. In many cryptographic applications of hash functions, it is desirable that the only way to produce a valid pair (x, y) is to first choose x, and then compute $y = h(x)$ by applying the function h to x. Other security requirements of hash functions are motivated by their applications in particular protocols, such as signature schemes (see Chapter 8). We now define three problems; if a hash function is to be considered secure, it should be the case that these three problems are difficult to solve.

Problem 5.1: Preimage

Instance: A hash function $h : \mathcal{X} \rightarrow \mathcal{Y}$ and an element $y \in \mathcal{Y}$.
Find: $x \in \mathcal{X}$ such that $h(x) = y$.

Given a (possible) message digest y, the problem **Preimage** asks if a an element $x \in \mathcal{X}$ can be found such that $h(x) = y$. Such a value x would be a ***preimage*** of y. If **Preimage** can be solved for a given $y \in \mathcal{Y}$, then the pair (x, y) is a valid pair. A hash function for which **Preimage** cannot be efficiently solved is often said to be ***one-way*** or ***preimage resistant***.

Problem 5.2: Second Preimage

Instance: A hash function $h : \mathcal{X} \rightarrow \mathcal{Y}$ and an element $x \in \mathcal{X}$.
Find: $x' \in \mathcal{X}$ such that $x' \neq x$ and $h(x') = h(x)$.

Given a message x, the problem **Second Preimage** asks if $x' \neq x$ can be found such that $h(x') = h(x)$. Here, we begin with x, which is a preimage of y, and we are seeking to find a value x' that would be a ***second preimage*** of y. Note that, if this can be done, then $(x', h(x))$ is a valid pair. A hash function for which **Second Preimage** cannot be efficiently solved is often said to be ***second preimage resistant***.

Problem 5.3: Collision

Instance: A hash function $h : \mathcal{X} \rightarrow \mathcal{Y}$.
Find: $x, x' \in \mathcal{X}$ such that $x' \neq x$ and $h(x') = h(x)$.

The problem **Collision** asks if any pair of distinct inputs x, x' can be found such that $h(x') = h(x)$. (Unsurpisingly, this is called a ***collision***.) A solution to this problem yields two valid pairs, (x, y) and (x', y), where $y = h(x) = h(x')$. There are various scenarios where we want to avoid such a situation from arising. A hash function for which **Collision** cannot be efficiently solved is often said to be ***collision resistant***.

Some of the questions we address in the next sections concern the difficulty of each of these three problems, as well as the relative difficulty of the three problems.

5.2.1 The Random Oracle Model

In this section, we describe a certain idealized model for a hash function, which attempts to capture the concept of an "ideal" hash function. If a hash function h is well designed, it should be the case that the only efficient way to determine the value $h(x)$ for a given x is to actually evaluate the function h at the value x. This should remain true even if many other values $h(x_1), h(x_2)$, etc., have already been computed.

To illustrate an example where the above property does not hold, suppose that the hash function $h : \mathbb{Z}_n \times \mathbb{Z}_n \rightarrow \mathbb{Z}_n$ is a linear function, say

$$h(x, y) = ax + by \bmod n,$$

$a, b \in \mathbb{Z}_n$ and $n \geq 2$ is a positive integer. Suppose that we are given the values

$$h(x_1, y_1) = z_1$$

and

$$h(x_2, y_2) = z_2.$$

Let $r, s \in \mathbb{Z}_n$; then we have that

$$\begin{aligned} h(rx_1 + sx_2 \bmod n, ry_1 + sy_2 \bmod n) &= a(rx_1 + sx_2) + b(ry_1 + sy_2) \bmod n \\ &= r(ax_1 + by_1) + s(ax_2 + by_2) \bmod n \\ &= rh(x_1, y_1) + sh(x_2, y_2) \bmod n. \end{aligned}$$

Therefore, given the value of function h at two points (x_1, y_1) and (x_2, y_2), we know its value at various other points, without actually having to evaluate h at those points (and note also that we do not even need to know the values of the constants a and b in order to apply the above-described technique).

The ***random oracle model***, which was introduced by Bellare and Rogaway, provides a mathematical model of an "ideal" hash function. In this model, a hash function $h : \mathcal{X} \rightarrow \mathcal{Y}$ is chosen randomly from $\mathcal{F}^{\mathcal{X},\mathcal{Y}}$, and we are only permitted ***oracle access*** to the function h. This means that we are not given a formula or an algorithm to compute values of the function h. Therefore, the only way to compute a value $h(x)$ is to query the oracle. This can be thought of as looking up the value $h(x)$ in a giant book of random numbers such that, for each possible x, there is a completely random value $h(x)$.[1]

Although a true random oracle does not exist in real life, we hope that a well-designed hash function will "behave" like a random oracle. So it is useful to study the random oracle model and its security with respect to the three problems introduced above. This is done in the next section.

As a consequence of the assumptions made in the random oracle model, it is obvious that the following independence property holds:

THEOREM 5.1 *Suppose that $h \in \mathcal{F}^{\mathcal{X},\mathcal{Y}}$ is chosen randomly, and let $\mathcal{X}_0 \subseteq \mathcal{X}$. Denote $|\mathcal{Y}| = M$. Suppose that the values $h(x)$ have been determined (by querying an oracle for h) if and only if $x \in \mathcal{X}_0$. Then $\mathbf{Pr}[h(x) = y] = 1/M$ for all $x \in \mathcal{X} \backslash \mathcal{X}_0$ and all $y \in \mathcal{Y}$.*

In the above theorem, the probability $\mathbf{Pr}[h(x) = y]$ is in fact a conditional probability that is computed over all functions h that take on the specified values for all $x \in \mathcal{X}_0$. Theorem 5.1 is a key property used in proofs involving the complexity of problems in the random oracle model.

[1]In fact, the book *A Million Random Digits with 100,000 Normal Deviates* was published by the RAND Corporation in 1955. This book could be viewed as an approximation to a random oracle, perhaps.

Algorithm 5.1: FIND-PREIMAGE(h, y, Q)

choose any $\mathcal{X}_0 \subseteq \mathcal{X}, |\mathcal{X}_0| = Q$
for each $x \in \mathcal{X}_0$
  do { **if** $h(x) = y$
    then return (x) }
return (failure)

5.2.2 Algorithms in the Random Oracle Model

In this section, we consider the complexity of the three problems defined in Section 5.2 in the random oracle model. An algorithm in the random oracle model can be applied to any hash function, since the algorithm needs to know nothing whatsoever about the hash function (except that a method must be specified to evaluate the hash function for arbitrary values of x).

The algorithms we present and analyze are ***randomized algorithms***; they can make random choices during their execution. A ***Las Vegas algorithm*** is a randomized algorithm that may fail to give an answer (i.e., it can terminate with the message "failure"), but if the algorithm does return an answer, then the answer must be correct.

Suppose $0 \leq \epsilon < 1$ is a real number. A randomized algorithm has ***worst-case success probability*** ϵ if, for every problem instance, the algorithm returns a correct answer with probability at least ϵ. A randomized algorithm has ***average-case success probability*** ϵ if the probability that the algorithm returns a correct answer, averaged over all problem instances of a specified size, is at least ϵ. Note that, in this latter situation, the probability that the algorithm returns a correct answer for a given problem instance can be greater than or less than ϵ.

In this section, we use the terminology (ϵ, Q)-***algorithm*** to denote a Las Vegas algorithm with average-case success probability ϵ, in which the number of oracle queries (i.e., evaluations of h) made by the algorithm is at most Q. The success probability ϵ is the average over all possible random choices of $h \in \mathcal{F}^{\mathcal{X},\mathcal{Y}}$, and all possible random choices of $x \in \mathcal{X}$ or $y \in \mathcal{Y}$, if x and/or y is specified as part of the problem instance.

We analyze the trivial algorithms, which evaluate $h(x)$ for Q values of $x \in \mathcal{X}$, in the random oracle model. These x-values are often chosen in a random way; however, it turns out that the complexity of such an algorithm is independent of the particular choice of the x-values because we are averaging over all functions $h \in \mathcal{F}^{\mathcal{X},\mathcal{Y}}$.

We first consider Algorithm 5.1, which attempts to solve **Preimage** by evaluating h at Q points.

THEOREM 5.2 *For any $\mathcal{X}_0 \subseteq \mathcal{X}$ with $|\mathcal{X}_0| = Q$, the average-case success probability of Algorithm 5.1 is $\epsilon = 1 - (1 - 1/M)^Q$.*

Algorithm 5.2: FIND-SECOND-PREIMAGE(h, x, Q)

$y \leftarrow h(x)$
choose $\mathcal{X}_0 \subseteq \mathcal{X}\backslash\{x\}, |\mathcal{X}_0| = Q - 1$
for each $x_0 \in \mathcal{X}_0$
do $\begin{cases} \textbf{if } h(x_0) = y \\ \quad \textbf{then return } (x_0) \end{cases}$
return (failure)

PROOF Let $y \in \mathcal{Y}$ be fixed. Let $\mathcal{X}_0 = \{x_1, \ldots, x_Q\}$. For $1 \leq i \leq Q$, let E_i denote the event "$h(x_i) = y$." It follows from Theorem 5.1 that the E_i's are independent events, and $\mathbf{Pr}[E_i] = 1/M$ for all $1 \leq i \leq Q$. Thus $\mathbf{Pr}[E_i{}^c] = 1 - 1/M$ for all $1 \leq i \leq Q$, where $E_i{}^c$ denotes the complement of the event E_i (i.e., the event "$h(x_i) \neq y$").

Therefore, it holds that

$$\begin{aligned} \mathbf{Pr}[E_1 \vee E_2 \vee \cdots \vee E_Q] &= 1 - \mathbf{Pr}[E_i{}^c \wedge E_2{}^c \wedge \cdots \wedge E_Q^c] \\ &= 1 - \left(1 - \frac{1}{M}\right)^Q, \end{aligned}$$

where "$\vee$" denotes the logical "or" and "$\wedge$" denotes the logical "and" of events.

The success probability of Algorithm 5.1, for any fixed y, is constant. Therefore, the success probability averaged over all $y \in \mathcal{Y}$ is identical, too. ∎

Note that the above success probability is approximately Q/M provided that Q is small compared to M.

We now present and analyze a very similar algorithm, Algorithm 5.2, that attempts to solve **Second Preimage**. The analysis of Algorithm 5.2 is similar to the previous algorithm. The only difference is that we require an "extra" application of h to compute $y = h(x)$ for the input value x.

THEOREM 5.3 *For any $\mathcal{X}_0 \subseteq \mathcal{X}\backslash\{x\}$ with $|\mathcal{X}_0| = Q - 1$, the success probability of Algorithm 5.2 is $\epsilon = 1 - (1 - 1/M)^{Q-1}$.*

Next, we look at an elementary algorithm for **Collision**. In Algorithm 5.3, the test to see if $y_x = y_{x'}$ for some $x' \neq x$ could be done efficiently by sorting the y_x's, for example. This algorithm is analyzed using a probability argument analogous to the standard "birthday paradox." The ***birthday paradox*** says that in a group of 23 randomly chosen people, at least two will share a birthday with probability at least 1/2. (Of course this is not actually a paradox, but it is probably counter-intuitive and surprising to many people.) This may not appear to be relevant to hash functions, but if we reformulate the problem, the connection will be clear. Suppose that the function h has as its domain the set of all living human beings,

Algorithm 5.3: FIND-COLLISION(h, Q)

```
choose X_0 ⊆ X, |X_0| = Q
for each x ∈ X_0
  do y_x ← h(x)
if y_x = y_x' for some x' ≠ x
  then return (x, x')
  else return (failure)
```

and for all x, $h(x)$ denotes the birthday of person x. Then the range of h consists of the 365 days in a year (366 days if we include February 29). Finding two people with the same birthday is the same thing as finding a collision for this particular hash function. In this setting, the birthday paradox is saying that Algorithm 5.3 has success probability at least $1/2$ when $Q = 23$ and $M = 365$.

We now analyze Algorithm 5.3 in general, in the random oracle model. This algorithm is analogous to throwing Q balls randomly into M bins and then checking to see if some bin contains at least two balls. (The Q balls correspond to the Q random x_i's, and the M bins correspond to the M possible elements of $\mathcal{Y}$.)

THEOREM 5.4 *For any $\mathcal{X}_0 \subseteq \mathcal{X}$ with $|\mathcal{X}_0| = Q$, the success probability of Algorithm 5.3 is*

$$\epsilon = 1 - \left(\frac{M-1}{M}\right)\left(\frac{M-2}{M}\right)\cdots\left(\frac{M-Q+1}{M}\right).$$

PROOF Let $\mathcal{X}_0 = \{x_1, \dots, x_Q\}$. For $1 \leq i \leq Q$, let E_i denote the event

$$\text{“}h(x_i) \notin \{h(x_1), \dots, h(x_{i-1})\}.\text{”}$$

We observe trivially that $\mathbf{Pr}[E_1] = 1$. Using induction, it follows from Theorem 5.1 that

$$\mathbf{Pr}[E_i | E_1 \wedge E_2 \wedge \cdots \wedge E_{i-1}] = \frac{M-i+1}{M},$$

for $2 \leq i \leq Q$. Therefore, we have that

$$\mathbf{Pr}[E_1 \wedge E_2 \wedge \cdots \wedge E_Q] = \left(\frac{M-1}{M}\right)\left(\frac{M-2}{M}\right)\cdots\left(\frac{M-Q+1}{M}\right).$$

The probability that there is at least one collision is $1 - \mathbf{Pr}[E_1 \wedge E_2 \wedge \cdots \wedge E_Q]$, so the desired result follows. ∎

The above theorem shows that the probability of finding no collisions is

$$\left(1-\frac{1}{M}\right)\left(1-\frac{2}{M}\right)\cdots\left(1-\frac{Q-1}{M}\right) = \prod_{i=1}^{Q-1}\left(1-\frac{i}{M}\right).$$

If x is a small real number, then $1 - x \approx e^{-x}$. This estimate is derived by taking the first two terms of the series expansion

$$e^{-x} = 1 - x + \frac{x^2}{2!} - \frac{x^3}{3!} \cdots.$$

Using this estimate, the probability of finding no collisions is approximately

$$\begin{aligned}\prod_{i=1}^{Q-1}\left(1 - \frac{i}{M}\right) &\approx \prod_{i=1}^{Q-1} e^{\frac{-i}{M}} \\ &= e^{-\sum_{i=1}^{Q-1}\frac{i}{M}} \\ &= e^{\frac{-Q(Q-1)}{2M}}.\end{aligned}$$

Consequently, we can estimate the probability of finding at least one collision to be

$$1 - e^{\frac{-Q(Q-1)}{2M}}.$$

If we denote this probability by ϵ, then we can solve for Q as a function of M and ϵ:

$$\begin{aligned}e^{\frac{-Q(Q-1)}{2M}} &\approx 1 - \epsilon \\ \frac{-Q(Q-1)}{2M} &\approx \ln(1-\epsilon) \\ Q^2 - Q &\approx 2M \ln \frac{1}{1-\epsilon}.\end{aligned}$$

If we ignore the term $-Q$, then we estimate that

$$Q \approx \sqrt{2M \ln \frac{1}{1-\epsilon}}.$$

If we take $\epsilon = .5$, then our estimate is

$$Q \approx 1.17\sqrt{M}.$$

So this says that hashing just over $\sqrt{M}$ random elements of X yields a collision with a probability of 50%. Note that a different choice of ϵ leads to a different constant factor, but Q will still be proportional to $\sqrt{M}$. The algorithm is a $(1/2, O(\sqrt{M}))$-algorithm.

We return to the example we mentioned earlier. Taking $M = 365$ in our estimate, we get $Q \approx 22.3$. Hence, as mentioned earlier, the probability is at least $1/2$ that there will be at least one duplicated birthday among 23 randomly chosen people.

The birthday attack imposes a lower bound on the sizes of secure message digests. A 40-bit message digest would be very insecure, since a collision could be found with probability $1/2$ with just over 2^{20} (about a million) random hashes. *SHA-1*, which was a standard for a number of years, has a message digest that is 160 bits (a birthday attack would require over 2^{80} hashes in this case). The most recent standard for hash functions, *SHA-3*, utilizes hash functions having message digests of sizes between 224 and 512 bits in length.

Algorithm 5.4: COLLISION-TO-SECOND-PREIMAGE( )

external ORACLE-2ND-PREIMAGE, h
comment: we consider the hash function h to be fixed

choose $x \in \mathcal{X}$ uniformly at random
if ORACLE-2ND-PREIMAGE$(x) = x'$
  then return (x, x')
  else return (failure)

5.2.3 Comparison of Security Criteria

In the random oracle model, we have seen that solving **Collision** is easier than solving **Preimage** or **Second Preimage**. It is interesting to consider the relative difficulty of these problems in a general setting. This is accomplished using the standard technique of ***reductions***.

The basic idea of a reduction (in the context of algorithms and complexity) is to use a hypothetical algorithm (i.e., an ***oracle***) that solves one problem as a subroutine in an algorithm to solve a second problem. In this situation, we say that we have a reduction *from* the second problem *to* the first problem. Then, if we have a specific algorithm that solves the first problem, we can use this algorithm, in the place of the oracle, to obtain an algorithm to solve the second problem.

Suppose we want to describe a reduction from a problem Π_2 to another problem Π_1. We would assume the existence of an oracle solving Π_1 and then use this oracle in an (efficient) algorithm to solve Π_2. Informally, the existence of such a reduction shows that if we can solve Π_1, then we can solve Π_2. So this establishes that solving Π_2 is no more difficult than solving Π_1. Equivalently, we are saying that if it is infeasible to solve Π_2, then it is infeasible to solve Π_1 (this is basically the contrapositive of the previous statement).

In this section, we will discuss some reductions among the three problems (**Preimage**, **Second Preimage**, and **Collision**) that could be applied to arbitrary hash functions. First, we observe that it is fairly easy to find a reduction from **Collision** to **Second Preimage**; this is accomplished in Algorithm 5.4.

We analyze Algorithm 5.4. as follows. Suppose that ORACLE-2ND-PREIMAGE is an (ϵ, Q)-algorithm that solves **Second Preimage** for a particular, fixed hash function h. If ORACLE-2ND-PREIMAGE returns a value x' when it is given input (h, x), then it must be the case that $x' \neq x$, because ORACLE-2ND-PREIMAGE is assumed to be a Las Vegas algorithm. As a consequence, it is clear that COLLISION-TO-SECOND-PREIMAGE is an (ϵ, Q)-algorithm that solves **Collision** for the same hash function h. That is, if we can solve **Second Preimage** with probability ϵ using Q queries, then we can also solve **Collision** with probability ϵ using Q queries. This reduction does not make any assumptions about the hash function h. As a

Algorithm 5.5: COLLISION-TO-PREIMAGE()

external ORACLE-PREIMAGE, h
comment: we consider the hash function h to be fixed

choose $x \in \mathcal{X}$ uniformly at random
$y \leftarrow h(x)$
if (ORACLE-PREIMAGE(y) $= x'$) **and** ($x' \neq x$)
  then return (x, x')
  else return (failure)

consequence of this reduction, we could say that the property of collision resistance implies the property of second preimage resistance.

We are now going to investigate the perhaps more interesting question of whether **Collision** can be reduced to **Preimage**. In other words, does collision resistance imply preimage resistance? We will prove that this is indeed the case, at least in some special situations. More specifically, we will prove that an arbitrary algorithm that solves **Preimage** with probability equal to 1 can be used to solve **Collision**.

This reduction can be accomplished with a fairly weak assumption on the relative sizes of the domain and range of the hash function h. We will assume that the hash function $h : \mathcal{X} \rightarrow \mathcal{Y}$, where $\mathcal{X}$ and $\mathcal{Y}$ are finite sets and $|\mathcal{X}| \geq 2|\mathcal{Y}|$. Now, suppose that ORACLE-PREIMAGE is a $(1, Q)$-algorithm for the **Preimage** problem. ORACLE-PREIMAGE accepts as input a message digest $y \in \mathcal{Y}$, and always finds an element ORACLE-PREIMAGE(y) $\in \mathcal{X}$ such that h(ORACLE-PREIMAGE(y)) $= y$ (in particular, this implies that h is surjective). We will analyze the reduction COLLISION-TO-PREIMAGE, which is presented as Algorithm 5.5.

We prove the following theorem.

THEOREM 5.5 *Suppose $h : \mathcal{X} \rightarrow \mathcal{Y}$ is a hash function where $|\mathcal{X}|$ and $|\mathcal{Y}|$ are finite and $|\mathcal{X}| \geq 2|\mathcal{Y}|$. Suppose* ORACLE-PREIMAGE *is a $(1, Q)$-algorithm for* ***Preimage****, for the fixed hash function h. Then* COLLISION-TO-PREIMAGE *is a $(1/2, Q+1)$-algorithm for* ***Collision****, for the fixed hash function h.*

PROOF Clearly COLLISION-TO-PREIMAGE is a probabilistic algorithm of the Las Vegas type, since it either finds a collision or returns "failure." Thus our main task is to compute the average-case probability of success.

For any $x \in \mathcal{X}$, define $x \sim x_1$ if $h(x) = h(x_1)$. It is easy to see that $\sim$ is an equivalence relation. Define $[x] = \{x_1 \in \mathcal{X} : x \sim x_1\}$. Each equivalence class $[x]$ consists of the inverse image of an element of $\mathcal{Y}$, i.e., for every equivalence class $[x]$, there exists a (unique) value $y \in \mathcal{Y}$ such that $[x] = h^{-1}(y)$. We assumed that ORACLE-PREIMAGE always finds a preimage of any element y, which means that $h^{-1}(y) \neq \emptyset$ for all $y \in \mathcal{Y}$. Therefore, the number of equivalence classes $[x]$ is equal to $|\mathcal{Y}|$. Denote the set of these $|\mathcal{Y}|$ equivalence classes by $\mathcal{C}$.

Now, suppose x is the random element of $\mathcal{X}$ chosen by the algorithm COLLISION-TO-PREIMAGE. For this x, there are $|[x]|$ possible x_1's that could be returned as the output of ORACLE-PREIMAGE. $|[x]| - 1$ of these x_1's are different from x and thus yield a collision. (Note that the algorithm ORACLE-PREIMAGE does not know the representative of the equivalence class $[x]$ that was initially chosen by algorithm COLLISION-TO-PREIMAGE.) So, given the element $x \in \mathcal{X}$, the probability of success is $(|[x]| - 1)/|[x]|$.

The probability of success of the algorithm COLLISION-TO-PREIMAGE is computed by averaging over all possible choices for x:

$$\begin{aligned}
\mathbf{Pr}[\text{success}] &= \frac{1}{|\mathcal{X}|} \sum_{x \in \mathcal{X}} \frac{|[x]| - 1}{|[x]|} \\
&= \frac{1}{|\mathcal{X}|} \sum_{C \in \mathcal{C}} \sum_{x \in C} \frac{|C| - 1}{|C|} \\
&= \frac{1}{|\mathcal{X}|} \sum_{C \in \mathcal{C}} (|C| - 1) \\
&= \frac{1}{|\mathcal{X}|} \left(\sum_{C \in \mathcal{C}} |C| - \sum_{C \in \mathcal{C}} 1 \right) \\
&= \frac{|\mathcal{X}| - |\mathcal{Y}|}{|\mathcal{X}|} \\
&\geq \frac{|\mathcal{X}| - |\mathcal{X}|/2}{|\mathcal{X}|} \\
&= \frac{1}{2}.
\end{aligned}$$

Note that we use the fact that $|\mathcal{X}| \geq 2|\mathcal{Y}|$ in the next-to-last line of the computation performed above.

In summary, we have constructed a Las Vegas algorithm with average-case success probability at least $1/2$. ∎

Informally, the two preceding theorems have shown that collision resistance implies both preimage resistance and second preimage resistance (under certain plausible assumptions). Thus, the focus in the design of hash functions is to achieve the property of collision resistance. In practice, if any collision is found for a given hash function, then that hash function is considered to have been completely "broken."

5.3 Iterated Hash Functions

So far, we have considered hash functions with a finite domain (i.e., compression functions). We now study a particular technique by which a compression function, say **compress**, can be extended to a hash function h having an infinite domain. A hash function h constructed by this method is called an ***iterated hash function***.

In this section, we restrict our attention to hash functions whose inputs and outputs are bitstrings (i.e., strings formed of zeroes and ones). We denote the length of a bitstring x by $|x|$, and the concatenation of bitstrings x and y is written as $x \parallel y$.

Suppose that **compress** : $\{0,1\}^{m+t} \rightarrow \{0,1\}^m$ is a compression function (where $t \geq 1$). We will construct an iterated hash function

$$h : \bigcup_{i=m+t+1}^{\infty} \{0,1\}^i \rightarrow \{0,1\}^\ell,$$

based on the compression function **compress**. The evaluation of h consists of the following three main steps:

preprocessing step

Given an input string x, where $|x| \geq m+t+1$, construct a string y, using a public algorithm, such that $|y| \equiv 0 \pmod t$. Denote

$$y = y_1 \parallel y_2 \parallel \cdots \parallel y_r,$$

where $|y_i| = t$ for $1 \leq i \leq r$.

processing step

Let IV be a public initial value that is a bitstring of length m. Then compute the following:

$$\begin{array}{lcl} z_0 & \leftarrow & \text{IV} \\ z_1 & \leftarrow & \mathbf{compress}(z_0 \parallel y_1) \\ z_2 & \leftarrow & \mathbf{compress}(z_1 \parallel y_2) \\ \vdots & \vdots & \vdots \\ z_r & \leftarrow & \mathbf{compress}(z_{r-1} \parallel y_r). \end{array}$$

This processing step is illustrated in Figure 5.1.

output transformation

Let $g : \{0,1\}^m \rightarrow \{0,1\}^\ell$ be a public function. Define $h(x) = g(z_r)$.

REMARK The output transformation is optional. If an output transformation is not desired, then define $h(x) = z_r$. ∎

A commonly used preprocessing step is to construct the string y in the following way:

$$y = x \parallel \mathbf{pad}(x),$$

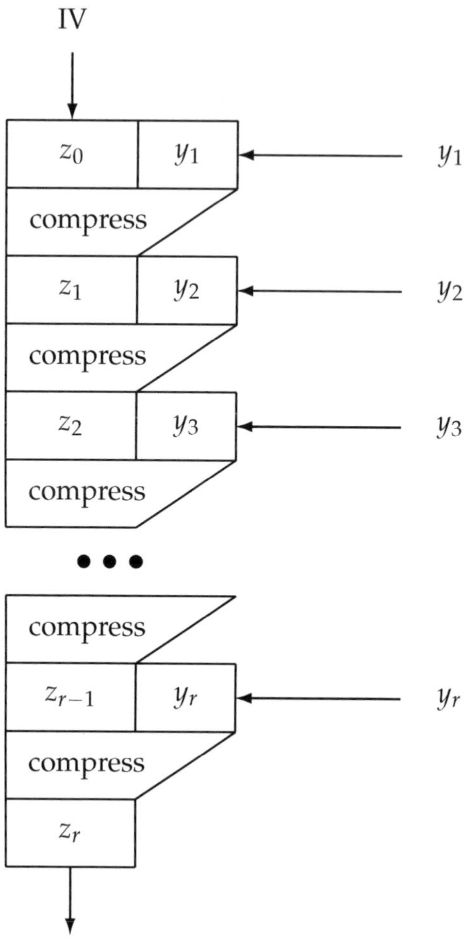

FIGURE 5.1: The processing step in an iterated hash function

where **pad**(x) is constructed from x using a ***padding function***. A padding function typically incorporates the value of $|x|$, and pads the result with additional bits (zeros, for example) so that the resulting string y has a length that is a multiple of t.

The preprocessing step must ensure that the mapping $x \mapsto y$ is an injection. (If the mapping $x \mapsto y$ is not one-to-one, then it may be possible to find $x \neq x'$ so that $y = y'$. Then $h(x) = h(x')$, and h would not be collision resistant.) Note also that $|y| = rt \geq |x|$ because of the required injective property.

Many hash functions commonly used in practice are in fact iterated hash functions and can be viewed as special cases of the generic construction described above. The Merkle-Damgård construction, which we discuss in the next section, is a construction of a certain kind of iterated hash function that permits a formal security proof to be given.

5.3.1 The Merkle-Damgård Construction

In this section, we present a particular method of constructing a hash function from a compression function. This construction has the property that the resulting hash function satisfies desirable security properties, such as collision resistance, provided that the compression function does. This technique is often called the ***Merkle-Damgård construction***.

Suppose **compress** : $\{0,1\}^{m+t} \rightarrow \{0,1\}^m$ is a collision resistant compression function, where $t \geq 1$. So **compress** takes $m + t$ input bits and produces m output bits. We will use **compress** to construct a collision resistant hash function $h : \mathcal{X} \rightarrow \{0,1\}^m$, where

$$\mathcal{X} = \bigcup_{i=m+t+1}^{\infty} \{0,1\}^i.$$

Thus, the hash function h takes any finite bitstring of length at least $m + t + 1$ and creates a message digest that is a bitstring of length m. We first consider the situation where $t \geq 2$ (the case $t = 1$ will be handled a bit later).

We will treat elements of $x \in \mathcal{X}$ as bitstrings. Suppose $|x| = n \geq m + t + 1$. We can express x as the concatenation

$$x = x_1 \parallel x_2 \parallel \cdots \parallel x_k,$$

where

$$|x_1| = |x_2| = \cdots = |x_{k-1}| = t - 1$$

and

$$|x_k| = t - 1 - d,$$

where $0 \leq d \leq t - 2$. Hence, we have that

$$k = \left\lceil \frac{n}{t-1} \right\rceil.$$

We define $h(x)$ to be the output of Algorithm 5.6.

Denote

$$y(x) = y_1 \parallel y_2 \parallel \cdots \parallel y_{k+1}.$$

Observe that y_k is formed from x_k by padding on the right with d zeroes, so that all the blocks y_i ($1 \leq i \leq k$) are of length $t - 1$. Also, y_{k+1} should be padded on the left with zeroes so that $|y_{k+1}| = t - 1$.

As was done in the generic construction described in Section 5.3, we hash x by first constructing $y(x)$, and then processing the blocks $y_1, y_2, \ldots, y_{k+1}$ in a particular fashion. y_{k+1} is defined in such a way that the mapping $x \mapsto y(x)$ is an injection, which we observed is necessary for the iterated hash function to be collision resistant.

Algorithm 5.6: MERKLE-DAMGÅRD(x)

external compress
comment: compress : $\{0,1\}^{m+t} \rightarrow \{0,1\}^m$, where $t \geq 2$

$n \leftarrow |x|$
$k \leftarrow \lceil n/(t-1) \rceil$
$d \leftarrow k(t-1) - n$
for $i \leftarrow 1$ **to** $k-1$
  do $y_i \leftarrow x_i$
$y_k \leftarrow x_k \parallel 0^d$
$y_{k+1} \leftarrow$ the binary representation of d
$z_1 \leftarrow 0^{m+1} \parallel y_1$
$g_1 \leftarrow$ **compress**(z_1)
for $i \leftarrow 1$ **to** k
  do $\begin{cases} z_{i+1} \leftarrow g_i \parallel 1 \parallel y_{i+1} \\ g_{i+1} \leftarrow \textbf{compress}(z_{i+1}) \end{cases}$
$h(x) \leftarrow g_{k+1}$
return $(h(x))$

The following theorem proves that, if a collision can be found for h, then a collision can be found for **compress**. In other words, h is collision resistant provided that **compress** is collision resistant.

THEOREM 5.6 *Suppose* **compress** : $\{0,1\}^{m+t} \rightarrow \{0,1\}^m$ *is a collision resistant compression function, where* $t \geq 2$. *Then the function*

$$h : \bigcup_{i=m+t+1}^{\infty} \{0,1\}^i \rightarrow \{0,1\}^m,$$

as constructed in Algorithm 5.6, is a collision resistant hash function.

PROOF Suppose that we can find $x \neq x'$ such that $h(x) = h(x')$. We will show how we can find a collision for **compress** in polynomial time.

Denote

$$y(x) = y_1 \parallel y_2 \parallel \cdots \parallel y_{k+1}$$

and

$$y(x') = y'_1 \parallel y'_2 \parallel \cdots \parallel y'_{\ell+1},$$

where x and x' are padded with d and d' 0's, respectively. Denote the g-values computed in the algorithm by $g_1, \ldots, g_{k+1}$ and $g'_1, \ldots, g'_{\ell+1}$, respectively.

We identify two cases, depending on whether $|x| \equiv |x'| \pmod{t-1}$ (or not).

case 1: $|x| \not\equiv |x'| \pmod{t-1}$.

Here $d \neq d'$ and $y_{k+1} \neq y'_{\ell+1}$. We have

$$\begin{aligned}\textbf{compress}(g_k \parallel 1 \parallel y_{k+1}) &= g_{k+1} \\ &= h(x) \\ &= h(x') \\ &= g'_{\ell+1} \\ &= \textbf{compress}(g'_\ell \parallel 1 \parallel y'_{\ell+1}),\end{aligned}$$

which is a collision for **compress** because $y_{k+1} \neq y'_{\ell+1}$.

case 2: $|x| \equiv |x'| \pmod{t-1}$.

It is convenient to split this case into two subcases:

case 2a: $|x| = |x'|$.

Here we have $k = \ell$ and $y_{k+1} = y'_{k+1}$. We begin as in case 1:

$$\begin{aligned}\textbf{compress}(g_k \parallel 1 \parallel y_{k+1}) &= g_{k+1} \\ &= h(x) \\ &= h(x') \\ &= g'_{k+1} \\ &= \textbf{compress}(g'_k \parallel 1 \parallel y'_{k+1}).\end{aligned}$$

If $g_k \neq g'_k$, then we find a collision for **compress**, so assume $g_k = g'_k$. Then we have

$$\begin{aligned}\textbf{compress}(g_{k-1} \parallel 1 \parallel y_k) &= g_k \\ &= g'_k \\ &= \textbf{compress}(g'_{k-1} \parallel 1 \parallel y'_k).\end{aligned}$$

Either we find a collision for **compress**, or $g_{k-1} = g'_{k-1}$ and $y_k = y'_k$. Assuming we do not find a collision, we continue working backwards, until finally we obtain

$$\begin{aligned}\textbf{compress}(0^{m+1} \parallel y_1) &= g_1 \\ &= g'_1 \\ &= \textbf{compress}(0^{m+1} \parallel y'_1).\end{aligned}$$

If $y_1 \neq y'_1$, then we find a collision for **compress**, so we assume $y_1 = y'_1$. But then $y_i = y'_i$ for $1 \leq i \leq k+1$, so $y(x) = y(x')$. But this implies $x = x'$, because the mapping $x \mapsto y(x)$ is an injection. We assumed $x \neq x'$, so we have a contradiction.

case 2b: $|x| \neq |x'|$.

Without loss of generality, assume $|x'| > |x|$, so $\ell > k$. This case proceeds in

Algorithm 5.7: MERKLE-DAMGÅRD2(x)

external compress
comment: compress : $\{0,1\}^{m+1} \rightarrow \{0,1\}^m$
$n \leftarrow |x|$
$y \leftarrow 11 \parallel f(x_1) \parallel f(x_2) \parallel \cdots \parallel f(x_n)$
denote $y = y_1 \parallel y_2 \parallel \cdots \parallel y_k$, where $y_i \in \{0,1\}, 1 \leq i \leq k$
$g_1 \leftarrow$ **compress**$(0^m \parallel y_1)$
for $i \leftarrow 1$ **to** $k-1$
  do $g_{i+1} \leftarrow$ **compress**$(g_i \parallel y_{i+1})$
return (g_k)

a similar fashion as case 2a. Assuming we find no collisions for **compress**, we eventually reach the situation where

$$\begin{aligned} \mathbf{compress}(0^{m+1} \parallel y_1) &= g_1 \\ &= g'_{\ell-k+1} \\ &= \mathbf{compress}(g'_{\ell-k} \parallel 1 \parallel y'_{\ell-k+1}). \end{aligned}$$

But the $(m+1)$st bit of

$$0^{m+1} \parallel y_1$$

is a 0 and the $(m+1)$st bit of

$$g'_{\ell-k} \parallel 1 \parallel y'_{\ell-k+1}$$

is a 1. So we find a collision for **compress**.

Since we have considered all possible cases, we have proven the desired conclusion. ∎

The construction presented in Algorithm 5.6 can be used only when $t \geq 2$. Let's now look at the situation where $t = 1$. We need to use a different construction for h. Suppose $|x| = n \geq m+2$. We first encode x in a special way. This will be done using the function f defined as follows:

$$\begin{aligned} f(0) &= 0 \\ f(1) &= 01. \end{aligned}$$

The construction of $h(x)$ is presented as Algorithm 5.7.

The encoding $x \mapsto y = y(x)$, as defined in Algorithm 5.7, satisfies two important properties:

1. If $x \neq x'$, then $y(x) \neq y(x')$ (i.e., $x \mapsto y(x)$ is an injection).

2. There do not exist two strings $x \neq x'$ and a string z such that $y(x) = z \parallel y(x')$. (In other words, no encoding is a ***postfix*** of another encoding. This is easily seen because each string $y(x)$ begins with 11, and there do not exist two consecutive 1's in the remainder of the string.)

THEOREM 5.7 *Suppose* **compress** : $\{0,1\}^{m+1} \to \{0,1\}^m$ *is a collision resistant compression function. Then the function*

$$h : \bigcup_{i=m+2}^{\infty} \{0,1\}^i \to \{0,1\}^m,$$

as constructed in Algorithm 5.7, is a collision resistant hash function.

PROOF Suppose that we can find $x \neq x'$ such that $h(x) = h(x')$. Denote

$$y(x) = y_1 y_2 \cdots y_k$$

and

$$y(x') = y_1' y_2' \cdots y_\ell'.$$

We consider two cases.

case 1: $k = \ell$.

As in Theorem 5.6, either we find a collision for **compress**, or we obtain $y = y'$. But this implies $x = x'$, a contradiction.

case 2: $k \neq \ell$.

Without loss of generality, assume $\ell > k$. This case proceeds in a similar fashion. Assuming we find no collisions for **compress**, we have the following sequence of equalities:

$$\begin{aligned} y_k &= y_\ell' \\ y_{k-1} &= y_{\ell-1}' \\ &\vdots \\ y_1 &= y_{\ell-k+1}'. \end{aligned}$$

But this contradicts the "postfix-free" property stated above.

We conclude that h is collision resistant. ∎

We summarize the two above-described constructions of hash functions, and the number of applications of **compress** needed to compute h, in the following theorem.

THEOREM 5.8 *Suppose* **compress** : $\{0,1\}^{m+t} \to \{0,1\}^m$ *is a collision resistant compression function, where* $t \geq 1$. *Then there exists a collision resistant hash function*

$$h : \bigcup_{i=m+t+1}^{\infty} \{0,1\}^i \to \{0,1\}^m.$$

The number of times **compress** *is computed in the evaluation of h is at most*

$$\begin{cases} 1 + \left\lceil \frac{n}{t-1} \right\rceil & \text{if } t \geq 2, \text{ and} \\ 2n + 2 & \text{if } t = 1, \end{cases}$$

where $|x| = n$.

5.3.2 Some Examples of Iterated Hash Functions

Many commonly used hash functions have been constructed using the Merkle-Damgård approach. The first of these was *MD4*, which was proposed by Rivest in 1990. Rivest then modified *MD4* to produce *MD5* in 1992. Next, *SHA* was proposed as a standard by NIST in 1993, and it was adopted as FIPS 180. *SHA-1* is a slight modification of *SHA*; it was published in 1995 as FIPS 180-1 (and *SHA* was subsequently referred to by the name *SHA-0*).

This progression of hash functions incorporated various modifications to improve the security of the later versions of the hash functions against attacks that were found against earlier versions. For example, collisions in the compression functions of *MD4* and *MD5* were discovered in the mid-1990s. It was shown in 1998 that *SHA-0* had a weakness that would allow collisions to be found in approximately 2^{61} steps (this attack is much more efficient than a birthday attack, which would require about 2^{80} steps).

In 2004, a collision for *SHA-0* was found by Joux and reported at CRYPTO 2004. Collisions for *MD5* and several other popular hash functions were also presented at CRYPTO 2004, by Wang, Feng, Lai, and Yu.

The first collision for *SHA-1* was found by Stevens, Bursztein, Karpman, Albertini, and Markov and announced on February 23, 2017. This attack was approxiately 100000 times faster than a brute-force "birthday paradox" search that would have required roughly 2^{80} trials.

SHA-2 includes four hash functions, which are known as *SHA-224*, *SHA-256*, *SHA-384*, and *SHA-512*. The suffixes "224", "256," "384," and "512" refer to the sizes of the message digests of these four hash functions. These hash functions are also iterated hash functions, but they have a more complex description than *SHA-1*. The last three of these four hash functions comprised the FIPS standard that was approved in 2002; *SHA-224* was added in 2004. It is probably fair to say that the *SHA-2* hash functions were not used nearly as frequently as *SHA-1*.

The most recent hash functions in the SHA family are known as *SHA-3*. These hash functions are based on a different design technique—known as the sponge construction—which will be discussed in the next section. *SHA-3* became a FIPS standard in August, 2015.

To close this section, we will now discuss *SHA-1* in a bit more detail, without giving a complete description of it (complete specifications of *SHA-1* and all the other FIPS standards are readily available on the internet). *SHA-1*, which creates a 160-bit message digest, provides a typical example of a hash standard prior to *SHA-3*. The padding scheme of *SHA-1* extends the input x by at most one extra

512-bit block. The compression function maps $160 + 512 = 672$ bits to 160 bits, where the 512 bits comprise a block of the message. *SHA-1* is built from word-oriented operations on bitstrings, where a word consists of 32 bits (or eight hexadecimal digits). The operations used in *SHA-1* are as follows:

$X \wedge Y$	bitwise "and" of X and Y
$X \vee Y$	bitwise "or" of X and Y
$X \oplus Y$	bitwise "x-or" of X and Y
$\neg X$	bitwise complement of X
$X + Y$	integer addition modulo 2^{32}
$\mathbf{ROTL}^s(X)$	circular left shift of X by s positions ($0 \leq s \leq 31$)

The point is that these operations are very efficient, but when a suitable sequence of these operations is performed, the output is quite unpredictable.

5.4 The Sponge Construction

SHA-3 is based on a design strategy called the ***sponge construction***. This technique was developed by Bertoni, Daemen, Peeters, and Van Assche. Instead of using a compression function, the basic "building block" is a function f that maps bitstrings of a fixed length to bitstrings of the same length. Typically f will be a bijection, so every bitstring will have a unique preimage. The sponge construction is quite versatile and can be used to construct various cryptographic tools. For hash functions, one of the useful features of the sponge construction is that it can produce output (i.e., a message digest) of arbitrary length.

Suppose that f operates on bitstrings of length b. That is, we have $f : \{0,1\}^b \rightarrow \{0,1\}^b$. The integer b is called the ***width***. We write b as the sum of two positive integers, say $b = r + c$, where r is the ***bitrate*** and c is the ***capacity***. The value of r affects the efficiency of the resulting sponge function, as a message will be processed r bits at a time. The value of c affects the resulting security of the sponge function. The security level against a certain kind of collision attack is intended to be roughly $2^{c/2}$. This is comparable to the security of a random oracle with a c-bit output (see Section 5.2.2).

The ***sponge function*** based on f is depicted in Figure 5.2.[2] This sponge function works as follows. The input will be a message M, which is a bitstring of arbitrary length. M is padded appropriately so that its length is a multiple of r. Then the padded message is split into ***blocks*** of length r.

Initially, the ***state*** is a bitstring of length b consisting of zeroes. The first block of the padded message is exclusive-ored with the first r bits of the state. Then the

[2]The diagram is taken from `http://sponge.noekeon.org` and is available under the Creative Commons Attribution License.

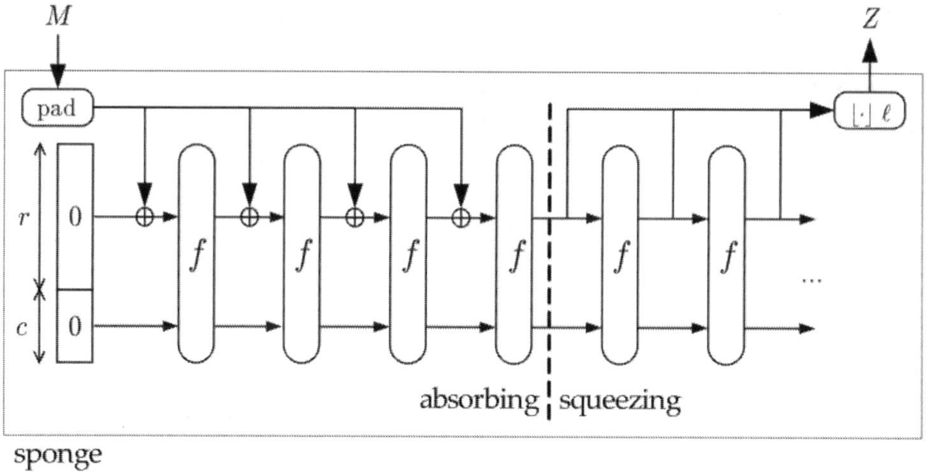

FIGURE 5.2: A sponge function

function f is applied, which updates the state. This process is then repeated with the remaining blocks of the padded message. Each block, in turn, is exclusive-ored with the first r bits of the current state and then the function f is applied to update the state. This constitutes the ***absorbing phase*** of the sponge function.

Following the absorbing phase, the ***squeezing phase*** is used to produce the output of the sponge function (i.e., the message digest). Suppose that ℓ ***output bits*** are desired. We begin by taking the first r bits of the current state; this forms an ***output block***. If $\ell > r$, then we apply f to the current state (which consists of $r + c$ bits) and take the first r output bits as another output block. This process is repeated until we have a total of at least ℓ bits. Finally, we truncate the concatenation of these output blocks (each of which has length r) to ℓ bits. This forms the desired message digest.

We can describe the absorbing process succinctly using mathematical notation as follows. Suppose the padded message is

$$M = m_1 \parallel \cdots \parallel m_k,$$

where $m_1, \dots, m_k \in \{0,1\}^r$. Define

$$x_0 = \underbrace{00\dots0}_{r} \quad \text{and} \quad y_0 = \underbrace{00\dots0}_{c}.$$

Then compute the following values:

$$\begin{aligned} f(x_0 \oplus m_1 \parallel y_0) &= x_1 \parallel y_1 \\ f(x_1 \oplus m_2 \parallel y_1) &= x_2 \parallel y_2 \\ \vdots \quad & \vdots \quad \vdots \\ f(x_{k-1} \oplus m_k \parallel y_{k-1}) &= x_k \parallel y_k, \end{aligned}$$

where $x_i \in \{0,1\}^r$ and $y_i \in \{0,1\}^c$ for all $i \geq 0$. The bitstring $x_k \parallel y_k$ is the output of the absorbing phase. If $\ell \leq r$, then the message digest Z just consists of the first ℓ bits of x_k. If $\ell > r$, then one or more additional applications of f are required to compute the message digest (see the description of the squeezing phase that was given above).

The security of a sponge function based on f is comparable to that of a random oracle that outputs c bits, assuming that f is a random function.[3] We are not going to provide a proof of this result, but we will informally discuss how to find a collision for a sponge function by evaluating the function f approximately $2^{c/2}$ times (it is a kind of birthday attack). This shows, roughly speaking, that the security of the sponge function cannot be higher than that of a random oracle that outputs c bits.

The collision we are going to find is an ***internal collision***, i.e., a collision in the b-bit state of the sponge function. Suppose we define

$$x_0 = \underbrace{00\ldots0}_{r} \quad \text{and} \quad y_0 = \underbrace{00\ldots0}_{c},$$

and we perform the following computations:

$$\begin{aligned} f(x_0 \parallel y_0) &= x_1 \parallel y_1 \\ f(x_0 \parallel y_1) &= x_2 \parallel y_2 \\ &\vdots \\ f(x_0 \parallel y_{k-1}) &= x_k \parallel y_k, \end{aligned}$$

terminating when we find a repeated y-value, say $y_k = y_h$, where $h < k$. As above, $x_i \in \{0,1\}^r$ and $y_i \in \{0,1\}^c$ for all $i \geq 0$. Observe that all evaluations of f have inputs that begin with r 0's. If we think of the values $y_1, \ldots, y_k$ outputted by f as being random bitstrings of length 2^c, then this is just a birthday attack, and we expect that the number of evaluations of f, which is denoted by k, is within a constant factor of $2^{c/2}$.

Now consider the following two messages (we are ignoring padding):

$$M = x_0 \parallel \cdots \parallel x_h$$

and

$$M' = x_0 \parallel \cdots \parallel x_k.$$

When we evaluate the sponge function with input M, we obtain

$$\begin{aligned} f(x_0 \oplus x_0 \parallel y_0) &= f(x_0 \parallel y_0) = x_1 \parallel y_1 \\ f(x_1 \oplus x_1 \parallel y_1) &= f(x_0 \parallel y_1) = x_2 \parallel y_2 \\ &\vdots \\ f(x_{h-1} \oplus x_{h-1} \parallel y_{h-1}) &= f(x_0 \parallel y_{h-1}) = x_h \parallel y_h \\ f(x_h \oplus x_h \parallel y_h) &= f(x_0 \parallel y_h) = x_{h+1} \parallel y_{h+1} \end{aligned}$$

[3]Here we are considering only the absorbing phase of the sponge function. The effect of the squeezing phase will be addressed a bit later.

Thus, $x_{h+1} \parallel y_{h+1} = f(x_0 \parallel y_h)$ is the output of the absorbing phase when the sponge function is computed with input M. Similarly, when the sponge function is evaluated with input M', the output of the absorbing phase is $x_{k+1} \parallel y_{k+1} = f(x_0 \parallel y_k)$. Since $y_h = y_k$, it must be the case that $x_{h+1} \parallel y_{h+1} = x_{k+1} \parallel y_{k+1}$. The two messages have the same output from the absorbing phase, and hence their message digests will be the same. This is a collision.

Now we look briefly at the squeezing phase, which creates an ℓ-bit message digest from a given sponge function. We always have the option of performing a birthday attack on the ℓ-bit message digest. Using this approach, we could generate an ***output collision*** by evaluating the sponge function roughly $2^{\ell/2}$ times. This does not require an internal collision to be generated first.

If our goal is simply to generate an output collision, the most efficient approach depends on the relationship between ℓ and c. If $c < \ell$, then the fastest method would be to first generate an internal collision, which would then yield an output collision. On the other hand, if $c > \ell$, then we would just generate an output collision directly. Overall, we would quantify the security of the sponge function (against a collision attack) as $\min\{2^{c/2}, 2^{\ell/2}\}$.

5.4.1 SHA-3

SHA-3 consists of four hash functions, which are named *SHA3-224*, *SHA3-256*, *SHA3-384*, and *SHA3-512*. Again, the suffixes denote the lengths of the message digests (i.e., the parameter ℓ in the discussion above). The *SHA-3* hash functions are derived from the hash function known as *Keccac*, which was proposed in the SHA-3 competition. The width, bitrate, and capacity for these functions are summarized in Table 5.1. Observe that all four of the hash functions in *SHA-3* produce message digests that are less than r bits in length.

The function f is a bijective function operating on a state that is a bitstring of length 1600. It consists of 24 rounds, each of which is composed of five simple steps (called sub-rounds). The actual operations performed are very efficient operations similar to the ones done in *SHA-1*. Note that the squeezing phase for any of the four versions of *SHA-3* does not require any applications of the function f.

There are two additional functions included in *SHA-3*. These functions, which are named *SHAKE128* and *SHAKE256*, are ***extendable output functions*** (which is abbreviated to ***XOF***). The difference between a hash function and an XOF is that an XOF has a variable-length output of d bits. It uses the same sponge construction, but it may employ additional applications of f in the squeezing phase in order to generate longer message digests. It is important to note, however, that when we generate longer message digests, the security is ultimately limited by the capacity, c.

In Table 5.1, we also list the security levels of these hash functions against the best-known attacks. The phrase "collision security" refers to the complexity of finding a collision; if the collision security is equal to t, this indicates that the attack requires approximately 2^t steps. The term "preimage security" has a similar meaning; however, it covers attacks to find either preimages or second preimages.

TABLE 5.1: Parameters and Security Levels for *SHA-3*

hash function	b	r	c	collision security	preimage security
SHA3-224	1600	1152	448	112	224
SHA3-256	1600	1088	512	128	256
SHA3-384	1600	832	768	192	384
SHA3-512	1600	576	1024	256	512
SHAKE128	1600	1344	256	$\min\{d/2, 128\}$	$\min\{d, 128\}$
SHAKE256	1600	1088	512	$\min\{d/2, 256\}$	$\min\{d, 256\}$

Note: d denotes the output length of *SHAKE128* or *SHAKE256*.

5.5 Message Authentication Codes

We now turn our attention to message authentication codes, which are keyed hash functions satisfying certain security properties. As we will see, the security properties required by a MAC are rather different than those required by an (unkeyed) hash function.

One common way of constructing a MAC is to incorporate a secret key into an unkeyed hash function, by including it as part of the message to be hashed. This must be done carefully, however, in order to prevent certain attacks from being carried out. We illustrate the possible pitfalls with a couple of simple examples.

As a first attempt, suppose we construct a keyed hash function h_K from an unkeyed iterated hash function, say h, by defining $\text{IV} = K$ and keeping its value secret. For simplicity, suppose also that h does not have a preprocessing step or an output transformation. Such a hash function requires that every input message x have length that is a multiple of t, where $\textbf{compress} : \{0,1\}^{m+t} \rightarrow \{0,1\}^m$ is the compression function used to build h. Further, the key K is an m-bit key.

We show how an adversary can construct a valid tag for a certain message, without knowing the secret key K, given any message x and its corresponding tag, $h_K(x)$. Let x' be any bitstring of length t, and consider the message $x \parallel x'$. The tag for this message, $h_K(x \parallel x')$, is computed to be

$$h_K(x \parallel x') = \textbf{compress}(h_K(x) \parallel x').$$

Since $h_K(x)$ and x' are both known, it is a simple matter for an adversary to compute $h_K(x \parallel x')$, even though K is secret. This is called a ***length extension attack***.

Even if messages are padded, a modification of the above attack can be carried out. For example, suppose that $y = x \parallel \textbf{pad}(x)$ in the preprocessing step. Note that $|y| = rt$ for some integer r. Let w be any bitstring of length t, and define

$$x' = x \parallel \textbf{pad}(x) \parallel w.$$

In the preprocessing step, we would compute

$$y' = x' \parallel \textbf{pad}(x') = x \parallel \textbf{pad}(x) \parallel w \parallel \textbf{pad}(x'),$$

where $|y'| = r't$ for some integer $r' > r$.

Consider the computation of $h_K(x')$. (This is the same as computing $h(x')$ when $\mathrm{IV} = K$.) In the processing step, it is clear that $z_r = h_K(x)$. It is therefore possible for an adversary to compute the following:

$$\begin{array}{rcl} z_{r+1} & \leftarrow & \textbf{compress}(h_K(x) \parallel y_{r+1}) \\ z_{r+2} & \leftarrow & \textbf{compress}(z_{r+1} \parallel y_{r+2}) \\ \vdots & \vdots & \vdots \\ z_{r'} & \leftarrow & \textbf{compress}(z_{r'-1} \parallel y_{r'}), \end{array}$$

and then

$$h_K(x') = z_{r'}.$$

Therefore the adversary can compute $h_K(x')$ even though he doesn't know the secret key K (and notice that the attack makes no assumptions about the length of the pad).

Keeping the above examples in mind, we formulate definitions of what it should mean for a MAC algorithm to be secure. As we saw, the objective of an adversary (Oscar) is to try to produce a message-tag pair (x, y) that is valid under a fixed but unknown key, K. The adversary might have some prior examples of message-tag pairs that are valid for the key K, say $(x_1, y_1), (x_2, y_2), \ldots, (x_Q, y_Q)$.

These Q message-tag pairs might be ones that Oscar observes being sent from Alice to Bob or from Alice to Bob. This scenario is often termed a ***known message attack***, which indicates that the messages $x_1, \ldots, x_Q$ are known to Oscar, but it was Alice or Bob who decided which messages to transmit.

An alternative scenario is when Oscar is permitted to choose the messages $x_1, \ldots, x_Q$ himself. Oscar is then allowed to ask Alice or Bob (or equivalently, a signing oracle) for the corresponding tags $y_1, \ldots, y_Q$. This variation is called a ***chosen message attack***.

In either scenario, the adversary obtains a list of message-tag pairs (all of which are valid under the same unknown key K):

$$(x_1, y_1), (x_2, y_2), \ldots, (x_Q, y_Q).$$

Later, when the adversary outputs the message-tag pair (x, y), it is required that x is a "new" message, i.e., $x \notin \{x_1, \ldots, x_Q\}$. If, in addition, (x, y) is a valid pair, then the pair is said to be a ***forgery***. If the probability that the adversary outputs a forgery is at least ϵ, then the adversary is said to be an (ϵ, Q)-***forger*** for the given MAC.

For an (ϵ, Q)-***forger***, we should specify whether it is a known message attack or a chosen message attack. In the case of a chosen message attack, the adversary can choose his queries (i.e., the messages) with the goal of maximizing the probability of a successful attack. In the case of a known message attack, the messages are beyond the control of the adversary. When we say that we have an (ϵ, Q)-forger in this setting, it means that the adversary should succeed with probability at least ϵ no matter what messages he observes.

Finally, the probability ϵ of a successful forgery could be taken to be either an average-case probability over all the possible keys, or the worst-case probability. To be concrete, in the following sections, we will generally take ϵ to be a worst-case probability. This means that the adversary can produce a forgery with probability at least ϵ, regardless of the secret key being used.

Using this terminology, the attacks described above are known-message $(1, 1)$-forgers.

We close this section by mentioning two obvious attacks on MACs. The first is a ***key guessing attack***, wherein the adversary chooses $K \in \mathcal{K}$ uniformly at random, and outputs the tag $h_K(x)$ for an arbitrary message x. This attack will succeed with probability $1/|\mathcal{K}|$. The second attack is a ***tag guessing attack***. Here, the adversary chooses the tag $y \in \mathcal{Y}$ uniformly at random and outputs y as the tag for an arbitrary message x. This attack will succeed with probability $1/|\mathcal{Y}|$.

5.5.1 Nested MACs and HMAC

A ***nested MAC*** builds a MAC algorithm from the composition of two (keyed) hash families. Suppose that $(\mathcal{X}, \mathcal{Y}, \mathcal{K}, \mathcal{G})$ and $(\mathcal{Y}, \mathcal{Z}, \mathcal{L}, \mathcal{H})$ are hash families. The ***composition*** of these hash families is the hash family $(\mathcal{X}, \mathcal{Z}, \mathcal{M}, \mathcal{G} \circ \mathcal{H})$ in which $\mathcal{M} = \mathcal{K} \times \mathcal{L}$ and

$$\mathcal{G} \circ \mathcal{H} = \{g \circ h : g \in \mathcal{G}, h \in \mathcal{H}\},$$

where $(g \circ h)_{(K,L)}(x) = h_L(g_K(x))$ for all $x \in \mathcal{X}$. In this construction, $\mathcal{Y}$ and $\mathcal{Z}$ are finite sets such that $|\mathcal{Y}| \geq |\mathcal{Z}|$; $\mathcal{X}$ could be finite or infinite. If $\mathcal{X}$ is finite, then $|\mathcal{X}| > |\mathcal{Y}|$.

Observe that a nested MAC is just the composition of two hash functions. We first apply a function that takes a message x as input and produces an output y. The second function takes input y and produces the message digest z. The first function is chosen from a hash family $\mathcal{G}$ and the second function is chosen from a hash family $\mathcal{H}$.

We are interested in finding situations under which we can guarantee that a nested MAC is secure, assuming that the two hash families from which it is constructed satisfy appropriate security requirements. All security results in this section will be assumed to refer to chosen message attacks. Roughly speaking, it can be shown that the nested MAC is secure provided that the following two conditions are satisfied:

1. $(\mathcal{Y}, \mathcal{Z}, \mathcal{L}, \mathcal{H})$ is secure as a MAC, given a fixed (unknown) key, and

2. $(\mathcal{X}, \mathcal{Y}, \mathcal{K}, \mathcal{G})$ is collision resistant, given a fixed (unknown) key.

Intuitively, we are building a secure "big MAC" (namely, the nested MAC) from the composition of a secure "little MAC" (namely, $(\mathcal{Y}, \mathcal{Z}, \mathcal{L}, \mathcal{H})$) and a certain kind of collision resistant keyed hash family (namely, $(\mathcal{X}, \mathcal{Y}, \mathcal{K}, \mathcal{G})$). Let's try to make the above conditions more precise, and then present a proof of a specific security result.

The security result will in fact be comparing the relative difficulties of certain types of attacks against the three hash families. We will be considering the following three adversaries:

- a forger for the little MAC (which carries out a "little MAC attack"),
- a collision-finder for the hash family $(\mathcal{X}, \mathcal{Y}, \mathcal{K}, \mathcal{G})$, when the key is secret (this is an "unknown-key collision attack"), and
- a forger for the nested MAC (which we term a "big MAC attack").

Here is a more careful description of each of the three adversaries: First, in a ***little MAC attack***, a key L is chosen and kept secret. The adversary is allowed to choose values for y and query a little MAC oracle for values of $h_L(y)$. Then the adversary attempts to output a pair (y', z) such that $z = h_L(y')$, where y' was not one of its previous queries.

In an ***unknown-key collision attack***, a key K is chosen and kept secret. The adversary is allowed to choose values for x and query a hash oracle for values of $g_K(x)$. Then the adversary attempts to output a pair x', x'' such that $x' \neq x''$ and $g_K(x') = g_K(x'')$.

Finally, in a ***big MAC attack***, a pair of keys, (K, L), is chosen and kept secret. The adversary is allowed to choose values for x and query a big MAC oracle for values of $h_L(g_K(x))$. Then the adversary attempts to output a pair (x', z) such that $z = h_L(g_K(x'))$, where x' was not one of its previous queries.

We will assume that there does not exist an $(\epsilon_1, Q+1)$-unknown-key collision attack for a randomly chosen function $g_K \in \mathcal{G}$. (If the key K were not secret, then this would correspond to our usual notion of collision resistance. Since we assume that K is secret, the problem facing the adversary is more difficult, and therefore we are making a weaker security assumption than collision resistance.) We also assume that there does not exist an (ϵ_2, Q)-little MAC attack for a randomly chosen function $h_L \in \mathcal{H}$, where L is secret. Finally, suppose that there exists an (ϵ, Q)-big MAC attack for a randomly chosen function $(g \circ h)_{(K,L)} \in \mathcal{G} \circ \mathcal{H}$, where (K, L) is secret.

With probability at least ϵ, the big MAC attack outputs a valid pair (x, z) after making at most Q queries to a big MAC oracle. Let $x_1, \ldots, x_Q$ denote the queries made by the adversary, and let $z_1, \ldots, z_Q$ be the corresponding responses made by the oracle. After the adversary has finished executing, we have the list of valid message-tag pairs $(x_1, z_1), \ldots, (x_Q, z_Q)$, as well as the possibly valid message-tag pair (x, z).

Suppose we now take the values $x_1, \ldots, x_Q$, and x, and make $Q+1$ queries to a hash oracle g_K. We obtain the list of values $y_1 = g_K(x_1), \ldots, y_Q = g_K(x_Q)$, and $y = g_K(x)$. Suppose it happens that $y \in \{y_1, \ldots, y_Q\}$, say $y = y_i$. Then we can output the pair x, x_i as a solution to **Collision**. This would be a successful unknown-key collision attack. On the other hand, if $y \notin \{y_1, \ldots, y_Q\}$, then we output the pair (y, z), which (possibly) is a valid pair for the little MAC. This would be a forgery constructed after (indirectly) obtaining Q answers to Q little MAC queries, namely $(y_1, z_1), \ldots, (y_Q, z_Q)$.

By the assumption we made, any unknown-key collision attack has probability at most ϵ_1 of succeeding. As well, we assumed that the big MAC attack has success probability at least ϵ. Therefore, the probability that (x, z) is a valid pair and $y \notin \{y_1, \ldots, y_Q\}$ is at least $\epsilon - \epsilon_1$. The success probability of any little MAC attack is at most ϵ_2, and the success probability of the little MAC attack described above is at least $\epsilon - \epsilon_1$. Hence, it follows that $\epsilon \leq \epsilon_1 + \epsilon_2$.

We have proven the following result.

THEOREM 5.9 *Suppose $(\mathcal{X}, \mathcal{Z}, \mathcal{M}, \mathcal{G} \circ \mathcal{H})$ is a nested MAC. Suppose there does not exist an $(\epsilon_1, Q+1)$-collision attack for a randomly chosen function $g_K \in \mathcal{G}$, when the key K is secret. Further, suppose that there does not exist an (ϵ_2, Q)-forger for a randomly chosen function $h_L \in \mathcal{H}$, where L is secret. Finally, suppose there exists an (ϵ, Q)-forger for the nested MAC, for a randomly chosen function $(g \circ h)_{(K,L)} \in \mathcal{G} \circ \mathcal{H}$. Then $\epsilon \leq \epsilon_1 + \epsilon_2$.*

HMAC is a nested MAC algorithm that was adopted as a FIPS standard in March, 2002. It constructs a MAC from an (unkeyed) hash function; we describe *HMAC* based on *SHA-1*. This version of *HMAC* uses a 512-bit key, denoted K. *ipad* and *opad* are 512-bit constants, defined in hexadecimal notation as follows:

$$\begin{aligned} ipad &= \texttt{3636}\cdots\texttt{36} \\ opad &= \texttt{5C5C}\cdots\texttt{5C} \end{aligned}$$

Let x be the message to be authenticated. Then the 160-bit MAC is defined as follows:

$$\text{HMAC}_K(x) = \text{SHA-1}((K \oplus opad) \parallel \text{SHA-1}((K \oplus ipad) \parallel x)).$$

Note that *HMAC* uses *SHA-1* with the value $K \oplus ipad$, which is prepended to x, used as the key. This application of *SHA-1* is assumed to be secure against an unknown-key collision attack. Now the key value $K \oplus opad$ is prepended to the previously constructed message digest, and *SHA-1* is applied again. This second computation of *SHA-1* requires only one application of the compression function, and we are assuming that *SHA-1* when used in this way is secure as a MAC. If these two assumptions are valid, then Theorem 5.9 says that *HMAC* is secure as a MAC.

We observe that *HMAC* is quite efficient. At first glance, we might think that it takes twice as long as evaluating the underlying hash function. However, as observed above, the second, "outer" hash takes a fixed-length, short bitstring as input. So the extra hash computation only takes constant time.

Upon the adoption of *SHA-3*, it may be the case that *HMAC* will become obsolete. The reason is that a MAC based on the sponge construction is not susceptible to the length extension attack described in Section 5.5. The simpler technique of prepending the key and then hashing using the sponge function would yield a secure MAC. This is the basis for a proposed MAC known as *KMAC*, which also includes an additional (variable) parameter to indicate the length of the tag.

Cryptosystem 5.1: *CBC-MAC*(x, K)

denote $x = x_1 \parallel \cdots \parallel x_n$
IV $\leftarrow 00 \cdots 0$
$y_0 \leftarrow$ IV
for $i \leftarrow 1$ **to** n
  do $y_i \leftarrow e_K(y_{i-1} \oplus x_i)$
return (y_n)

5.5.2 CBC-MAC

One of the more popular ways to construct a MAC is to use a block cipher in CBC mode with a fixed (public) initialization vector. In CBC mode, recall from Section 4.7 that each ciphertext block y_i is x-ored with the next plaintext block, x_{i+1}, before being encrypted with the secret key K. More formally, we start with an initialization vector, denoted by IV, and define $y_0 =$ IV. Then we construct $y_1, y_2, \ldots$ using the rule

$$y_i = e_K(y_{i-1} \oplus x_i),$$

for all $i \geq 1$.

Suppose that $(\mathcal{P}, \mathcal{C}, \mathcal{K}, \mathcal{E}, \mathcal{D})$ is an (endomorphic) cryptosystem, where $\mathcal{P} = \mathcal{C} = \{0,1\}^t$. Let IV be the bitstring consisting of t zeroes, and let $K \in \mathcal{K}$ be a secret key. Finally, let $x = x_1 \parallel \cdots \parallel x_n$ be a bitstring of length tn (for some positive integer n), where each x_i is a bitstring of length t. We compute CBC-MAC(x, K) as shown in Algorithm 5.1. Basically, we "encrypt" the plaintext in CBC mode and we only retain the last ciphertext block, which we define to be the tag.

The best known general attack on *CBC-MAC* is a birthday (i.e., collision-finding) chosen message attack. We describe this attack now. Basically, we allow the adversary to request tags on a large number of messages. If a duplicated tag is found, then the adversary can construct one additional message and request its tag. Finally, the adversary can produce a new message and its corresponding tag (i.e., a forgery), even though he does not know the secret key. The attack works for messages of any prespecified fixed length, $n \geq 3$.

In preparation for the attack, let $n \geq 3$ be an integer and let $x_3, \ldots, x_n$ be fixed bitstrings of length t. Let $Q \approx 1.17 \times 2^{t/2}$ be an integer, and choose any Q distinct bitstrings of length t, which we denote $x_1^1, \ldots, x_1^Q$. Next, let $x_2^1, \ldots, x_2^Q$ be randomly chosen bitstrings of length t. Finally, for $1 \leq i \leq Q$ and for $3 \leq k \leq n$, define $x_k^i = x_k$, and then define

$$x^i = x_1^i \parallel \cdots \parallel x_n^i$$

for $1 \leq i \leq Q$. Note that $x^i \neq x^j$ if $i \neq j$, because $x_1^i \neq x_1^j$.

The attack can now be carried out. First the adversary requests the tags for the messages $x^1, x^2, \ldots, x^Q$. In the computation of the tag of each x^i using Cryptosystem 5.1, values $y_0^i, \ldots, y_n^i$ are computed, and y_n^i is the resulting tag. Now suppose that x^i and x^j have identical tags, i.e., $y_n^i = y_n^j$. This happens if and only if $y_2^i = y_2^j$.

Observe that $y_2^i = e_K(y_1^i \oplus x_2^i)$ and $y_2^j = e_K(y_1^j \oplus x_2^j)$. The values y_2^i and y_2^j are both encryptions using the same key K. If we regard e_K as a random function, then a collision of the form $e_K(y_1^i \oplus x_2^i) = e_K(y_1^j \oplus x_2^j)$ will occur with probability 1/2 after $Q \approx 1.17 \times 2^{t/2}$ encryptions have been performed (this is basically a birthday attack).

We are assuming that $y_2^i = y_2^j$. This happens if and only if

$$y_1^i \oplus x_2^i = y_1^j \oplus x_2^j.$$

Let x_δ be any nonzero bitstring of length t. Define

$$v = x_1^i \parallel (x_2^i \oplus x_\delta) \parallel \cdots \parallel x_n^i$$

and

$$w = x_1^j \parallel (x_2^j \oplus x_\delta) \parallel \cdots \parallel x_n^j.$$

Then the adversary requests the tag for the message v. It is not difficult to see that v and w have identical tags, so the adversary is able to construct the tag for the message w even though he does not know the key K. This attack produces a $(1/2, O(2^{t/2}))$-forger.

It is known that *CBC-MAC* is secure if the underlying encryption satisfies appropriate security properties. That is, if certain plausible but unproved assumptions about the randomness of an encryption scheme are true, then *CBC-MAC* will be secure.

5.5.3 Authenticated Encryption

We have been using encryption to provide secrecy and a MAC to provide data integrity. It is often desirable to combine encryption with a MAC, so that the properties of secrecy and data integrity are achieved simultaneously. The resulting combined process is often called ***authenticated encryption***. There are at least three ways in which we could consider combining encryption with a MAC. In each of these methods, we will use two independent keys, one for the MAC and one for the encryption scheme.

MAC-and-encrypt

Given a message x, compute a tag $z = h_{K_1}(x)$ and a ciphertext $y = e_{K_2}(x)$. The pair (y, z) is transmitted. The receiver would decrypt y, obtaining x, and then verify the correctness of the tag z on x.

MAC-then-encrypt

Here the tag $z = h_{K_1}(x)$ would be computed first. Then the plaintext and tag would both be encrypted, yielding $y = e_{K_2}(x \parallel z)$. The ciphertext y would be transmitted. The receiver will decrypt y, obtaining x and z, and then verify the correctness of the tag z on x.

encrypt-then-MAC

Here the first step is to encrypt x, producing a ciphertext $y = e_{K_2}(x)$. Then a tag is created for the ciphertext y, namely, $z = h_{K_1}(y)$. The pair (y, z) is transmitted. The receiver will first verify the correctness of the tag z on y. Then, provided that the tag is valid, the receiver will decrypt y to obtain x.

Of the three methods presented above, encrypt-then-MAC is usually preferred. A security result due to Bellare and Namprempre says that this method of authenticated encryption is secure provided that that the two component schemes are secure. On the other hand, there exist instantiations of MAC-then-encrypt and MAC-and-encrypt that are insecure, even though the component schemes are secure.

Aside from security considerations, encrypt-then-MAC also has an advantage from the point of view of efficiency. In the case where the transmitted data has been modified, the tag will be invalid and the decryption operation will not be necessary. In contrast, in the cases of MAC-then-encrypt and MAC-and-encrypt, both the decryption and tag verifications are required, even when the data has been modified.

The *CCM mode* of operation provides authenticated encryption using a type of MAC-then-encrypt approach ("CCM" is an abbreviation for "Counter with CBC-MAC"). CCM mode, which is actually a NIST standard, computes a tag using *CBC-MAC*. This is then followed by an encryption in counter mode. Let K be the encryption key and let $x = x_1 \parallel \cdots \parallel x_n$ be the plaintext. As in counter mode, we choose a counter, *ctr*. Then we construct the sequence of counters $T_0, T_1, T_2, \ldots, T_n$, defined as follows:

$$T_i = ctr + i \bmod 2^m$$

for $0 \leq i \leq n$, where m is the block length of the cipher. We encrypt the plaintext blocks $x_1, x_2, \ldots, x_n$ by computing

$$y_i = x_i \oplus e_K(T_i),$$

for all $i \geq 1$. Then we compute $temp = \text{CBC-MAC}(x, K)$ and $y' = T_0 \oplus temp$. The ciphertext consists of the string $y = y_1 \parallel \cdots \parallel y_n \parallel y'$.

To decrypt and verify y, one would first decrypt $y_1 \parallel \cdots \parallel y_n$ using counter mode decryption with the counter sequence $T_1, T_2, \ldots, T_n$, obtaining the plaintext string x. The second step is to compute $\text{CBC-MAC}(x, K)$ and see if it is equal to $y' \oplus T_0$. The ciphertext is rejected if this condition does not hold.

GCM also provides authenticated encryption ("GCM" is an abbreviation for "Galois/Counter mode"). A detailed description of *GCM* is given in NIST Special Publication 800-38D; we give a brief summary of how it works here. See Figure 5.3 for a diagram illustrating Galois/Counter mode.[4]

Encryption is done in counter mode using a 128-bit *AES* key. The initial value

[4] This image or file is a work of a United States Department of Commerce employee, taken or made as part of that person's official duties. As a work of the U.S. federal government, the image is in the public domain.

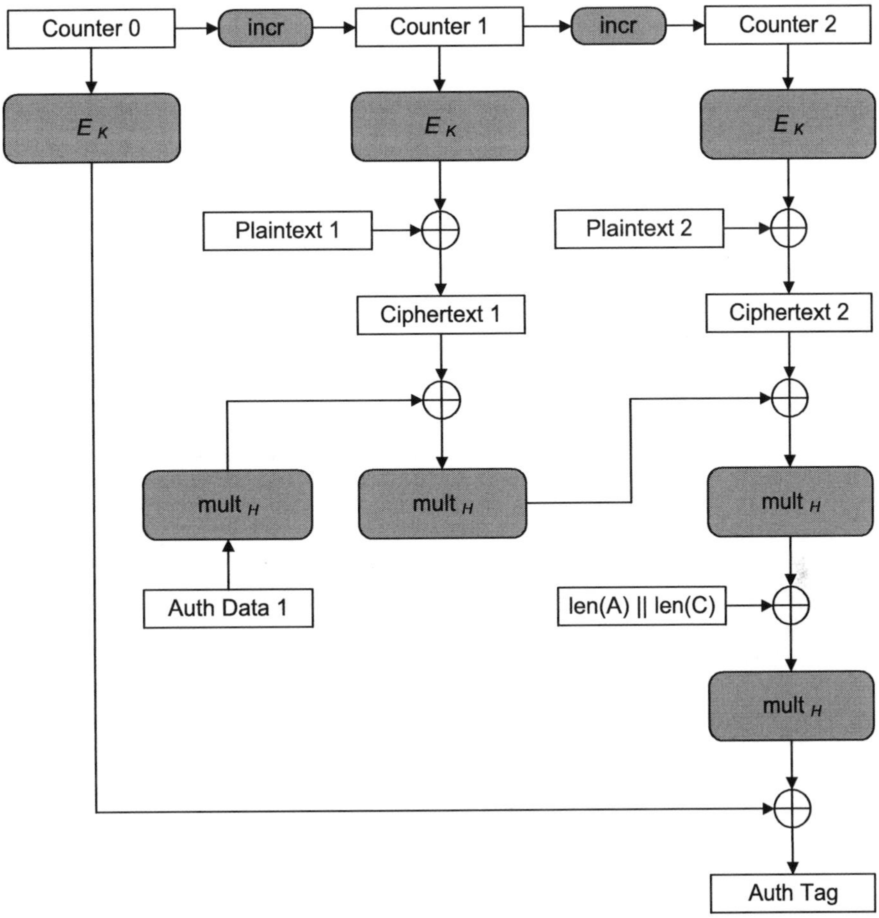

FIGURE 5.3: Galois/Counter mode

of the 128-bit counter (which is denoted by Counter 0) is derived from an IV that is typically 96 bits in length. The IV is transmitted along with the ciphertext, and it should be changed every time a new encryption is performed. Computation of the authentication tag requires performing multiplications by a constant value H in the finite field $\mathbb{F}_{2^{128}}$. The value of H is determined by encrypting Counter 0.

"Auth data 1" is authenticated data that is not encrypted (so it can be transmitted in unencrypted form), but which is incorporated into the authentication tag. Starting with this authenticated data, we successively multiply by H and x-or with a ciphertext block, repeating these two operations until all the ciphertext blocks have been processed. A final x-or is done with a block of data that records the length of the authenticated data as well as the length of the ciphertext, followed by a final multiplication by h and an x-or with the encryption of Counter 0.

5.6 Unconditionally Secure MACs

In this section, we study unconditionally secure MACs, where we assume that the adversary has infinite computing power. However, we will assume that any given key is used to produce only one authentication tag. Also, we will be analyzing the security of these types of codes against known message attacks.

For $Q \in \{0,1\}$, we define the ***deception probability*** Pd_Q to be the probability ϵ that an adversary can create a successful forgery after observing Q valid message-tag pairs. The attack when $Q = 0$ is termed ***impersonation*** and the attack when $Q = 1$ is termed ***substitution***. In an impersonation attack, the adversary (Oscar) creates a message and a tag, hoping that the tag is valid under the key that is being used by Alice and Bob (which is not known to Oscar). In a substitution attack, Oscar sees one valid message-tag pair, intercepts it, and then replaces it with another message-tag pair that he hopes is valid.

For simplicity, we assume that K is chosen uniformly at random from $\mathcal{K}$. In an impersonation attack, Oscar's success probability ϵ may depend on the particular message-tag pair (x, y) that he observes. So there could be different probabilities $\epsilon(x, y)$ for different message-tag pairs (x, y). There are various ways in which we could define Pd_1 as a function of these values $\epsilon(x, y)$. For the purposes of our discussion, we will define Pd_1 to be the maximum of the relevant values $\epsilon(x, y)$. Thus, when we prove an upper bound $Pd_1 \leq \epsilon$, we are saying that Oscar's success probability is at most ϵ, regardless of the message-tag pair that he observes prior to making his substitution.

We illustrate the above concepts by considering a small example of an unconditionally secure MAC. As usual, each function $h_K : \mathcal{X} \rightarrow \mathcal{Y}$.

Example 5.1 Suppose

$$\mathcal{X} = \mathcal{Y} = \mathbb{Z}_3$$

and

$$\mathcal{K} = \mathbb{Z}_3 \times \mathbb{Z}_3.$$

For each $K = (a, b) \in \mathcal{K}$ and each $x \in \mathcal{X}$, define

$$h_{(a,b)}(x) = ax + b \bmod 3,$$

and then define

$$\mathcal{H} = \{h_{(a,b)} : (a, b) \in \mathbb{Z}_3 \times \mathbb{Z}_3\}.$$

Each of the nine keys will be used with probability $1/9$.

It will be useful to study the ***authentication matrix*** of the hash family $(\mathcal{X}, \mathcal{Y}, \mathcal{K}, \mathcal{H})$, which tabulates all the values $h_{(a,b)}(x)$ as follows. For each key $(a, b) \in \mathcal{K}$ and for each $x \in \mathcal{X}$, place the authentication tag $h_{(a,b)}(x)$ in row (a, b) and column x of a $|\mathcal{K}| \times |\mathcal{X}|$ matrix, say M. The matrix M is presented in Table 5.2.

Let's first consider an impersonation attack. Oscar can pick any message x, and

TABLE 5.2: An authentication matrix

key	0	1	2
(0,0)	0	0	0
(0,1)	1	1	1
(0,2)	2	2	2
(1,0)	0	1	2
(1,1)	1	2	0
(1,2)	2	0	1
(2,0)	0	2	1
(2,1)	1	0	2
(2,2)	2	1	0

then attempt to guess the "correct" authentication tag. Denote by K_0 the actual key being used (which is unknown to Oscar). Oscar will succeed in creating a forgery if he guesses the tag $y_0 = h_{K_0}(x)$. However, for any $x \in \mathcal{X}$ and $y \in \mathcal{Y}$, it is easy to verify that there are exactly three (out of nine) keys $K \in \mathcal{K}$ such that $h_K(x) = y$. (In other words, each symbol occurs three times in each column of the authentication matrix.) Thus, any message-tag pair (x, y) will be a valid pair with probability $1/3$. Hence, it follows that $Pd_0 = 1/3$.

Substitution is a bit more complicated to analyze. As a specific case, suppose Oscar observes the valid pair $(0, 0)$ being transmitted from Alice to Bob. This gives Oscar some information about the key: he knows that

$$K_0 \in \{(0,0), (1,0), (2,0)\}.$$

Now suppose Oscar outputs the message-tag pair $(1, 1)$ as a (possible) forgery. The pair $(1, 1)$ is a forgery if and only if $K_0 = (1, 0)$. The (conditional) probability that K_0 is the key, given that $(0, 0)$ is a valid pair, is $1/3$, since the key is known to be in the set $\{(0,0), (1,0), (2,0)\}$.

A similar analysis can be done for any valid pair (x, y) and for any substitution (x', y') (where $x' \neq x$) that Oscar outputs as his (possible) forgery. In general, knowledge of any valid pair (x, y) restricts the key to three possibilities. Then, for each choice of a message-tag pair (x', y') (where $x' \neq x$), it can be verified that there is one key (out of the three possible keys) under which y' is the correct authentication tag for x'. Hence, it follows that $Pd_1 = 1/3$. □

We now discuss how to compute the deception probabilities for an arbitrary message authentication code by examining its authentication matrix. (Recall that we are assuming that keys are chosen uniformly at random. This makes the analysis simpler than it would otherwise be.)

First, we consider Pd_0. As above, let K_0 denote the key chosen by Alice and Bob. For $x \in \mathcal{X}$ and $y \in \mathcal{Y}$, define **payoff**(x, y) to be the probability that the message-tag pair (x, y) is valid. It is not difficult to see that

$$\begin{aligned}\textbf{payoff}(x,y) &= \textbf{Pr}[y = h_{K_0}(x)] \\ &= \frac{|\{K \in \mathcal{K} : h_K(x) = y\}|}{|\mathcal{K}|}.\end{aligned}$$

That is, $\textbf{payoff}(x, y)$ is computed by counting the number of rows of the authentication matrix that have entry y in column x, and dividing the result by the number of possible keys.

In order to maximize his chance of success, Oscar will choose a message-tag pair (x, y) such that $\textbf{payoff}(x, y)$ is a maximum. Hence, we have the following formula:

$$Pd_0 = \max\{\textbf{payoff}(x,y) : x \in \mathcal{X}, y \in \mathcal{Y}\}. \tag{5.1}$$

Now, we turn our attention to substitution. As stated earlier, we analyze Pd_1 in the known message setting, where Oscar observes a message-tag pair (x, y) in the communication channel and replaces it with another pair (x', y'), where $x \neq x'$.

Suppose we fix $x \in \mathcal{X}$ and $y \in \mathcal{Y}$ such that (x, y) is a valid pair. This means that there is at least one key K such that $h_K(x) = y$. Now let $x' \in \mathcal{X}$, where $x' \neq x$. Define $\textbf{payoff}(x', y'; x, y)$ to be the (conditional) probability that (x', y') is a valid pair, given that (x, y) is a valid pair. As before, let K_0 denote the key chosen by Alice and Bob. Then we can compute the following:

$$\begin{aligned}\textbf{payoff}(x',y';x,y) &= \textbf{Pr}[y' = h_{K_0}(x') | y = h_{K_0}(x)] \\ &= \frac{\textbf{Pr}[y' = h_{K_0}(x') \wedge y = h_{K_0}(x)]}{\textbf{Pr}[y = h_{K_0}(x)]} \\ &= \frac{|\{K \in \mathcal{K} : h_K(x') = y', h_K(x) = y\}|}{|\{K \in \mathcal{K} : h_K(x) = y\}|}.\end{aligned}$$

The numerator of this fraction is the number of rows of the authentication matrix that have the value y in column x, and also have the value y' in column x'; the denominator is the number of rows that have the value y in column x. Note that the denominator is non-zero because we are assuming that (x, y) is a valid pair under at least one key.

Suppose we define

$$\mathcal{V} = \{(x,y) : |\{K \in \mathcal{K} : h_K(x) = y\}| \geq 1\}.$$

Observe that $\mathcal{V}$ is just the set of all message-tag pairs (x, y) that are valid pairs under at least one key. This is the set of all the messages that Oscar could possibly observe in the channel. Then the following formula can be used to compute Pd_1:

$$Pd_1 = \max_{(x,y) \in \mathcal{V}} \left\{ \max_{(x',y'), x' \neq x} \{\textbf{payoff}(x',y';x,y)\} \right\}. \tag{5.2}$$

Some explanation would be helpful, as this is a complicated formula. The quantity **payoff**$(x', y'; x, y)$ denotes the probability that a substitution of (x, y) with (x', y') will be accepted. After observing a message-tag pair (x, y), Oscar will choose (x', y') to maximize **payoff**$(x', y'; x, y)$. Then, as we have discussed at the beginning of this section, Pd_1 is defined to be the maximum success probability over all possible observed message-tag pairs (x, y). (So, no matter which message-tag pair Oscar observes, he cannot succeed in his deception with probability greater than Pd_1.)

Referring again to Example 5.1, we have that **payoff**$(x, y) = 1/3$ for all x, y and **payoff**$(x', y'; x, y) = 1/3$ for all x, y, x', y' (where $x \neq x'$).

5.6.1 Strongly Universal Hash Families

Strongly universal hash families are used in several areas of cryptography. We begin with a definition of these important objects.

Definition 5.2: Suppose that $(\mathcal{X}, \mathcal{Y}, \mathcal{K}, \mathcal{H})$ is an (N, M) hash family. This hash family is ***strongly universal*** provided that the following condition is satisfied for every $x, x' \in \mathcal{X}$ such that $x \neq x'$, and for every $y, y' \in \mathcal{Y}$:

$$|\{K \in \mathcal{K} : h_K(x) = y, h_K(x') = y'\}| = \frac{|\mathcal{K}|}{M^2}.$$

As an example, the reader can verify that the hash family in Example 5.1 is a strongly universal $(3, 3)$-hash family.

Here is a bit of intuition to motivate Definition 5.2. Suppose we fix x and x', where $x' \neq x$. There are M^2 possible choices for the ordered pair (y, y'). The definition is saying that the number of hash functions in the family $\mathcal{H}$ that map x to y and also map x' to y' is independent of the choice of y and y'. Since there are $|\mathcal{K}|$ hash functions in total, this number must equal $|\mathcal{K}|/M^2$.

Strongly universal hash families immediately yield authentication codes in which Pd_0 and Pd_1 can easily be computed. We prove a theorem on the values of these deception probabilities after stating and proving a simple lemma about strongly universal hash families.

LEMMA 5.10 *Suppose that $(\mathcal{X}, \mathcal{Y}, \mathcal{K}, \mathcal{H})$ is a strongly universal (N, M)-hash family. Then*

$$|\{K \in \mathcal{K} : h_K(x) = y\}| = \frac{|\mathcal{K}|}{M},$$

for every $x \in \mathcal{X}$ and for every $y \in \mathcal{Y}$.

PROOF Let $x, x' \in \mathcal{X}$ and $y \in \mathcal{Y}$ be fixed, where $x \neq x'$. Then we have the following:

$$\begin{aligned}|\{K \in \mathcal{K} : h_K(x) = y\}| &= \sum_{y' \in \mathcal{Y}} |\{K \in \mathcal{K} : h_K(x) = y, h_K(x') = y'\}| \\ &= \sum_{y' \in \mathcal{Y}} \frac{|\mathcal{K}|}{M^2} \\ &= \frac{|\mathcal{K}|}{M}.\end{aligned}$$

∎

THEOREM 5.11 *Suppose that $(\mathcal{X}, \mathcal{Y}, \mathcal{K}, \mathcal{H})$ is a strongly universal (N, M)-hash family. Then $(\mathcal{X}, \mathcal{Y}, \mathcal{K}, \mathcal{H})$ is an authentication code with $Pd_0 = Pd_1 = 1/M$.*

PROOF We proved in Lemma 5.10 that

$$|\{K \in \mathcal{K} : h_K(x) = y\}| = \frac{|\mathcal{K}|}{M},$$

for every $x \in \mathcal{X}$ and for every $y \in \mathcal{Y}$. Therefore $\textbf{payoff}(x, y) = 1/M$ for every $x \in \mathcal{X}, y \in \mathcal{Y}$, and hence $Pd_0 = 1/M$.

Now let $x, x' \in \mathcal{X}$ such that $x \neq x'$, and let $y, y' \in \mathcal{Y}$. Note that $\mathcal{V} = \{(x, y) : x \in \mathcal{X}, y \in \mathcal{Y}\}$. We have that

$$\begin{aligned}\textbf{payoff}(x', y'; x, y) &= \frac{|\{K \in \mathcal{K} : h_K(x') = y', h_K(x) = y\}|}{|\{K \in \mathcal{K} : h_K(x) = y\}|} \\ &= \frac{|\mathcal{K}|/M^2}{|\mathcal{K}|/M} \\ &= \frac{1}{M}.\end{aligned}$$

Therefore $Pd_1 = 1/M$. ∎

We now give a construction of strongly universal hash families. This construction generalizes Example 5.1.

THEOREM 5.12 *Let p be prime. For $a, b \in \mathbb{Z}_p$, define $f_{(a,b)} : \mathbb{Z}_p \to \mathbb{Z}_p$ by the rule*

$$f_{(a,b)}(x) = ax + b \bmod p.$$

Then $(\mathbb{Z}_p, \mathbb{Z}_p, \mathbb{Z}_p \times \mathbb{Z}_p, \{f_{(a,b)} : a, b \in \mathbb{Z}_p\})$ is a strongly universal (p, p)-hash family.

PROOF Suppose that $x, x', y, y' \in \mathbb{Z}_p$, where $x \neq x'$. We will show that there is a unique key $(a, b) \in \mathbb{Z}_p \times \mathbb{Z}_p$ such that $ax + b \equiv y \pmod p$ and $ax' + b \equiv y' \pmod p$. This is not difficult, as (a, b) is the solution of a system of two linear equations in two unknowns over $\mathbb{Z}_p$. Specifically,

$$\begin{aligned}a &= (y' - y)(x' - x)^{-1} \bmod p, \quad \text{and} \\ b &= y - x(y' - y)(x' - x)^{-1} \bmod p.\end{aligned}$$

(Note that $(x' - x)^{-1} \bmod p$ exists because $x \not\equiv x' \pmod p$ and p is prime.) ∎

5.6.2 Optimality of Deception Probabilities

In this section, we prove some lower bounds on deception probabilities of unconditionally secure MACs, which show that the authentication codes derived from strongly universal hash families have minimum possible deception probabilities.

Suppose $(\mathcal{X}, \mathcal{Y}, \mathcal{K}, \mathcal{H})$ is an (N, M)-hash family. Suppose we fix a message $x \in \mathcal{X}$. Then we can compute as follows:

$$\begin{aligned}\sum_{y \in \mathcal{Y}} \mathbf{payoff}(x,y) &= \sum_{y \in \mathcal{Y}} \frac{|\{K \in \mathcal{K} : h_K(x) = y\}|}{|\mathcal{K}|} \\ &= \frac{|\mathcal{K}|}{|\mathcal{K}|} \\ &= 1.\end{aligned}$$

Hence, for every $x \in \mathcal{X}$, there exists an authentication tag y (depending on x), such that

$$\mathbf{payoff}(x,y) \geq \frac{1}{M}.$$

The following theorem is an easy consequence of the above computations.

THEOREM 5.13 *Suppose $(\mathcal{X}, \mathcal{Y}, \mathcal{K}, \mathcal{H})$ is an (N, M)-hash family. Then $Pd_0 \geq 1/M$. Further, $Pd_0 = 1/M$ if and only if*

$$|\{K \in \mathcal{K} : h_K(x) = y\}| = \frac{|\mathcal{K}|}{M} \tag{5.3}$$

for every $x \in \mathcal{X}, y \in \mathcal{Y}$.

Now, we turn our attention to substitution. Suppose that we fix $x, x' \in \mathcal{X}$ and $y \in \mathcal{Y}$, where $x \neq x'$ and $(x,y) \in \mathcal{V}$. We have the following:

$$\begin{aligned}\sum_{y' \in \mathcal{Y}} \mathbf{payoff}(x', y'; x, y) &= \sum_{y' \in \mathcal{Y}} \frac{|\{K \in \mathcal{K} : h_K(x') = y', h_K(x) = y\}|}{|\{K \in \mathcal{K} : h_K(x) = y\}|} \\ &= \frac{|\{K \in \mathcal{K} : h_K(x) = y\}|}{|\{K \in \mathcal{K} : h_K(x) = y\}|} \\ &= 1.\end{aligned}$$

Hence, for each $(x,y) \in \mathcal{V}$ and for each x' such that $x' \neq x$, there exists an authentication tag y' such that

$$\mathbf{payoff}(x', y'; x, y) \geq \frac{1}{M}.$$

We have proven the following theorem.

THEOREM 5.14 *Suppose $(\mathcal{X}, \mathcal{Y}, \mathcal{K}, \mathcal{H})$ is an (N, M)-hash family. Then $Pd_1 \geq 1/M$.*

With a bit more work, we can determine necessary and sufficient conditions such that $Pd_1 = 1/M$.

THEOREM 5.15 *Suppose $(\mathcal{X}, \mathcal{Y}, \mathcal{K}, \mathcal{H})$ is an (N, M)-hash family. Then $Pd_1 = 1/M$ if and only if the hash family is strongly universal.*

PROOF We proved already in Theorem 5.11 that $Pd_1 = 1/M$ if the hash family is strongly universal. We need to prove the converse now; so, we assume that $Pd_1 = 1/M$.

We will show first that $\mathcal{V} = \mathcal{X} \times \mathcal{Y}$. Let $(x', y') \in \mathcal{X} \times \mathcal{Y}$; we will show that $(x', y') \in \mathcal{V}$. Let $x \in \mathcal{X}$, $x \neq x'$. Choose any $y \in \mathcal{Y}$ such that $(x, y) \in \mathcal{V}$. From the discussion preceding Theorem 5.14, it is clear that

$$\frac{|\{K \in \mathcal{K} : h_K(x') = y', h_K(x) = y\}|}{|\{K \in \mathcal{K} : h_K(x) = y\}|} = \frac{1}{M} \tag{5.4}$$

for every $x, x' \in \mathcal{X}$, $x' \neq x$, $y, y' \in \mathcal{Y}$ such that $(x, y) \in \mathcal{V}$. Therefore

$$|\{K \in \mathcal{K} : h_K(x') = y', h_K(x) = y\}| > 0,$$

and hence

$$|\{K \in \mathcal{K} : h_K(x') = y'\}| > 0.$$

This proves that $(x', y') \in \mathcal{V}$, and hence $\mathcal{V} = \mathcal{X} \times \mathcal{Y}$.

Now, let's look again at (5.4). Let $x, x' \in \mathcal{X}$, $x \neq x'$, and let $y, y' \in \mathcal{Y}$. We know that $(x, y) \in \mathcal{V}$ and $(x', y') \in \mathcal{V}$, so we can interchange the roles of (x, y) and (x', y') in (5.4). This yields

$$|\{K \in \mathcal{K} : h_K(x) = y\}| = |\{K \in \mathcal{K} : h_K(x') = y'\}|$$

for all such x, x', y, y'. Hence, the quantity

$$|\{K \in \mathcal{K} : h_K(x) = y\}|$$

is a constant. (In other words, the number of occurrences of any symbol y in any column x of the authentication matrix x is a constant.) Now, we can return one last time to (5.4), and it follows that the quantity

$$|\{K \in \mathcal{K} : h_K(x') = y', h_K(x) = y\}|$$

is also a constant. Therefore the hash family is strongly universal. ∎

The following corollary establishes that $Pd_0 = 1/M$ whenever $Pd_1 = 1/M$.

COROLLARY 5.16 *Suppose $(\mathcal{X}, \mathcal{Y}, \mathcal{K}, \mathcal{H})$ is an (N, M)-hash family such that $Pd_1 = 1/M$. Then $Pd_0 = 1/M$.*

PROOF Under the stated hypotheses, Theorem 5.15 says that $(\mathcal{X}, \mathcal{Y}, \mathcal{K}, \mathcal{H})$ is strongly universal. Then $Pd_0 = 1/M$ from Theorem 5.11. ∎

5.7 Notes and References

Concepts such as preimage resistance and collision resistance have been discussed for some time; see [166] for further details.

The random oracle model was introduced by Bellare and Rogaway in [22]; the analyses in Section 5.2.2 are based on Stinson [192].

The material from Section 5.3 is based on Damgård [64]. Similar methods were discovered by Merkle [137].

Rivest's *MD4* and *MD5* hashing algorithms are described in [170] and [171], respectively. They are of course obsolete, but they are still of historical interest because of their influence on *SHA-1* in particular. The state-of-the-art for finding collisions in MD5 is described in [204].

FIPS publication 180-4 [151] includes descriptions of *SHA-1* as well as the *SHA-2* family of hash functions. The *SHA-3* hash functions are presented in FIPS publication 202 [152].

Sponge functions were first described in [26]. The *Keccak* submission for *SHA-3* is found in the document [27].

The computations that were used to find the *SHA-1* collision are discussed in Stevens, Bursztein, Karpman, Albertini, and Markov [191].

Security proofs for several types of MACs have been published. For a detailed examination of the security of *HMAC*, see Koblitz and Menezes [113]. Bellare, Kilian, and Rogaway [15] showed that *CBC-MAC* is secure.

A modification of *CBC-MAC* known as *CMAC* is presented in the NIST special publication 800-38B [77]. *CMAC* is closely based on *OMAC*, which is due to Iwata and Kurosawa [99]. *HMAC* was adopted as a standard; see FIPS publication 198-1 [150].

Bellare and Namprempre [16] study the security of methods of composing authentication and encryption. CCM mode is described in NIST special publication 800-38C [78] and GCM mode can be found in NIST special publication 800-38D [79].

Unconditionally secure authentication codes were invented in 1974 by Gilbert, MacWilliams, and Sloane [87]. Much of the theory of unconditionally secure authentication codes was developed by Simmons, who proved many fundamental results in the area; Simmons [182] is a good survey.

Universal hash families were introduced by Carter and Wegman [55, 199]. Their paper [199] was the first to apply strongly universal hash families to authentication. We also note that universal hash families are used in the construction of efficient computationally secure MACs; one such MAC is called *UMAC*, which is described in Black *et al.* [32].

Exercises

5.1 Define a toy hash function $h : (\mathbb{Z}_2)^7 \rightarrow (\mathbb{Z}_2)^4$ by the rule $h(x) = xA$ where all operations are modulo 2 and

$$A = \begin{pmatrix} 1 & 0 & 0 & 0 \\ 1 & 1 & 0 & 0 \\ 1 & 1 & 1 & 0 \\ 1 & 1 & 1 & 1 \\ 0 & 1 & 1 & 1 \\ 0 & 0 & 1 & 1 \\ 0 & 0 & 0 & 1 \end{pmatrix}.$$

Find all preimages of $(0, 1, 0, 1)$.

5.2 Suppose $h : \mathcal{X} \rightarrow \mathcal{Y}$ is an (N, M)-hash function. For any $y \in \mathcal{Y}$, let

$$h^{-1}(y) = \{x : h(x) = y\}$$

and denote $s_y = |h^{-1}(y)|$. Define

$$S = |\{\{x_1, x_2\} : h(x_1) = h(x_2)\}|.$$

Note that S counts the number of unordered pairs in $\mathcal{X}$ that collide under h.

(a) Prove that

$$\sum_{y \in \mathcal{Y}} s_y = N,$$

so the mean of the s_y's is '

$$\overline{s} = \frac{N}{M}.$$

(b) Prove that

$$S = \sum_{y \in \mathcal{Y}} \binom{s_y}{2} = \frac{1}{2} \sum_{y \in \mathcal{Y}} {s_y}^2 - \frac{N}{2}.$$

(c) Prove that

$$\sum_{y \in \mathcal{Y}} (s_y - \overline{s})^2 = 2S + N - \frac{N^2}{M}.$$

(d) Using the result proved in part (c), prove that

$$S \geq \frac{1}{2} \left(\frac{N^2}{M} - N \right).$$

Further, show that equality is attained if and only if

$$s_y = \frac{N}{M}$$

for every $y \in \mathcal{Y}$.

5.3 As in Exercise 5.2, suppose $h : \mathcal{X} \rightarrow \mathcal{Y}$ is an (N, M)-hash function, and let

$$h^{-1}(y) = \{x : h(x) = y\}$$

for any $y \in \mathcal{Y}$. Let ϵ denote the probability that $h(x_1) = h(x_2)$, where x_1 and x_2 are random (not necessarily distinct) elements of $\mathcal{X}$. Prove that

$$\epsilon \geq \frac{1}{M},$$

with equality if and only if

$$|h^{-1}(y)| = \frac{N}{M}$$

for every $y \in \mathcal{Y}$.

5.4 Suppose that $h : \mathcal{X} \rightarrow \mathcal{Y}$ is an (N, M)-hash function, let

$$h^{-1}(y) = \{x : h(x) = y\}$$

and let $s_y = |h^{-1}(y)|$ for any $y \in \mathcal{Y}$. Suppose that we try to solve **Preimage** for the function h, using Algorithm 5.1, assuming that we have only oracle access for h. For a given $y \in \mathcal{Y}$, suppose that $\mathcal{X}_0$ is chosen to be a random subset of $\mathcal{X}$ having cardinality q.

(a) Prove that the success probability of Algorithm 5.1, given y, is

$$1 - \frac{\binom{N-s_y}{q}}{\binom{N}{q}}.$$

(b) Prove that the average success probabilty of Algorithm 5.1 (over all $y \in \mathcal{Y}$) is

$$1 - \frac{1}{M} \sum_{y \in \mathcal{Y}} \frac{\binom{N-s_y}{q}}{\binom{N}{q}}.$$

(c) In the case $q = 1$, show that the success probability in part (b) is $1/M$.

5.5 Suppose that $h : \mathcal{X} \rightarrow \mathcal{Y}$ is an (N, M)-hash function, let

$$h^{-1}(y) = \{x : h(x) = y\}$$

and let $s_y = |h^{-1}(y)|$ for any $y \in \mathcal{Y}$. Suppose that we try to solve **Second Preimage** for the function h, using Algorithm 5.2, assuming that we have only oracle access for h. For a given $x \in \mathcal{Y}$, suppose that $\mathcal{X}_0$ is chosen to be a random subset of $\mathcal{X} \backslash \{x\}$ having cardinality $q - 1$.

(a) Prove that the success probability of Algorithm 5.2, given x, is

$$1 - \frac{\binom{N-s_y}{q-1}}{\binom{N-1}{q-1}}.$$

(b) Prove that the average success probabilty of Algorithm 5.2 (over all $x \in \mathcal{X}$) is

$$1 - \frac{1}{N} \sum_{y \in \mathcal{Y}} \frac{s_y \binom{N-s_y}{q-1}}{\binom{N-1}{q-1}}.$$

(c) In the case $q = 2$, show that the success probability in part (b) is

$$\frac{\sum_{y \in \mathcal{Y}} {s_y}^2}{N(N-1)} - \frac{1}{N-1}.$$

5.6 (This exercise is based on an example from the *Handbook of Applied Cryptography* by A.J. Menezes, P.C. Van Oorschot, and S.A. Vanstone.) Suppose g is a collision resistant hash function that takes an arbitrary bitstring as input and produces an n-bit message digest. Define a hash function h as follows:

$$h(x) = \begin{cases} 0 \parallel x & \text{if } x \text{ is a bitstring of length } n \\ 1 \parallel g(x) & \text{otherwise.} \end{cases}$$

(a) Prove that h is collision resistant.

(b) Prove that h is not preimage resistant. More precisely, show that preimages (for the function h) can easily be found for half of the possible message digests.

5.7 If we define a hash function (or compression function) h that will hash an n-bit binary string to an m-bit binary string, we can view h as a function from $\mathbb{Z}_{2^n}$ to $\mathbb{Z}_{2^m}$. It is tempting to define h using integer operations modulo 2^m. We show in this exercise that some simple constructions of this type are insecure and should therefore be avoided.

(a) Suppose that $n = m > 1$ and $h : \mathbb{Z}_{2^m} \rightarrow \mathbb{Z}_{2^m}$ is defined as

$$h(x) = x^2 + ax + b \bmod 2^m.$$

Prove that it is (usually) easy to solve **Second Preimage** for any $x \in \mathbb{Z}_{2^m}$ without having to solve a quadratic equation.

HINT Show that it is possible to find a linear function $g(x)$ such that $h(g(x)) = h(x)$ for all x. This solves **Second Preimage** for any x such that $g(x) \neq x$.

(b) Suppose that $n > m$ and $h : \mathbb{Z}_{2^n} \rightarrow \mathbb{Z}_{2^m}$ is defined to be a polynomial of degree d:

$$h(x) = \sum_{i=0}^{d} a_i x^i \bmod 2^m,$$

where $a_i \in \mathbb{Z}$ for $0 \leq i \leq d$. Prove that it is easy to solve **Second Preimage** for any $x \in \mathbb{Z}_{2^n}$ without having to solve a polynomial equation.

HINT Make use of the fact that $h(x)$ is defined using reduction modulo 2^m, but the domain of h is $\mathbb{Z}_{2^n}$, where $n > m$.

5.8 Suppose that $f : \{0,1\}^m \rightarrow \{0,1\}^m$ is a preimage resistant bijection. Define $h : \{0,1\}^{2m} \rightarrow \{0,1\}^m$ as follows. Given $x \in \{0,1\}^{2m}$, write

$$x = x' \parallel x''$$

where $x', x'' \in \{0,1\}^m$. Then define

$$h(x) = f(x' \oplus x'').$$

Prove that h is not second preimage resistant.

5.9 For $M = 365$ and $15 \leq q \leq 30$, compare the exact value of ϵ given by the formula in the statement of Theorem 5.4 with the estimate for ϵ derived after the proof of that theorem.

5.10 Suppose that messages are designated as "safe" or "dangerous" and an adversary is trying to find a collision of one safe and one dangerous message under a hash function h. That is, the adversary is trying to find a safe message x and a dangerous message x' such that $h(x) = h(x')$. An obvious attack would be to choose a set $\mathcal{X}_0$ of Q safe messages and a set $\mathcal{X}_0'$ of Q' dangerous messages, and test the QQ' resulting ordered pairs $(x, x') \in \mathcal{X}_0 \times \mathcal{X}_0'$ to see if a collision occurs. We analyze the success of this approach in the random oracle model, assuming that there are M possible message digests.

(a) For a fixed value $x \in \mathcal{X}_0$, determine an upper bound on the probability that $h(x) \neq h(x')$ for all $x' \in \mathcal{X}_0'$.

(b) Using the result from (a), determine an upper bound on the probability that $h(x) \neq h(x')$ for all $x \in \mathcal{X}_0$ and all $x' \in \mathcal{X}_0'$.

(c) Show that there is a 50% probability of finding at least one collision using this method if $QQ' \approx cM$, for a suitable positive constant c.

5.11 Suppose $h : \mathcal{X} \rightarrow \mathcal{Y}$ is a hash function where $|\mathcal{X}|$ and $|\mathcal{Y}|$ are finite and $|\mathcal{X}| \geq 2|\mathcal{Y}|$. Suppose that h is a ***balanced hash function*** (i.e.,

$$|h^{-1}(y)| = \frac{|\mathcal{X}|}{|\mathcal{Y}|}$$

for all $y \in \mathcal{Y}$). Finally, suppose ORACLE-PREIMAGE is an (ϵ, Q)-algorithm for **Preimage**, for the fixed hash function h. Prove that COLLISION-TO-PREIMAGE is an $(\epsilon/2, Q+1)$-algorithm for **Collision**, for the fixed hash function h.

5.12 Suppose $h_1 : \{0,1\}^{2m} \rightarrow \{0,1\}^m$ is a collision resistant hash function.

(a) Define $h_2 : \{0,1\}^{4m} \rightarrow \{0,1\}^m$ as follows:

1. Write $x \in \{0,1\}^{4m}$ as $x = x_1 \parallel x_2$, where $x_1, x_2 \in \{0,1\}^{2m}$.
2. Define $h_2(x) = h_1(h_1(x_1) \parallel h_1(x_2))$.

Prove that h_2 is collision resistant (i.e., given a collision for h_2, show how to find a collision for h_1).

(b) For an integer $i \geq 2$, define a hash function $h_i : \{0,1\}^{2^i m} \rightarrow \{0,1\}^m$ recursively from h_{i-1}, as follows:

1. Write $x \in \{0,1\}^{2^i m}$ as $x = x_1 \parallel x_2$, where $x_1, x_2 \in \{0,1\}^{2^{i-1} m}$.
2. Define $h_i(x) = h_1(h_{i-1}(x_1) \parallel h_{i-1}(x_2))$.

Prove that h_i is collision resistant.

5.13 In this exercise, we consider a simplified version of the Merkle-Damgård construction. Suppose

$$\mathbf{compress} : \{0,1\}^{m+t} \rightarrow \{0,1\}^m,$$

where $t \geq 1$, and suppose that

$$x = x_1 \parallel x_2 \parallel \cdots \parallel x_k,$$

where

$$|x_1| = |x_2| = \cdots = |x_k| = t.$$

We study the following iterated hash function:

Algorithm 5.8: SIMPLIFIED MERKLE-DAMGÅRD (x, k, t)

external compress
$z_1 \leftarrow 0^m \parallel x_1$
$g_1 \leftarrow \mathbf{compress}(z_1)$
for $i \leftarrow 1$ **to** $k - 1$
  do $\begin{cases} z_{i+1} \leftarrow g_i \parallel x_{i+1} \\ g_{i+1} \leftarrow \mathbf{compress}(z_{i+1}) \end{cases}$
$h(x) \leftarrow g_k$
return $(h(x))$

Suppose that **compress** is collision resistant, and suppose further that **compress** is ***zero preimage resistant***, which means that it is hard to find $z \in \{0,1\}^{m+t}$ such that $\mathbf{compress}(z) = 0^m$. Under these assumptions, prove that h is collision resistant.

5.14 Message authentication codes are often constructed using block ciphers in CBC mode. Here we consider the construction of a message authentication code using a block cipher in CFB mode. Given a sequence of plaintext blocks,

$x_1, \dots, x_n$, suppose we define the initialization vector IV to be x_1. Then encrypt the sequence $x_2, \dots, x_n$ using key K in CFB mode, obtaining the ciphertext sequence $y_1, \dots, y_{n-1}$ (note that there are only $n-1$ ciphertext blocks). Finally, define the MAC to be $e_K(y_{n-1})$. Prove that this MAC actually turns out to be identical to CBC-MAC, as presented in Section 5.5.2.

5.15 Suppose that $(\mathcal{P}, \mathcal{C}, \mathcal{K}, \mathcal{E}, \mathcal{D})$ is a cryptosystem with $\mathcal{P} = \mathcal{C} = \{0,1\}^m$. Let $n \geq 2$ be a fixed integer, and define a hash family $(\mathcal{X}, \mathcal{Y}, \mathcal{K}, \mathcal{H})$, where $\mathcal{X} = (\{0,1\}^m)^n$ and $\mathcal{Y} = \{0,1\}^m$, as follows:

$$h_K(x_1, \dots, x_n) = e_K(x_1) \oplus \cdots \oplus e_K(x_n).$$

Suppose that $(x_1, \dots, x_n)$ is an arbitrary message. Show how an adversary can then determine $h_K(x_1, \dots, x_n) = e_K(x_1)$ by using at most one oracle query. (This is called a ***selective forgery***, because a specific message is given to the adversary and the adversary is then required to find the tag for the given message.)

HINT The proof is divided into three mutually exclusive cases as follows:

case 1 In this case, we assume that not all of the x_i's are identical. Here, one oracle query suffices.

case 2 In this case, we assume n is even and $x_1 = \cdots = x_n$. Here, no oracle queries are required.

case 3 In this case, we assume $n \geq 3$ is odd and $x_1 = \cdots = x_n$. Here, one oracle query suffices.

5.16 (a) Suppose that the hash family $(\mathcal{X}, \mathcal{Y}, \mathcal{K}, \mathcal{H})$ is a secure MAC algorithm. The tag for a message $x \in \mathcal{X}$ is $h_K(x)$. Suppose we instead computed the tag to be $x \parallel h_K(x)$. Would the resulting MAC algorithm still be considered secure? Explain.

(b) Discuss why the general strategy of MAC-and-encrypt should be avoided.

HINT Consider modifying a secure MAC algorithm as described in part (a) and examine the impact of this change in the context of MAC-and-encrypt.

5.17 Suppose that $(\mathcal{X}, \mathcal{Y}, \mathcal{K}, \mathcal{H})$ is a strongly universal (N, M)-hash family.

(a) If $|\mathcal{K}| = M^2$, show that there exists a $(1,2)$-forger for this hash family (i.e., $Pd_2 = 1$).

(b) (This generalizes the result proven in part (a).) Denote $\lambda = |\mathcal{K}|/M^2$. Prove there exists a $(1/\lambda, 2)$-forger for this hash family (i.e., $Pd_2 \geq 1/\lambda$).

5.18 Compute Pd_0 and Pd_1 for the following authentication code, represented in matrix form:

key	1	2	3	4
1	1	1	2	3
2	1	2	3	1
3	2	1	3	1
4	2	3	1	2
5	3	2	1	3
6	3	3	2	1

5.19 Let p be an odd prime. For $a, b \in \mathbb{Z}_p$, define $f_{(a,b)} : \mathbb{Z}_p \to \mathbb{Z}_p$ by the rule

$$f_{(a,b)}(x) = (x + a)^2 + b \bmod p.$$

Prove that $(\mathbb{Z}_p, \mathbb{Z}_p, \mathbb{Z}_p \times \mathbb{Z}_p, \{f_{(a,b)} : a, b \in \mathbb{Z}_p\})$ is a strongly universal (p, p)-hash family.

5.20 Let $k \geq 1$ be an integer. An (N, M) hash family, $(\mathcal{X}, \mathcal{Y}, \mathcal{K}, \mathcal{H})$, is ***strongly** k-**universal*** provided that the following condition is satisfied for all choices of k distinct elements $x_1, x_2, \ldots, x_k \in \mathcal{X}$ and for all choices of k (not necessarily distinct) elements $y_1, \ldots, y_k \in \mathcal{Y}$:

$$|\{K \in \mathcal{K} : h_K(x_i) = y_i \text{ for } 1 \leq i \leq k\}| = \frac{|\mathcal{K}|}{M^k}.$$

(a) Prove that a strongly k-universal hash family is strongly ℓ-universal for all ℓ such that $1 \leq \ell \leq k$.

(b) Let p be prime and let $k \geq 1$ be an integer. For all k-tuples $(a_0, \ldots, a_{k-1}) \in (\mathbb{Z}_p)^k$, define $f_{(a_0,\ldots,a_{k-1})} : \mathbb{Z}_p \to \mathbb{Z}_p$ by the rule

$$f_{(a_0,\ldots,a_{k-1})}(x) = \sum_{i=0}^{k-1} a_i x^i \bmod p.$$

Prove that $\left(\mathbb{Z}_p, \mathbb{Z}_p, (\mathbb{Z}_p)^k, \{f_{(a_0,\ldots,a_{k-1})} : (a_0, \ldots, a_{k-1}) \in (\mathbb{Z}_p)^k\}\right)$ is a strongly k-universal (p, p) hash family.

HINT Use the fact that any degree d polynomial over a field has at most d roots.

Chapter 6

The RSA Cryptosystem and Factoring Integers

In this chapter, we discuss the *RSA Cryptosystem*, which was the first example of a public-key cryptosystem to be discovered, along with related mathematical concepts including algorithms for factoring large integers.

6.1 Introduction to Public-key Cryptography

In the classical model of cryptography that we have been studying up until now, Alice and Bob secretly choose the key K. K then gives rise to an encryption rule e_K and a decryption rule d_K. In the cryptosystems we have seen so far, d_K is either the same as e_K, or easily derived from it. A cryptosystem of this type is known as a ***secret-key cryptosystem*** or, alternatively, a ***symmetric-key cryptosystem***. Usually, in such a cryptosystem, the exposure of either of e_K or d_K renders the system insecure.

One drawback of a secret-key system is that it requires the prior communication of the key K between Alice and Bob, using a secure channel, before any ciphertext is transmitted. In practice, this may be very difficult to achieve. For example, suppose Alice and Bob live far away from each other and they decide that they want to communicate electronically, using email. In a situation such as this, Alice and Bob may not have access to a reasonable secure channel.

The idea behind a ***public-key cryptosystem*** is that it might be possible to find a cryptosystem where it is computationally infeasible to determine d_K given e_K. If so, then the encryption rule e_K is a ***public key***, the value of which can be made known to everyone (hence the term public-key system). The advantage of a public-key system is that Alice (or anyone else) can send an encrypted message to Bob (without the prior communication of a shared secret key) by using the public encryption rule e_K. Bob will be the only person that can decrypt the ciphertext, using the decryption rule d_K, which is called the ***private key***.

Consider the following analogy: Alice places an object in a metal box, and then locks it with a combination lock left there by Bob. Bob is the only person who can open the box since only he knows the combination.

When Alice wants to encrypt a message to send to Bob, it is essential that the public encryption key that Alice is using is actually Bob's public key. In practice, public keys are authenticated using certificates, which are discussed in Section 8.6.

An alternative approach is to use an identity-based encryption scheme; see Section 13.1.

The idea of a public-key cryptosystem was put forward by Diffie and Hellman in 1976. Then, in 1977, Rivest, Shamir, and Adleman invented the well-known *RSA Cryptosystem*, which we study in this chapter. Several public-key systems have since been proposed, whose security rests on different computational problems. Of these, the most important are the *RSA Cryptosystem* (and variations of it), in which the security is based on the difficulty of factoring large integers; and the *ElGamal Cryptosystem* (and variations such as *Elliptic Curve Cryptosystems*) in which the security is based on the discrete logarithm problem. We discuss the *RSA Cryptosystem* and its variants in this chapter, while the *ElGamal Cryptosystem* is studied in Chapter 7. A variety of other public-key cryptosystems are presented in Chapter 9.

It should be mentioned that all known examples of secure public-key cryptosystems are much slower than commonly-used secret-key cryptosystems such as *AES*. So, in practice, public-key cryptosystems are almost never used to encrypt "long" messages; their main use is in encrypting short keys used in secret-key cryptosystems. We could encrypt data using a *AES*, and then encrypt the AES key using a public-key cryptosystem. This process is known as ***hybrid cryptography***. That is, the following two-step process is used in hybrid cryptography:

1. Alice first chooses a key L for a secret-key cryptosystem and computes $y = e_L(x)$.

2. Alice then encrypts L using Bob's public key $e_{K_{Bob}}$ for a public-key cryptosystem, obtaining $z = e_{K_{Bob}}(L)$.

The ciphertext y and the encrypted key z would both be transmitted to Bob. When Bob receives y and z, he decrypts the ciphertext as follows:

1. Bob first decrypts z using his private key $d_{K_{Bob}}$, obtaining $L = d_{K_{Bob}}(z)$.

2. Bob then uses L to decrypt y, obtaining the plaintext $x = d_L(y)$.

Prior to Diffie and Hellman, the idea of public-key cryptography had already been proposed by James Ellis in January 1970, in a paper entitled *The possibility of non-secret encryption*. (The phrase "non-secret encryption" can be read as "public-key cryptography.") James Ellis was a member of the Communication-Electronics Security Group (CESG), which is a special section of the British Government Communications Headquarters (GCHQ). This paper was not published in the open literature, and was one of five papers released by the GCHQ officially in December 1997. Also included in these five papers was a 1973 paper written by Clifford Cocks, entitled *A note on non-secret encryption*, in which a public-key cryptosystem is described that is essentially the same as the *RSA Cryptosystem*.

One very important observation is that a public-key cryptosystem can never provide unconditional security. This is because an opponent, on observing a ciphertext y, can encrypt each possible plaintext in turn using the public encryption rule e_K until he finds the unique x such that $y = e_K(x)$. This x is the decryption of y. Consequently, we study the computational security of public-key systems.

It is helpful conceptually to think of a public-key system in terms of an abstraction called a "trapdoor one-way function." We informally define this notion now.

Bob's public encryption function, e_K, should be easy to compute. We have just noted that computing the inverse function (i.e., decrypting) should be hard (for anyone other than Bob). Recall from Section 5.2 that a function that is easy to compute but hard to invert is often called a one-way function. In the context of encryption, we desire that e_K be an injective one-way function so that decryption can be performed. Unfortunately, although there are many injective functions that are believed to be one-way, there currently do not exist such functions that can be proved to be one-way.

Here is an example of a function that is believed to be one-way. Suppose n is the product of two large primes p and q, and let b be a positive integer. Then define $f : \mathbb{Z}_n \rightarrow \mathbb{Z}_n$ to be

$$f(x) = x^b \bmod n.$$

(If $\gcd(b, \phi(n)) = 1$, then this is in fact an RSA encryption function; we will have much more to say about it later.)

If we are to construct a public-key cryptosystem, then it is not sufficient to find an injective one-way function. We do not want e_K to be one-way from Bob's point of view, because he needs to be able to decrypt messages that he receives in an efficient way. Thus, it is necessary that Bob possesses a ***trapdoor***, which consists of secret information that permits easy inversion of e_K. That is, Bob can decrypt efficiently because he has some extra secret knowledge, namely, K, which provides him with the decryption function d_K. So, we say that a function is a ***trapdoor one-way function*** if it is a one-way function, but it becomes easy to invert with the knowledge of a certain trapdoor.

Let's consider the function $f(x) = x^b \bmod n$ considered above. We will see in Section 6.3 that the inverse function f^{-1} has a similar form: $f(x) = x^a \bmod n$ for an appropriate value of a. The trapdoor is an efficient method for computing the correct exponent a (as a function of b), which makes use of the factorization of n.

It is often convenient to specify a family of trapdoor one-way functions, say $\mathcal{F}$. Then a function $f \in \mathcal{F}$ is chosen at random and used as the public encryption function; the inverse function, f^{-1}, is the private decryption function. This is analogous to choosing a random key from a specified keyspace, as we did with secret-key cryptosystems.

The rest of this chapter is organized as follows. Section 6.2 introduces several important number-theoretic results. In Section 6.3, we begin our study of the *RSA Cryptosystem*. Section 6.4 presents some important methods of primality testing. Section 6.5 is a short section on the existence of square roots modulo n. Then we present several algorithms for factoring in Section 6.6. Section 6.7 considers other attacks against the *RSA Cryptosystem*, and the *Rabin Cryptosystem* is described in Section 6.8. Semantic security of RSA-like cryptosystems is the topic of Section 6.9.

6.2 More Number Theory

Before describing how the *RSA Cryptosystem* works, we need to discuss some more facts concerning modular arithmetic and number theory. Two fundamental tools that we require are the EUCLIDEAN ALGORITHM and the Chinese remainder theorem.

6.2.1 The Euclidean Algorithm

We already observed in Chapter 2 that $\mathbb{Z}_n$ is a ring for any positive integer n. We also proved there that $b \in \mathbb{Z}_n$ has a multiplicative inverse if and only if $\gcd(b, n) = 1$, and that the number of positive integers less than n and relatively prime to n is $\phi(n)$.

The set of residues modulo n that are relatively prime to n is denoted $\mathbb{Z}_n{}^*$. It is not hard to see that $\mathbb{Z}_n{}^*$ forms an abelian group under multiplication. We already have stated that multiplication modulo n is associative and commutative, and that 1 is the multiplicative identity. Any element in $\mathbb{Z}_n{}^*$ will have a multiplicative inverse (which is also in $\mathbb{Z}_n{}^*$). Finally, $\mathbb{Z}_n{}^*$ is closed under multiplication since xy is relatively prime to n whenever x and y are relatively prime to n (prove this!).

At this point, we know that any $b \in \mathbb{Z}_n{}^*$ has a multiplicative inverse, b^{-1}, but we do not yet have an efficient algorithm to compute b^{-1}. Such an algorithm exists; it is called the EXTENDED EUCLIDEAN ALGORITHM. However, we first describe the EUCLIDEAN ALGORITHM, in its basic form, which can be used to compute the greatest common divisor of two positive integers, say a and b. The EUCLIDEAN ALGORITHM sets r_0 to be a and r_1 to be b, and performs the following sequence of divisions:

$$\begin{array}{rcll} r_0 & = & q_1 r_1 + r_2, & 0 < r_2 < r_1 \\ r_1 & = & q_2 r_2 + r_3, & 0 < r_3 < r_2 \\ \vdots & \vdots & \vdots & \vdots \\ r_{m-2} & = & q_{m-1} r_{m-1} + r_m, & 0 < r_m < r_{m-1} \\ r_{m-1} & = & q_m r_m. & \end{array}$$

A pseudocode description of the EUCLIDEAN ALGORITHM is presented as Algorithm 6.1.

REMARK We will make use of the list $(q_1, \ldots, q_m)$ that is computed during the execution of Algorithm 6.1 in a later section of this chapter. ∎

Algorithm 6.1: EUCLIDEAN ALGORITHM(a, b)

$r_0 \leftarrow a$
$r_1 \leftarrow b$
$m \leftarrow 1$
while $r_m \neq 0$

$$\textbf{do} \begin{cases} q_m \leftarrow \lfloor \frac{r_{m-1}}{r_m} \rfloor \\ r_{m+1} \leftarrow r_{m-1} - q_m r_m \\ m \leftarrow m + 1 \end{cases}$$

$m \leftarrow m - 1$
return $(q_1, \ldots, q_m; r_m)$
comment: $r_m = \gcd(a, b)$

In Algorithm 6.1, it is not hard to show that

$$\gcd(r_0, r_1) = \gcd(r_1, r_2) = \cdots = \gcd(r_{m-1}, r_m) = r_m.$$

Hence, it follows that $\gcd(r_0, r_1) = r_m$.

Since the EUCLIDEAN ALGORITHM computes greatest common divisors, it can be used to determine if a positive integer $b < n$ has a multiplicative inverse modulo n, by calling EUCLIDEAN ALGORITHM(n, b) and checking to see if $r_m = 1$. However, it does not compute the value of $b^{-1} \bmod n$ (if it exists).

Now, suppose we define two sequences of numbers,

$$t_0, t_1, \ldots, t_m \quad \text{and} \quad s_0, s_1, \ldots, s_m,$$

according to the following recurrences (where the q_j's are defined as in Algorithm 6.1):

$$t_j = \begin{cases} 0 & \text{if } j = 0 \\ 1 & \text{if } j = 1 \\ t_{j-2} - q_{j-1} t_{j-1} & \text{if } j \geq 2 \end{cases}$$

and

$$s_j = \begin{cases} 1 & \text{if } j = 0 \\ 0 & \text{if } j = 1 \\ s_{j-2} - q_{j-1} s_{j-1} & \text{if } j \geq 2. \end{cases}$$

Then we have the following useful result.

THEOREM 6.1 *For $0 \leq j \leq m$, we have that $r_j = s_j r_0 + t_j r_1$, where the r_j's are defined as in Algorithm 6.1, and the s_j's and t_j's are defined in the above recurrence.*

PROOF The proof is by induction on j. The assertion is trivially true for $j = 0$ and

$j = 1$. Assume the assertion is true for $j = i - 1$ and $i - 2$, where $i \geq 2$; we will prove the assertion is true for $j = i$. By induction, we have that

$$r_{i-2} = s_{i-2}r_0 + t_{i-2}r_1$$

and

$$r_{i-1} = s_{i-1}r_0 + t_{i-1}r_1.$$

Now, we compute:

$$\begin{aligned} r_i &= r_{i-2} - q_{i-1}r_{i-1} \\ &= s_{i-2}r_0 + t_{i-2}r_1 - q_{i-1}(s_{i-1}r_0 + t_{i-1}r_1) \\ &= (s_{i-2} - q_{i-1}s_{i-1})r_0 + (t_{i-2} - q_{i-1}t_{i-1})r_1 \\ &= s_ir_0 + t_ir_1. \end{aligned}$$

Hence, the result is true, for all integers $j \geq 0$, by induction. ∎

In Algorithm 6.2, we present the EXTENDED EUCLIDEAN ALGORITHM, which takes two integers a and b as input and computes integers r, s, and t such that $r = \gcd(a, b)$ and $sa + tb = r$. In this version of the algorithm, we do not keep track of all the q_j's, r_j's, s_j's, and t_j's; it suffices to record only the "last" two terms in each of these sequences at any point in the algorithm.

The next corollary is an immediate consequence of Theorem 6.1.

COROLLARY 6.2 *Suppose* $\gcd(r_0, r_1) = 1$. *Then* $r_1^{-1} \bmod r_0 = t_m \bmod r_0$.

PROOF From Theorem 6.1, we have that

$$1 = \gcd(r_0, r_1) = s_mr_0 + t_mr_1.$$

Reducing this equation modulo r_0, we obtain

$$t_mr_1 \equiv 1 \pmod{r_0}.$$

The result follows. ∎

We present a small example to illustrate, in which we show the values of all the s_j's, t_j's, q_j's, and r_j's.

Example 6.1 Suppose we wish to calculate $28^{-1} \bmod 75$ using Algorithm 6.2. Then we compute:

i	r_i	q_i	s_i	t_i
0	75		1	0
1	28	2	0	1
2	19	1	1	−2
3	9	2	−1	3
4	1	9	3	−8

Algorithm 6.2: EXTENDED EUCLIDEAN ALGORITHM(a, b)

$a_0 \leftarrow a$
$b_0 \leftarrow b$
$t_0 \leftarrow 0$
$t \leftarrow 1$
$s_0 \leftarrow 1$
$s \leftarrow 0$
$q \leftarrow \lfloor \frac{a_0}{b_0} \rfloor$
$r \leftarrow a_0 - qb_0$
while $r > 0$

$$\textbf{do} \begin{cases} temp \leftarrow t_0 - qt \\ t_0 \leftarrow t \\ t \leftarrow temp \\ temp \leftarrow s_0 - qs \\ s_0 \leftarrow s \\ s \leftarrow temp \\ a_0 \leftarrow b_0 \\ b_0 \leftarrow r \\ q \leftarrow \lfloor \frac{a_0}{b_0} \rfloor \\ r \leftarrow a_0 - qb_0 \end{cases}$$

$r \leftarrow b_0$
return (r, s, t)
comment: $r = \gcd(a, b)$ and $sa + tb = r$

Therefore, we have found that $3 \times 75 - 8 \times 28 = 1$. Applying Corollary 6.2, we see that

$$28^{-1} \bmod 75 = -8 \bmod 75 = 67.$$

□

The EXTENDED EUCLIDEAN ALGORITHM immediately yields the value b^{-1} modulo a (if it exists). In fact, the multiplicative inverse $b^{-1} \bmod a = t \bmod a$; this follows immediately from Corollary 6.2. However, a more efficient way to compute multiplicative inverses is to remove the s's from Algorithm 6.2, and to reduce the t's modulo a during each iteration of the main loop. We obtain Algorithm 6.3 as a result.

6.2.2 The Chinese Remainder Theorem

The Chinese remainder theorem is really a method of solving certain systems of congruences. Suppose $m_1, \ldots, m_r$ are pairwise relatively prime positive integers (that is, $\gcd(m_i, m_j) = 1$ if $i \neq j$). Suppose $a_1, \ldots, a_r$ are integers, and consider the

Algorithm 6.3: MULTIPLICATIVE INVERSE(a, b)

```
a_0 ← a
b_0 ← b
t_0 ← 0
t ← 1
q ← ⌊a_0 / b_0⌋
r ← a_0 − q b_0
while r > 0
   do { temp ← (t_0 − q t) mod a
        t_0 ← t
        t ← temp
        a_0 ← b_0
        b_0 ← r
        q ← ⌊a_0 / b_0⌋
        r ← a_0 − q b_0 }
if b_0 ≠ 1
  then b has no inverse modulo a
  else return (t)
```

following system of congruences:

$$\begin{aligned} x &\equiv a_1 \pmod{m_1} \\ x &\equiv a_2 \pmod{m_2} \\ &\vdots \\ x &\equiv a_r \pmod{m_r}. \end{aligned}$$

The Chinese remainder theorem asserts that this system has a unique solution modulo $M = m_1 \times m_2 \times \cdots \times m_r$. We will prove this result in this section, and also describe an efficient algorithm for solving systems of congruences of this type.

It is convenient to study the "projection function" $\chi : \mathbb{Z}_M \rightarrow \mathbb{Z}_{m_1} \times \cdots \times \mathbb{Z}_{m_r}$, which we define as follows:

$$\chi(x) = (x \bmod m_1, \ldots, x \bmod m_r).$$

Example 6.2 Suppose $r = 2$, $m_1 = 5$ and $m_2 = 3$, so $M = 15$. Then the function χ has the following values:

$$\begin{array}{lll} \chi(0) = (0,0) & \chi(1) = (1,1) & \chi(2) = (2,2) \\ \chi(3) = (3,0) & \chi(4) = (4,1) & \chi(5) = (0,2) \\ \chi(6) = (1,0) & \chi(7) = (2,1) & \chi(8) = (3,2) \\ \chi(9) = (4,0) & \chi(10) = (0,1) & \chi(11) = (1,2) \\ \chi(12) = (2,0) & \chi(13) = (3,1) & \chi(14) = (4,2). \end{array}$$

□

Proving the Chinese remainder theorem amounts to proving that the function χ is a bijection. In Example 6.2 this is easily seen to be the case. In fact, we will be able to give an explicit general formula for the inverse function χ^{-1}.

For $1 \leq i \leq r$, define

$$M_i = \frac{M}{m_i}.$$

Then it is not difficult to see that

$$\gcd(M_i, m_i) = 1$$

for $1 \leq i \leq r$. Next, for $1 \leq i \leq r$, define

$$y_i = M_i^{-1} \bmod m_i.$$

(This inverse exists because $\gcd(M_i, m_i) = 1$, and it can be found using Algorithm 6.3.) Note that

$$M_i y_i \equiv 1 \pmod{m_i}$$

for $1 \leq i \leq r$.

Now, define a function $\rho : \mathbb{Z}_{m_1} \times \cdots \times \mathbb{Z}_{m_r} \rightarrow \mathbb{Z}_M$ as follows:

$$\rho(a_1, \ldots, a_r) = \sum_{i=1}^{r} a_i M_i y_i \bmod M.$$

We will show that the function $\rho = \chi^{-1}$, i.e., it provides an explicit formula for solving the original system of congruences.

Denote $X = \rho(a_1, \ldots, a_r)$, and let $1 \leq j \leq r$. Consider a term $a_i M_i y_i$ in the above summation, reduced modulo m_j: If $i = j$, then

$$a_i M_i y_i \equiv a_i \pmod{m_i}$$

because

$$M_i y_i \equiv 1 \pmod{m_i}.$$

On the other hand, if $i \neq j$, then

$$a_i M_i y_i \equiv 0 \pmod{m_j}$$

because $m_j \mid M_i$ in this case. Thus, we have that

$$\begin{aligned} X &\equiv \sum_{i=1}^{r} a_i M_i y_i \pmod{m_j} \\ &\equiv a_j \pmod{m_j}. \end{aligned}$$

Since this is true for all j, $1 \leq j \leq r$, X is a solution to the system of congruences.

At this point, we need to show that the solution X is unique modulo M. But this can be done by simple counting. The function χ is a function from a domain of cardinality M to a range of cardinality M. We have just proved that χ is a surjective (i.e., onto) function. Hence, χ must also be injective (i.e., one-to-one), since the domain and range have the same cardinality. It follows that χ is a bijection and $\chi^{-1} = \rho$. Note also that χ^{-1} is a linear function of its arguments $a_1, \dots, a_r$.

Here is a bigger example to illustrate.

Example 6.3 Suppose $r = 3$, $m_1 = 7$, $m_2 = 11$, and $m_3 = 13$. Then $M = 1001$. We compute $M_1 = 143$, $M_2 = 91$, and $M_3 = 77$, and then $y_1 = 5$, $y_2 = 4$, and $y_3 = 12$. Then the function $\chi^{-1} : \mathbb{Z}_7 \times \mathbb{Z}_{11} \times \mathbb{Z}_{13} \rightarrow \mathbb{Z}_{1001}$ is the following:

$$\chi^{-1}(a_1, a_2, a_3) = (715a_1 + 364a_2 + 924a_3) \bmod 1001.$$

For example, if $x \equiv 5 \pmod 7$, $x \equiv 3 \pmod{11}$, and $x \equiv 10 \pmod{13}$, then this formula tells us that

$$\begin{aligned} x &= (715 \times 5 + 364 \times 3 + 924 \times 10) \bmod 1001 \\ &= 13907 \bmod 1001 \\ &= 894. \end{aligned}$$

This can be verified by reducing 894 modulo 7, 11, and 13. □

For future reference, we record the results of this section as a theorem.

THEOREM 6.3 (Chinese remainder theorem) *Suppose $m_1, \dots, m_r$ are pairwise relatively prime positive integers, and suppose $a_1, \dots, a_r$ are integers. Then the system of r congruences $x \equiv a_i \pmod{m_i}$ $(1 \leq i \leq r)$ has a unique solution modulo $M = m_1 \times \cdots \times m_r$, which is given by*

$$x = \sum_{i=1}^{r} a_i M_i y_i \bmod M,$$

where $M_i = M/m_i$ and $y_i = M_i^{-1} \bmod m_i$, for $1 \leq i \leq r$.

6.2.3 Other Useful Facts

We next mention another result from elementary group theory, called Lagrange's theorem, that will be relevant in our treatment of the *RSA Cryptosystem*. Let G be a (finite) multiplicative group. The ***order*** of G is the number of elements in G. The ***order*** of an element $g \in G$ is defined to be the smallest positive integer m such that $g^m = 1$. The following result is fairly simple, but we will not prove it here.

THEOREM 6.4 (Lagrange) *Suppose G is a multiplicative group of order n, and $g \in G$. Then the order of g divides n.*

For our purposes, the following corollaries are essential.

COROLLARY 6.5 *If* $b \in \mathbb{Z}_n{}^*$*, then* $b^{\phi(n)} \equiv 1 \pmod{n}$.

PROOF $\mathbb{Z}_n{}^*$ is a multiplicative group of order $\phi(n)$. ∎

COROLLARY 6.6 (Fermat) *Suppose p is prime and* $b \in \mathbb{Z}_p$*. Then* $b^p \equiv b \pmod{p}$.

PROOF If p is prime, then $\phi(p) = p - 1$. So, for $b \not\equiv 0 \pmod{p}$, the result follows from Corollary 6.5. For $b \equiv 0 \pmod{p}$, the result is also true since $0^p \equiv 0 \pmod{p}$. ∎

At this point, we know that if p is prime, then $\mathbb{Z}_p{}^*$ is a group of order $p - 1$, and any element in $\mathbb{Z}_p{}^*$ has order dividing $p - 1$. In fact, if p is prime, then the group $\mathbb{Z}_p{}^*$ is a ***cyclic group***: there exists an element $\alpha \in \mathbb{Z}_p{}^*$ having order equal to $p - 1$. We will not prove this very important fact, but we do record it for future reference:

THEOREM 6.7 *If p is prime, then* $\mathbb{Z}_p{}^*$ *is a cyclic group.*

An element α having order $p - 1$ modulo p is called a ***primitive element modulo*** p. Observe that α is a primitive element modulo p if and only if

$$\{\alpha^i : 0 \leq i \leq p-2\} = \mathbb{Z}_p{}^*.$$

Now, suppose p is prime and α is a primitive element modulo p. Any element $\beta \in \mathbb{Z}_p{}^*$ can be written as $\beta = \alpha^i$, where $0 \leq i \leq p - 2$, in a unique way. It is not difficult to prove that the order of $\beta = \alpha^i$ is

$$\frac{p-1}{\gcd(p-1, i)}.$$

Thus β is itself a primitive element if and only if $\gcd(p - 1, i) = 1$. It follows that the number of primitive elements modulo p is $\phi(p - 1)$.

We do a small example to illustrate.

Example 6.4 Suppose $p = 13$. The results proven above establish that there are exactly four primitive elements modulo 13. First, by computing successive powers of 2, we can verify that 2 is a primitive element modulo 13:

$$\begin{array}{rclrcl}
2^0 \bmod 13 & = & 1 & 2^1 \bmod 13 & = & 2\\
2^2 \bmod 13 & = & 4 & 2^3 \bmod 13 & = & 8\\
2^4 \bmod 13 & = & 3 & 2^5 \bmod 13 & = & 6\\
2^6 \bmod 13 & = & 12 & 2^7 \bmod 13 & = & 11\\
2^8 \bmod 13 & = & 9 & 2^9 \bmod 13 & = & 5\\
2^{10} \bmod 13 & = & 10 & 2^{11} \bmod 13 & = & 7.
\end{array}$$

The element 2^i is primitive if and only if $\gcd(i, 12) = 1$, i.e., if and only if $i = 1, 5, 7$, or 11. Hence, the primitive elements modulo 13 are $2, 6, 7$, and 11. □

In the above example, we computed all the powers of 2 in order to verify that it was a primitive element modulo 13. If p is a large prime, however, it would take a long time to compute $p-1$ powers of an element $\alpha \in \mathbb{Z}_p{}^*$. Fortunately, if the factorization of $p-1$ is known, then we can verify whether $\alpha \in \mathbb{Z}_p{}^*$ is a primitive element much more quickly, by making use of the following result.

THEOREM 6.8 *Suppose that $p > 2$ is prime and $\alpha \in \mathbb{Z}_p{}^*$. Then α is a primitive element modulo p if and only if $\alpha^{(p-1)/q} \not\equiv 1 \pmod p$ for all primes q such that $q \mid (p-1)$.*

PROOF If α is a primitive element modulo p, then $\alpha^i \not\equiv 1 \pmod p$ for all i such that $1 \leq i \leq p-2$, so the result follows.

Conversely, suppose that $\alpha \in \mathbb{Z}_p{}^*$ is not a primitive element modulo p. Let d be the order of α. Then $d \mid (p-1)$ by Lagrange's theorem, and $d < p-1$ because α is not primitive. Then $(p-1)/d$ is an integer exceeding 1. Let q be a prime divisor of $(p-1)/d$. Then d is a divisor of the integer $(p-1)/q$. Since $\alpha^d \equiv 1 \pmod p$ and $d \mid (p-1)/q$, it follows that $\alpha^{(p-1)/q} \equiv 1 \pmod p$. ∎

The factorization of 12 is $12 = 2^2 \times 3$. Therefore, in the previous example, we could verify that 2 is a primitive element modulo 13 by verifying that $2^6 \not\equiv 1 \pmod{13}$ and $2^4 \not\equiv 1 \pmod{13}$.

6.3 The RSA Cryptosystem

We can now describe the *RSA Cryptosystem*. This cryptosystem uses computations in $\mathbb{Z}_n$, where n is the product of two distinct odd primes p and q. For such an integer n, note that $\phi(n) = (p-1)(q-1)$. The formal description is given as Cryptosystem 6.1.

Let's verify that encryption and decryption are inverse operations. Since

$$ab \equiv 1 \pmod{\phi(n)},$$

we have that

$$ab = t\phi(n) + 1$$

for some integer $t \geq 1$. Suppose that $x \in \mathbb{Z}_n{}^*$; then we have

$$\begin{aligned}(x^b)^a &\equiv x^{t\phi(n)+1} \pmod n\\ &\equiv (x^{\phi(n)})^t x \pmod n\\ &\equiv 1^t x \pmod n\\ &\equiv x \pmod n,\end{aligned}$$

as desired. We leave it as an Exercise to show that $(x^b)^a \equiv x \pmod n$ if $x \in \mathbb{Z}_n \backslash \mathbb{Z}_n{}^*$.

Here is a small (insecure) example of the *RSA Cryptosystem*.

Cryptosystem 6.1: *RSA Cryptosystem*

Let $n = pq$, where p and q are primes. Let $\mathcal{P} = \mathcal{C} = \mathbb{Z}_n$, and define

$$\mathcal{K} = \{(n, p, q, a, b) : ab \equiv 1 \pmod{\phi(n)}\}.$$

For $K = (n, p, q, a, b)$, define

$$e_K(x) = x^b \bmod n$$

and

$$d_K(y) = y^a \bmod n$$

($x, y \in \mathbb{Z}_n$). The values n and b comprise the public key, and the values p, q, and a form the private key.

Example 6.5 Suppose Bob chooses $p = 101$ and $q = 113$. Then $n = 11413$ and $\phi(n) = 100 \times 112 = 11200$. Since $11200 = 2^6 5^2 7$, an integer b can be used as an encryption exponent if and only if b is not divisible by 2, 5, or 7. (In practice, however, Bob will not factor $\phi(n)$. He will verify that $\gcd(\phi(n), b) = 1$ using Algorithm 6.3. If this is the case, then he will compute b^{-1} at the same time.) Suppose Bob chooses $b = 3533$. Then

$$b^{-1} \bmod 11200 = 6597.$$

Hence, Bob's secret decryption exponent is $a = 6597$.

Bob publishes $n = 11413$ and $b = 3533$ in a directory. Now, suppose Alice wants to encrypt the plaintext 9726 to send to Bob. She will compute

$$9726^{3533} \bmod 11413 = 5761$$

and send the ciphertext 5761 over the channel. When Bob receives the ciphertext 5761, he uses his secret decryption exponent to compute

$$5761^{6597} \bmod 11413 = 9726.$$

(At this point, the encryption and decryption operations might appear to be very complicated, but we will discuss efficient algorithms for these operations in the next section.) □

The security of the *RSA Cryptosystem* is based on the belief that the encryption function $e_K(x) = x^b \bmod n$ is a one-way function, so it will be computationally infeasible for an opponent to decrypt a ciphertext. The trapdoor that allows Bob to decrypt a ciphertext is the knowledge of the factorization $n = pq$. Since Bob knows this factorization, he can compute $\phi(n) = (p-1)(q-1)$, and then compute the decryption exponent a using the EXTENDED EUCLIDEAN ALGORITHM. We will say more about the security of the *RSA Cryptosystem* later on.

6.3.1 Implementing RSA

There are many aspects of the *RSA Cryptosystem* to discuss, including the details of setting up the cryptosystem, the efficiency of encrypting and decrypting, and security issues. In order to set up the system, Bob uses the RSA PARAMETER GENERATION algorithm, presented informally as Algorithm 6.4. How Bob carries out the steps of this algorithm will be discussed later in this chapter.

Algorithm 6.4: RSA PARAMETER GENERATION

1. Generate two large primes, p and q, such that $p \neq q$
2. $n \leftarrow pq$ and $\phi(n) \leftarrow (p-1)(q-1)$
3. Choose a random b $(1 < b < \phi(n))$ such that $\gcd(b, \phi(n)) = 1$
4. $a \leftarrow b^{-1} \bmod \phi(n)$
5. The public key is (n, b) and the private key is (p, q, a).

One obvious attack on the *RSA Cryptosystem* is for a cryptanalyst to attempt to factor n. If this can be done, it is a simple matter to compute $\phi(n) = (p-1)(q-1)$ and then compute the decryption exponent a from b exactly as Bob did. (It has been conjectured that breaking the *RSA Cryptosystem* is polynomially equivalent[1] to factoring n, but this remains unproved.)

If the *RSA Cryptosystem* is to be secure, it is certainly necessary that $n = pq$ must be large enough that factoring it will be computationally infeasible. As of the writing of this book, factoring algorithms are able to factor RSA moduli having up to 768 bits in their binary representation (for more information on factoring, see Section 6.6). It is currently recommended that, to be on the safe side, one should choose each of p and q to be 1024-bit primes; then n will be a 2048-bit modulus. Factoring a number of this size is well beyond the capability of the best current factoring algorithms.

Leaving aside for the moment the question of how to find 1024-bit primes, let us look now at the arithmetic operations of encryption and decryption. An encryption (or decryption) involves performing one exponentiation modulo n. Since n is very large, we must use multiprecision arithmetic to perform computations in $\mathbb{Z}_n$, and the time required will depend on the number of bits in the binary representation of n.

Suppose that x and y are positive integers having k and ℓ bits respectively in their binary representations; i.e., $k = \lfloor \log_2 x \rfloor + 1$ and $\ell = \lfloor \log_2 y \rfloor + 1$. Assume that $k \geq \ell$. Using standard "grade-school" arithmetic techniques, it is not difficult to obtain big-oh upper bounds on the amount of time to perform various operations on x and y. We summarize these results now (and we do not claim that these are the best possible bounds).

[1]Two problems are said to be ***polynomially equivalent*** if the existence of a polynomial-time algorithm for either problem implies the existence of a polynomial-time algorithm for the other problem.

- $x + y$ can be computed in time $O(k)$
- $x - y$ can be computed in time $O(k)$
- xy can be computed in time $O(k\ell)$
- $\lfloor x/y \rfloor$ can be computed in time $O(\ell(k - \ell))$. Note that $O(k\ell)$ is a weaker bound.
- $\gcd(x, y)$ can be computed in time $O(k^3)$.

In reference to the last item, a gcd can be computed using Algorithm 6.1. It can be shown that the number of iterations required in the EUCLIDEAN ALGORITHM is $O(k)$ (see the Exercises). Each iteration performs a long division requiring time $O(k^2)$; so, the complexity of a gcd computation is seen to be $O(k^3)$. (Actually, a more careful analysis can be used to show that the complexity is, in fact, $O(k^2)$.)

Now we turn to modular arithmetic, i.e., operations in $\mathbb{Z}_n$. Suppose that n is a k-bit integer, and $0 \leq m_1, m_2 \leq n - 1$. Also, let c be a positive integer. We have the following:

- Computing $(m_1 + m_2) \bmod n$ can be done in time $O(k)$.
- Computing $(m_1 - m_2) \bmod n$ can be done in time $O(k)$.
- Computing $(m_1 m_2) \bmod n$ can be done in time $O(k^2)$.
- Computing $(m_1)^{-1} \bmod n$ can be done in time $O(k^3)$ (provided that this inverse exists).
- Computing $(m_1)^c \bmod n$ can be done in time $O((\log c) \times k^2)$.

Most of the above results are not hard to prove. The first three operations (modular addition, subtraction, and multiplication) can be accomplished by doing the corresponding integer operation and then performing a single reduction modulo n. Modular inversion (i.e., computing multiplicative inverses) is done using Algorithm 6.3. The complexity is analyzed in a similar fashion as a gcd computation.

We now consider ***modular exponentiation***, i.e., computation of a function of the form $x^c \bmod n$. Both the encryption and the decryption operations in the *RSA Cryptosystem* are modular exponentiations. Computation of $x^c \bmod n$ can be done using $c - 1$ modular multiplications; however, this is very inefficient if c is large. Note that c might be as big as $\phi(n) - 1$, which is almost as big as n and exponentially large compared to k.

The well-known SQUARE-AND-MULTIPLY ALGORITHM reduces the number of modular multiplications required to compute $x^c \bmod n$ to at most 2ℓ, where ℓ is the number of bits in the binary representation of c. It follows that $x^c \bmod n$ can be computed in time $O(\ell k^2)$. If we assume that $c < n$ (as it is in the definition of the *RSA Cryptosystem*), then we see that RSA encryption and decryption can both be done in time $O((\log n)^3)$, which is a polynomial function of the number of bits in one plaintext (or ciphertext) character.

Algorithm 6.5: SQUARE-AND-MULTIPLY(x, c, n)

$z \leftarrow 1$
for $i \leftarrow \ell - 1$ **downto** 0
$$\textbf{do} \begin{cases} z \leftarrow z^2 \bmod n \\ \textbf{if } c_i = 1 \\ \quad \textbf{then } z \leftarrow (z \times x) \bmod n \end{cases}$$
return (z)

The SQUARE-AND-MULTIPLY ALGORITHM assumes that the exponent c is represented in binary notation, say

$$c = \sum_{i=0}^{\ell-1} c_i 2^i,$$

where $c_i = 0$ or 1, $0 \leq i \leq \ell - 1$. The algorithm to compute $z = x^c \bmod n$ is presented as Algorithm 6.5.

The proof of correctness of this algorithm is left as an Exercise. It is easy to count the number of modular multiplications in the algorithm. There are always ℓ squarings performed. The number of modular multiplications of the type $z \leftarrow (z \times x) \bmod n$ is equal to the number of 1's in the binary representation of c. This is an integer between 0 and ℓ. Thus, the total number of modular multiplications is at least ℓ and at most 2ℓ, as stated above.

We will illustrate the use of the SQUARE-AND-MULTIPLY ALGORITHM by returning to Example 6.5.

Example 6.5 (Cont.) Recall that $n = 11413$, and the public encryption exponent is $b = 3533$. The binary representation of 3533 is 110111001101. Alice encrypts the plaintext 9726 by computing $9726^{3533} \bmod 11413$, using the SQUARE-AND-MULTIPLY ALGORITHM, as shown in Figure 6.1. Hence, as stated earlier, the ciphertext is 5761. □

So far, we have discussed the RSA encryption and decryption operations. Regarding RSA PARAMETER GENERATION, methods to construct the primes p and q (Step 1) will be discussed in the next section. Step 2 is straightforward and can be done in time $O((\log n)^2)$. Steps 3 and 4 utilize Algorithm 6.3, which has complexity $O((\log n)^2)$.

6.4 Primality Testing

In setting up the *RSA Cryptosystem*, it is necessary to generate large "random primes." The way this is done is to generate large random numbers, and then

i	b_i	z
11	1	$1^2 \times 9726 = 9726$
10	1	$9726^2 \times 9726 = 2659$
9	0	$2659^2 = 5634$
8	1	$5634^2 \times 9726 = 9167$
7	1	$9167^2 \times 9726 = 4958$
6	1	$4958^2 \times 9726 = 7783$
5	0	$7783^2 = 6298$
4	0	$6298^2 = 4629$
3	1	$4629^2 \times 9726 = 10185$
2	1	$10185^2 \times 9726 = 105$
1	0	$105^2 = 11025$
0	1	$11025^2 \times 9726 = 5761$

FIGURE 6.1: Exponentiation using the SQUARE-AND-MULTIPLY ALGORITHM

test them for primality. In 2002, it was proven by Agrawal, Kayal, and Saxena that there is a polynomial-time deterministic algorithm for primality testing. This was a major breakthrough that solved a longstanding open problem. However, in practice, primality testing is still done mainly by using a randomized polynomial-time Monte Carlo algorithm such as the SOLOVAY-STRASSEN ALGORITHM or the MILLER-RABIN ALGORITHM, both of which we will present in this section. These algorithms are fast (i.e., an integer n can be tested in time that is polynomial in $\log_2 n$, the number of bits in the binary representation of n), but there is a possibility that the algorithm may claim that n is prime when it is not. However, by running the algorithm enough times, the error probability can be reduced below any desired threshold. (We will discuss this in more detail a bit later.)

The other pertinent question is how many random integers (of a specified size) will need to be tested until we find one that is prime. Suppose we define $\pi(N)$ to be the number of primes that are less than or equal to N. A famous result in number theory, called the ***Prime number theorem***, states that $\pi(N)$ is approximately $N/\ln N$. Hence, if an integer p is chosen at random between 1 and N, then the probability that it is prime is about $1/\ln N$. For a 2048 bit modulus $n = pq$, p and q will be chosen to be 1024-bit primes. A random 1024-bit integer will be prime with probability approximately $1/\ln 2^{1024} \approx 1/710$. That is, on average, given 710 random 1024-bit integers p, one of them will be prime (of course, if we restrict our attention to odd integers, the probability doubles, to about $1/355$). So we can in fact generate sufficiently large random numbers that are "probably prime," and hence parameter generation for the *RSA Cryptosystem* is indeed practical. We proceed to describe how this is done.

A ***decision problem*** is a problem in which a question is to be answered "yes" or "no." Recall that a randomized algorithm is any algorithm that uses random numbers (in contrast, an algorithm that does not use random numbers is called a ***deterministic algorithm***). The following definitions pertain to randomized algorithms for decision problems.

Definition 6.1: A ***yes-biased Monte Carlo algorithm*** is a randomized algorithm for a decision problem in which a "yes" answer is (always) correct, but a "no" answer may be incorrect. A ***no-biased Monte Carlo algorithm*** is defined in the obvious way. We say that a yes-biased Monte Carlo algorithm has ***error probability*** equal to ϵ if, for any instance in which the answer is "yes," the algorithm will give the (incorrect) answer "no" with probability at most ϵ. (This probability is computed over all possible random choices made by the algorithm when it is run with a given input.)

REMARK A Las Vegas algorithm may not give an answer, but any answer it gives is correct. In contrast, a Monte Carlo algorithm always gives an answer, but the answer may be incorrect. ∎

The decision problem called **Composites** is presented as Problem 6.1.

Problem 6.1: Composites

Instance: A positive integer $n \geq 2$.
Question: Is n composite?

Note that an algorithm for a decision problem only has to answer "yes" or "no." In particular, in the case of the problem **Composites**, we do not require the algorithm to find a factorization in the case that n is composite.

We will first describe the SOLOVAY-STRASSEN ALGORITHM, which is a yes-biased Monte Carlo algorithm for **Composites** with error probability $1/2$. Hence, if the algorithm answers "yes," then n is composite; conversely, if n is composite, then the algorithm answers "yes" with probability at least $1/2$.

Although the MILLER-RABIN ALGORITHM (which we will discuss later) is faster than the SOLOVAY-STRASSEN ALGORITHM, we first look at the SOLOVAY-STRASSEN ALGORITHM because it is easier to understand conceptually and because it involves some number-theoretic concepts that will be useful in later chapters of the book. We begin by developing some further background from number theory before describing the algorithm.

6.4.1 Legendre and Jacobi Symbols

Definition 6.2: Suppose p is an odd prime and a is an integer. a is defined to be a ***quadratic residue*** modulo p if $a \not\equiv 0 \pmod{p}$ and the congruence $y^2 \equiv a \pmod{p}$ has a solution $y \in \mathbb{Z}_p$. a is defined to be a ***quadratic non-residue*** modulo p if $a \not\equiv 0 \pmod{p}$ and a is not a quadratic residue modulo p.

Example 6.6 In $\mathbb{Z}_{11}$, we have that $1^2 = 1, 2^2 = 4, 3^2 = 9, 4^2 = 5, 5^2 = 3, 6^2 = 3, 7^2 = 5, 8^2 = 9, 9^2 = 4$, and $(10)^2 = 1$. Therefore the quadratic residues modulo 11 are $1, 3, 4, 5$, and 9, and the quadratic non-residues modulo 11 are $2, 6, 7, 8$, and 10.
□

Suppose that p is an odd prime and a is a quadratic residue modulo p. Then there exists $y \in \mathbb{Z}_p{}^*$ such that $y^2 \equiv a \pmod p$. Clearly, $(-y)^2 \equiv a \pmod p$, and $y \not\equiv -y \pmod p$ because p is odd and $y \neq 0$. Now consider the quadratic congruence $x^2 - a \equiv 0 \pmod p$. This congruence can be factored as $(x-y)(x+y) \equiv 0 \pmod p$, which is the same thing as saying that $p \mid (x-y)(x+y)$. Now, because p is prime, it follows that $p \mid (x-y)$ or $p \mid (x+y)$. In other words, $x \equiv \pm y \pmod p$, and we conclude that there are exactly two solutions (modulo p) to the congruence $x^2 - a \equiv 0 \pmod p$. Moreover, these two solutions are negatives of each other modulo p.

We now study the problem of determining whether an integer a is quadratic residue modulo p. The decision problem **Quadratic Residues** (Problem 6.2) is defined in the obvious way. Notice that this problem just asks for a "yes" or "no" answer: it does not require us to compute square roots in the case when a is a quadratic residue modulo p.

Problem 6.2: Quadratic Residues

Instance: An odd prime p, and an integer a.
Question: Is a a quadratic residue modulo p?

We prove a result, known as Euler's criterion, that will give rise to a polynomial-time deterministic algorithm for **Quadratic Residues**.

THEOREM 6.9 (Euler's Criterion) *Let p be an odd prime. Then a is a quadratic residue modulo p if and only if*

$$a^{(p-1)/2} \equiv 1 \pmod p.$$

PROOF First, suppose $a \equiv y^2 \pmod p$. Recall from Corollary 6.6 that if p is prime, then $a^{p-1} \equiv 1 \pmod p$ for any $a \not\equiv 0 \pmod p$. Thus we have

$$\begin{aligned} a^{(p-1)/2} &\equiv (y^2)^{(p-1)/2} \pmod p \\ &\equiv y^{p-1} \pmod p \\ &\equiv 1 \pmod p. \end{aligned}$$

Conversely, suppose $a^{(p-1)/2} \equiv 1 \pmod p$. Let b be a primitive element modulo p. Then $a \equiv b^i \pmod p$ for some positive integer i. Then we have

$$\begin{aligned} a^{(p-1)/2} &\equiv (b^i)^{(p-1)/2} \pmod p \\ &\equiv b^{i(p-1)/2} \pmod p. \end{aligned}$$

Since b has order $p-1$, it must be the case that $p-1$ divides $i(p-1)/2$. Hence, i is even, and then the square roots of a are $\pm b^{i/2} \bmod p$. ∎

Theorem 6.9 yields a polynomial-time algorithm for **Quadratic Residues**, by using the SQUARE-AND-MULTIPLY ALGORITHM for exponentiation modulo p. The complexity of the algorithm will be $O((\log p)^3)$.

We now need to give some further definitions from number theory.

Definition 6.3: Suppose p is an odd prime. For any integer a, define the ***Legendre symbol*** $\left(\frac{a}{p}\right)$ as follows:

$$\left(\frac{a}{p}\right) = \begin{cases} 0 & \text{if } a \equiv 0 \pmod{p} \\ 1 & \text{if } a \text{ is a quadratic residue modulo } p \\ -1 & \text{if } a \text{ is a quadratic non-residue modulo } p. \end{cases}$$

We have already seen that $a^{(p-1)/2} \equiv 1 \pmod{p}$ if and only if a is a quadratic residue modulo p. If a is a multiple of p, then it is clear that $a^{(p-1)/2} \equiv 0 \pmod{p}$. Finally, if a is a quadratic non-residue modulo p, then $a^{(p-1)/2} \equiv -1 \pmod{p}$ because

$$(a^{(p-1)/2})^2 \equiv a^{p-1} \equiv 1 \pmod{p}$$

and $a^{(p-1)/2} \not\equiv 1 \pmod{p}$. Hence, we have the following result, which provides an efficient algorithm to evaluate Legendre symbols:

THEOREM 6.10 *Suppose p is an odd prime. Then*

$$\left(\frac{a}{p}\right) \equiv a^{(p-1)/2} \pmod{p}.$$

Next, we define a generalization of the Legendre symbol.

Definition 6.4: Suppose n is an odd positive integer, and the prime power factorization of n is

$$n = \prod_{i=1}^{k} p_i^{e_i}.$$

Let a be an integer. The ***Jacobi symbol*** $\left(\frac{a}{n}\right)$ is defined to be

$$\left(\frac{a}{n}\right) = \prod_{i=1}^{k} \left(\frac{a}{p_i}\right)^{e_i}.$$

Example 6.7 Consider the Jacobi symbol $\left(\frac{6278}{9975}\right)$. The prime power factorization of

Algorithm 6.6: SOLOVAY-STRASSEN(n)

choose a random integer a such that $1 \leq a \leq n-1$
$x \leftarrow \left(\frac{a}{n}\right)$
if $x = 0$
  then return ("n is composite")
$y \leftarrow a^{(n-1)/2} \pmod n$
if $x \equiv y \pmod n$
  then return ("n is prime")
  else return ("n is composite")

9975 is $9975 = 3 \times 5^2 \times 7 \times 19$. Thus we have

$$\begin{aligned}
\left(\frac{6278}{9975}\right) &= \left(\frac{6278}{3}\right)\left(\frac{6278}{5}\right)^2\left(\frac{6278}{7}\right)\left(\frac{6278}{19}\right) \\
&= \left(\frac{2}{3}\right)\left(\frac{3}{5}\right)^2\left(\frac{6}{7}\right)\left(\frac{8}{19}\right) \\
&= (-1)(-1)^2(-1)(-1) \\
&= -1.
\end{aligned}$$

□

Suppose $n > 1$ is odd. If n is prime, then $\left(\frac{a}{n}\right) \equiv a^{(n-1)/2} \pmod n$ for any a. On the other hand, if n is composite, it may or may not be the case that $\left(\frac{a}{n}\right) \equiv a^{(n-1)/2} \pmod n$. If this congruence holds, then n is called an ***Euler pseudo-prime*** to the base a. For example, 91 is an Euler pseudo-prime to the base 10, because

$$\left(\frac{10}{91}\right) = -1 \equiv 10^{45} \pmod{91}.$$

It can be shown that, for any odd composite n, n is an Euler pseudo-prime to the base a for at most half of the integers $a \in \mathbb{Z}_n{}^*$ (see the Exercises). It is also easy to see that $\left(\frac{a}{n}\right) = 0$ if and only if $\gcd(a, n) > 1$ (therefore, if $1 \leq a \leq n-1$ and $\left(\frac{a}{n}\right) = 0$, it must be the case that n is composite).

6.4.2 The Solovay-Strassen Algorithm

We present the SOLOVAY-STRASSEN ALGORITHM, as Algorithm 6.6. The facts proven in the previous section show that this is is a yes-biased Monte Carlo algorithm with error probability at most $1/2$.

At this point it is not clear that Algorithm 6.6 is a polynomial-time algorithm. We already know how to evaluate $a^{(n-1)/2} \bmod n$ in time $O((\log n)^3)$, but how do we compute Jacobi symbols efficiently? It might appear to be necessary to first factor n, since the Jacobi symbol $\left(\frac{a}{n}\right)$ is defined in terms of the factorization of n. But, if we could factor n, we would already know if it is prime; so this approach ends up in a vicious circle.

Fortunately, we can evaluate a Jacobi symbol without factoring n by using some results from number theory, the most important of which is a generalization of the law of quadratic reciprocity (property 4 below). We now enumerate these properties without proof:

1. If n is a positive odd integer and $m_1 \equiv m_2 \pmod{n}$, then
$$\left(\frac{m_1}{n}\right) = \left(\frac{m_2}{n}\right).$$

2. If n is a positive odd integer, then
$$\left(\frac{2}{n}\right) = \begin{cases} 1 & \text{if } n \equiv \pm 1 \pmod{8} \\ -1 & \text{if } n \equiv \pm 3 \pmod{8}. \end{cases}$$

3. If n is a positive odd integer, then
$$\left(\frac{m_1 m_2}{n}\right) = \left(\frac{m_1}{n}\right)\left(\frac{m_2}{n}\right).$$
In particular, if $m = 2^k t$ and t is odd, then
$$\left(\frac{m}{n}\right) = \left(\frac{2}{n}\right)^k \left(\frac{t}{n}\right).$$

4. Suppose m and n are positive odd integers. Then
$$\left(\frac{m}{n}\right) = \begin{cases} -\left(\frac{n}{m}\right) & \text{if } m \equiv n \equiv 3 \pmod{4} \\ \left(\frac{n}{m}\right) & \text{otherwise.} \end{cases}$$

Example 6.8 As an illustration of the application of these properties, we evaluate the Jacobi symbol $\left(\frac{7411}{9283}\right)$ in Figure 6.2. Notice that we successively apply properties 4, 1, 3, and 2 (in this order) in this computation. □

In general, by applying these four properties in the same manner as was done in the example above, it is possible to compute a Jacobi symbol $\left(\frac{a}{n}\right)$ in polynomial time. The only arithmetic operations that are required are modular reductions and factoring out powers of two. Note that if an integer is represented in binary notation, then factoring out powers of two amounts to determining the number of trailing zeroes. So, the complexity of the algorithm is determined by the number of modular reductions that must be done. It is not difficult to show that at most $O(\log n)$ modular reductions are performed, each of which can be done in time $O((\log n)^2)$. This shows that the complexity is $O((\log n)^3)$, which is polynomial

$$\begin{aligned}
\left(\frac{7411}{9283}\right) &= -\left(\frac{9283}{7411}\right) && \text{by property 4}\\
&= -\left(\frac{1872}{7411}\right) && \text{by property 1}\\
&= -\left(\frac{2}{7411}\right)^4\left(\frac{117}{7411}\right) && \text{by property 3}\\
&= -\left(\frac{117}{7411}\right) && \text{by property 2}\\
&= -\left(\frac{7411}{117}\right) && \text{by property 4}\\
&= -\left(\frac{40}{117}\right) && \text{by property 1}\\
&= -\left(\frac{2}{117}\right)^3\left(\frac{5}{117}\right) && \text{by property 3}\\
&= \left(\frac{5}{117}\right) && \text{by property 2}\\
&= \left(\frac{117}{5}\right) && \text{by property 4}\\
&= \left(\frac{2}{5}\right) && \text{by property 1}\\
&= -1 && \text{by property 2.}
\end{aligned}$$

FIGURE 6.2: Evaluation of a Jacobi symbol

in $\log n$. (In fact, the complexity can be shown to be $O((\log n)^2)$ by more precise analysis.)

Suppose that we have generated a random number n and tested it for primality using the SOLOVAY-STRASSEN ALGORITHM. If we have run the algorithm m times, what is our confidence that n is prime? It is tempting to conclude that the probability that such an integer n is prime is $1 - 2^{-m}$. This conclusion is often stated in both textbooks and technical articles, but it cannot be inferred from the given data.

We need to be careful about our use of probabilities. Suppose we define the following random variables: **a** denotes the event

"a random odd integer n of a specified size is composite,"

and **b** denotes the event

"the algorithm answers 'n is prime' m times in succession."

It is certainly the case that the probability $\mathbf{Pr}[\mathbf{b}|\mathbf{a}] \leq 2^{-m}$. However, the probability that we are really interested is $\mathbf{Pr}[\mathbf{a}|\mathbf{b}]$, which is usually not the same as $\mathbf{Pr}[\mathbf{b}|\mathbf{a}]$.

We can compute $\mathbf{Pr}[\mathbf{a}|\mathbf{b}]$ using Bayes' theorem (Theorem 3.1). In order to do this, we need to know $\mathbf{Pr}[\mathbf{a}]$. Suppose $N \leq n \leq 2N$. Applying the Prime number theorem, the number of (odd) primes between N and $2N$ is approximately

$$\frac{2N}{\ln 2N} - \frac{N}{\ln N} \approx \frac{N}{\ln N} \approx \frac{n}{\ln n}.$$

There are $N/2 \approx n/2$ odd integers between N and $2N$, so we estimate

$$\mathbf{Pr}[\mathbf{a}] \approx 1 - \frac{2}{\ln n}.$$

Then we can compute as follows:

$$\begin{aligned}
\mathbf{Pr}[\mathbf{a}|\mathbf{b}] &= \frac{\mathbf{Pr}[\mathbf{b}|\mathbf{a}]\,\mathbf{Pr}[\mathbf{a}]}{\mathbf{Pr}[\mathbf{b}]} \\
&= \frac{\mathbf{Pr}[\mathbf{b}|\mathbf{a}]\,\mathbf{Pr}[\mathbf{a}]}{\mathbf{Pr}[\mathbf{b}|\mathbf{a}]\,\mathbf{Pr}[\mathbf{a}] + \mathbf{Pr}[\mathbf{b}|\overline{\mathbf{a}}]\,\mathbf{Pr}[\overline{\mathbf{a}}]} \\
&\approx \frac{\mathbf{Pr}[\mathbf{b}|\mathbf{a}]\,(1 - \frac{2}{\ln n})}{\mathbf{Pr}[\mathbf{b}|\mathbf{a}]\,(1 - \frac{2}{\ln n}) + \frac{2}{\ln n}} \\
&= \frac{\mathbf{Pr}[\mathbf{b}|\mathbf{a}](\ln n - 2)}{\mathbf{Pr}[\mathbf{b}|\mathbf{a}](\ln n - 2) + 2} \\
&\leq \frac{2^{-m}(\ln n - 2)}{2^{-m}(\ln n - 2) + 2} \\
&= \frac{\ln n - 2}{\ln n - 2 + 2^{m+1}}.
\end{aligned}$$

Note that in this computation, $\overline{\mathbf{a}}$ denotes the event

"a random odd integer n is prime."

It is interesting to compare the two quantities $(\ln n - 2)/(\ln n - 2 + 2^{m+1})$ and 2^{-m} as a function of m. Suppose that $n \approx 2^{1024} \approx e^{710}$, since these are the sizes of primes p and q used to construct an RSA modulus. Then the first function is roughly $708/(708 + 2^{m+1})$. We tabulate the two functions for some values of m in Table 6.1.

Although $708/(708 + 2^{m+1})$ approaches zero exponentially quickly, it does not do so as quickly as 2^{-m}. In practice, however, one would take m to be something like 50 or 100, which will reduce the probability of error to a very small quantity.

6.4.3 The Miller-Rabin Algorithm

We now present another Monte Carlo algorithm for **Composites**, which is called the MILLER-RABIN ALGORITHM (this is also known as the ***strong pseudo-prime test***). This algorithm is presented as Algorithm 6.7.

Algorithm 6.7 is clearly a polynomial-time algorithm: an elementary analysis shows that its complexity is $O((\log n)^3)$, as is the SOLOVAY-STRASSEN ALGORITHM. In fact, the MILLER-RABIN ALGORITHM performs better in practice than the SOLOVAY-STRASSEN ALGORITHM.

TABLE 6.1: Error probabilities for the SOLOVAY-STRASSEN ALGORITHM

m	2^{-m}	bound on error probability
1	.500	.994
2	.250	.989
5	$.312 \times 10^{-1}$	.917
10	$.977 \times 10^{-3}$	.257
20	$.954 \times 10^{-6}$	$.337 \times 10^{-3}$
30	$.931 \times 10^{-9}$	$.330 \times 10^{-6}$
50	$.888 \times 10^{-15}$	$.314 \times 10^{-12}$
100	$.789 \times 10^{-30}$	$.279 \times 10^{-27}$

We show now that this algorithm cannot answer "n is composite" if n is prime, i.e., the algorithm is yes-biased.

THEOREM 6.11 *The* MILLER-RABIN ALGORITHM *for* ***Composites*** *is a yes-biased Monte Carlo algorithm.*

PROOF We will prove this by assuming that Algorithm 6.7 answers "n is composite" for some prime integer n, and obtain a contradiction. Since the algorithm answers "n is composite," it must be the case that $a^m \not\equiv 1 \pmod{n}$. Now consider the sequence of values b tested in the algorithm. Since b is squared in each iteration of the **for** loop, we are testing the values $a^m, a^{2m}, \ldots, a^{2^{k-1}m}$. Since the algorithm answers "n is composite," we conclude that

$$a^{2^i m} \not\equiv -1 \pmod{n}$$

for $0 \leq i \leq k-1$.

Now, using the assumption that n is prime, Fermat's theorem (Corollary 6.6) tells us that

$$a^{2^k m} \equiv 1 \pmod{n}$$

since $n - 1 = 2^k m$. Then $a^{2^{k-1}m}$ is a square root of 1 modulo n. Because n is prime, there are only two square roots of 1 modulo n, namely, $\pm 1 \bmod n$. We have that

$$a^{2^{k-1}m} \not\equiv -1 \pmod{n},$$

so it follows that

$$a^{2^{k-1}m} \equiv 1 \pmod{n}.$$

Then $a^{2^{k-2}m}$ must be a square root of 1. By the same argument,

$$a^{2^{k-2}m} \equiv 1 \pmod{n}.$$

Repeating this argument, we eventually obtain

$$a^m \equiv 1 \pmod{n},$$

Algorithm 6.7: MILLER-RABIN(n)

write $n-1 = 2^k m$, where m is odd
choose a random integer $a, 1 \leq a \leq n-1$
$b \leftarrow a^m \bmod n$
if $b \equiv 1 \pmod{n}$
  then return ("n is prime")
for $i \leftarrow 0$ **to** $k-1$
  do { **if** $b \equiv -1 \pmod{n}$
    then return ("n is prime")
    else $b \leftarrow b^2 \bmod n$ }
return ("n is composite")

which is a contradiction, since the algorithm would have answered "n is prime" in this case. ∎

It remains to consider the error probability of the MILLER-RABIN ALGORITHM. Although we will not prove it here, the error probability can be shown to be at most 1/4.

6.5 Square Roots Modulo n

In this section, we briefly discuss several useful results related to the existence of square roots modulo n. Throughout this section, we will suppose that n is odd and $\gcd(n, a) = 1$. The first question we will consider is the number of solutions $y \in \mathbb{Z}_n$ to the congruence $y^2 \equiv a \pmod{n}$. We already know from Section 6.4 that this congruence has either zero or two solutions if n is prime, depending on whether $\left(\frac{a}{n}\right) = -1$ or $\left(\frac{a}{n}\right) = 1$.

Our next theorem extends this characterization to (odd) prime powers. A proof is outlined in the Exercises.

THEOREM 6.12 *Suppose that p is an odd prime, e is a positive integer, and* $\gcd(a, p) = 1$. *Then the congruence $y^2 \equiv a \pmod{p^e}$ has no solutions if $\left(\frac{a}{p}\right) = -1$, and two solutions (modulo p^e) if $\left(\frac{a}{p}\right) = 1$.*

Notice that Theorem 6.12 tells us that the existence of square roots of a modulo p^e can be determined by evaluating the Legendre symbol $\left(\frac{a}{p}\right)$.

It is not difficult to extend Theorem 6.12 to the case of an arbitrary odd integer n. The following result is basically an application of the Chinese remainder theorem.

THEOREM 6.13 *Suppose that $n > 1$ is an odd integer having factorization*

$$n = \prod_{i=1}^{\ell} p_i^{e_i},$$

where the p_i's are distinct primes and the e_i's are positive integers. Suppose further that $\gcd(a, n) = 1$. Then the congruence $y^2 \equiv a \pmod{n}$ has 2^ℓ solutions modulo n if $\left(\frac{a}{p_i}\right) = 1$ for all $i \in \{1, \dots, \ell\}$, and no solutions, otherwise.

PROOF It is clear that $y^2 \equiv a \pmod{n}$ if and only if $y^2 \equiv a \pmod{p_i^{e_i}}$ for all $i \in \{1, \dots, \ell\}$. If $\left(\frac{a}{p_i}\right) = -1$ for some i, then the congruence $y^2 \equiv a \pmod{p_i^{e_i}}$ has no solutions, and hence $y^2 \equiv a \pmod{n}$ has no solutions.

Now suppose that $\left(\frac{a}{p_i}\right) = 1$ for all $i \in \{1, \dots, \ell\}$. It follows from Theorem 6.12 that each congruence $y^2 \equiv a \pmod{p_i^{e_i}}$ has two solutions modulo $p_i^{e_i}$, say $y \equiv b_{i,1}$ or $b_{i,2} \pmod{p_i^{e_i}}$. For $1 \leq i \leq \ell$, let $b_i \in \{b_{i,1}, b_{i,2}\}$. Then the system of congruences $y \equiv b_i \pmod{p_i^{e_i}}$ $(1 \leq i \leq \ell)$ has a unique solution modulo n, which can be found using the Chinese remainder theorem. There are 2^ℓ ways to choose the ℓ-tuple $(b_1, \dots, b_\ell)$, and therefore there are 2^ℓ solutions modulo n to the congruence $y^2 \equiv a \pmod{n}$. ∎

Suppose that $x^2 \equiv y^2 \equiv a \pmod{n}$, where $\gcd(a, n) = 1$. Let $z = xy^{-1} \bmod n$. It follows that $z^2 \equiv 1 \pmod{n}$. Conversely, if $z^2 \equiv 1 \pmod{n}$, then $(xz)^2 \equiv x^2 \pmod{n}$ for any x. It is therefore possible to obtain all 2^ℓ square roots of an element $a \in \mathbb{Z}_n{}^*$ by taking all 2^ℓ products of one given square root of a with the 2^ℓ square roots of 1. We will make use of this observation later in this chapter.

6.6 Factoring Algorithms

The most obvious way to attack the *RSA Cryptosystem* is to attempt to factor the public modulus. There is a huge amount of literature on factoring algorithms, and a thorough treatment would require more pages than we have in this book. We will just try to give a brief overview here, including an informal discussion of the best current factoring algorithms and their use in practice. The three algorithms that are most effective on very large numbers are the QUADRATIC SIEVE, the ELLIPTIC CURVE FACTORING ALGORITHM, and the NUMBER FIELD SIEVE. Other well-known algorithms that were precursors include Pollard's rho-method and $p - 1$ algorithm, Williams' $p + 1$ algorithm, the continued fraction algorithm, and, of course, trial division.

Throughout this section, we suppose that the integer n that we wish to factor is odd. If n is composite, then it is easy to see that n has a prime factor $p \leq \lfloor\sqrt{n}\rfloor$. Therefore, the simple method of ***trial division***, which consists of dividing n by every odd integer up to $\lfloor\sqrt{n}\rfloor$, suffices to determine if n is prime or composite. If

Algorithm 6.8: POLLARD $p-1$ FACTORING ALGORITHM(n, B)

```
a ← 2
for j ← 2 to B
  do a ← a^j mod n
d ← gcd(a − 1, n)
if 1 < d < n
  then return (d)
  else return ("failure")
```

$n < 10^{12}$, say, this is a perfectly reasonable factorization method, but for larger n we generally need to use more sophisticated techniques.

When we say that we want to factor n, we could ask for a complete factorization into primes, or we might be content with finding any non-trivial factor. In most of the algorithms we study, we are just searching for an arbitrary non-trivial factor. In general, we obtain factorizations of the form $n = n_1 n_2$, where $1 < n_1 < n$ and $1 < n_2 < n$. If we desire a complete factorization of n into primes, we could test n_1 and n_2 for primality using a randomized primality test, and then factor one or both of them further if they are not prime.

6.6.1 The Pollard $p-1$ Algorithm

As an example of a simple algorithm that can sometimes be applied to larger integers, we describe the POLLARD $p-1$ ALGORITHM, which dates from 1974. This algorithm, presented as Algorithm 6.8, has two inputs: the (odd) integer n to be factored, and a prespecified "bound," B.

Here is what is taking place in the POLLARD $p-1$ ALGORITHM: Suppose p is a prime divisor of n, and suppose that $q \leq B$ for every prime power $q \mid (p-1)$. Then it must be the case that

$$(p-1) \mid B!$$

At the end of the **for** loop, we have that

$$a \equiv 2^{B!} \pmod{n}.$$

Since $p \mid n$, it must be the case that

$$a \equiv 2^{B!} \pmod{p}.$$

Now,

$$2^{p-1} \equiv 1 \pmod{p}$$

by Fermat's theorem. Since $(p-1) \mid B!$, it follows that

$$a \equiv 1 \pmod{p},$$

and hence $p \mid (a-1)$. Since we also have that $p \mid n$, we see that $p \mid d$, where $d = \gcd(a-1, n)$. The integer d will be a non-trivial divisor of n (unless $a = 1$). Having found a non-trivial factor d, we would then proceed to attempt to factor d and n/d if they are expected to be composite.

Here is an example to illustrate.

Example 6.9 Suppose $n = 15770708441$. If we apply Algorithm 6.8 with $B = 180$, then we find that $a = 11620221425$ and d is computed to be 135979. In fact, the complete factorization of n into primes is

$$15770708441 = 135979 \times 115979.$$

In this example, the factorization succeeds because 135978 has only "small" prime factors:

$$135978 = 2 \times 3 \times 131 \times 173.$$

Hence, by taking $B \geq 173$, it will be the case that $135978 \mid B!$, as desired. □

In the POLLARD $p-1$ ALGORITHM, there are $B-1$ modular exponentiations, each requiring at most $2 \log_2 B$ modular multiplications using the SQUARE-AND-MULTIPLY ALGORITHM. The gcd can be computed in time $O((\log n)^3)$ using the EXTENDED EUCLIDEAN ALGORITHM. Hence, the complexity of the algorithm is $O(B \log B (\log n)^2 + (\log n)^3)$. If the integer B is $O((\log n)^i)$ for some fixed integer i, then the algorithm is indeed a polynomial-time algorithm (as a function of $\log n$); however, for such a choice of B the probability of success will be very small. On the other hand, if we increase the size of B drastically, say to $\sqrt{n}$, then the algorithm is guaranteed to be successful, but it will be no faster than trial division.

Thus, the drawback of this method is that it requires n to have a prime factor p such that $p-1$ has only "small" prime factors. It would be very easy to construct an RSA modulus $n = pq$ that would resist factorization by this method. One would start by finding a large prime p_1 such that $p = 2p_1 + 1$ is also prime, and a large prime q_1 such that $q = 2q_1 + 1$ is also prime (using one of the Monte Carlo primality testing algorithms discussed in Section 6.4). Then the RSA modulus $n = pq$ will be resistant to factorization using the $p-1$ method.

The more powerful elliptic curve algorithm, developed by Lenstra in the mid-1980s, is in fact a generalization of the POLLARD $p-1$ ALGORITHM. The success of the elliptic curve method depends on the more likely situation that an integer "close to" p has only "small" prime factors. Whereas the $p-1$ method depends on a relation that holds in the group $\mathbb{Z}_p$, the elliptic curve method involves groups defined on elliptic curves modulo p.

6.6.2 The Pollard Rho Algorithm

Let p be the smallest prime divisor of n. Suppose there exist two integers $x, x' \in \mathbb{Z}_n$, such that $x \neq x'$ and $x \equiv x' \pmod{p}$. Then $p \leq \gcd(x - x', n) < n$, so we obtain a non-trivial factor of n by computing a greatest common divisor. (Note that the value of p does not need to be known ahead of time in order for this method to work.)

Suppose we try to factor n by first choosing a random subset $X \subseteq \mathbb{Z}_n$, and then computing $\gcd(x - x', n)$ for all distinct values $x, x' \in X$. This method will be successful if and only if the mapping $x \mapsto x \bmod p$ yields at least one collision for $x \in X$. This situation can be analyzed using the birthday paradox described in Section 5.2.2: if $|X| \approx 1.17\sqrt{p}$, then there is a 50% probability that there is at least one collision, and hence a non-trivial factor of n will be found. However, in order to find a collision of the form $x \bmod p = x' \bmod p$, we need to compute $\gcd(x - x', n)$. (We cannot explicitly compute the values $x \bmod p$ for $x \in X$, and sort the resulting list, as suggested in Section 5.2.2, because the value of p is not known.) This means that we would expect to compute more than $\binom{|X|}{2} > p/2$ greatest common divisors before finding a factor of n.

The POLLARD RHO ALGORITHM incorporates a variation of this technique that requires fewer gcd computations and less memory. Suppose that the function f is a polynomial with integer coefficients, e.g., $f(x) = x^2 + a$, where a is a small constant ($a = 1$ is a commonly used value). Let's assume that the mapping $x \mapsto f(x) \bmod p$ behaves like a random mapping. (It is, of course, not "random," which means that what we are presenting is a heuristic analysis rather than a rigorous proof.) Let $x_1 \in \mathbb{Z}_n$, and consider the sequence $x_1, x_2, \ldots$, where

$$x_j = f(x_{j-1}) \bmod n,$$

for all $j \geq 2$. Let m be an integer, and define $X = \{x_1, \ldots, x_m\}$. To simplify matters, suppose that X consists of m distinct residues modulo n. Hopefully it will be the case that X is a random subset of m elements of $\mathbb{Z}_n$.

We are looking for two distinct values $x_i, x_j \in X$ such that $\gcd(x_j - x_i, n) > 1$. Each time we compute a new term x_j in the sequence, we could compute $\gcd(x_j - x_i, n)$ for all $i < j$. However, it turns out that we can reduce the number of gcd computations greatly. We describe how this can be done.

Suppose that $x_i \equiv x_j \pmod{p}$. Using the fact that f is a polynomial with integer coefficients, we have that $f(x_i) \equiv f(x_j) \pmod{p}$. Recall that $x_{i+1} = f(x_i) \bmod n$ and $x_{j+1} = f(x_j) \bmod n$. Then

$$x_{i+1} \bmod p = (f(x_i) \bmod n) \bmod p = f(x_i) \bmod p,$$

because $p \mid n$. Similarly,

$$x_{j+1} \bmod p = f(x_j) \bmod p.$$

Therefore, $x_{i+1} \equiv x_{j+1} \pmod{p}$. Repeating this argument, we obtain the following important result:

If $x_i \equiv x_j \pmod{p}$, then $x_{i+\delta} \equiv x_{j+\delta} \pmod{p}$ for all integers $\delta \geq 0$.

Denoting $\ell = j - i$, it follows that $x_{i'} \equiv x_{j'} \pmod{p}$ if $j' > i' \geq i$ and $j' - i' \equiv 0 \pmod{\ell}$.

Suppose that we construct a graph G on vertex set $\mathbb{Z}_p$, where for all $i \geq 1$, we have a directed edge from $x_i \bmod p$ to $x_{i+1} \bmod p$. There must exist a first pair x_i, x_j with $i < j$ such that $x_i \equiv x_j \pmod{p}$. By the observation made above, it is easily seen that the graph G consists of a "tail"

$$x_1 \bmod p \rightarrow x_2 \bmod p \rightarrow \cdots \rightarrow x_i \bmod p\ ,$$

and an infinitely repeated cycle of length ℓ, having vertices

$$x_i \bmod p \rightarrow x_{i+1} \bmod p \rightarrow \cdots \rightarrow x_j \bmod p = x_i \bmod p.$$

Thus G looks like the Greek letter ρ, which is the reason for the name "rho algorithm."

We illustrate the above with an example.

Example 6.10 Suppose that $n = 7171 = 71 \times 101$, $f(x) = x^2 + 1$, and $x_1 = 1$. The sequence of x_i's begins as follows:

1	2	5	26	677	6557	4105
6347	4903	2218	219	4936	4210	4560
4872	375	4377	4389	2016	5471	88

The above values, when reduced modulo 71, are as follows:

1	2	5	26	38	25	58
28	4	17	6	37	21	16
44	20	46	58	28	4	17

The first collision in the above list is

$$x_7 \bmod 71 = x_{18} \bmod 71 = 58.$$

Therefore the graph G consists of a tail of length seven and a cycle of length 11. □

We have already mentioned that our goal is to discover two terms $x_i \equiv x_j \pmod{p}$ with $i < j$, by computing a greatest common divisor. It is not necessary that we discover the first occurrence of a collision of this type. In order to simplify and improve the algorithm, we restrict our search for collisions by taking $j = 2i$. The resulting algorithm is presented as Algorithm 6.9.

This algorithm is not hard to analyze. If $x_i \equiv x_j \pmod{p}$, then it is also the case that $x_{i'} = x_{2i'} \pmod{p}$ for all i' such that $i' \equiv 0 \pmod{\ell}$ and $i' \geq i$. Among the ℓ consecutive integers $i, \ldots, j-1$, there must be one that is divisible by ℓ. Therefore the smallest value i' that satisfies the two conditions above is at most $j - 1$. Hence, the number of iterations required to find a factor p is at most j, which is expected to be at most $\sqrt{p}$.

Algorithm 6.9: POLLARD RHO FACTORING ALGORITHM(n, x_1)

external f
$x \leftarrow x_1$
$x' \leftarrow f(x) \bmod n$
$p \leftarrow \gcd(x - x', n)$
while $p = 1$
do {
  comment: in the ith iteration, $x = x_i$ and $x' = x_{2i}$
  $x \leftarrow f(x) \bmod n$
  $x' \leftarrow f(x') \bmod n$
  $x' \leftarrow f(x') \bmod n$
  $p \leftarrow \gcd(x - x', n)$
}
if $p = n$
  then return ("failure")
  else return (p)

In Example 6.10, the first collision modulo 71 occurs for $i = 7, j = 18$. The smallest integer $i' \geq 7$ that is divisible by 11 is $i' = 11$. Therefore Algorithm 6.9 will discover the factor 71 of n when it computes $\gcd(x_{11} - x_{22}, n) = 71$.

In general, since $p < \sqrt{n}$, the expected complexity of the algorithm is $O(n^{1/4})$ (ignoring logarithmic factors). However, we again emphasize that this is a heuristic analysis, and not a mathematical proof. On the other hand, the actual performance of the algorithm in practice is similar to this estimate.

It is possible that Algorithm 6.9 could fail to find a nontrivial factor of n. This happens if and only if the first values x and x' that satisfy $x \equiv x' \pmod{p}$ actually satisfy $x \equiv x' \pmod{n}$ (this is equivalent to $x = x'$, because x and x' are reduced modulo n). We would estimate heuristically that the probability of this situation occurring is roughly p/n, which is quite small when n is large, because $p < \sqrt{n}$. If the algorithm does fail in this way, it is a simple matter to run it again with a different initial value or a different choice for the function f.

The reader might wish to run Algorithm 6.9 on a larger value of n. When $n = 15770708441$ (the same value of n considered in Example 6.9), $x_1 = 1$ and $f(x) = x^2 + 1$, it can be verified that $x_{422} = 2261992698$, $x_{211} = 7149213937$, and

$$\gcd(x_{422} - x_{211}, n) = 135979.$$

6.6.3 Dixon's Random Squares Algorithm

Many factoring algorithms are based on the following very simple idea. Suppose we can find $x \not\equiv \pm y \pmod{n}$ such that $x^2 \equiv y^2 \pmod{n}$. Then

$$n \mid (x - y)(x + y)$$

but neither of $x - y$ or $x + y$ is divisible by n. It therefore follows that $\gcd(x + y, n)$ is a non-trivial factor of n (and similarly, $\gcd(x - y, n)$ is also a non-trivial factor of n).

As an example, it is easy to verify that $10^2 \equiv 32^2 \pmod{77}$. By computing $\gcd(10 + 32, 77) = 7$, we discover the factor 7 of 77.

The RANDOM SQUARES ALGORITHM uses a ***factor base***, which is a set $\mathcal{B}$ of the b smallest primes, for an appropriate value b. We first obtain several integers z such that all the prime factors of $z^2 \bmod n$ occur in the factor base $\mathcal{B}$. (How this is done will be discussed a bit later.) The idea is to then take the product of a subset of these z's in such a way that every prime in the factor base is used an even number of times. This then gives us a congruence of the desired type $x^2 \equiv y^2 \pmod{n}$, which (we hope) will lead to a factorization of n.

We illustrate with a carefully contrived example.

Example 6.11 Suppose $n = 15770708441$ (this was the same n that we used in Example 6.9). Let $b = 6$; then $\mathcal{B} = \{2, 3, 5, 7, 11, 13\}$. Consider the three congruences:

$$\begin{aligned} 8340934156^2 &\equiv 3 \times 7 \pmod{n} \\ 12044942944^2 &\equiv 2 \times 7 \times 13 \pmod{n} \\ 2773700011^2 &\equiv 2 \times 3 \times 13 \pmod{n}. \end{aligned}$$

If we take the product of these three congruences, then we have

$$(8340934156 \times 12044942944 \times 2773700011)^2 \equiv (2 \times 3 \times 7 \times 13)^2 \pmod{n}.$$

Reducing the expressions inside the parentheses modulo n, we have

$$9503435785^2 \equiv 546^2 \pmod{n}.$$

Then, using the EUCLIDEAN ALGORITHM, we compute

$$\gcd(9503435785 - 546, 15770708441) = 115759,$$

finding the factor 115759 of n. □

Suppose $\mathcal{B} = \{p_1, \ldots, p_b\}$ is the factor base. Let c be slightly larger than b (say $c = b + 4$), and suppose we have obtained c congruences:

$$z_j^2 \equiv p_1^{\alpha_{1j}} \times p_2^{\alpha_{2j}} \cdots \times p_b^{\alpha_{bj}} \pmod{n},$$

$1 \leq j \leq c$. For each j, consider the vector

$$a_j = (\alpha_{1j} \bmod 2, \ldots, \alpha_{bj} \bmod 2) \in (\mathbb{Z}_2)^b.$$

If we can find a subset of the a_j's that sum modulo 2 to the vector $(0, \ldots, 0)$, then the product of the corresponding z_j's will use each factor in $\mathcal{B}$ an even number of times.

We illustrate by returning to Example 6.11, where there exists a dependence, even though $c < b$ in this example.

Example 6.11 (Cont.) The three vectors a_1, a_2, a_3 are as follows:

$$\begin{aligned} a_1 &= (0,1,0,1,0,0) \\ a_2 &= (1,0,0,1,0,1) \\ a_3 &= (1,1,0,0,0,1). \end{aligned}$$

It is easy to see that

$$a_1 + a_2 + a_3 = (0,0,0,0,0,0) \bmod 2.$$

This gives rise to the congruence we saw earlier that successfully factored n. □

Observe that finding a subset of the c vectors $a_1, \dots, a_c$ that sums modulo 2 to the all-zero vector is nothing more than finding a linear dependence (over $\mathbb{Z}_2$) of these vectors. Provided $c > b$, such a linear dependence must exist, and it can be found easily using the standard method of Gaussian elimination. The reason why we take $c > b + 1$ is that there is no guarantee that any given congruence $x^2 \equiv y^2 \pmod{n}$ will yield the factorization of n. However, we argue heuristically that $x \equiv \pm y \pmod{n}$ at most 50% of the time, as follows. Suppose that $x^2 \equiv y^2 \equiv a \pmod{n}$, where $\gcd(a, n) = 1$. Theorem 6.13 tells us that a has 2^ℓ square roots modulo n, where ℓ is the number of prime divisors of n. If $\ell \geq 2$, then a has at least four square roots. Hence, if we assume that x and y are "random," we can then conclude that $x \equiv \pm y \pmod{n}$ with probability $2/2^\ell \leq 1/2$.

Now, if $c > b + 1$, we can obtain several such congruences of the form $x^2 \equiv y^2 \pmod{n}$ (arising from different linear dependencies among the a_j's). Hopefully, at least one of the resulting congruences will yield a congruence of the form $x^2 \equiv y^2 \bmod n$ where $x \not\equiv \pm y \pmod{n}$, and a non-trivial factor of n will be obtained.

We now discuss how to obtain integers z such that the values $z^2 \bmod n$ factor completely over a given factor base $\mathcal{B}$. There are several methods of doing this. One way is simply to choose the z's at random; this approach yields the so-called RANDOM SQUARES ALGORITHM. However, it is particularly useful to try integers of the form $j + \lceil\sqrt{kn}\rceil$, $j = 0, 1, 2, \dots$, $k = 1, 2, \dots$. These integers tend to be small when squared and reduced modulo n, and hence they have a higher than average probability of factoring over $\mathcal{B}$. Another useful trick is to try integers of the form $z = \lfloor\sqrt{kn}\rfloor$. When squared and reduced modulo n, these integers are a bit less than n. This means that $-z^2 \bmod n$ is small and can perhaps be factored over $\mathcal{B}$. Therefore, if we include -1 in $\mathcal{B}$, we can factor $z^2 \bmod n$ over $\mathcal{B}$.

We illustrate these techniques with a small example.

Example 6.12 Suppose that $n = 1829$ and $\mathcal{B} = \{-1, 2, 3, 5, 7, 11, 13\}$. We compute $\sqrt{n} = 42.77$, $\sqrt{2n} = 60.48$, $\sqrt{3n} = 74.07$, and $\sqrt{4n} = 85.53$. Suppose we take $z = 42, 43, 60, 61, 74, 75, 85, 86$. We obtain several factorizations of $z^2 \bmod n$ over

$\mathcal{B}$. In the following table, all congruences are modulo n:

$$\begin{array}{lllllll} {z_1}^2 & \equiv & 42^2 & \equiv & -65 & \equiv & (-1) \times 5 \times 13 \\ {z_2}^2 & \equiv & 43^2 & \equiv & 20 & \equiv & 2^2 \times 5 \\ {z_3}^2 & \equiv & 61^2 & \equiv & 63 & \equiv & 3^2 \times 7 \\ {z_4}^2 & \equiv & 74^2 & \equiv & -11 & \equiv & (-1) \times 11 \\ {z_5}^2 & \equiv & 85^2 & \equiv & -91 & \equiv & (-1) \times 7 \times 13 \\ {z_6}^2 & \equiv & 86^2 & \equiv & 80 & \equiv & 2^4 \times 5. \end{array}$$

We therefore have six factorizations, which yield six vectors in $(\mathbb{Z}_2)^7$. This is not enough to guarantee a dependence relation, but it turns out to be sufficient in this particular case. The six vectors are as follows:

$$\begin{array}{lll} a_1 & = & (1,0,0,1,0,0,1) \\ a_2 & = & (0,0,0,1,0,0,0) \\ a_3 & = & (0,0,0,0,1,0,0) \\ a_4 & = & (1,0,0,0,0,1,0) \\ a_5 & = & (1,0,0,0,1,0,1) \\ a_6 & = & (0,0,0,1,0,0,0). \end{array}$$

Clearly $a_2 + a_6 = (0,0,0,0,0,0,0)$; however, the reader can check that this dependence relation does not yield a factorization of n. A dependence relation that does work is

$$a_1 + a_2 + a_3 + a_5 = (0,0,0,0,0,0,0).$$

The congruence that we obtain is

$$(42 \times 43 \times 61 \times 85)^2 \equiv (2 \times 3 \times 5 \times 7 \times 13)^2 \pmod{1829}.$$

This simplifies to give

$$1459^2 \equiv 901^2 \pmod{1829}.$$

It is then straightforward to compute

$$\gcd(1459 + 901, 1829) = 59,$$

and thus we have obtained a nontrivial factor of n. □

An important general question is how large the factor base should be (as a function of the integer n that we are attempting to factor) and what the complexity of the algorithm is. In general, there is a trade-off: if $b = |\mathcal{B}|$ is large, then it is more likely that an integer $z^2 \bmod n$ factors over $\mathcal{B}$. But the larger b is, the more congruences we need to accumulate before we are able to find a dependence relation. A good choice for b can be determined with the help of some results from number theory. We discuss some of the main ideas now. This will be a heuristic analysis in which we will be assuming that the integers z are chosen randomly.

Suppose that n and m are positive integers. We say that the integer n is ***m*-smooth** provided that every prime factor of n is less than or equal to m. $\Psi(n,m)$ is defined to be the number of positive integers less than or equal to n that are m-smooth. An important result in number theory says that, if $n \gg m$, then

$$\frac{\Psi(n,m)}{n} \approx \frac{1}{u^u},$$

where $u = \log n / \log m$. Observe that $\Psi(n,m)/n$ represents the probability that a random integer in the set $\{1,\ldots,n\}$ is m-smooth.

Suppose that $n \approx 2^r$ and $m \approx 2^s$. Then

$$u = \frac{\log n}{\log m} \approx \frac{r}{s}.$$

Division of an r-bit integer by an s-bit integer can be done in time $O(rs)$. From this, it is possible to show that we can determine if an integer in the set $\{1,\ldots,n\}$ is m-smooth in time $O(rsm)$ if we assume that $r < m$ (see the Exercises).

Our factor base $\mathcal{B}$ consists of all the primes less than or equal to m. Therefore, applying the Prime number theorem, we have that

$$|\mathcal{B}| = b = \pi(m) \approx \frac{m}{\ln m}.$$

We need to find slightly more than b m-smooth squares modulo n in order for the algorithm to succeed. We expect to test bu^u integers in order to find b of them that are m-smooth. Therefore, the expected time to find the necessary m-smooth squares is $O(bu^u \times rsm)$. We have that b is $O(m/s)$, so the running time of the first part of the algorithm is $O(rm^2u^u)$.

In the second part of the algorithm, we need to reduce the associated matrix modulo 2, construct our congruence of the form $x^2 \equiv y^2 \pmod{n}$, and apply the EUCLIDEAN ALGORITHM. It can be checked without too much difficulty that these steps can be done in time that is polynomial in r and m, say $O(r^i m^j)$, where i and j are positive integers. (On average, this second part of the algorithm must be done at most twice, because the probability that a congruence does not provide a factor of n is at most $1/2$. This contributes a constant factor of at most 2, which is absorbed into the big-oh.)

At this point, we know that the total running time of the algorithm can be written in the form $O(rm^2u^u + r^i m^j)$. Recall that $n \approx 2^r$ is given, and we are trying to choose $m \approx 2^s$ to optimize the running time. It turns out that a good choice for m is to take $s \approx \sqrt{r \log_2 r}$. Then

$$u \approx \frac{r}{s} \approx \sqrt{\frac{r}{\log_2 r}}.$$

Now we can compute

$$\begin{aligned}\log_2 u^u &= u \log_2 u \\ &\approx \sqrt{\frac{r}{\log_2 r}} \log_2 \left(\sqrt{\frac{r}{\log_2 r}}\right) \\ &< \sqrt{\frac{r}{\log_2 r}} \log_2 \sqrt{r} \\ &= \sqrt{\frac{r}{\log_2 r}} \times \frac{\log_2 r}{2} \\ &= \frac{\sqrt{r \log_2 r}}{2}.\end{aligned}$$

It follows that

$$u^u \leq 2^{0.5\sqrt{r \log_2 r}}.$$

We also have that

$$m \approx 2^{\sqrt{r \log_2 r}}$$

and

$$r = 2^{\log_2 r}.$$

Hence, the total running time can be expressed in the form

$$O\left(2^{\log_2 r + 2\sqrt{r \log_2 r} + 0.5\sqrt{r \log_2 r}} + 2^{i \log_2 r + j\sqrt{r \log_2 r}}\right),$$

which is easily seen to be

$$O\left(2^{c\sqrt{r \log_2 r}}\right)$$

for some constant c. Using the fact that $r \approx \log_2 n$, we obtain a running time of

$$O\left(2^{c\sqrt{\log_2 n \log_2 \log_2 n}}\right).$$

Often the running time is expressed in terms of logarithms and exponentials to the base e. A more precise analysis, using an optimal choice of m, leads to the following commonly stated expected running time:

$$O\left(e^{(1+o(1))\sqrt{\ln n \ln \ln n}}\right).$$

The notation $o(1)$ denotes a function of n that approaches 0 as $n \to \infty$.

6.6.4 Factoring Algorithms in Practice

One specific, well-known algorithm that has been widely used in practice is the QUADRATIC SIEVE due to Pomerance. The name "quadratic sieve" comes from a sieving procedure (which we will not describe here) that is used to determine the values $z^2 \bmod n$ that factor over $\mathcal{B}$. The NUMBER FIELD SIEVE is a more recent factoring algorithm from the late 1980s. It also factors n by constructing a congruence

$x^2 \equiv y^2 \pmod{n}$, but it does so by means of computations in rings of algebraic integers. In recent years, the number field sieve has become the algorithm of choice for factoring large integers.

The asymptotic running times of the QUADRATIC SIEVE, ELLIPTIC CURVE, and NUMBER FIELD SIEVE factoring algorithms are as follows:

quadratic sieve	$O\left(e^{(1+o(1))\sqrt{\ln n \ln \ln n}}\right)$
elliptic curve	$O\left(e^{(1+o(1))\sqrt{2 \ln p \ln \ln p}}\right)$
number field sieve	$O\left(e^{(1.92+o(1))(\ln n)^{1/3}(\ln \ln n)^{2/3}}\right)$

In the above, p denotes the smallest prime factor of n. In the worst case, $p \approx \sqrt{n}$ and the asymptotic running times of the QUADRATIC SIEVE and ELLIPTIC CURVE algorithms are essentially the same. But in such a situation, QUADRATIC SIEVE is faster than ELLIPTIC CURVE. The ELLIPTIC CURVE ALGORITHM is more useful if the prime factors of n are of differing size. One very large number that was factored using the ELLIPTIC CURVE ALGORITHM was the Fermat number $2^{2^{11}} + 1$, which was factored in 1988 by Brent.

For factoring RSA moduli (where $n = pq$, p, q are prime, and p and q are roughly the same size), the QUADRATIC SIEVE was the most-used algorithm up until the mid-1990s. The number field sieve is the most recently developed of the three algorithms. Its advantage over the other algorithms is that its asymptotic running time is faster than either QUADRATIC SIEVE or ELLIPTIC CURVE. The NUMBER FIELD SIEVE has proven to be faster for numbers having more than about 125–130 digits. An early use of the NUMBER FIELD SIEVE was in 1990, when Lenstra, Lenstra, Manasse, and Pollard factored $2^{2^9} + 1$ into three primes having 7, 49, and 99 digits.

Some notable milestones in factoring have included the following factorizations. In 1983, the QUADRATIC SIEVE successfully factored a 69-digit number that was a (composite) factor of $2^{251} - 1$ (a computation that was done by Davis, Holdridge, and Simmons). Progress continued throughout the 1980s, and by 1989, numbers having up to 106 digits were factored by this method by Lenstra and Manasse, by distributing the computations to hundreds of widely separated workstations (they called this approach "factoring by electronic mail").

In the early 1990s, RSA published a series of "challenge" numbers for factoring algorithms on the Internet. The numbers were known as RSA-100, RSA-110, ..., RSA-500, where each number RSA-d is a d-digit integer that is the product of two primes of approximately the same length. Several of the smaller challenges were factored, culminating in the factorization of RSA-220 in May, 2016.

RSA put forward a new factoring challenge in 2001, where the size of the numbers involved were measured in bits rather than decimal digits. There were eight numbers in the newer RSA challenge: RSA-576, RSA-640, RSA-704, RSA-768, RSA-896, RSA-1024, RSA-1536 and RSA-2048. There were prizes for factoring these numbers, ranging from \$10,000 to \$200,000, but the challenge was terminated by RSA in 2007. Nevertheless, people have continued to try to solve additional challenges on the list. The largest of these challenges to be factored to date was RSA-768, which was factored in December, 2009. This factorization was accomplished using the NUMBER FIELD SIEVE.

6.7 Other Attacks on RSA

In this section, we address the following question: are there possible attacks on the *RSA Cryptosystem* other than factoring n? For example, it is at least conceivable that there could exist a method of decrypting RSA ciphertexts that does not involve finding the factorization of the modulus n.

6.7.1 Computing $\phi(n)$

We first observe that computing $\phi(n)$ is no easier than factoring n. For, if n and $\phi(n)$ are known, and n is the product of two primes p, q, then n can be easily factored, by solving the two equations

$$\begin{aligned} n &= pq \\ \phi(n) &= (p-1)(q-1) \end{aligned}$$

for the two "unknowns" p and q. This is easily accomplished, as follows. If we substitute $q = n/p$ into the second equation, we obtain a quadratic equation in the unknown value p:

$$p^2 - (n - \phi(n) + 1)p + n = 0. \tag{6.1}$$

The two roots of equation (6.1) will be p and q, the factors of n. Hence, if a cryptanalyst can learn the value of $\phi(n)$, then he can factor n and break the system. In other words, computing $\phi(n)$ is no easier than factoring n.

Here is an example to illustrate.

Example 6.13 Suppose $n = 84773093$, and the adversary has learned that $\phi(n) = 84754668$. This information gives rise to the following quadratic equation:

$$p^2 - 18426p + 84773093 = 0.$$

This can be solved by the quadratic formula, yielding the two roots 9539 and 8887. These are the two factors of n. □

6.7.2 The Decryption Exponent

We will now prove the very interesting result that, if the decryption exponent a is known, then n can be factored in polynomial time by means of a randomized algorithm. Therefore we can say that computing a is (essentially) no easier than factoring n. (However, this does not rule out the possibility of breaking the *RSA Cryptosystem* without computing a.) Notice that this result is of much more than theoretical interest. It tells us that if a is revealed (accidentally or otherwise), then it is not sufficient for Bob to choose a new encryption exponent; he must also choose a new modulus n.

The algorithm we are going to describe is a randomized algorithm of the Las Vegas type (see Section 5.2.2 for the definition). Here, we consider Las Vegas algorithms having worst-case success probability at least $1 - \epsilon$. Therefore, for any problem instance, the algorithm may fail to give an answer with probability at most ϵ.

If we have such a Las Vegas algorithm, then we simply run the algorithm over and over again until it finds an answer. The probability that the algorithm will return "no answer" m times in succession is ϵ^m. It follows that the average (i.e., expected) number of times the algorithm must be run in order to obtain an answer is $1/(1-\epsilon)$ (see the Exercises).

We will describe a Las Vegas algorithm that will factor n with probability at least $1/2$ when given the values a, b, and n as input. Hence, if the algorithm is run m times, then n will be factored with probability at least $1 - 1/2^m$.

The algorithm is based on certain facts concerning square roots of 1 modulo n, where $n = pq$ is the product of two distinct odd primes. $x^2 \equiv 1 \pmod{p}$ and Theorem 6.13 asserts that there are four square roots of 1 modulo n. Two of these square roots are $\pm 1 \bmod n$; these are called the ***trivial square roots*** of 1 modulo n. The other two square roots are called ***non-trivial square roots***; they are also negatives of each other modulo n.

Here is a small example to illustrate.

Example 6.14 Suppose $n = 403 = 13 \times 31$. The four square roots of 1 modulo 403 are $1, 92, 311$, and 402. The square root 92 is obtained by solving the system

$$\begin{aligned} x &\equiv 1 \pmod{13}, \\ x &\equiv -1 \pmod{31}. \end{aligned}$$

using the Chinese remainder theorem. The other non-trivial square root is $403 - 92 = 311$. It is the solution to the system

$$\begin{aligned} x &\equiv -1 \quad \pmod{13}, \\ x &\equiv 1 \quad \pmod{31}. \end{aligned}$$

□

Suppose x is a non-trivial square root of 1 modulo n. Then

$$x^2 \equiv 1^2 \pmod{n}$$

but

$$x \not\equiv \pm 1 \pmod{n}.$$

Then, as in the Random squares factoring algorithm, we can find the factors of n by computing $\gcd(x+1,n)$ and $\gcd(x-1,n)$. In Example 6.14 above,

$$\gcd(93,403) = 31$$

and

$$\gcd(312,403) = 13.$$

Algorithm 6.10 attempts to factor n by finding a non-trivial square root of 1 modulo n. Before analyzing the algorithm, we first do an example to illustrate its application.

Example 6.15 Suppose $n = 89855713$, $b = 34986517$, and $a = 82330933$, and the random value $w = 5$. We have

$$ab - 1 = 2^3 \times 360059073378795.$$

Then

$$w^r \bmod n = 85877701.$$

It happens that

$$85877701^2 \equiv 1 \pmod{n}.$$

Therefore the algorithm will return the value

$$x = \gcd(85877702, n) = 9103.$$

This is one factor of n; the other is $n/9103 = 9871$. □

Let's now proceed to the analysis of Algorithm 6.10. First, observe that if we are lucky enough to choose w to be a multiple of p or q, then we can factor n immediately. If w is relatively prime to n, then we compute $w^r, w^{2r}, w^{4r}, \ldots$, by successive squaring, until

$$w^{2^t r} \equiv 1 \pmod{n}$$

for some t. Since

$$ab - 1 = 2^s r \equiv 0 \pmod{\phi(n)},$$

we know that $w^{2^s r} \equiv 1 \pmod{n}$. Hence, the **while** loop terminates after at most s iterations. At the end of the **while** loop, we have found a value v_0 such that $(v_0)^2 \equiv 1 \pmod{n}$ but $v_0 \not\equiv 1 \pmod{n}$. If $v_0 \equiv -1 \pmod{n}$, then the algorithm fails; otherwise, v_0 is a non-trivial square root of 1 modulo n and we are able to factor n.

The main task facing us now is to prove that the algorithm succeeds with probability at least $1/2$. There are two ways in which the algorithm can fail to factor n:

1. $w^r \equiv 1 \pmod{n}$, or

2. $w^{2^t r} \equiv -1 \pmod{n}$ for some t, $0 \le t \le s-1$.

Algorithm 6.10: RSA-FACTOR(n, a, b)

comment: we are assuming that $ab \equiv 1 \pmod{\phi(n)}$

write $ab - 1 = 2^s r, r$ odd
choose w at random such that $1 \leq w \leq n-1$
$x \leftarrow \gcd(w, n)$
if $1 < x < n$
  then return (x)
comment: x is a factor of n

$v \leftarrow w^r \bmod n$
if $v \equiv 1 \pmod{n}$
  then return ("failure")
while $v \not\equiv 1 \pmod{n}$
  do $\begin{cases} v_0 \leftarrow v \\ v \leftarrow v^2 \bmod n \end{cases}$
if $v_0 \equiv -1 \pmod{n}$
  then return ("failure")
  else $\begin{cases} x \leftarrow \gcd(v_0 + 1, n) \\ \textbf{return } (x) \end{cases}$
comment: x is a factor of n

We have $s+1$ congruences to consider. If a random value w is a solution to at least one of these $s+1$ congruences, then it is a "bad" choice, and the algorithm fails. So we proceed by counting the number of solutions to each of these congruences.

First, consider the congruence $w^r \equiv 1 \pmod{n}$. The way to analyze a congruence such as this is to consider solutions modulo p and modulo q separately, and then combine them using the Chinese remainder theorem. Observe that $x \equiv 1 \pmod{n}$ if and only if $x \equiv 1 \pmod{p}$ and $x \equiv 1 \pmod{q}$.

So, we first consider $w^r \equiv 1 \pmod{p}$. Since p is prime, $\mathbb{Z}_p{}^*$ is a cyclic group by Theorem 6.7. Let g be a primitive element modulo p. We can write $w = g^u$ for a unique integer $u, 0 \leq u \leq p-2$. Then we have

$$\begin{aligned} w^r &\equiv 1 \pmod{p}, \\ g^{ur} &\equiv 1 \pmod{p}, \text{and hence} \\ (p-1) &\mid ur. \end{aligned}$$

Let us write

$$p - 1 = 2^i p_1$$

where p_1 is odd, and

$$q - 1 = 2^j q_1$$

where q_1 is odd. Since

$$\phi(n) = (p-1)(q-1) \mid (ab-1) = 2^s r,$$

we have that

$$2^{i+j} p_1 q_1 \mid 2^s r.$$

Hence

$$i + j \leq s$$

and

$$p_1 q_1 \mid r.$$

Now, the condition $(p-1) \mid ur$ becomes $2^i p_1 \mid ur$. Since $p_1 \mid r$ and r is odd, it is necessary and sufficient that $2^i \mid u$. Hence, $u = k2^i$, $0 \leq k \leq p_1 - 1$, and the number of solutions to the congruence $w^r \equiv 1 \pmod{p}$ is p_1.

By an identical argument, the congruence $w^r \equiv 1 \pmod{q}$ has exactly q_1 solutions. We can combine any solution modulo p with any solution modulo q to obtain a unique solution modulo n, using the Chinese remainder theorem. Consequently, the number of solutions to the congruence $w^r \equiv 1 \pmod{n}$ is $p_1 q_1$.

The next step is to consider a congruence $w^{2^t r} \equiv -1 \pmod{n}$ for a fixed value t (where $0 \leq t \leq s-1$). Again, we first look at the congruence modulo p and then modulo q (note that $w^{2^t r} \equiv -1 \pmod{n}$ if and only if $w^{2^t r} \equiv -1 \pmod{p}$ and $w^{2^t r} \equiv -1 \pmod{q}$). First, consider $w^{2^t r} \equiv -1 \pmod{p}$. Writing $w = g^u$, as above, we get

$$g^{u2^t r} \equiv -1 \pmod{p}.$$

Since $g^{(p-1)/2} \equiv -1 \pmod{p}$, we have that

$$\begin{aligned} u2^t r &\equiv \frac{p-1}{2} \pmod{p-1} \\ (p-1) &\mid \left(u2^t r - \frac{p-1}{2}\right) \\ 2(p-1) &\mid (u2^{t+1} r - (p-1)). \end{aligned}$$

Since $p - 1 = 2^i p_1$, we get

$$2^{i+1} p_1 \mid (u2^{t+1} r - 2^i p_1).$$

Taking out a common factor of p_1, this becomes

$$2^{i+1} \mid \left(\frac{u2^{t+1} r}{p_1} - 2^i\right).$$

Now, if $t \geq i$, then there can be no solutions since $2^{i+1} \mid 2^{t+1}$ but $2^{i+1} \nmid 2^i$. On the other hand, if $t \leq i-1$, then u is a solution if and only if u is an odd multiple of 2^{i-t-1} (note that r/p_1 is an odd integer). So, the number of solutions in this case is

$$\frac{p-1}{2^{i-t-1}} \times \frac{1}{2} = 2^t p_1.$$

By similar reasoning, the congruence $w^{2^t r} \equiv -1 \pmod{q}$ has no solutions if $t \geq j$, and $2^t q_1$ solutions if $t \leq j-1$. Using the Chinese remainder theorem, we see that the number of solutions of $w^{2^t r} \equiv -1 \pmod{n}$ is

$$\begin{array}{rl} 0 & \text{if } t \geq \min\{i,j\} \\ 2^{2t} p_1 q_1 & \text{if } t \leq \min\{i,j\} - 1. \end{array}$$

Now, t can range from 0 to $s-1$. Without loss of generality, suppose $i \leq j$; then the number of solutions is 0 if $t \geq i$. The total number of "bad" choices for w is at most

$$\begin{aligned} p_1 q_1 + p_1 q_1 (1 + 2^2 + 2^4 + \cdots + 2^{2i-2}) &= p_1 q_1 \left(1 + \frac{2^{2i}-1}{3}\right) \\ &= p_1 q_1 \left(\frac{2}{3} + \frac{2^{2i}}{3}\right). \end{aligned}$$

Recall that $p - 1 = 2^i p_1$ and $q - 1 = 2^j q_1$. Now, $j \geq i \geq 1$, so $p_1 q_1 < n/4$. We also have that

$$2^{2i} p_1 q_1 \leq 2^{i+j} p_1 q_1 = (p-1)(q-1) < n.$$

Hence, we obtain

$$\begin{aligned} p_1 q_1 \left(\frac{2}{3} + \frac{2^{2i}}{3}\right) &< \frac{n}{6} + \frac{n}{3} \\ &= \frac{n}{2}. \end{aligned}$$

Since at most $(n-1)/2$ choices for w are "bad," it follows that at least $(n-1)/2$ choices are "good" and hence the probability of success of the algorithm is at least $1/2$.

6.7.3 Wiener's Low Decryption Exponent Attack

As always, suppose that $n = pq$ where p and q are prime; then $\phi(n) = (p-1)(q-1)$. In this section, we present an attack, due to M. Wiener, that succeeds in computing the secret decryption exponent, a, whenever the following hypotheses are satisfied:

$$3a < n^{1/4} \quad \text{and} \quad q < p < 2q. \tag{6.2}$$

If n has ℓ bits in its binary representation, then the attack will work when a has fewer than $\ell/4 - 1$ bits in its binary representation and p and q are not too far apart.

Note that Bob might be tempted to choose his decryption exponent to be small in order to speed up decryption. If he uses Algorithm 6.5 to compute $y^a \bmod n$, then the running time of decryption will be reduced by roughly 75% if he chooses a value of a that satisfies (6.2). The results we prove in this section show that this method of reducing decryption time should be avoided.

Since $ab \equiv 1 \pmod{\phi(n)}$, it follows that there is an integer t such that

$$ab - t\phi(n) = 1.$$

Since $n = pq > q^2$, we have that $q < \sqrt{n}$. Hence,

$$0 < n - \phi(n) = p + q - 1 < 2q + q - 1 < 3q < 3\sqrt{n}.$$

Now, we see that

$$\begin{aligned} \left| \frac{b}{n} - \frac{t}{a} \right| &= \left| \frac{ba - tn}{an} \right| \\ &= \left| \frac{1 + t(\phi(n) - n)}{an} \right| \\ &< \frac{3t\sqrt{n}}{an} \\ &= \frac{3t}{a\sqrt{n}}. \end{aligned}$$

Since $t < a$, we have that $3t < 3a < n^{1/4}$, and hence

$$\left| \frac{b}{n} - \frac{t}{a} \right| < \frac{1}{an^{1/4}}.$$

Finally, since $3a < n^{1/4}$, we have that

$$\left| \frac{b}{n} - \frac{t}{a} \right| < \frac{1}{3a^2}.$$

Therefore the fraction t/a is a very close approximation to the fraction b/n. From the theory of continued fractions, it is known that any approximation of b/n that is this close must be one of the convergents of the continued fraction expansion of b/n (see Theorem 6.14). This expansion can be obtained from the EUCLIDEAN ALGORITHM, as we describe now.

A (finite) ***continued fraction*** is an m-tuple of non-negative integers, say

$$[q_1, \ldots, q_m],$$

which is shorthand for the following expression:

$$q_1 + \cfrac{1}{q_2 + \cfrac{1}{q_3 + \cdots + \frac{1}{q_m}}}.$$

Suppose a and b are positive integers such that $\gcd(a, b) = 1$, and suppose that the output of Algorithm 6.1 is the m-tuple $(q_1, \ldots, q_m)$. Then it is not hard to see that $a/b = [q_1, \ldots, q_m]$. We say that $[q_1, \ldots, q_m]$ is the ***continued fraction expansion*** of a/b in this case. Now, for $1 \leq j \leq m$, define $C_j = [q_1, \ldots, q_j]$. C_j is said to be the

jth ***convergent*** of $[q_1, \ldots, q_m]$. Each C_j can be written as a rational number c_j/d_j, where the c_j's and d_j's satisfy the following recurrences:

$$c_j = \begin{cases} 1 & \text{if } j = 0 \\ q_1 & \text{if } j = 1 \\ q_j c_{j-1} + c_{j-2} & \text{if } j \geq 2 \end{cases}$$

and

$$d_j = \begin{cases} 0 & \text{if } j = 0 \\ 1 & \text{if } j = 1 \\ q_j d_{j-1} + d_{j-2} & \text{if } j \geq 2. \end{cases}$$

Example 6.16 We compute the continued fraction expansion of 34/99. The EUCLIDEAN ALGORITHM proceeds as follows:

$$\begin{aligned} 34 &= 0 \times 99 + 34 \\ 99 &= 2 \times 34 + 31 \\ 34 &= 1 \times 31 + 3 \\ 31 &= 10 \times 3 + 1 \\ 3 &= 3 \times 1. \end{aligned}$$

Hence, the continued fraction expansion of 34/99 is $[0, 2, 1, 10, 3]$, i.e.,

$$\frac{34}{99} = 0 + \frac{1}{2 + \frac{1}{1 + \frac{1}{10 + \frac{1}{3}}}}.$$

The convergents of this continued fraction are as follows:

$$\begin{aligned} [0] &= 0 \\ [0, 2] &= 1/2 \\ [0, 2, 1] &= 1/3 \\ [0, 2, 1, 10] &= 11/32, \quad \text{and} \\ [0, 2, 1, 10, 3] &= 34/99. \end{aligned}$$

The reader can verify that these convergents can be computed using the recurrence relations given above. □

The convergents of a continued fraction expansion of a rational number satisfy many interesting properties. For our purposes, the most important property is the following.

THEOREM 6.14 *Suppose that* $\gcd(a, b) = \gcd(c, d) = 1$ *and*

$$\left| \frac{a}{b} - \frac{c}{d} \right| < \frac{1}{2d^2}.$$

Then c/d *is one of the convergents of the continued fraction expansion of* a/b.

```
Algorithm 6.11: WIENER'S ALGORITHM(n, b)

(q_1, ..., q_m; r_m) ← EUCLIDEAN ALGORITHM(b, n)
c_0 ← 1
c_1 ← q_1
d_0 ← 0
d_1 ← 1
for j ← 2 to m
      ⎧ c_j ← q_j c_{j-1} + c_{j-2}
      ⎪ d_j ← q_j d_{j-1} + d_{j-2}
      ⎪ n' ← (d_j b - 1)/c_j
      ⎪ comment: n' = φ(n) if c_j/d_j is the correct convergent
  do  ⎨ if n' is an integer
      ⎪              ⎧ let p and q be the roots of the equation
      ⎪              ⎪    x^2 - (n - n' + 1)x + n = 0
      ⎪       then   ⎨ if p and q are positive integers less than n
      ⎩              ⎩   then return (p, q)
return ("failure")
```

Now we can apply this result to the *RSA Cryptosystem*. We already observed that, if condition (6.2) holds, then the unknown fraction t/a is a close approximation to b/n. Theorem 6.14 tells us that t/a must be one of the convergents of the continued fraction expansion of b/n. Since the value of b/n is public information, it is a simple matter to compute its convergents. All we need is a method to test each convergent to see if it is the "right" one.

But this is also not difficult to do. If t/a is a convergent of b/n, then we can compute the value of $\phi(n)$ to be $\phi(n) = (ab-1)/t$. Once n and $\phi(n)$ are known, we can factor n by solving the quadratic equation (6.1) for p. We do not know ahead of time which convergent of b/n will yield the factorization of n, so we try each one in turn, until the factorization of n is found. If we do not succeed in factoring n by this method, then it must be the case that the hypotheses (6.2) are not satisfied.

A pseudocode description of WIENER'S ALGORITHM is presented as Algorithm 6.11.

We present an example to illustrate.

Example 6.17 Suppose that $n = 160523347$ and $b = 60728973$. The continued fraction expansion of b/n is

$$[0, 2, 1, 1, 1, 4, 12, 102, 1, 1, 2, 3, 2, 2, 36].$$

The first few convergents are

$$0, \frac{1}{2}, \frac{1}{3}, \frac{2}{5}, \frac{3}{8}, \frac{14}{37}.$$

Cryptosystem 6.2: *Rabin Cryptosystem*

Let $n = pq$, where p and q are primes and $p, q \equiv 3 \pmod 4$. Let $\mathcal{P} = \mathcal{C} = \mathbb{Z}_n{}^*$, and define

$$\mathcal{K} = \{(n, p, q)\}.$$

For $K = (n, p, q)$, define

$$e_K(x) = x^2 \bmod n$$

and

$$d_K(y) = \sqrt{y} \bmod n.$$

The value n is the public key, while p and q are the private key.

The reader can verify that the first five convergents do not produce a factorization of n. However, the convergent $14/37$ yields

$$n' = \frac{37 \times 60728973 - 1}{14} = 160498000.$$

Now, if we solve the equation

$$x^2 - 25348x + 160523347 = 0,$$

then we find the roots $x = 12347, 13001$. Therefore we have discovered the factorization

$$160523347 = 12347 \times 13001.$$

Notice that for the modulus $n = 160523347$, WIENER'S ALGORITHM will work for

$$a < \frac{n^{1/4}}{3} \approx 37.52.$$

□

6.8 The Rabin Cryptosystem

In this section, we describe the *Rabin Cryptosystem*, which is computationally secure against a chosen-plaintext attack provided that the modulus $n = pq$ cannot be factored. Therefore, the *Rabin Cryptosystem* provides an example of a provably secure cryptosystem: assuming that the problem of factoring is computationally infeasible, the *Rabin Cryptosystem* is secure. We present the *Rabin Cryptosystem* as Cryptosystem 6.2.

REMARK The requirement that $p, q \equiv 3 \pmod 4$ can be omitted. As well, the cryptosystem still "works" if we take $\mathcal{P} = \mathcal{C} = \mathbb{Z}_n$ instead of $\mathbb{Z}_n{}^*$. However, the more restrictive description we use simplifies some aspects of computation and analysis of the cryptosystem. ∎

One drawback of the *Rabin Cryptosystem* is that the encryption function e_K is not an injection, so decryption cannot be done in an unambiguous fashion. We prove this as follows. Suppose that y is a valid ciphertext; this means that $y = x^2 \bmod n$ for some $x \in \mathbb{Z}_n{}^*$. Theorem 6.13 proves that there are four square roots of y modulo n, which are the four possible plaintexts that encrypt to y. In general, there will be no way for Bob to distinguish which of these four possible plaintexts is the "right" plaintext, unless the plaintext contains sufficient redundancy to eliminate three of these four possible values.

Let us look at the decryption problem from Bob's point of view. He is given a ciphertext y and wants to determine x such that

$$x^2 \equiv y \pmod n.$$

This is a quadratic equation in $\mathbb{Z}_n$ in the unknown x, and decryption requires extracting square roots modulo n. This is equivalent to solving the two congruences

$$z^2 \equiv y \pmod p$$

and

$$z^2 \equiv y \pmod q.$$

We can use Euler's criterion to determine if y is a quadratic residue modulo p (and modulo q). In fact, y will be a quadratic residue modulo p and modulo q if encryption was performed correctly. Unfortunately, Euler's criterion does not help us find the square roots of y; it yields only an answer "yes" or "no."

When $p \equiv 3 \pmod 4$, there is a simple formula to compute square roots of quadratic residues modulo p. Suppose y is a quadratic residue modulo p, where $p \equiv 3 \pmod 4$. Then we have that

$$\begin{aligned}(\pm y^{(p+1)/4})^2 &\equiv y^{(p+1)/2} \pmod p\\ &\equiv y^{(p-1)/2} y \pmod p\\ &\equiv y \pmod p.\end{aligned}$$

Here we have again made use of Euler's criterion, which says that if y is a quadratic residue modulo p, then $y^{(p-1)/2} \equiv 1 \pmod p$. Hence, the two square roots of y modulo p are $\pm y^{(p+1)/4} \bmod p$. In a similar fashion, the two square roots of y modulo q are $\pm y^{(q+1)/4} \bmod q$. It is then straightforward to obtain the four square roots of y modulo n using the Chinese remainder theorem.

REMARK For $p \equiv 1 \pmod 4$, there is no known polynomial-time deterministic algorithm to compute square roots of quadratic residues modulo p. (There is a polynomial-time Las Vegas algorithm, however.) This is why we stipulated that $p, q \equiv 3 \pmod 4$ in the definition of the *Rabin Cryptosystem*. ∎

Example 6.18 Let's illustrate the encryption and decryption procedures for the *Rabin Cryptosystem* with a toy example. Suppose $n = 77 = 7 \times 11$. Then the encryption function is

$$e_K(x) = x^2 \bmod 77$$

and the decryption function is

$$d_K(y) = \sqrt{y} \bmod 77.$$

Suppose Bob wants to decrypt the ciphertext $y = 23$. It is first necessary to find the square roots of 23 modulo 7 and modulo 11. Since 7 and 11 are both congruent to 3 modulo 4, we use our formula:

$$23^{(7+1)/4} \equiv 2^2 \equiv 4 \pmod 7$$

and

$$23^{(11+1)/4} \equiv 1^3 \equiv 1 \pmod{11}.$$

Using the Chinese remainder theorem, we compute the four square roots of 23 modulo 77 to be $\pm 10, \pm 32 \bmod 77$. Therefore, the four possible plaintexts are $x = 10, 32, 45$, and 67. It can be verified that each of these plaintexts yields the value 23 when squared and reduced modulo 77. This proves that 23 is indeed a valid ciphertext. □

6.8.1 Security of the Rabin Cryptosystem

We now discuss the (provable) security of the *Rabin Cryptosystem*. The security proof uses a Turing reduction, which is defined in Definition 6.5.

A Turing reduction $\mathbf{G} \propto_T \mathbf{H}$ does not necessarily yield a polynomial-time algorithm to solve **G**. It actually proves the truth of the following implication:

If there exists a polynomial-time algorithm to solve **H**, then there exists a polynomial-time algorithm to solve **G**.

This is because any algorithm SOLVEH that solves **H** can be "plugged into" the algorithm SOLVEG, thereby producing an algorithm that solves **G**. Clearly this resulting algorithm will be a polynomial-time algorithm if SOLVEH is a polynomial-time algorithm.

We will provide an explicit example of a Turing reduction: We will prove that a decryption oracle RABIN DECRYPT can be incorporated into a Las Vegas algorithm, Algorithm 6.12, that factors the modulus n with probability at least $1/2$. In other words, we show that **Factoring** $\propto_T$ **Rabin decryption**, where the Turing reduction is itself a randomized algorithm. In Algorithm 6.12, we assume that n is the product of two distinct primes p and q; and RABIN DECRYPT is an oracle that performs

Definition 6.5: Suppose that **G** and **H** are problems. A ***Turing reduction*** from **G** to **H** is an algorithm SOLVEG with the following properties:

1. SOLVEG assumes the existence of an arbitrary algorithm SOLVEH that solves the problem **H**.
2. SOLVEG can call the algorithm SOLVEH and make use of any values it outputs, but SOLVEG cannot make any assumption about the actual computations performed by SOLVEH (in other words, SOLVEH is an oracle that is treated as a "black box").
3. SOLVEG is a polynomial-time algorithm, when each call to the oracle is regarded as taking $O(1)$ time. (Note that the complexity of SOLVEG takes into account all the computations that are done "outside" the oracle.)
4. SOLVEG correctly solves the problem **G**.

If there is a Turing reduction from **G** to **H**, we denote this by writing $\mathbf{G} \propto_T \mathbf{H}$.

Algorithm 6.12: RABIN ORACLE FACTORING(n)

external RABIN DECRYPT
choose a random integer $r \in \mathbb{Z}_n{}^*$
$y \leftarrow r^2 \bmod n$
$x \leftarrow$ RABIN DECRYPT(y)
if $x \equiv \pm r \pmod{n}$
  then return ("failure")
  else $\begin{cases} p \leftarrow \gcd(x+r, n) \\ q \leftarrow n/p \\ \textbf{return } (\text{"}n = p \times q\text{"}) \end{cases}$

Rabin decryption, returning one of the four possible plaintexts corresponding to a given ciphertext.

There are several points of explanation needed. First, observe that y is a valid ciphertext and RABIN DECRYPT(y) will return one of four possible plaintexts as the value of x. In fact, it holds that $x \equiv \pm r \pmod{n}$ or $x \equiv \pm \omega r \pmod{n}$, where ω is one of the non-trivial square roots of 1 modulo n. In the second case, we have $x^2 \equiv r^2 \pmod{n}$, $x \not\equiv \pm r \pmod{n}$. Hence, computation of $\gcd(x+r, n)$ must yield either p or q, and the factorization of n is accomplished.

Let's compute the probability of success of this algorithm, over all choices for the random value $r \in \mathbb{Z}_n{}^*$. For a residue $r \in \mathbb{Z}_n{}^*$, define

$$[r] = \{\pm r \bmod n, \pm \omega r \bmod n\}.$$

Clearly any two residues in $[r]$ yield the same y-value in Algorithm 6.12, and the value of x that is output by the oracle RABIN DECRYPT is also in $[r]$. We have already observed that Algorithm 6.12 succeeds if and only if $x \equiv \pm \omega r \pmod{n}$. The oracle does not know which of four possible r-values was used to construct y, and r was chosen at random before the oracle RABIN DECRYPT is called. Hence, the probability that $x \equiv \pm \omega r \pmod{n}$ is $1/2$. We conclude that the probability of success of Algorithm 6.12 is $1/2$.

We have shown that the *Rabin Cryptosystem* is provably secure against a chosen plaintext attack. However, the system is completely insecure against a chosen ciphertext attack. In fact, Algorithm 6.12 can be used to break the *Rabin Cryptosystem* in a chosen ciphertext attack! In the chosen ciphertext attack, the (hypothetical) oracle RABIN DECRYPT is replaced by an actual decryption algorithm. (Informally, the security proof says that a decryption oracle can be used to factor n; and a chosen ciphertext attack assumes that a decryption oracle exists. Together, these break the cryptosystem!)

6.9 Semantic Security of RSA

To this point in the text, we have assumed that an adversary trying to break a cryptosystem is actually trying to determine the secret key (in the case of a secret-key cryptosystem) or the private key (in the case of a public-key cryptosystem). If Oscar can do this, then the system is completely broken. However, it is possible that the goal of an adversary is somewhat less ambitious. Even if Oscar cannot find the secret or private key, he still may be able to gain more information than we would like. If we want to be assured that the cryptosystem is "secure," we should take into account these more modest goals that an adversary might have.

Here is a short list of potential adversarial goals:

total break

The adversary is able to determine Bob's private key (in the case of a public-key cryptosystem) or the secret key (in the case of a secret-key cryptosystem). Therefore he can decrypt any ciphertext that has been encrypted using the given key.

partial break

With some non-negligible probability, the adversary is able to decrypt a previously unseen ciphertext (without knowing the key). Or, the adversary can determine some specific information about the plaintext, given the ciphertext.

distinguishability of ciphertexts

With some probability exceeding 1/2, the adversary is able to distinguish between encryptions of two given plaintexts, or between an encryption of a given plaintext and a random string.

In the next sections, we will consider some possible attacks against RSA-like cryptosystems that achieve some of these types of goals. We also describe how to construct a public-key cryptosystem in which the adversary cannot (in polynomial time) distinguish ciphertexts, provided that certain computational assumptions hold. Such cryptosystems are said to achieve ***semantic security***. Achieving semantic security is quite difficult, because we are providing protection against a very weak, and therefore easy to achieve, adversarial goal.

6.9.1 Partial Information Concerning Plaintext Bits

A weakness of some cryptosystems is the fact that partial information about the plaintext might be "leaked" by the ciphertext. This represents a type of partial break of the system, and it happens, in fact, in the *RSA Cryptosystem*. Suppose we are given a ciphertext, $y = x^b \bmod n$, where x is the plaintext. Since $\gcd(b, \phi(n)) = 1$, it must be the case that b is odd. Therefore the Jacobi symbol

$$\left(\frac{y}{n}\right) = \left(\frac{x}{n}\right)^b = \left(\frac{x}{n}\right).$$

Hence, given the ciphertext y, anyone can efficiently compute $\left(\frac{x}{n}\right)$ without decrypting the ciphertext. In other words, an RSA encryption "leaks" some information concerning the plaintext x, namely, the value of the Jacobi symbol $\left(\frac{x}{n}\right)$.

In this section, we consider some other specific types of partial information that could be leaked by a cryptosystem:

1. given $y = e_K(x)$, compute $\mathbf{parity}(y)$, where $\mathbf{parity}(y)$ denotes the low-order bit of x (i.e., $\mathbf{parity}(y) = 0$ if x is even and $\mathbf{parity}(y) = 1$ if x is odd).

2. given $y = e_K(x)$, compute $\mathbf{half}(y)$, where $\mathbf{half}(y) = 0$ if $0 \leq x < n/2$ and $\mathbf{half}(y) = 1$ if $n/2 < x \leq n-1$.

We will prove that the *RSA Cryptosystem* does not leak these types of information provided that RSA encryption is secure. More precisely, we show that the problem of RSA decryption can be Turing reduced to the problem of computing $\mathbf{half}(y)$. This means that the existence of a polynomial-time algorithm that computes $\mathbf{half}(y)$ implies the existence of a polynomial-time algorithm for RSA decryption. In other words, computing certain partial information about the plaintext, namely $\mathbf{half}(y)$, is no easier than decrypting the ciphertext to obtain the whole plaintext.

We will now show how to compute $x = d_K(y)$, given a hypothetical algorithm (oracle) HALF which computes $\mathbf{half}(y)$. The algorithm is presented as Algorithm 6.13.

Algorithm 6.13: ORACLE RSA DECRYPTION(n, b, y)

```
external HALF
k ← ⌊log₂ n⌋
for i ← 0 to k
  do { h_i ← HALF(n, b, y)
       y ← (y × 2^b) mod n
lo ← 0
hi ← n
for i ← 0 to k
  do { mid ← (hi + lo)/2
       if h_i = 1
         then lo ← mid
         else hi ← mid
return (⌊hi⌋)
```

We explain what is happening in the above algorithm. First, we note that the RSA encryption function satisfies the following (multiplicative) homomorphic property in $\mathbb{Z}_n$:

$$e_K(x_1)e_K(x_2) = e_K(x_1x_2).$$

Now, using the fact that

$$y = e_K(x) = x^b \bmod n,$$

it is easily seen in the ith iteration of the first **for** loop that

$$h_i = \mathbf{half}(y \times (e_K(2))^i) = \mathbf{half}(e_K(x \times 2^i)),$$

for $0 \leq i \leq \lfloor \log_2 n \rfloor$. We observe that

$$\begin{aligned}
\mathbf{half}(e_K(x)) = 0 \quad &\Leftrightarrow \quad x \in \left[0, \frac{n}{2}\right)\\
\mathbf{half}(e_K(2x)) = 0 \quad &\Leftrightarrow \quad x \in \left[0, \frac{n}{4}\right) \cup \left[\frac{n}{2}, \frac{3n}{4}\right)\\
\mathbf{half}(e_K(4x)) = 0 \quad &\Leftrightarrow \quad x \in \left[0, \frac{n}{8}\right) \cup \left[\frac{n}{4}, \frac{3n}{8}\right) \cup \left[\frac{n}{2}, \frac{5n}{8}\right) \cup \left[\frac{3n}{4}, \frac{7n}{8}\right),
\end{aligned}$$

and so on. Hence, we can find x by a binary search technique, which is done in the second **for** loop. Here is a small example to illustrate.

Example 6.19 Suppose $n = 1457$, $b = 779$, and we have a ciphertext $y = 722$. Then suppose, using our oracle HALF, that we obtain the following values for h_i:

i	0	1	2	3	4	5	6	7	8	9	10
h_i	1	0	1	0	1	1	1	1	1	0	0

Then the binary search proceeds as shown in Figure 6.3. Hence, the plaintext is $x = \lfloor 999.55 \rfloor = 999$. □

FIGURE 6.3: Binary search for RSA decryption

i	lo	mid	hi
0	0.00	728.50	1457.00
1	728.50	1092.75	1457.00
2	728.50	910.62	1092.75
3	910.62	1001.69	1092.75
4	910.62	956.16	1001.69
5	956.16	978.92	1001.69
6	978.92	990.30	1001.69
7	990.30	996.00	1001.69
8	996.00	998.84	1001.69
9	998.84	1000.26	1001.69
10	998.84	999.55	1000.26
	998.84	999.55	999.55

The complexity of Algorithm 6.13 is easily seen to be

$$O((\log n)^3) + O(\log n) \times \text{the complexity of } \textsc{Half}.$$

Therefore we will obtain a polynomial-time algorithm for RSA decryption if HALF is a polynomial-time algorithm.

It is a simple matter to observe that computing $\mathbf{parity}(y)$ is polynomially equivalent to computing $\mathbf{half}(y)$. This follows from the following two easily proved identities involving RSA encryption (see the exercises):

$$\mathbf{half}(y) = \mathbf{parity}((y \times e_K(2)) \bmod n) \tag{6.3}$$

$$\mathbf{parity}(y) = \mathbf{half}((y \times e_K(2^{-1})) \bmod n), \tag{6.4}$$

and from the above-mentioned multiplicative rule, $e_K(x_1)e_K(x_2) = e_K(x_1x_2)$. Hence, from the results proved above, it follows that the existence of a polynomial-time algorithm to compute **parity** implies the existence of a polynomial-time algorithm for RSA decryption.

We have provided evidence that computing **parity** or **half** is difficult, provided that RSA decryption is difficult. However, the proofs we have presented do not rule out the possibility that it might be possible to find an efficient algorithm that computes **parity** or **half** with 75% accuracy, say. There are also many other types of plaintext information that could possibly be considered, and we certainly cannot deal with all possible types of information using separate proofs. Therefore the results of this section only provide evidence of security against certain types of attacks.

6.9.2 Obtaining Semantic Security

What we really want is to find a method of designing a cryptosystem that allows us to prove (assuming some plausible computational assumptions) that no information of any kind regarding the plaintext is revealed in polynomial time by examining the ciphertext. It can be shown that this is equivalent to showing that an adversary cannot distinguish ciphertexts. Therefore, we consider the problem of **Ciphertext Distinguishability**, which is defined as follows:

Problem 6.3: Ciphertext Distinguishability

Instance: An encryption function $f : X \rightarrow X$; two plaintexts $x_1, x_2 \in X$; and a ciphertext $y = f(x_i)$, where $i \in \{1,2\}$.
Question: Is $i = 1$?

Problem 6.3 is of course trivial if the encryption function f is deterministic, since it suffices to compute $f(x_1)$ and $f(x_2)$ and see which one yields the ciphertext y. Hence, if **Ciphertext Distinguishability** is going to be computationally infeasible, then it will be necessary for the encryption process to be randomized. In this section, we present one concrete method to realize this objective.

We consider Cryptosystem 6.3. This system is based on an arbitrary ***trapdoor one-way permutation***, which is a (bijective) trapdoor one-way function from a set X to itself. If $f : X \rightarrow X$ is a trapdoor one-way permutation, then the inverse permutation is denoted, as usual, by f^{-1}. f is the encryption function, and f^{-1} is the decryption function of the public-key cryptosystem.

In the case of the *RSA Cryptosystem*, we would take $n = pq$, $X = \mathbb{Z}_n$, $f(x) = x^b \bmod n$ and $f^{-1}(x) = x^a \bmod n$, where $ab \equiv 1 \pmod{\phi(n)}$. Cryptosystem 6.3 also employs a certain random function, G. Actually, G will be modeled by a random oracle, which was defined in Section 5.2.1.

We observe that Cryptosystem 6.3 is quite efficient: it requires little additional computation as compared to the underlying public-key cryptosystem based on f. In practice, the function G could be built from a secure hash function in a very efficient manner. The main drawback of Cryptosystem 6.3 is the data expansion: m bits of plaintext are encrypted to yield $k + m$ bits of ciphertext. If f is based on the RSA encryption function, for example, then it will be necessary to take $k \geq 2048$ in order for the system to be secure, using a 2048-bit RSA modulus.

An intuitive argument that Cryptosystem 6.3 is semantically secure in the random oracle model goes as follows: In order to determine any information about the plaintext x, we need to have some information about $G(r)$. Assuming that G is a random oracle, the only way to ascertain any information about the value of $G(r)$ is to first compute $r = f^{-1}(y_1)$. (It is not sufficient to compute some partial information about r; it is necessary to have complete information about r in order to obtain any information about $G(r)$.) However, if f is one-way, then r cannot be computed in a reasonable amount of time by an adversary who does not know the trapdoor, f^{-1}.

The preceding argument might be fairly convincing, but it is not a proof. If we are going to massage this argument into a proof, we need to describe a reduction, from the problem of inverting the function f to the problem of **Ciphertext Distinguishability**. When f is randomized, as in Cryptosystem 6.3, it may not be feasible

Cryptosystem 6.3: *Semantically Secure Public-key Cryptosystem*

Let m, k be positive integers; let $\mathcal{F}$ be a family of trapdoor one-way permutations such that $f : \{0,1\}^k \rightarrow \{0,1\}^k$ for all $f \in \mathcal{F}$; and let $G : \{0,1\}^k \rightarrow \{0,1\}^m$ be a random oracle. Let $\mathcal{P} = \{0,1\}^m$ and $\mathcal{C} = \{0,1\}^k \times \{0,1\}^m$, and define

$$\mathcal{K} = \{(f, f^{-1}, G) : f \in \mathcal{F}\}.$$

For $K = (f, f^{-1}, G)$, let $r \in \{0,1\}^k$ be chosen randomly, and define

$$e_K(x) = (y_1, y_2) = (f(r), G(r) \oplus x),$$

where $y_1 \in \{0,1\}^k$, $x, y_2 \in \{0,1\}^m$. Further, define

$$d_K(y_1, y_2) = G(f^{-1}(y_1)) \oplus y_2$$

($y_1 \in \{0,1\}^k$, $y_2 \in \{0,1\}^m$). The functions f and G are the public key; the function f^{-1} is the private key.

to solve Problem 6.3 if there are sufficiently many possible encryptions of a given plaintext.

We are going to describe a reduction that is more general than the Turing reductions considered previously. We will assume the existence of an algorithm DISTINGUISH that solves the problem of **Ciphertext Distinguishability** for two plaintexts x_1 and x_2, and then we will modify this algorithm in such a way that we obtain an algorithm to invert f. The algorithm DISTINGUISH need not be a "perfect" algorithm; we will only require that it gives the right answer with some probability $1/2 + \epsilon$, where $\epsilon > 0$ (i.e., it is more accurate than a random guess of "1" or "2"). DISTINGUISH is allowed to query the random oracle, and therefore it can compute encryptions of plaintexts. In other words, we are assuming a chosen plaintext attack.

As mentioned above, we will prove that Cryptosystem 6.3 is semantically secure in the random oracle model. The main features of this model (which we introduced in Section 5.2.1), and the reduction we describe, are as follows.

1. G is assumed to be a random oracle, so the only way to determine any information about a value $G(r)$ is to call the function G with input r.

2. We construct a new algorithm INVERT, by modifying the algorithm DISTINGUISH, which will invert randomly chosen elements y with probability bounded away from 0 (i.e., given a value $y = f(x)$ where x is chosen randomly, the algorithm INVERT will find x with some specified probability).

3. The algorithm INVERT will replace the random oracle by a specific func-

```
Algorithm 6.14: INVERT(y)

 external f
 global RList, GList, ℓ
 procedure SIMG(r)
  i ← 1
  found ← false
  while i ≤ ℓ and not found
       ⎧if RList[i] = r
    do ⎨  then found ← true
       ⎩  else i ← i + 1
  if found
    then return (GList[i])
  if f(r) = y
         ⎧let j ∈ {1,2} be chosen at random
    then ⎨
         ⎩g ← y2 ⊕ xj
    else let g be chosen at random
  ℓ ← ℓ + 1
  RList[ℓ] ← r
  GList[ℓ] ← g
  return (g)

 main
  y1 ← y
  choose y2 at random
  ℓ ← 0
  insert the code for DISTINGUISH(x1, x2, (y1, y2)) here
  for i ← 1 to ℓ
       ⎧if f(RList[i]) = y
    do ⎨
       ⎩  then return (RList[i])
  return ("failure")
```

tion that we will describe, SIMG, all of whose outputs are random numbers. SIMG is a perfect simulation of a random oracle.

The algorithm INVERT is presented as Algorithm 6.14.

Given two plaintexts x_1 and x_2, DISTINGUISH solves the **Ciphertext Distinguishability** problem with probability $1/2 + \epsilon$. The input to INVERT is the element y to be inverted; the objective is to output $f^{-1}(y)$. INVERT begins by constructing a ciphertext (y_1, y_2) in which $y_1 = y$ and y_2 is random. INVERT runs the algorithm DISTINGUISH on the ciphertext (y_1, y_2), attempting to determine if it is an encryption of x_1 or of x_2. DISTINGUISH will query the simulated oracle, SIMG, at various times during its execution. The following points summarize the operation of SIMG:

1. SIMG maintains a list, denoted *RList*, of all inputs r for which it is queried during the execution of DISTINGUISH; and the corresponding list, denoted *GList*, of outputs SIMG(r).

2. If an input r satisfies $f(r) = y$, then SIMG(r) is defined so that (y_1, y_2) is a valid encryption of one of x_1 or x_2 (chosen at random).

3. If the oracle was previously queried with input r, then SIMG(r) is already defined.

4. Otherwise, the value for SIMG(r) is chosen randomly.

Observe that, for any possible plaintext $x_0 \in X$, (y_1, y_2) is a valid encryption of x_0 if and only if

$$\text{SIMG}(f^{-1}(y_1)) = y_2 \oplus x_0.$$

In particular, (y_1, y_2) can be a valid encryption of either of x_1 or x_2, provided that SIMG$(f^{-1}(y_1))$ is defined appropriately. The description of the algorithm SIMG ensures that (y_1, y_2) is a valid encryption of one of x_1 or x_2.

Eventually, the algorithm DISTINGUISH will terminate with an answer "1" or "2," which may or may not be correct. At this point, the algorithm INVERT examines the list *RList* to see if any r in the *RList* satisfies $f(r) = y$. If such a value r is found, then it is the desired value $f^{-1}(y)$, and the algorithm INVERT succeeds (INVERT fails if $f^{-1}(y)$ is not discovered in *RList*).

It is in fact possible to make algorithm INVERT more efficient by observing that the function SIMG checks to see if $y = f(r)$ for every r that it is queried with. Once it is discovered, within the function SIMG, that $y = f(r)$, we can terminate the algorithm INVERT immediately, returning the value r as its output. It is not necessary to keep running the algorithm DISTINGUISH to its conclusion. However, the analysis of the success probability, which we are going to do next, is a bit easier to understand for Algorithm 6.14 as we have presented it. (The reader might want to verify that the above-mentioned modification of INVERT will not change its success probability.)

We now proceed to compute a lower bound on the success probability of the algorithm INVERT. We do this by examining the success probability of DISTINGUISH. We are assuming that the success probability of DISTINGUISH is at least $1/2 + \epsilon$ when it interacts with a random oracle. In the algorithm INVERT, DISTINGUISH interacts with the simulated random oracle, SIMG. Clearly SIMG is completely indistinguishable from a true random oracle for all inputs, except possibly for the input $r = f^{-1}(y)$. However, if $f(r) = y$ and (y, y_2) is a valid encryption of x_1 or x_2, then it must be the case that SIMG$(r) = y_2 \oplus x_1$ or SIMG$(r) = y_2 \oplus x_2$. SIMG is choosing randomly from these two possible alternatives. Therefore, the output it produces is indistinguishable from a true random oracle for the input $r = f^{-1}(y)$, as well. Consequently, the success probability of DISTINGUISH is at least $1/2 + \epsilon$ when it interacts with the simulated random oracle, SIMG.

We now calculate the success probability of DISTINGUISH, conditioned on whether (or not) $f^{-1}(y) \in RList$:

$$\begin{aligned}&\mathbf{Pr}[\text{DISTINGUISH succeeds}] = \\ &\quad \mathbf{Pr}[\text{DISTINGUISH succeeds} \mid f^{-1}(y) \in RList]\, \mathbf{Pr}[f^{-1}(y) \in RList] + \\ &\quad \mathbf{Pr}[\text{DISTINGUISH succeeds} \mid f^{-1}(y) \notin RList]\, \mathbf{Pr}[f^{-1}(y) \notin RList].\end{aligned}$$

It is clear that

$$\mathbf{Pr}[\text{DISTINGUISH succeeds} \mid f^{-1}(y) \notin RList] = 1/2,$$

because there is no way to distinguish an encryption of x_1 from an encryption of x_2 if the value of SIMG$(f^{-1}(y))$ is not determined. Now, using the fact that

$$\mathbf{Pr}[\text{DISTINGUISH succeeds} \mid f^{-1}(y) \in RList] \leq 1,$$

we obtain the following:

$$\begin{aligned}\frac{1}{2} + \epsilon &\leq \mathbf{Pr}[\text{DISTINGUISH succeeds}] \\ &\leq \mathbf{Pr}[f^{-1}(y) \in RList] + \frac{1}{2}\mathbf{Pr}[f^{-1}(y) \notin RList] \\ &\leq \mathbf{Pr}[f^{-1}(y) \in RList] + \frac{1}{2}.\end{aligned}$$

Therefore, it follows that

$$\mathbf{Pr}[f^{-1}(y) \in RList] \geq \epsilon.$$

Since

$$\mathbf{Pr}[\text{INVERSE succeeds}] = \mathbf{Pr}[f^{-1}(y) \in RList],$$

it follows that

$$\mathbf{Pr}[\text{INVERSE succeeds}] \geq \epsilon.$$

It is straightforward to consider the running time of INVERT as compared to that of DISTINGUISH. Suppose that t_1 is the running time of DISTINGUISH, t_2 is the time required to evaluate the function f, and q denotes the number of oracle queries made by DISTINGUISH. Then it is not difficult to see that the running time of INVERT is $t_1 + O(q^2 + qt_2)$.

It is easy to see that there must be some data expansion in any semantically secure cryptosystem due to the fact that encryption is randomized. However, there are more efficient provably secure schemes than Cryptosystem 6.3. The most important of these is *Optimal Asymmetric Encryption Padding* (or *OAEP*), which is widely used in practice. The data expansion of *OAEP* is considerably less than that of Cryptosystem 6.3. In *OAEP*, an m-bit plaintext is encrypted to form a $(k_0 + m)$-bit ciphertext, where k_0 is the security parameter. The complexity of breaking *OAEP*, under certain computational assumptions, is approximately 2^{k_0}.

The adjective "optimal" in *Optimal Asymmetric Encryption Padding* refers to the message expansion, which is k_0 bits. Each plaintext has 2^{k_0} possible valid encryptions. One way of solving the problem of **Ciphertext Distinguishability** for *OAEP* would be simply to compute all the possible encryptions of one of the two given plaintexts, say x_1, and check to see if the given ciphertext y is obtained. The complexity of this algorithm is 2^{k_0}. It is therefore clear that the message expansion of the cryptosystem must be at least as big as the logarithm to the base 2 of the amount of computation time of an algorithm that solves the **Ciphertext Distinguishability** problem.

6.10 Notes and References

The idea of public-key cryptography was introduced in the open literature by Diffie and Hellman in 1976. Although [71] is the most cited reference, the conference paper [70] actually appeared a bit earlier. The *RSA Cryptosystem* was discovered by Rivest, Shamir, and Adleman [172].

The Solovay-Strassen test was first described in [188]. The Miller-Rabin test was given in [138] and [168]. Our discussion of error probabilities is motivated by observations of Brassard and Bratley [45] (see also [12]).

One recommended cryptography textbook that emphasizes number theory is Koblitz [111] (note that Example 6.12 is taken from Koblitz's book). Bressoud and Wagon [47] is a good elementary textbook on number-theoretic concepts relevant to *RSA*, including factoring and primality testing. We also recommend Galbraith [84] for a thorough treatment of mathematical topics that are useful for the study of public-key cryptography in general. We should also mention Lenstra and Lenstra [121], which is a monograph on the number field sieve. Finally, the factorization of a 768-bit RSA modulus is described in [108].

The material in Sections 6.7.2 and 6.9.1 is based on the treatment by Salomaa [173, pp. 143–154] (the factorization of n, given the decryption exponent, was first presented in [67]; the results on partial information revealed by RSA ciphertexts is from [90]). Wiener's attack can be found in [201]. Ten years later, an improvement was found by Boneh and Durfee; it is described in [41]. A thorough treatment of various types of attacks on *RSA* can be found in the books by Hinek [95] and Yan [205].

The *Rabin Cryptosystem* was described in Rabin [167]. Other provably secure RSA-type cryptosystems in which decryption is unambiguous have been found by Williams [202] and Kurosawa, Ito, and Takeuchi [117].

Partial information leaked by RSA ciphertexts was studied in Alexi, Chor, Goldreich, and Schnorr [3]. The concept of semantic security is due to Goldwasser and Micali [89]. Cryptosystem 6.3 was presented by Bellare and Rogaway in [22] and *Optimal Asymmetric Encryption Padding* was first described in [19].

Exercises 5.15–5.17 give some examples of protocol failures. For an influential, pioneering article on this subject, see Moore [141].

Exercises

6.1 In Algorithm 6.1, prove that

$$\gcd(r_0, r_1) = \gcd(r_1, r_2) = \cdots = \gcd(r_{m-1}, r_m) = r_m$$

and, hence, $r_m = \gcd(a, b)$.

6.2 Suppose that $a > b$ in Algorithm 6.1.

(a) Prove that $r_i \geq 2r_{i+2}$ for all i such that $0 \leq i \leq m-2$.

(b) Prove that m is $O(\log a)$.

(c) Prove that m is $O(\log b)$.

6.3 Use the EXTENDED EUCLIDEAN ALGORITHM to compute the following multiplicative inverses:

(a) $17^{-1} \bmod 101$

(b) $357^{-1} \bmod 1234$

(c) $3125^{-1} \bmod 9987$.

6.4 Compute $\gcd(57, 93)$, and find integers s and t such that $57s + 93t = \gcd(57, 93)$.

6.5 Suppose $\chi : \mathbb{Z}_{105} \rightarrow \mathbb{Z}_3 \times \mathbb{Z}_5 \times \mathbb{Z}_7$ is defined as

$$\chi(x) = (x \bmod 3, x \bmod 5, x \bmod 7).$$

Give an explicit formula for the function χ^{-1}, and use it to compute $\chi^{-1}(2, 2, 3)$.

6.6 Solve the following system of congruences:

$$\begin{aligned} x &\equiv 12 \pmod{25} \\ x &\equiv 9 \pmod{26} \\ x &\equiv 23 \pmod{27}. \end{aligned}$$

6.7 Solve the following system of congruences:

$$\begin{aligned} 13x &\equiv 4 \pmod{99} \\ 15x &\equiv 56 \pmod{101}. \end{aligned}$$

HINT First use the EXTENDED EUCLIDEAN ALGORITHM, and then apply the Chinese remainder theorem.

6.8 Use Theorem 6.8 to find the smallest primitive element modulo 97.

6.9 How many primitive elements are there modulo 1041817?

6.10 Suppose that $p = 2q + 1$, where p and q are odd primes. Suppose further that $\alpha \in \mathbb{Z}_p{}^*$, $\alpha \not\equiv \pm 1 \pmod p$. Prove that α is a primitive element modulo p if and only if $\alpha^q \equiv -1 \pmod p$.

6.11 Suppose that $n = pq$, where p and q are distinct odd primes and $ab \equiv 1 \pmod{(p-1)(q-1)}$. The RSA encryption operation is $e(x) = x^b \bmod n$ and the decryption operation is $d(y) = y^a \bmod n$. We proved that $d(e(x)) = x$ if $x \in \mathbb{Z}_n{}^*$. Prove that the same statement is true for any $x \in \mathbb{Z}_n$.

HINT Use the fact that $x_1 \equiv x_2 \pmod{pq}$ if and only if $x_1 \equiv x_2 \pmod p$ and $x_1 \equiv x_2 \pmod q$. This follows from the Chinese remainder theorem.

6.12 For $n = pq$, where p and q are distinct odd primes, define

$$\lambda(n) = \frac{(p-1)(q-1)}{\gcd(p-1, q-1)}.$$

Suppose that we modify the *RSA Cryptosystem* by requiring that $ab \equiv 1 \pmod{\lambda(n)}$.

(a) Prove that encryption and decryption are still inverse operations in this modified cryptosystem.

(b) If $p = 37$, $q = 79$, and $b = 7$, compute a in this modified cryptosystem, as well as in the original *RSA Cryptosystem*.

6.13 Two samples of RSA ciphertext are presented in Tables 6.2 and 6.3. Your task is to decrypt them. The public parameters of the system are $n = 18923$ and $b = 1261$ (for Table 6.2) and $n = 31313$ and $b = 4913$ (for Table 6.3). This can be accomplished as follows. First, factor n (which is easy because it is so small). Then compute the exponent a from $\phi(n)$, and, finally, decrypt the ciphertext. Use the SQUARE-AND-MULTIPLY ALGORITHM to exponentiate modulo n.

In order to translate the plaintext back into ordinary English text, you need to know how alphabetic characters are "encoded" as elements in $\mathbb{Z}_n$. Each element of $\mathbb{Z}_n$ represents three alphabetic characters as in the following examples:

$$\begin{array}{llll} DOG & \rightarrow & 3 \times 26^2 + 14 \times 26 + 6 & = & 2398 \\ CAT & \rightarrow & 2 \times 26^2 + 0 \times 26 + 19 & = & 1371 \\ ZZZ & \rightarrow & 25 \times 26^2 + 25 \times 26 + 25 & = & 17575. \end{array}$$

You will have to invert this process as the final step in your program.

TABLE 6.2: RSA ciphertext

12423	11524	7243	7459	14303	6127	10964	16399
9792	13629	14407	18817	18830	13556	3159	16647
5300	13951	81	8986	8007	13167	10022	17213
2264	961	17459	4101	2999	14569	17183	15827
12693	9553	18194	3830	2664	13998	12501	18873
12161	13071	16900	7233	8270	17086	9792	14266
13236	5300	13951	8850	12129	6091	18110	3332
15061	12347	7817	7946	11675	13924	13892	18031
2620	6276	8500	201	8850	11178	16477	10161
3533	13842	7537	12259	18110	44	2364	15570
3460	9886	8687	4481	11231	7547	11383	17910
12867	13203	5102	4742	5053	15407	2976	9330
12192	56	2471	15334	841	13995	17592	13297
2430	9741	11675	424	6686	738	13874	8168
7913	6246	14301	1144	9056	15967	7328	13203
796	195	9872	16979	15404	14130	9105	2001
9792	14251	1498	11296	1105	4502	16979	1105
56	4118	11302	5988	3363	15827	6928	4191
4277	10617	874	13211	11821	3090	18110	44
2364	15570	3460	9886	9988	3798	1158	9872
16979	15404	6127	9872	3652	14838	7437	2540
1367	2512	14407	5053	1521	297	10935	17137
2186	9433	13293	7555	13618	13000	6490	5310
18676	4782	11374	446	4165	11634	3846	14611
2364	6789	11634	4493	4063	4576	17955	7965
11748	14616	11453	17666	925	56	4118	18031
9522	14838	7437	3880	11476	8305	5102	2999
18628	14326	9175	9061	650	18110	8720	15404
2951	722	15334	841	15610	2443	11056	2186

The first plaintext was taken from *The Diary of Samuel Marchbanks*, by Robertson Davies, 1947, and the second was taken from *Lake Wobegon Days*, by Garrison Keillor, 1985.

6.14 A common way to speed up RSA decryption incorporates the Chinese remainder theorem, as follows. Suppose that $d_K(y) = y^d \bmod n$ and $n = pq$. Define $d_p = d \bmod (p-1)$ and $d_q = d \bmod (q-1)$; and let $M_p = q^{-1} \bmod p$ and $M_q = p^{-1} \bmod q$. Then consider the following algorithm:

TABLE 6.3: RSA ciphertext

6340	8309	14010	8936	27358	25023	16481	25809
23614	7135	24996	30590	27570	26486	30388	9395
27584	14999	4517	12146	29421	26439	1606	17881
25774	7647	23901	7372	25774	18436	12056	13547
7908	8635	2149	1908	22076	7372	8686	1304
4082	11803	5314	107	7359	22470	7372	22827
15698	30317	4685	14696	30388	8671	29956	15705
1417	26905	25809	28347	26277	7897	20240	21519
12437	1108	27106	18743	24144	10685	25234	30155
23005	8267	9917	7994	9694	2149	10042	27705
15930	29748	8635	23645	11738	24591	20240	27212
27486	9741	2149	29329	2149	5501	14015	30155
18154	22319	27705	20321	23254	13624	3249	5443
2149	16975	16087	14600	27705	19386	7325	26277
19554	23614	7553	4734	8091	23973	14015	107
3183	17347	25234	4595	21498	6360	19837	8463
6000	31280	29413	2066	369	23204	8425	7792
25973	4477	30989					

Algorithm 6.15: CRT-OPTIMIZED RSA DECRYPTION(n, d_p, d_q, M_p, M_q, y)

$x_p \leftarrow y^{d_p} \bmod p$
$x_q \leftarrow y^{d_q} \bmod q$
$x \leftarrow M_p q x_p + M_q p x_q \bmod n$
return (x)

Algorithm 6.15 replaces an exponentiation modulo n by modular exponentiations modulo p and q. If p and q are ℓ-bit integers and exponentiation modulo an ℓ-bit integer takes time $c\ell^3$, then the time to perform the required exponentiation(s) is reduced from $c(2\ell)^3$ to $2c\ell^3$, a savings of 75%. The final step, involving the Chinese remainder theorem, requires time $O(\ell^2)$ if d_p, d_q, M_p, and M_q have been pre-computed.

(a) Prove that the value x returned by Algorithm 6.15 is, in fact, $y^d \bmod n$.

(b) Given that $p = 1511$, $q = 2003$, and $d = 1234577$, compute d_p, d_q, M_p, and M_q.

(c) Given the above values of p, q, and d, decrypt the ciphertext $y = 152702$ using Algorithm 6.15.

6.15 Prove that the *RSA Cryptosystem* is insecure against a chosen ciphertext attack. In particular, given a ciphertext y, describe how to choose a ciphertext

$\hat{y} \neq y$, such that knowledge of the plaintext $\hat{x} = d_K(\hat{y})$ allows $x = d_K(y)$ to be computed.

HINT Use the multiplicative property of the *RSA Cryptosystem*, i.e., that

$$e_K(x_1)e_K(x_2) \bmod n = e_K(x_1x_2 \bmod n).$$

6.16 This exercise exhibits what is called a ***protocol failure***. It provides an example where ciphertext can be decrypted by an opponent, without determining the key, if a cryptosystem is used in a careless way. The moral is that it is not sufficient to use a "secure" cryptosystem in order to guarantee "secure" communication.

Suppose Bob has an *RSA Cryptosystem* with a large modulus n for which the factorization cannot be found in a reasonable amount of time. Suppose Alice sends a message to Bob by representing each alphabetic character as an integer between 0 and 25 (i.e., $A \leftrightarrow 0$, $B \leftrightarrow 1$, etc.), and then encrypting each residue modulo 26 as a separate plaintext character.

(a) Describe how Oscar can easily decrypt a message that is encrypted in this way.

(b) Illustrate this attack by decrypting the following ciphertext (which was encrypted using an *RSA Cryptosystem* with $n = 18721$ and $b = 25$) without factoring the modulus:

$$365, 0, 4845, 14930, 2608, 2608, 0.$$

6.17 This exercise illustrates another example of a protocol failure (due to Simmons) involving the *RSA Cryptosystem*; it is called the ***common modulus protocol failure***. Suppose Bob has an *RSA Cryptosystem* with modulus n and encryption exponent b_1, and Charlie has an *RSA Cryptosystem* with (the same) modulus n and encryption exponent b_2. Suppose also that $\gcd(b_1, b_2) = 1$. Now, consider the situation that arises if Alice encrypts the same plaintext x to send to both Bob and Charlie. Thus, she computes $y_1 = x^{b_1} \bmod n$ and $y_2 = x^{b_2} \bmod n$, and then she sends y_1 to Bob and y_2 to Charlie. Suppose Oscar intercepts y_1 and y_2, and performs the computations indicated in Algorithm 6.16.

Algorithm 6.16: RSA COMMON MODULUS DECRYPTION(n, b_1, b_2, y_1, y_2)

$c_1 \leftarrow {b_1}^{-1} \bmod b_2$
$c_2 \leftarrow (c_1b_1 - 1)/b_2$
$x_1 \leftarrow {y_1}^{c_1}({y_2}^{c_2})^{-1} \bmod n$
return (x_1)

(a) Prove that the value x_1 computed in Algorithm 6.16 is in fact Alice's plaintext, x. Thus, Oscar can decrypt the message Alice sent, even though the cryptosystem may be "secure."

(b) Illustrate the attack by computing x by this method if $n = 18721$, $b_1 = 43$, $b_2 = 7717$, $y_1 = 12677$, and $y_2 = 14702$.

6.18 We give yet another protocol failure involving the *RSA Cryptosystem*. Suppose that three users in a network, say Bob, Bart, and Bert, all have public encryption exponents $b = 3$. Let their moduli be denoted by n_1, n_2, n_3, and assume that n_1, n_2, and n_3, are pairwise relatively prime. Now suppose Alice encrypts the same plaintext x to send to Bob, Bart, and Bert. That is, Alice computes $y_i = x^3 \bmod n_i$, $1 \leq i \leq 3$. Describe how Oscar can compute x, given y_1, y_2, and y_3, without factoring any of the moduli.

6.19 A plaintext x is said to be a ***fixed plaintext*** if $e_K(x) = x$. Show that, for the *RSA Cryptosystem*, the number of fixed plaintexts $x \in \mathbb{Z}_n{}^*$ is equal to

$$\gcd(b-1, p-1) \times \gcd(b-1, q-1).$$

HINT Consider the following system of two congruences:

$$\begin{aligned} e_K(x) &\equiv x \pmod{p}, \\ e_K(x) &\equiv x \pmod{q}. \end{aligned}$$

6.20 Suppose **A** is a deterministic algorithm that is given as input an RSA modulus n, an encryption exponent b, and a ciphertext y. **A** will either decrypt y or return no answer. Supposing that there are $\epsilon(n-1)$ nonzero ciphertexts which **A** is able to decrypt, show how to use **A** as an oracle in a Las Vegas decryption algorithm having success probability ϵ.

6.21 Write a program to evaluate Jacobi symbols using the four properties presented in Section 6.4. The program should not do any factoring, other than dividing out powers of two. Test your program by computing the following Jacobi symbols:

$$\left(\frac{610}{987}\right), \left(\frac{20964}{1987}\right), \left(\frac{1234567}{11111111}\right).$$

6.22 For $n = 837, 851$, and 1189, find the number of bases b such that n is an Euler pseudo-prime to the base b.

6.23 The purpose of this question is to prove that the error probability of the Solovay-Strassen primality test is at most $1/2$. Let $\mathbb{Z}_n{}^*$ denote the group of units modulo n. Define

$$G(n) = \left\{ a : a \in \mathbb{Z}_n{}^*, \left(\frac{a}{n}\right) \equiv a^{(n-1)/2} \pmod{n} \right\}.$$

(a) Prove that $G(n)$ is a subgroup of $\mathbb{Z}_n{}^*$. Hence, by Lagrange's theorem, if $G(n) \neq \mathbb{Z}_n{}^*$, then

$$|G(n)| \leq \frac{|\mathbb{Z}_n{}^*|}{2} \leq \frac{n-1}{2}.$$

(b) Suppose $n = p^k q$, where p and q are odd, p is prime, $k \geq 2$, and $\gcd(p, q) = 1$. Let $a = 1 + p^{k-1}q$. Prove that

$$\left(\frac{a}{n}\right) \not\equiv a^{(n-1)/2} \pmod{n}.$$

HINT Use the binomial theorem to compute $a^{(n-1)/2}$.

(c) Suppose $n = p_1 \dots p_s$, where the p_i's are distinct odd primes. Suppose $a \equiv u \pmod{p_1}$ and $a \equiv 1 \pmod{p_2 p_3 \cdots p_s}$, where u is a quadratic non-residue modulo p_1 (note that such an a exists by the Chinese remainder theorem). Prove that

$$\left(\frac{a}{n}\right) \equiv -1 \pmod{n},$$

but

$$a^{(n-1)/2} \equiv 1 \pmod{p_2 p_3 \cdots p_s},$$

so

$$a^{(n-1)/2} \not\equiv -1 \pmod{n}.$$

(d) If n is odd and composite, prove that $|G(n)| \leq (n-1)/2$.

(e) Summarize the above: prove that the error probability of the Solovay-Strassen primality test is at most $1/2$.

6.24 Suppose we have a Las Vegas algorithm with failure probability ϵ.

(a) Prove that the probability of first achieving success on the nth trial is $p_n = \epsilon^{n-1}(1-\epsilon)$.

(b) The average (expected) number of trials to achieve success is

$$\sum_{n=1}^{\infty} (n \times p_n).$$

Show that this average is equal to $1/(1-\epsilon)$.

(c) Let δ be a positive real number less than 1. Show that the number of iterations required in order to reduce the probability of failure to at most δ is

$$\left\lceil \frac{\log_2 \delta}{\log_2 \epsilon} \right\rceil.$$

6.25 Suppose throughout this question that p is an odd prime and $\gcd(a, p) = 1$.

(a) Suppose that $i \geq 2$ and $b^2 \equiv a \pmod{p^{i-1}}$. Prove that there is a unique $x \in \mathbb{Z}_{p^i}$ such that $x^2 \equiv a \pmod{p^i}$ and $x \equiv b \pmod{p^{i-1}}$. Describe how this x can be computed efficiently.

(b) Illustrate your method in the following situation: starting with the congruence $6^2 \equiv 17 \pmod{19}$, find square roots of 17 modulo 19^2 and modulo 19^3.

(c) For all $i \geq 1$, prove that the number of solutions to the congruence $x^2 \equiv a \pmod{p^i}$ is either 0 or 2.

6.26 Using various choices for the bound, B, attempt to factor 262063 and 9420457 using the $p-1$ method. How big does B have to be in each case to be successful?

6.27 Factor 262063, 9420457, and 181937053 using the POLLARD RHO ALGORITHM, if the function f is defined to be $f(x) = x^2 + 1$. How many iterations are needed to factor each of these three integers?

6.28 Suppose we want to factor the integer $n = 256961$ using the RANDOM SQUARES ALGORITHM. Using the factor base

$$\{-1, 2, 3, 5, 7, 11, 13, 17, 19, 23, 29, 31\},$$

test the integers $z^2 \bmod n$ for $z = 500, 501, \ldots,$ until a congruence of the form $x^2 \equiv y^2 \pmod{n}$ is obtained and the factorization of n is found.

6.29 In the RANDOM SQUARES ALGORITHM, we need to test a positive integer $w \leq n-1$ to see if it factors completely over the factor base $\mathcal{B} = \{p_1, \ldots, p_B\}$ consisting of the B smallest prime numbers. Recall that $p_B = m \approx 2^s$ and $n \approx 2^r$.

(a) Prove that this can be done using at most $B + r$ divisions of an integer having at most r bits by an integer having at most s bits.

(b) Assuming that $r < m$, prove that the complexity of this test is $O(rsm)$.

6.30 In this exercise, we show that parameter generation for the *RSA Cryptosystem* should take care to ensure that $q - p$ is not too small, where $n = pq$ and $q > p$.

(a) Suppose that $q - p = 2d > 0$, and $n = pq$. Prove that $n + d^2$ is a perfect square.

(b) Given an integer n that is the product of two odd primes, and given a small positive integer d such that $n + d^2$ is a perfect square, show how this information can be used to factor n.

(c) Use this technique to factor $n = 2189284635403183$.

6.31 Suppose Bob has carelessly revealed his decryption exponent to be $a = 14039$ in an *RSA Cryptosystem* with public key $n = 36581$ and $b = 4679$. Implement the randomized algorithm to factor n given this information. Test your algorithm with the "random" choices $w = 9983$ and $w = 13461$. Show all computations.

6.32 Compute the continued fraction expansion of 144/89.

6.33 If $q_1, \ldots, q_m$ is the sequence of quotients obtained in applying the EUCLIDEAN ALGORITHM with input r_0, r_1, prove that the continued fraction $[q_1, \ldots, q_m] = r_0/r_1$.

6.34 Suppose that $n = 317940011$ and $b = 77537081$ in the *RSA Cryptosystem*. Using WIENER'S ALGORITHM, attempt to factor n.

6.35 Consider the modification of the *Rabin Cryptosystem* in which $e_K(x) = x(x + B) \bmod n$, where $B \in \mathbb{Z}_n$ is part of the public key. Supposing that $p = 199$, $q = 211$, $n = pq$, and $B = 1357$, perform the following computations.

(a) Compute the encryption $y = e_K(32767)$.

(b) Determine the four possible decryptions of this given ciphertext y.

6.36 The security of the *Rabin Cryptosystem* is established by showing that if $x^2 \equiv y^2 \pmod{n}$, and $x \not\equiv \pm y \pmod{n}$, then $\gcd(x - y, n) = p$ or q, where n is the product of two primes p and q. In this question, we consider a variation where n is the product of three primes.

In what follows, you can assume that $x, y, z \in \mathbb{Z}_n{}^*$.

(a) If n is the product of three primes, and we are given x and y such that that $x^2 \equiv y^2 \pmod{n}$, and $x \not\equiv \pm y \pmod{n}$, show that it is easy to compute at least one prime factor of n using a gcd computation.

(b) Suppose n is the product of three primes, and we are given x, y, and z such that that $x^2 \equiv y^2 \equiv z^2 \pmod{n}$, $x \not\equiv \pm y \pmod{n}$, $x \not\equiv \pm z \pmod{n}$, and $y \not\equiv \pm z \pmod{n}$. Prove that we can compute all three prime factors of n by means of gcd computations.

HINT You can use the Chinese remainder theorem to prove this.

6.37 Prove Equations (6.3) and (6.4) relating the functions **half** and **parity**.

6.38 Prove that Cryptosystem 6.3 is not semantically secure against a chosen ciphertext attack. Given x_1, x_2, a ciphertext (y_1, y_2) that is an encryption of x_i ($i = 1$ or 2), and given a decryption oracle DECRYPT for Cryptosystem 6.3, describe an algorithm to determine whether $i = 1$ or $i = 2$. You are allowed to call the algorithm DECRYPT with any input except for the given ciphertext (y_1, y_2), and it will output the corresponding plaintext.

Chapter 7

Public-Key Cryptography and Discrete Logarithms

The theme of this chapter concerns public-key cryptosystems based on the **Discrete Logarithm** problem. The first and best-known of these is the *ElGamal Cryptosystem*. The **Discrete Logarithm** problem forms the basis of numerous cryptographic protocols; thus we devote a considerable amount of time to discussion of this important problem. In later sections of this chapter, we give treatments of some other ElGamal-type systems based on finite fields and elliptic curves.

7.1 Introduction

The *ElGamal Cryptosystem* is based on the **Discrete Logarithm** problem. We begin by describing this problem in the setting of a finite multiplicative group $(G, \cdot)$. For an element $\alpha \in G$ having order n, define

$$\langle \alpha \rangle = \{\alpha^i : 0 \leq i \leq n-1\}.$$

It is easy to see that $\langle \alpha \rangle$ is a subgroup of G and that $\langle \alpha \rangle$ is cyclic of order n. The subgroup $\langle \alpha \rangle$ is called the ***subgroup generated by*** α.

An often-used example is to take G to be the multiplicative group of a finite field $\mathbb{Z}_p$ (where p is prime) and to let α be a primitive element modulo p. In this situation, we have that $n = |\langle \alpha \rangle| = p - 1$. Another frequently used setting is to take α to be an element having prime order q in the multiplicative group $\mathbb{Z}_p{}^*$ (where p is prime and $p - 1 \equiv 0 \pmod{q}$). Such an α can be obtained by raising a primitive element in $\mathbb{Z}_p{}^*$ to the $(p-1)/q$th power.

In Problem 7.1, we define a general version of the **Discrete Logarithm** problem in a subgroup $\langle \alpha \rangle$ of a group $(G, \cdot)$.

The utility of the **Discrete Logarithm** problem in a cryptographic setting is that finding discrete logarithms is (probably) difficult, but the inverse operation of exponentiation can be computed efficiently by using the square-and-multiply method (Algorithm 6.5). Stated another way, exponentiation is a one-way function in suitable groups G.

Problem 7.1: Discrete Logarithm

Instance: A multiplicative group $(G, \cdot)$, an element $\alpha \in G$ having order n, and an element $\beta \in \langle\alpha\rangle$.
Question: Find the unique integer a, $0 \leq a \leq n-1$, such that

$$\alpha^a = \beta.$$

We will denote this integer a by $\log_\alpha \beta$; it is called the ***discrete logarithm*** of β.

7.1.1 The ElGamal Cryptosystem

ElGamal proposed a public-key cryptosystem that is based on the **Discrete Logarithm** problem in $(\mathbb{Z}_p{}^*, \cdot)$. This system is presented as Cryptosystem 7.1.

The encryption operation in the *ElGamal Cryptosystem* is randomized, since the ciphertext depends on both the plaintext x and on the random value k chosen by Alice. Hence, there will be many ciphertexts ($p-1$, in fact) that are encryptions of the same plaintext.

Informally, this is how the *ElGamal Cryptosystem* works: The plaintext x is "masked" by multiplying it by β^k, yielding y_2. The value α^k is also transmitted as part of the ciphertext. Bob, who knows the private key, a, can compute β^k from α^k. Then he can "remove the mask" by dividing y_2 by β^k to obtain x.

A small example will illustrate the computations performed in the *ElGamal Cryptosystem*.

Example 7.1 Suppose $p = 2579$ and $\alpha = 2$. It can be verified that α is a primitive element modulo p. Let $a = 765$, so

$$\beta = 2^{765} \bmod 2579 = 949.$$

Now, suppose that Alice wishes to send the message $x = 1299$ to Bob. Say $k = 853$ is the random integer she chooses. Then she computes

$$\begin{aligned} y_1 &= 2^{853} \bmod 2579 \\ &= 435, \end{aligned}$$

and

$$\begin{aligned} y_2 &= 1299 \times 949^{853} \bmod 2579 \\ &= 2396. \end{aligned}$$

When Bob receives the ciphertext $y = (435, 2396)$, he computes

$$\begin{aligned} x &= 2396 \times (435^{765})^{-1} \bmod 2579 \\ &= 1299, \end{aligned}$$

which was the plaintext that Alice encrypted. □

Cryptosystem 7.1: *ElGamal Public-key Cryptosystem in* $\mathbb{Z}_p{}^*$

Let p be a prime such that the **Discrete Logarithm** problem in $(\mathbb{Z}_p{}^*, \cdot)$ is infeasible, and let $\alpha \in \mathbb{Z}_p{}^*$ be a primitive element. Let $\mathcal{P} = \mathbb{Z}_p{}^*$, $\mathcal{C} = \mathbb{Z}_p{}^* \times \mathbb{Z}_p{}^*$, and define

$$\mathcal{K} = \{(p, \alpha, a, \beta) : \beta \equiv \alpha^a \pmod{p}\}.$$

The values p, α, and β are the public key, and a is the private key.

For $K = (p, \alpha, a, \beta)$, and for a (secret) random number $k \in \mathbb{Z}_{p-1}$, define

$$e_K(x, k) = (y_1, y_2),$$

where

$$y_1 = \alpha^k \bmod p$$

and

$$y_2 = x\beta^k \bmod p.$$

For $y_1, y_2 \in \mathbb{Z}_p{}^*$, define

$$d_K(y_1, y_2) = y_2(y_1{}^a)^{-1} \bmod p.$$

Clearly the *ElGamal Cryptosystem* will be insecure if Oscar can compute the value $a = \log_\alpha \beta$, for then Oscar can decrypt ciphertexts exactly as Bob does. Hence, a necessary condition for the *ElGamal Cryptosystem* to be secure is that the **Discrete Logarithm** problem in $\mathbb{Z}_p{}^*$ is infeasible. This is generally regarded as being the case if p is carefully chosen and α is a primitive element modulo p. In particular, there is no known polynomial-time algorithm for this version of the **Discrete Logarithm** problem. However, for a secure setting, it is recommended that p should have at least 2048 bits in its binary representation, and $p - 1$ should have at least one "large" prime factor (see Section 7.6 for more details).

The rest of this chapter is organized as follows. Section 7.2 presents some algorithms to solve the **Discrete Logarithm** problem. Section 7.3 derives lower bounds for so-called "generic" algorithms for this problem. Section 7.4 introduces finite fields, and Section 7.5 gives the basic concepts of elliptic curves and pairings. Section 7.6 discusses discrete logarithm algorithms in practice and Section 7.7 addresses some additional security considerations for *ElGamal Cryptosystems*.

7.2 Algorithms for the Discrete Logarithm Problem

Throughout this section, we assume that $(G, \cdot)$ is a multiplicative group and $\alpha \in G$ has order n. Hence the **Discrete Logarithm** problem can be phrased in the following form: Given $\beta \in \langle\alpha\rangle$, find the unique exponent a, $0 \leq a \leq n-1$, such that $\alpha^a = \beta$.

We begin by analyzing some elementary algorithms that can be used to solve the **Discrete Logarithm** problem. In our analyses, we will assume that computing a product of two elements in the group G requires constant (i.e., $O(1)$) time.

First, we observe that the **Discrete Logarithm** problem can be solved by exhaustive search in $O(n)$ time and $O(1)$ space, simply by computing $\alpha, \alpha^2, \alpha^3, \dots$, until $\beta = \alpha^a$ is found. (Each term α^i in the above list is computed by multiplying the previous term α^{i-1} by α, and hence the total time required is $O(n)$.)

Another approach is to precompute all possible values α^i, and then sort the list of ordered pairs (i, α^i) with respect to their second coordinates. Then, given β, we can perform a binary search of the sorted list in order to find the value a such that $\alpha^a = \beta$. This requires precomputation time $O(n)$ to compute the n powers of α, and time $O(n \log n)$ to sort the list of size n. (The sorting step can be done in time $O(n \log n)$ if an efficient sorting algorithm, such as the QUICKSORT algorithm, is used.) If we neglect logarithmic factors, as is usually done in the analysis of these algorithms, the precomputation time is $O(n)$. The time for a binary search of a sorted list of size n is $O(\log n)$. If we (again) ignore the logarithmic term, then we see that we can solve the **Discrete Logarithm** problem in $O(1)$ time with $O(n)$ precomputation and $O(n)$ memory.

7.2.1 Shanks' Algorithm

The first non-trivial algorithm we describe is a time-memory trade-off due to Shanks. SHANKS' ALGORITHM is presented in Algorithm 7.1. It can be seen that this algorithm constructs two lists, each of size $\sqrt{n}$, and then searches for a "collision." A collision allows the desired discrete logarithm to be computed.

Here are some details to justify the correctness of the algorithm. Observe that if $(j, y) \in L_1$ and $(i, y) \in L_2$, then

$$\alpha^{mj} = y = \beta\alpha^{-i},$$

so

$$\alpha^{mj+i} = \beta,$$

as desired. Conversely, for any $\beta \in \langle\alpha\rangle$, we have that $0 \leq \log_\alpha \beta \leq n-1$. If we divide $\log_\alpha \beta$ by the integer m, then we can express $\log_\alpha \beta$ in the form

$$\log_\alpha \beta = mj + i,$$

where $0 \leq j, i \leq m-1$. The fact that $j \leq m-1$ follows because

$$\log_\alpha \beta \leq n-1 \leq m^2 - 1 = m(m-1) + m - 1.$$

Algorithm 7.1: SHANKS(G, n, α, β)

1. $m \leftarrow \lceil \sqrt{n} \rceil$
2. **for** $j \leftarrow 0$ **to** $m - 1$
   do compute α^{mj}
3. Sort the m ordered pairs (j, α^{mj}) with respect to their second coordinates, obtaining a list L_1
4. **for** $i \leftarrow 0$ **to** $m - 1$
   do compute $\beta\alpha^{-i}$
5. Sort the m ordered pairs $(i, \beta\alpha^{-i})$ with respect to their second coordinates, obtaining a list L_2
6. Find a pair $(j, y) \in L_1$ and a pair $(i, y) \in L_2$ (i.e., find two pairs having identical second coordinates)
7. $\log_\alpha \beta \leftarrow (mj + i) \bmod n$

Hence, the search in step 6 will be successful. (However, if it happened that $\beta \notin \langle \alpha \rangle$, then step 6 will not be successful.)

It is not difficult to implement the algorithm to run in $O(m)$ time with $O(m)$ memory (neglecting logarithmic factors). Here are a few details: Step 2 can be performed by first computing α^m, and then computing its powers by successively multiplying by α^m. The total time for this step is $O(m)$. In a similar fashion, step 4 also takes time $O(m)$. Steps 3 and 5 can be done in time $O(m \log m)$ using an efficient sorting algorithm. Finally, step 6 can be done with one (simultaneous) pass through each of the two lists L_1 and L_2; so it requires time $O(m)$.

We also note that steps 2 and 3 can be precomputed, if desired (this will not affect the asymptotic running time, however).

Here is a small example to illustrate SHANKS' ALGORITHM.

Example 7.2 Suppose we wish to find $\log_3 525$ in $(\mathbb{Z}_{809}{}^*, \cdot)$. Note that 809 is prime and 3 is a primitive element in $\mathbb{Z}_{809}{}^*$, so we have $\alpha = 3$, $n = 808$, $\beta = 525$, and $m = \lceil \sqrt{808} \rceil = 29$. Then

$$\alpha^{29} \bmod 809 = 99.$$

First, we compute the ordered pairs $(j, 99^j \bmod 809)$ for $0 \leq j \leq 28$. We obtain the list

$$\begin{array}{ccccc} (0,1) & (1,99) & (2,93) & (3,308) & (4,559) \\ (5,329) & (6,211) & (7,664) & (8,207) & (9,268) \\ (10,644) & (11,654) & (12,26) & (13,147) & (14,800) \\ (15,727) & (16,781) & (17,464) & (18,632) & (19,275) \\ (20,528) & (21,496) & (22,564) & (23,15) & (24,676) \\ (25,586) & (26,575) & (27,295) & (28,81) & \end{array}$$

which is then sorted to produce L_1.

The second list contains the ordered pairs $(i, 525 \times (3^i)^{-1} \bmod 809)$, $0 \leq j \leq 28$. It is as follows:

$$\begin{array}{ccccc} (0,525) & (1,175) & (2,328) & (3,379) & (4,396) \\ (5,132) & (6,44) & (7,554) & (8,724) & (9,511) \\ (10,440) & (11,686) & (12,768) & (13,256) & (14,355) \\ (15,388) & (16,399) & (17,133) & (18,314) & (19,644) \\ (20,754) & (21,521) & (22,713) & (23,777) & (24,259) \\ (25,356) & (26,658) & (27,489) & (28,163) & \end{array}$$

After sorting this list, we get L_2.

Now, if we proceed simultaneously through the two sorted lists (searching for a common second co-ordinate), we find that $(10,644)$ is in L_1 and $(19,644)$ is in L_2. Hence, we can compute

$$\begin{aligned} \log_3 525 &= (29 \times 10 + 19) \bmod 808 \\ &= 309. \end{aligned}$$

As a check, it can be verified that $3^{309} \equiv 525 \pmod{809}$. □

7.2.2 The Pollard Rho Discrete Logarithm Algorithm

We previously discussed the POLLARD RHO ALGORITHM for factoring in Section 6.6.2. There is a corresponding algorithm for finding discrete logarithms, which we describe now. As before, let $(G, \cdot)$ be a group and let $\alpha \in G$ be an element having order n. Let $\beta \in \langle\alpha\rangle$ be the element whose discrete logarithm we want to find. Since $\langle\alpha\rangle$ is cyclic of order n, we can treat $\log_\alpha \beta$ as an element of $\mathbb{Z}_n$.

As with the rho algorithm for factoring, we form a sequence $x_1, x_2, \ldots$ by iteratively applying a random-looking function, f. Once we obtain two elements x_i and x_j in the sequence such that $x_i = x_j$ and $i < j$, we can (hopefully) compute $\log_\alpha \beta$. Just as we did in the case of the factoring algorithm, we will seek a collision of the form $x_i = x_{2i}$, in order to save time and memory.

Let $S_1 \cup S_2 \cup S_3$ be a partition of G into three subsets of roughly equal size. We define a function $f : \langle\alpha\rangle \times \mathbb{Z}_n \times \mathbb{Z}_n \rightarrow \langle\alpha\rangle \times \mathbb{Z}_n \times \mathbb{Z}_n$ as follows:

$$f(x,a,b) = \begin{cases} (\beta x, a, b+1) & \text{if } x \in S_1 \\ (x^2, 2a, 2b) & \text{if } x \in S_2 \\ (\alpha x, a+1, b) & \text{if } x \in S_3. \end{cases}$$

Further, each of the triples (x, a, b) that we form is required to have the property that $x = \alpha^a \beta^b$. We begin with an initial triple having this property, say $(1, 0, 0)$. Observe that $f(x, a, b)$ satisfies the desired property if (x, a, b) does. So we define

$$(x_i, a_i, b_i) = \begin{cases} (1,0,0) & \text{if } i = 0 \\ f(x_{i-1}, a_{i-1}, b_{i-1}) & \text{if } i \geq 1. \end{cases}$$

We compare the triples (x_{2i}, a_{2i}, b_{2i}) and (x_i, a_i, b_i) until we find a value of $i \geq 1$ such that $x_{2i} = x_i$. When this occurs, we have that

$$\alpha^{a_{2i}} \beta^{b_{2i}} = \alpha^{a_i} \beta^{b_i}.$$

If we denote $c = \log_\alpha \beta$, then it must be the case that

$$\alpha^{a_{2i} + c b_{2i}} = \alpha^{a_i + c b_i}.$$

Since α has order n, it follows that

$$a_{2i} + c b_{2i} \equiv a_i + c b_i \pmod{n}.$$

This can be rewritten as

$$c(b_{2i} - b_i) \equiv a_i - a_{2i} \pmod{n}.$$

If $\gcd(b_{2i} - b_i, n) = 1$, then we can solve for c as follows:

$$c = (a_i - a_{2i})(b_{2i} - b_i)^{-1} \bmod n.$$

We illustrate the application of the above algorithm with an example. Notice that we take care to ensure that $1 \notin S_2$ (since we would obtain $x_i = (1, 0, 0)$ for all integers $i \geq 0$ if $1 \in S_2$).

Example 7.3 The integer $p = 809$ is prime, and it can be verified that the element $\alpha = 89$ has order $n = 101$ in $\mathbb{Z}_{809}{}^*$. The element $\beta = 618$ is in the subgroup $\langle \alpha \rangle$; we will compute $\log_\alpha \beta$.

Suppose we define the sets S_1, S_2, and S_3 as follows:

$$\begin{aligned} S_1 &= \{x \in \mathbb{Z}_{809} : x \equiv 1 \pmod 3\} \\ S_2 &= \{x \in \mathbb{Z}_{809} : x \equiv 0 \pmod 3\} \\ S_3 &= \{x \in \mathbb{Z}_{809} : x \equiv 2 \pmod 3\}. \end{aligned}$$

For $i = 1, 2, \ldots$, we obtain triples (x_{2i}, a_{2i}, b_{2i}) and (x_i, a_i, b_i) as follows:

i	(x_i, a_i, b_i)	(x_{2i}, a_{2i}, b_{2i})
1	$(618, 0, 1)$	$(76, 0, 2)$
2	$(76, 0, 2)$	$(113, 0, 4)$
3	$(46, 0, 3)$	$(488, 1, 5)$
4	$(113, 0, 4)$	$(605, 4, 10)$
5	$(349, 1, 4)$	$(422, 5, 11)$
6	$(488, 1, 5)$	$(683, 7, 11)$
7	$(555, 2, 5)$	$(451, 8, 12)$
8	$(605, 4, 10)$	$(344, 9, 13)$
9	$(451, 5, 10)$	$(112, 11, 13)$
10	$(422, 5, 11)$	$(422, 11, 15)$

Algorithm 7.2: POLLARD RHO DISCRETE LOG ALGORITHM(G, n, α, β)

```
procedure f(x, a, b)
  if x ∈ S1
    then f ← (βx, a, (b + 1) mod n)
    else if x ∈ S2
    then f ← (x², 2a mod n, 2b mod n)
    else f ← (αx, (a + 1) mod n, b)
  return (f)

main
  define the partition G = S1 ∪ S2 ∪ S3
  (x, a, b) ← f(1, 0, 0)
  (x', a', b') ← f(x, a, b)
  while x ≠ x'
       ⎧ (x, a, b) ← f(x, a, b)
    do ⎨ (x', a', b') ← f(x', a', b')
       ⎩ (x', a', b') ← f(x', a', b')
  if gcd(b' − b, n) ≠ 1
    then return ("failure")
    else return ((a − a')(b' − b)^{-1} mod n)
```

The first collision in the above list is $x_{10} = x_{20} = 422$. The equation to be solved is

$$c = (5 - 11)(15 - 11)^{-1} \bmod 101 = (-6 \times 4^{-1}) \bmod 101 = 49.$$

Therefore, $\log_{89} 618 = 49$ in the group $\mathbb{Z}_{809}{}^*$. □

The POLLARD RHO ALGORITHM for discrete logarithms is presented as Algorithm 7.2. In this algorithm, we assume, as usual, that $\alpha \in G$ has order n and $\beta \in \langle\alpha\rangle$.

In the situation where $\gcd(b' - b, n) > 1$, Algorithm 7.2 terminates with the output "failure." The situation may not be so bleak, however. If $\gcd(b' - b, n) = d$, then it is not hard to show that the congruence $c(b' - b) \equiv a - a' \pmod{n}$ has exactly d possible solutions. If d is not too large, then it is relatively straightforward to find the d solutions to the congruence and check to see which one is the correct one.

Algorithm 7.2 can be analyzed in a similar fashion as the Pollard rho factoring algorithm. Under reasonable assumptions concerning the randomness of the function f, we expect to be able to compute discrete logarithms in cyclic groups of order n in $O(\sqrt{n})$ iterations of the algorithm.

7.2.3 The Pohlig-Hellman Algorithm

The next algorithm we study is the POHLIG-HELLMAN ALGORITHM. Suppose that

$$n = \prod_{i=1}^{k} p_i^{c_i},$$

where the p_i's are distinct primes. The value $a = \log_\alpha \beta$ is determined (uniquely) modulo n. We first observe that if we can compute $a \bmod p_i^{c_i}$ for each i, $1 \leq i \leq k$, then we can compute $a \bmod n$ by the Chinese remainder theorem. So, let's suppose that q is prime,

$$n \equiv 0 \pmod{q^c}$$

and

$$n \not\equiv 0 \pmod{q^{c+1}}.$$

We will show how to compute the value

$$x = a \bmod q^c,$$

where $0 \leq x \leq q^c - 1$. We can express x in radix q representation as

$$x = \sum_{i=0}^{c-1} a_i q^i,$$

where $0 \leq a_i \leq q - 1$ for $0 \leq i \leq c - 1$. Also, observe that we can express a as

$$a = x + sq^c$$

for some integer s. Hence, we have that

$$a = \sum_{i=0}^{c-1} a_i q^i + sq^c.$$

The first step of the algorithm is to compute a_0. The main observation used in the algorithm is the following:

$$\beta^{n/q} = \alpha^{a_0 n/q}. \tag{7.1}$$

We prove that equation (7.1) holds as follows:

$$\begin{aligned}
\beta^{n/q} &= (\alpha^a)^{n/q} \\
&= (\alpha^{a_0 + a_1 q + \cdots + a_{c-1} q^{c-1} + sq^c})^{n/q} \\
&= (\alpha^{a_0 + Kq})^{n/q} \qquad \text{where } K \text{ is an integer} \\
&= \alpha^{a_0 n/q} \alpha^{Kn} \\
&= \alpha^{a_0 n/q}.
\end{aligned}$$

Using equation (7.1), it is a simple matter to determine a_0. This can be done, for example, by computing

$$\gamma = \alpha^{n/q}, \gamma^2, \dots,$$

until

$$\gamma^i = \beta^{n/q}$$

for some $i \leq q - 1$. When this happens, we know that $a_0 = i$.

Now, if $c = 1$, we're done. Otherwise $c > 1$, and we proceed to determine $a_1, \dots, a_{c-1}$. This is done in a similar fashion as the computation of a_0. Denote $\beta_0 = \beta$, and define

$$\beta_j = \beta\alpha^{-(a_0 + a_1 q + \dots + a_{j-1} q^{j-1})}$$

for $1 \leq j \leq c - 1$. We make use of the following generalization of equation (7.1):

$$\beta_j^{n/q^{j+1}} = \alpha^{a_j n/q}. \tag{7.2}$$

Observe that equation (7.2) reduces to equation (7.1) when $j = 0$.

The proof of equation (7.2) is similar to that of equation (7.1):

$$\begin{aligned}
\beta_j^{n/q^{j+1}} &= (\alpha^{a-(a_0 + a_1 q + \dots + a_{j-1} q^{j-1})})^{n/q^{j+1}} \\
&= (\alpha^{a_j q^j + \dots + a_{c-1} q^{c-1} + s q^c})^{n/q^{j+1}} \\
&= (\alpha^{a_j q^j + K_j q^{j+1}})^{n/q^{j+1}} \qquad \text{where } K_j \text{ is an integer} \\
&= \alpha^{a_j n/q} \alpha^{K_j n} \\
&= \alpha^{a_j n/q}.
\end{aligned}$$

Hence, given β_j, it is straightforward to compute a_j from equation (7.2).

To complete the description of the algorithm, it suffices to observe that β_{j+1} can be computed from β_j by means of a simple recurrence relation, once a_j is known:

$$\beta_{j+1} = \beta_j \alpha^{-a_j q^j}. \tag{7.3}$$

Therefore, we can compute $a_0, \beta_1, a_1, \beta_2, \dots, \beta_{c-1}, a_{c-1}$ by alternately applying equations (7.2) and (7.3).

A pseudo-code description of the POHLIG-HELLMAN ALGORITHM is given as Algorithm 7.3. To summarize the operation of this algorithm, α is an element of order n in a multiplicative group G, q is prime,

$$n \equiv 0 \pmod{q^c}$$

and

$$n \not\equiv 0 \pmod{q^{c+1}}.$$

The algorithm calculates $a_0, \dots, a_{c-1}$, where

$$\log_\alpha \beta \bmod q^c = \sum_{i=0}^{c-1} a_i q^i.$$

We illustrate the Pohlig-Hellman algorithm with a small example.

Algorithm 7.3: POHLIG-HELLMAN$(G, n, \alpha, \beta, q, c)$

$j \leftarrow 0$
$\beta_j \leftarrow \beta$
while $j \leq c - 1$
do $\begin{cases} \delta \leftarrow \beta_j^{\,n/q^{j+1}} \\ \text{find } i \text{ such that } \delta = \alpha^{in/q} \\ a_j \leftarrow i \\ \beta_{j+1} \leftarrow \beta_j \alpha^{-a_j q^j} \\ j \leftarrow j+1 \end{cases}$
return $(a_0, \dots, a_{c-1})$

Example 7.4 Suppose $p = 29$ and $\alpha = 2$. p is prime and α is a primitive element modulo p, and we have that

$$n = p - 1 = 28 = 2^2 7^1.$$

Suppose $\beta = 18$, so we want to determine $a = \log_2 18$. We proceed by first computing $a \bmod 4$ and then computing $a \bmod 7$.

We start by setting $q = 2$ and $c = 2$ and applying Algorithm 7.3. We find that $a_0 = 1$ and $a_1 = 1$. Hence, $a \equiv 3 \pmod 4$.

Next, we apply Algorithm 7.3 with $q = 7$ and $c = 1$. We find that $a_0 = 4$, so $a \equiv 4 \pmod 7$.

Finally, solving the system

$$\begin{aligned} a &\equiv 3 \pmod 4 \\ a &\equiv 4 \pmod 7 \end{aligned}$$

using the Chinese remainder theorem, we get $a \equiv 11 \pmod{28}$. That is, we have computed $\log_2 18 = 11$ in $\mathbb{Z}_{29}$. □

Let's consider the complexity of Algorithm 7.3. It is not difficult to see that a straightforward implementation of this algorithm runs in time $O(cq)$. This can be improved, however, by observing that each computation of a value i such that $\delta = \alpha^{in/q}$ can be viewed as the solution of a particular instance of the **Discrete Logarithm** problem. To be specific, we have that $\delta = \alpha^{in/q}$ if and only if

$$i = \log_{\alpha^{n/q}} \delta.$$

The element $\alpha^{n/q}$ has order q, and therefore each i can be computed (using SHANKS' ALGORITHM, for example) in time $O(\sqrt{q})$. The complexity of Algorithm 7.3 can therefore be reduced to $O(c\sqrt{q})$.

7.2.4 The Index Calculus Method

The algorithms in the three previous sections can be applied to any group. The algorithm we describe in this section, the INDEX CALCULUS ALGORITHM, is more specialized: it applies to the particular situation of finding discrete logarithms in $\mathbb{Z}_p{}^*$ when p is prime and α is a primitive element modulo p. In this situation, the INDEX CALCULUS ALGORITHM is faster than the algorithms previously considered.

The ***Index calculus algorithm*** for computing discrete logarithms bears considerable resemblance to many of the best factoring algorithms. The method uses a ***factor base***, which, as in Section 6.6.3, is a set $\mathcal{B}$ of "small" primes. Suppose $\mathcal{B} = \{p_1, p_2, \ldots, p_B\}$. The first step (a preprocessing step) is to find the logarithms of the B primes in the factor base. The second step is to compute the discrete logarithm of a desired element β, using the knowledge of the discrete logarithms of the elements in the factor base.

Let C be a bit bigger than B; say $C = B + 10$. In the precomputation phase, we will construct C congruences modulo p, which have the following form:

$$\alpha^{x_j} \equiv p_1{}^{a_{1j}} p_2{}^{a_{2j}} \ldots p_B{}^{a_{Bj}} \pmod{p},$$

for $1 \leq j \leq C$. Notice that these congruences can be written equivalently as

$$x_j \equiv a_{1j} \log_\alpha p_1 + \cdots + a_{Bj} \log_\alpha p_B \pmod{p-1},$$

$1 \leq j \leq C$. Given C congruences in the B "unknowns" $\log_\alpha p_i$ $(1 \leq i \leq B)$, we try to solve the system of linear congruences, hoping that there is a unique solution modulo $p-1$. If this is the case, then we can compute the logarithms of the elements in the factor base.

How do we generate the C congruences of the desired form? One elementary way is to take a random value x, compute $\alpha^x \bmod p$, and then determine if $\alpha^x \bmod p$ has all its factors in $\mathcal{B}$ (using trial division, for example).

Now, supposing that we have already successfully carried out the precomputation step, we compute a desired logarithm $\log_\alpha \beta$ by means of a Las Vegas type randomized algorithm. Choose a random integer s $(1 \leq s \leq p-2)$ and compute

$$\gamma = \beta\alpha^s \bmod p.$$

Now attempt to factor γ over the factor base $\mathcal{B}$. If this can be done, then we obtain a congruence of the form

$$\beta\alpha^s \equiv p_1{}^{c_1} p_2{}^{c_2} \ldots p_B{}^{c_B} \pmod{p}.$$

This can be written equivalently as

$$\log_\alpha \beta + s \equiv c_1 \log_\alpha p_1 + \cdots + c_B \log_\alpha p_B \pmod{p-1}.$$

Since all terms in the above congruence are now known, except for $\log_\alpha \beta$, we can easily solve for $\log_\alpha \beta$.

Here is a small, very artificial, example to illustrate the two steps in the algorithm.

Example 7.5 The integer $p = 10007$ is prime. Suppose that $\alpha = 5$ is the primitive element used as the base of logarithms modulo p. Suppose we take $\mathcal{B} = \{2, 3, 5, 7\}$ as the factor base. Of course $\log_5 5 = 1$, so there are three logs of factor base elements to be determined.

Some examples of "lucky" exponents that might be chosen are 4063, 5136, and 9865.

With $x = 4063$, we compute

$$5^{4063} \bmod 10007 = 42 = 2 \times 3 \times 7.$$

This yields the congruence

$$\log_5 2 + \log_5 3 + \log_5 7 \equiv 4063 \pmod{10006}.$$

Similarly, since

$$5^{5136} \bmod 10007 = 54 = 2 \times 3^3$$

and

$$5^{9865} \bmod 10007 = 189 = 3^3 \times 7,$$

we obtain two more congruences:

$$\log_5 2 + 3 \log_5 3 \equiv 5136 \pmod{10006}$$

and

$$3 \log_5 3 + \log_5 7 \equiv 9865 \pmod{10006}.$$

We now have three congruences in three unknowns, and there happens to be a unique solution modulo 10006, namely $\log_5 2 = 6578$, $\log_5 3 = 6190$ and $\log_5 7 = 1301$.

Now, let's suppose that we wish to find $\log_5 9451$. Suppose we choose the "random" exponent $s = 7736$, and compute

$$9451 \times 5^{7736} \bmod 10007 = 8400.$$

Since $8400 = 2^4 3^1 5^2 7^1$ factors over $\mathcal{B}$, we obtain

$$\begin{aligned} \log_5 9451 &= (4 \log_5 2 + \log_5 3 + 2 \log_5 5 + \log_5 7 - s) \bmod 10006 \\ &= (4 \times 6578 + 6190 + 2 \times 1 + 1301 - 7736) \bmod 10006 \\ &= 6057. \end{aligned}$$

To verify, we can check that $5^{6057} \equiv 9451 \pmod{10007}$. □

Heuristic analyses of various versions of the INDEX CALCULUS ALGORITHM have been done. Under reasonable assumptions, such as those considered in the analysis of DIXON'S ALGORITHM in Section 6.6.3, the asymptotic running time of the precomputation phase is

$$O\left(e^{(1+o(1))\sqrt{\ln p \ln \ln p}}\right),$$

and the time to find a particular discrete logarithm is

$$O\left(e^{(1/2+o(1))\sqrt{\ln p \ln \ln p}}\right).$$

7.3 Lower Bounds on the Complexity of Generic Algorithms

In this section, we turn our attention to an interesting lower bound on the complexity of the **Discrete Logarithm** problem. Several of the algorithms we have described for the **Discrete Logarithm** problem can be applied in any group. An algorithm of this type is called a ***generic algorithm***, because it does not depend on any property of the representation of the group. Examples of generic algorithms for the **Discrete Logarithm** problem include SHANKS' ALGORITHM, the POLLARD RHO ALGORITHM and the POHLIG-HELLMAN ALGORITHM. On the other hand, the INDEX CALCULUS ALGORITHM studied in the previous section is not generic. This algorithm involves treating elements of ${\mathbb{Z}_p}^*$ as integers, and then computing their factorizations into primes. Clearly this is something that cannot be done in an arbitrary group.

Another example of a non-generic algorithm for a particular group is provided by studying the **Discrete Logarithm** problem in the additive group $(\mathbb{Z}_n, +)$. (We defined the **Discrete Logarithm** problem in a multiplicative group, but this was done solely to establish a consistent notation for the algorithms we presented.) Suppose that $\gcd(\alpha, n) = 1$, so α is a generator of $\mathbb{Z}_n$. Since the group operation is addition modulo n, an "exponentiation" operation, α^a, corresponds to multiplication by a modulo n. Hence, in this setting, the **Discrete Logarithm** problem is to find the integer a such that

$$\alpha a \equiv \beta \pmod{n}.$$

Since $\gcd(\alpha, n) = 1$, α has a multiplicative inverse modulo n, and we can compute $\alpha^{-1} \bmod n$ easily using the EXTENDED EUCLIDEAN ALGORITHM. Then we can solve for a, obtaining

$$\log_\alpha \beta = \beta\alpha^{-1} \bmod n.$$

This algorithm is of course very fast; its complexity is polynomial in $\log n$.

An even more trivial algorithm can be used to solve the **Discrete Logarithm** problem in $(\mathbb{Z}_n, +)$ when $\alpha = 1$. In this situation, we have that $\log_1 \beta = \beta$ for all $\beta \in \mathbb{Z}_n$.

The **Discrete Logarithm** problem, by definition, takes place in a cyclic (sub)group of order n. It is well known, and almost trivial to prove, that all cyclic groups of order n are isomorphic. By the discussion above, we know how to compute discrete logarithms quickly in the additive group $(\mathbb{Z}_n, +)$. This suggests that we might be able to solve the **Discrete Logarithm** problem in any subgroup $\langle\alpha\rangle$ of order n of any group G by "reducing" the problem to the the easily solved formulation in $(\mathbb{Z}_n, +)$.

Let us think about how (in theory, at least) this could be done. The statement that $\langle\alpha\rangle$ is isomorphic to $(\mathbb{Z}_n, +)$ means that there is a bijection

$$\phi : \langle\alpha\rangle \rightarrow \mathbb{Z}_n$$

such that

$$\phi(xy) = (\phi(x) + \phi(y)) \bmod n$$

for all $x, y \in \langle\alpha\rangle$. It follows easily that

$$\phi(\alpha^a) = a\phi(\alpha) \bmod n,$$

so we have that

$$\beta = \alpha^a \Leftrightarrow a\phi(\alpha) \equiv \phi(\beta) \pmod{n}.$$

Hence, solving for a as described above (using the EXTENDED EUCLIDEAN ALGORITHM), we have that

$$\log_\alpha \beta = \phi(\beta)(\phi(\alpha))^{-1} \bmod n.$$

Consequently, if we have an efficient method of computing the isomorphism ϕ, then we would have an efficient algorithm to compute discrete logarithms in $\langle\alpha\rangle$. The catch is that there is no known general method to efficiently compute the isomorphism ϕ for an arbitrary subgroup $\langle\alpha\rangle$ of an arbitrary group G, even though we know the two groups in question are isomorphic. In fact, it is not hard to see that computing discrete logarithms in $\langle\alpha\rangle$ is equivalent to finding an explicit isomorphism between $\langle\alpha\rangle$ and $(\mathbb{Z}_n, +)$. Hence, this approach seems to lead to a dead end.

In view of the fact that an extremely efficient algorithm exists for the **Discrete Logarithm** problem in $(\mathbb{Z}_n, +)$, it is perhaps surprising that there is a nontrivial lower bound on the complexity of the general problem. However, a result of Shoup provides a lower bound on the complexity of generic algorithms for the **Discrete Logarithm** problem. Recall that Shanks' and the rho algorithms have the property that their complexity (in terms of the number of group operations required to run the algorithm) is roughly $\sqrt{n}$, where n is the order of the (sub)group in which the discrete logarithm is being sought. Shoup's result establishes that these algorithms are essentially optimal within the class of generic algorithms.

We begin by giving a precise description of what we mean by a generic algorithm. We consider a cyclic group or subgroup of order n, which is therefore isomorphic to $(\mathbb{Z}_n, +)$. We will study generic algorithms for the **Discrete Logarithm** problem in $(\mathbb{Z}_n, +)$. (As we shall see, the particular group that is used is irrelevant in the context of generic algorithms; the choice of $(\mathbb{Z}_n, +)$ is arbitrary.)

An ***encoding*** of $(\mathbb{Z}_n, +)$ is any injective mapping $\sigma : \mathbb{Z}_n \rightarrow S$, where S is a finite set. The encoding function specifies how group elements are represented. Any discrete logarithm problem in a (sub)group of cardinality n of an arbitrary group G can be specified by defining a suitable encoding function. For example, consider the multiplicative group $(\mathbb{Z}_p{}^*, \cdot)$, and let α be a primitive element in $\mathbb{Z}_p{}^*$. Let $n = p - 1$, and define the encoding function σ as follows: $\sigma(i) = \alpha^i \bmod p$, $0 \leq i \leq n-1$. Then it should be clear that solving the **Discrete Logarithm** problem in $(\mathbb{Z}_p{}^*, \cdot)$ with respect to the primitive element α is equivalent to solving the **Discrete Logarithm** problem in $(\mathbb{Z}_n, +)$ with generator 1 under the encoding σ.

A generic algorithm is one that works for any encoding. In particular, a generic algorithm must work correctly when the encoding function σ is a random injective function. For example, when $S = \mathbb{Z}_n$, we could take σ to be a random permutation of $\mathbb{Z}_n$. This is very similar to the random oracle model, where a hash function is regarded as a random function in order to define an idealized model in which formal security proofs can be given.

We suppose that we have a random encoding, σ, for the group $(\mathbb{Z}_n, +)$. In this group, the discrete logarithm of any element a to the base 1 is just a, of course. Given the encoding function σ, the encoding $\sigma(1)$ of the generator, and an encoding of an arbitrary group element $\sigma(a)$, a generic algorithm is trying to compute the value of a. In order to perform operations in this group when group elements are encoded using the function σ, we hypothesize the existence of an oracle (or subroutine) to perform this task.

Given encodings of two group elements, say $\sigma(i)$ and $\sigma(j)$, it should be possible to compute the encodings $\sigma((i+j) \bmod n)$ and $\sigma((i-j) \bmod n)$. This is necessary if we are going to add and subtract group elements, and we assume that our oracle will do this for us. By combining operations of the above type, it is possible to compute arbitrary linear combinations of the form $\sigma((ci \pm dj) \bmod n)$, where $c, d \in \mathbb{Z}_n$. However, using the fact that $-j \equiv n - j \pmod{n}$, we observe that we only need to be able to compute linear combinations of the form $\sigma((ci + dj) \bmod n)$. We will assume that the oracle can directly compute linear combinations of this form in one unit of time.

Group operations of the type described above are the only ones allowed in a generic algorithm. That is, we assume that we have some method of performing group operations on encoded elements, but we cannot do any more than that. Now let us consider how a generic algorithm, say GENLOG, can go about trying to compute a discrete logarithm. The input to the algorithm GENLOG consists of $\sigma_1 = \sigma(1)$ and $\sigma_2 = \sigma(a)$, where $a \in \mathbb{Z}_n$ is chosen randomly. GENLOG will be successful if and only if it outputs the value a. (We will assume that n is prime, in order to simplify the analysis.)

GENLOG will use the oracle to generate a sequence of m, say, encodings of linear combinations of 1 and a. The execution of GENLOG can be specified by a list of ordered pairs $(c_i, d_i) \in \mathbb{Z}_n \times \mathbb{Z}_n$, $1 \leq m$. (We can assume that these m ordered pairs are distinct.) For each ordered pair (c_i, d_i), the oracle computes the encoding $\sigma_i = \sigma((c_i + d_i a) \bmod n)$. Note that we can define $(c_1, d_1) = (1, 0)$ and $(c_2, d_2) = (0, 1)$ so our notation is consistent with the input to the algorithm.

In this way, the algorithm GENLOG obtains a list of encoded group elements, $(\sigma_1, \ldots, \sigma_m)$. Because the encoding function σ is injective, it follows immediately that $c_i + d_i a \equiv c_j + d_j a \pmod{n}$ if and only if $\sigma_i = \sigma_j$. This provides a method to possibly compute the value of the unknown a: Suppose that $\sigma_i = \sigma_j$ for two integers $i \neq j$. If $d_i = d_j$, then $c_i = c_j$ and the two ordered pairs (c_i, d_i) and (c_j, d_j) are the same. Since we are assuming the ordered pairs are distinct, it follows that $d_i \neq d_j$. Because n is prime, we can compute a as follows:

$$a = (c_i - c_j)(d_j - d_i)^{-1} \bmod n.$$

(Recall that we used a similar method of computing the value of a discrete logarithm in the POLLARD RHO ALGORITHM.)

Suppose first that the algorithm GENLOG chooses a set

$$\mathcal{C} = \{(c_i, d_i) : 1 \le i \le m\} \subseteq \mathbb{Z}_n \times \mathbb{Z}_n$$

of m distinct ordered pairs all at once, at the beginning of the algorithm. Such an algorithm is called a ***non-adaptive algorithm*** (SHANKS' ALGORITHM is an example of a non-adaptive algorithm). Then the list of m corresponding encodings is obtained from the oracle. Define $\mathbf{Good}(\mathcal{C})$ to consist of all elements $a \in \mathbb{Z}_n$ that are the solution of an equation $a = (c_i - c_j)(d_j - d_i)^{-1} \bmod n$ with $i \neq j$, $i, j \in \{1, \ldots, m\}$. By what we have said above, we know that the value of a can be computed by GENLOG if and only if $a \in \mathbf{Good}(\mathcal{C})$. It is clear that $|\mathbf{Good}(\mathcal{C})| \le \binom{m}{2}$, so there are at most $\binom{m}{2}$ elements for which GENLOG can compute the discrete logarithm after having obtained a sequence of m encoded group elements corresponding to the ordered pairs in $\mathcal{C}$. The probability that $a \in \mathbf{Good}(\mathcal{C})$ is at most $\binom{m}{2}/n$.

If $a \notin \mathbf{Good}(\mathcal{C})$, then the best strategy for the algorithm GENLOG is to guess the value of a by choosing a random value in $\mathbb{Z}_n \backslash \mathbf{Good}(\mathcal{C})$. Denote $g = |\mathbf{Good}(\mathcal{C})|$. Then, by conditioning on whether or not $a \in \mathbf{Good}(\mathcal{C})$, we can compute a bound on the success probability of the algorithm. Suppose we define A to be the event $a \in \mathbf{Good}(\mathcal{C})$ and we let B denote the event "the algorithm returns the correct value of a." Then we have that

$$\begin{aligned}
\mathbf{Pr}[B] &= \mathbf{Pr}[B|A] \times \mathbf{Pr}[A] + \mathbf{Pr}[B|\overline{A}] \times \mathbf{Pr}[\overline{A}] \\
&= 1 \times \frac{g}{n} + \frac{1}{n-g} \times \frac{n-g}{n} \\
&= \frac{g+1}{n} \\
&\le \frac{\binom{m}{2}+1}{n}.
\end{aligned}$$

If the algorithm always gives the correct answer, then $\mathbf{Pr}[B] = 1$. In this case, it is easy to see that m is $\Omega(\sqrt{n})$.

A generic discrete logarithm algorithm is not required to choose all the ordered pairs in $\mathcal{C}$ at the beginning of the algorithm, of course. It can choose later pairs after seeing what encodings of previous linear combinations look like (i.e., we allow the algorithm to be an ***adaptive algorithm***). However, it can be shown that this does not improve the success probability of the algorithm.

Let GENLOG be an adaptive generic algorithm for the **Discrete Logarithm** problem. For $1 \le i \le m$, let $\mathcal{C}_i$ consist of the first i ordered pairs, for which the oracle computes the corresponding encodings $\sigma_1, \ldots, \sigma_i$. The set $\mathcal{C}_i$ and the list $\sigma_1, \ldots, \sigma_i$ represent all the information available to GENLOG at time i of its execution.

Now, it can be proven that the value of a can be computed at time i if $a \in \mathbf{Good}(\mathcal{C}_i)$. Furthermore, if $a \notin \mathbf{Good}(\mathcal{C}_i)$, then a is equally likely to take on any given value in the set $\mathbb{Z}_n \backslash \mathbf{Good}(\mathcal{C}_i)$.

From these facts, it can be shown that adaptive generic algorithms have the same success probability as non-adaptive ones. It follows that $\Omega(\sqrt{n})$ is a lower bound on the complexity of any generic algorithm for the **Discrete Logarithm** problem in a (sub)group of prime order n.

7.4 Finite Fields

The *ElGamal Cryptosystem* can be implemented in any group where the **Discrete Logarithm** problem is infeasible. We used the multiplicative group $\mathbb{Z}_p{}^*$ in the description of Cryptosystem 7.1, but other groups are also suitable candidates. Two such classes of groups are

1. the multiplicative group of the finite field $\mathbb{F}_{p^n}$
2. the group of an elliptic curve defined over a finite field.

We will discuss these two classes of groups in the next sections.

We have already mentioned the fact that $\mathbb{Z}_p$ is a field if p is prime. However, there are other examples of finite fields not of this form. In fact, there is a finite field with q elements if $q = p^n$ where p is prime and $n \geq 1$ is an integer. We will now describe very briefly how to construct such a field. First, we need several definitions.

Definition 7.1: Suppose p is prime. Define $\mathbb{Z}_p[x]$ to be the set of all polynomials in the indeterminate x. By defining addition and multiplication of polynomials in the usual way (and reducing coefficients modulo p), we construct a ring.
For $f(x), g(x) \in \mathbb{Z}_p[x]$, we say that $f(x)$ ***divides*** $g(x)$ (notation: $f(x) \mid g(x)$) if there exists $q(x) \in \mathbb{Z}_p[x]$ such that

$$g(x) = q(x)f(x).$$

For $f(x) \in \mathbb{Z}_p[x]$, define $\mathbf{deg}(f)$, the ***degree*** of f, to be the highest exponent in a term of f.
Suppose $f(x), g(x), h(x) \in \mathbb{Z}_p[x]$, and $\mathbf{deg}(f) = n \geq 1$. We define

$$g(x) \equiv h(x) \pmod{f(x)}$$

if

$$f(x) \mid (g(x) - h(x)).$$

Notice the resemblance of the definition of congruence of polynomials to that of congruence of integers.

We are now going to define a ring of polynomials "modulo $f(x)$." This ring is denoted by $\mathbb{Z}_p[x]/(f(x))$. The construction of $\mathbb{Z}_p[x]/(f(x))$ from $\mathbb{Z}_p[x]$ is based on the idea of congruences modulo $f(x)$ and is analogous to the construction of $\mathbb{Z}_m$ from $\mathbb{Z}$.

Suppose $\mathbf{deg}(f) = n$. If we divide $g(x)$ by $f(x)$, we obtain a (unique) ***quotient*** $q(x)$ and ***remainder*** $r(x)$, where

$$g(x) = q(x)f(x) + r(x) \tag{7.4}$$

and

$$\mathbf{deg}(r) < n. \tag{7.5}$$

This can be done by usual long division of polynomials. Hence any polynomial in $\mathbb{Z}_p[x]$ is congruent modulo $f(x)$ to a unique polynomial of degree at most $n-1$.

Now we define the elements of $\mathbb{Z}_p[x]/(f(x))$ to be the p^n polynomials in $\mathbb{Z}_p[x]$ of degree at most $n-1$. Addition and multiplication in $\mathbb{Z}_p[x]/(f(x))$ is defined as in $\mathbb{Z}_p[x]$, followed by a reduction modulo $f(x)$. Equipped with these operations, $\mathbb{Z}_p[x]/(f(x))$ is a ring.

Recall that $\mathbb{Z}_m$ is a field if and only if m is prime, and multiplicative inverses can be found using the Euclidean algorithm. A similar situation holds for $\mathbb{Z}_p[x]/(f(x))$. The analog of primality for polynomials is irreducibility, which we define as follows:

Definition 7.2: A polynomial $f(x) \in \mathbb{Z}_p[x]$ is said to be ***irreducible*** if there do not exist polynomials $f_1(x), f_2(x) \in \mathbb{Z}_p[x]$ such that

$$f(x) = f_1(x)f_2(x),$$

where $\mathbf{deg}(f_1) > 0$ and $\mathbf{deg}(f_2) > 0$.

A very important fact is that $\mathbb{Z}_p[x]/(f(x))$ is a field if and only if $f(x)$ is irreducible. Further, multiplicative inverses in $\mathbb{Z}_p[x]/(f(x))$ can be computed using a straightforward modification of the EXTENDED EUCLIDEAN ALGORITHM. We do not give a formal description of the EXTENDED EUCLIDEAN ALGORITHM FOR POLYNOMIALS, but we illustrate the basic idea with an example.

Example 7.6 The polynomial $x^5 + x^2 + 1$ is irreducible over $\mathbb{Z}_2[x]$. Suppose we wish to calculate the inverse of $x^4 + x^3 + 1$ in $\mathbb{Z}_2[x]/(x^5 + x^2 + 1)$ using the EXTENDED EUCLIDEAN ALGORITHM FOR POLYNOMIALS. We basically follow the same steps as in Algorithm 6.2. The only modification is that we are now performing long division of polynomials at each step, obtaining quotients and remainders that satisfy (7.4) and (7.5).

We compute the following:

i	r_i	q_i	s_i	t_i
0	x^5+x^2+1		1	0
1	x^4+x^3+1	$x+1$	0	1
2	x^3+x^2+x	x	1	$x+1$
3	x^2+1	$x+1$	x	x^2+x+1
4	1	x^2+1	x^2+x+1	x^3+x

Therefore, we have found that

$$(x^2+x+1)(x^5+x^2+1)+(x^3+x)(x^4+x^3+1)=1.$$

This implies that x^3+x is the inverse of x^4+x^3+1 in $\mathbb{Z}_2[x]/(x^5+x^2+1)$. □

We now provide an example to illustrate the construction of a finite field using the techniques described above.

Example 7.7 Let's construct a field having eight elements. This can be done by finding an irreducible polynomial of degree three in $\mathbb{Z}_2[x]$. It is sufficient to consider the polynomials having constant term equal to 1, since any polynomial with constant term 0 is divisible by x and hence is reducible. There are four such polynomials:

$$\begin{aligned} f_1(x) &= x^3+1 \\ f_2(x) &= x^3+x+1 \\ f_3(x) &= x^3+x^2+1 \\ f_4(x) &= x^3+x^2+x+1. \end{aligned}$$

Now, $f_1(x)$ is reducible because

$$x^3+1=(x+1)(x^2+x+1)$$

(remember that all coefficients are to be reduced modulo 2). Also, f_4 is reducible because

$$x^3+x^2+x+1=(x+1)(x^2+1).$$

However, $f_2(x)$ and $f_3(x)$ are both irreducible, and either one can be used to construct a field having eight elements.

Let us use $f_2(x)$, and thus construct the field $\mathbb{Z}_2[x]/(x^3+x+1)$. The eight field elements are the eight polynomials 0, 1, x, $x+1$, x^2, x^2+1, x^2+x, and x^2+x+1.

To compute a product of two field elements, we multiply the two polynomials together, and reduce modulo x^3+x+1 (i.e., divide by x^3+x+1 and find the remainder polynomial). Since we are dividing by a polynomial of degree three, the remainder will have degree at most two and hence it is an element of the field.

For example, to compute $(x^2+1)(x^2+x+1)$ in $\mathbb{Z}_2[x]/(x^3+x+1)$, we first compute the product in $\mathbb{Z}_2[x]$, which is x^4+x^3+x+1. Then we divide by x^3+x+1, obtaining the expression

$$x^4+x^3+x+1=(x+1)(x^3+x+1)+x^2+x.$$

Hence, in the field $\mathbb{Z}_2[x]/(x^3+x+1)$, we have that

$$(x^2+1)(x^2+x+1)=x^2+x.$$

Below, we present a complete multiplication table for the non-zero field elements. To save space, we write a polynomial $a_2x^2 + a_1x + a_0$ as the ordered triple $a_2a_1a_0$.

	001	010	011	100	101	110	111
001	001	010	011	100	101	110	111
010	010	100	110	011	001	111	101
011	011	110	101	111	100	001	010
100	100	011	111	110	010	101	001
101	101	001	100	010	111	011	110
110	110	111	001	101	011	010	100
111	111	101	010	001	110	100	011

The multiplicative group of the non-zero polynomials in the field is a cyclic group of order seven. Since 7 is prime, it follows that any field element other than 0 or 1 is a generator of this group, i.e., a primitive element of the field.

For example, if we compute the powers of x, we obtain

$$\begin{aligned} x^1 &= x \\ x^2 &= x^2 \\ x^3 &= x+1 \\ x^4 &= x^2+x \\ x^5 &= x^2+x+1 \\ x^6 &= x^2+1 \\ x^7 &= 1, \end{aligned}$$

which comprise all the non-zero field elements. □

It remains to discuss existence and uniqueness of fields of this type. It can be shown that there is at least one irreducible polynomial of any given degree $n \geq 1$ in $\mathbb{Z}_p[x]$. Hence, there is a finite field with p^n elements for all primes p and all integers $n \geq 1$. There are usually many irreducible polynomials of degree n in $\mathbb{Z}_p[x]$. But the finite fields constructed from any two irreducible polynomials of degree n can be shown to be isomorphic. Thus there is a unique finite field of any size p^n (p prime, $n \geq 1$), which is denoted by $\mathbb{F}_{p^n}$. In the case $n = 1$, the resulting field $\mathbb{F}_p$ is the same thing as $\mathbb{Z}_p$. Finally, it can be shown that there does not exist a finite field with r elements unless $r = p^n$ for some prime p and some integer $n \geq 1$.

The field $\mathbb{Z}_p[x]/(f(x))$ contains the set of constant polynomials in $\mathbb{Z}_p[x]$, namely, the polynomials of degree zero, together with 0. Under addition and multiplication, these behave like the elements of $\mathbb{Z}_p$, since the sum of two constant polynomials is a constant polynomial, and the product of two constant polynomials is a constant polynomial. Thus we can regard $\mathbb{Z}_p$ to be contained in $\mathbb{F}_{p^n}$, and we say that $\mathbb{Z}_p$ is a ***subfield*** of $\mathbb{F}_{p^n}$ or, alternatively, that $\mathbb{F}_{p^n}$ is an ***extension field*** of $\mathbb{Z}_p$ of degree n. More generally, it can be shown that $\mathbb{F}_{p^n}$ contains a unique subfield isomorphic to $\mathbb{F}_{p^d}$ for each d that divides n. Given a field $\mathbb{F}_{p^n}$ and an irreducible polynomial $g(x) \in \mathbb{F}_{p^n}[x]$ of degree k, it holds that $\mathbb{F}_{p^n}[x]/(g(x))$ is the field $\mathbb{F}_{p^{kn}}$, which is an extension of $\mathbb{F}_{p^n}$ of degree k.

We have already noted that the multiplicative group $\mathbb{Z}_p{}^*$ (p prime) is a cyclic group of order $p-1$. In fact, the multiplicative group of any finite field is cyclic: $\mathbb{F}_{p^n}\backslash\{0\}$ is a cyclic group of order $p^n - 1$. This provides further examples of cyclic groups in which the **Discrete Logarithm** problem can be studied.

The ***characteristic*** of a field $\mathbb{F}_q$ is the smallest integer s such that the sum of s copies of "1" is equal to 0. Since any finite field $\mathbb{F}_q$ has $q = p^n$, and $\mathbb{F}_{p^n}$ is an extension of $\mathbb{Z}_p$, it follows that the characteristic of $\mathbb{F}_{p^n}$ is p.

In practice, the finite fields $\mathbb{F}_{2^n}$ have been most studied. Any generic algorithm works in a field $\mathbb{F}_{2^n}$, of course. More importantly, however, the INDEX CALCULUS ALGORITHM can be modified in a straightforward manner to work in these fields. Recall that the main steps in the INDEX CALCULUS ALGORITHM involve trying to factor elements in $\mathbb{Z}_p$ over a given factor base that consists of small primes. The analog of a factor base in $\mathbb{Z}_2[x]$ is a set of irreducible polynomials of low degree. The idea then is to try to factor elements in $\mathbb{F}_{2^n}$ into polynomials in the given factor base. The reader can easily fill in the details.

Once the appropriate modifications have been made, the precomputation time of the INDEX CALCULUS ALGORITHM in $\mathbb{F}_{2^n}$ turns out to be

$$O\left(e^{(1.405+o(1))n^{1/3}(\ln n)^{2/3}}\right),$$

and the time to find an individual discrete logarithm is

$$O\left(e^{(1.098+o(1))n^{1/3}(\ln n)^{2/3}}\right).$$

This algorithm was successfully used by Thomé in 2001 to compute discrete logarithms in $\mathbb{F}_{2^{607}}{}^*$. For values of $n > 1024$, the **Discrete Logarithm** problem in $\mathbb{F}_{2^n}{}^*$ was considered to be infeasible at that time, provided that $2^n - 1$ has at least one "large" prime factor (in order to thwart a Pohlig-Hellman attack). However, in 2013, Joux introduced a new variant of the INDEX CALCULUS ALGORITHM that improves the efficiency of discrete logarithm calculations in $\mathbb{F}_{2^n}$, especially for composite values of n. This approach has since been refined by various researchers, and in 2014, Robert Granger, Thorsten Kleinjung, and Jens Zumbrägel announced they had used this method to successfully compute discrete logarithms in $\mathbb{F}_{2^{9234}}{}^*$. We discuss Joux's algorithm in the next section.

7.4.1 Joux's Index Calculus for Fields of Small Characteristic

Joux's approach to solving the **Discrete Logarithm** problem in $(\mathbb{F}_{2^n}{}^*, \cdot)$ relies on the observation that, if $2^n = (q^2)^k$, then $\mathbb{F}_{2^n}$ can be viewed as a degree k extension of $\mathbb{F}_{q^2}$. (For this algorithm to be effective, we require k to be roughly the same size as q; if n does not have a suitable factor k, then it may be necessary to work in a small extension of $\mathbb{F}_{2^n}$ whose degree can be factored in the desired way.) The elements of $\mathbb{F}_{2^n} = \mathbb{F}_{(q^2)^k}$ are viewed as polynomials over $\mathbb{F}_{q^2}$, and Joux takes as a factor base all polynomials over $\mathbb{F}_{q^2}$ that have degree at most 2.

Where Joux's algorithm differs from the traditional INDEX CALCULUS ALGORITHM is in the process for obtaining relations among the elements of the factor base. Rather than taking random polynomials and testing to see whether they can be expressed as a product of small factors, Joux proposes an explicit method for constructing polynomials that have the desired form. The starting point is the observation that the elements of $\mathbb{F}_q{}^*$ form a subgroup of $\mathbb{F}_{q^2}{}^*$ of order $q-1$ under multiplication. This implies that any element $\alpha \in \mathbb{F}_q{}^*$ satisfies $\alpha^{q-1} = 1$, and hence $\alpha^q = \alpha$. Therefore, the q elements of $\mathbb{F}_q$ are all roots of the polynomial $x^q - x \in \mathbb{F}_{q^2}[x]$, and so we have

$$x^q - x = \prod_{\alpha \in \mathbb{F}_q} (x - \alpha). \tag{7.6}$$

Equation (7.6) shows that $x^q - x$ is a product of elements of the factor base. In order to use this as a relation between elements of the factor base, we need $x^q - x$ to have a suitably low degree, when considered as an element of $\mathbb{F}_{(q^2)^k}$. This depends on the choice of irreducible polynomial used to represent elements of $\mathbb{F}_{(q^2)^k}$ as polynomials over $\mathbb{F}_{q^2}$. If $I \in \mathbb{F}_{q^2}[x]$ is an irreducible polynomial that divides $h_1 x^q - h_0$ for some $h_0, h_1 \in \mathbb{F}_{q^2}[x]$, then $x^q = h_0/h_1$ in $\mathbb{F}_{q^2}[x]/(I)$. If we can find a suitable I, h_0, and h_1 for which h_0/h_1 is a polynomial of sufficiently small degree, then $x^q - x$ is an element of our factor base, as desired. Joux gives heuristic arguments to suggest that, in general, suitable irreducible polynomials do exist.

Once we have a single relation given by (7.6), we can use a clever trick to generate further relations. Let a, b, c, and d be elements of $\mathbb{F}_{q^2}$ with $ad - bc \neq 0$. We apply a change of variables to both sides of (7.6) in which we replace x by $(ax+b)/(cx+d)$. If we then multiply both sides of the resulting expression by $(cx+d)^{q+1}$, then we obtain another polynomial expression that gives us a relation between linear elements of the factor base. In some cases, for example when we have $a, b, c, d \in \mathbb{F}_q$, the resulting expression may not differ from the original expression. However, by repeating this process for a sufficient number of choices of $a, b, c, d \in \mathbb{F}_{q^2}$, we can expect to obtain the desired number of relations. A slightly different transformation can be used in a similar way to obtain relations between the quadratic elements of the factor base.

Once we have a sufficient number of relations, we can find the discrete logarithms of all the elements in the factor base through linear algebra; this step is directly analogous to the corresponding step in the traditional index calculus approach. Unlike the traditional method, however, generating relations and calculating the discrete logarithms of the elements of the factor base can be carried out (under certain heuristic assumptions) in randomized polynomial time. The costly part of this approach is actually using these discrete logarithm values to recover the discrete logarithm of an arbitrary field element. This involves a ***descent phase*** in which the element whose logarithm we wish to calculate is expressed in terms of polynomials of successively lower degree, until a point is reached at which the known logarithms of the elements of the factor base can be used. Barbulescu, Gaudry, Joux, and Thomé have described an approach to this descent phase that gives rise to an overall complexity for computing discrete logarithms in $\mathbb{F}_{2^n}{}^*$ of $2^{O((\log n)^2)}$, which is referred to as ***quasi-polynomial*** complexity.

Example 7.8 Consider the field $\mathbb{F}_{2^{12}}$, and observe that $2^{12} = (4^2)^3$. Set $q = 4$ and $k = 3$. Let w be a primitive element of $\mathbb{F}_{16}$. The elements of $\mathbb{F}_4 \subseteq \mathbb{F}_{16}$ are $\{0, 1, w^5, w^{10}\}$.

The polynomial $x^3 - w$ is irreducible over $\mathbb{F}_{16}$. Thus we have $\mathbb{F}_{2^{12}} = \mathbb{F}_{16}[x]/(x^3 - w)$, and in this field we have $x^4 - wx = 0$, so $x^4 = wx$.

Using (7.6) we derive the following relation:

$$\begin{aligned} x(x-1)(x-w^5)(x-w^{10}) &= x^4 - x, \\ &= (w-1)x. \end{aligned} \tag{7.7}$$

As an example of how we can derive new relations among the elements of the factor base, let $a = 0$, $b = 1$, $c = wx$, and $d = 0$. Replacing x by $1/(wx)$ in (7.7) gives

$$\left(\frac{1}{wx}\right)\left(\frac{1-wx}{wx}\right)\left(\frac{1-w^6x}{wx}\right)\left(\frac{1-w^{11}x}{wx}\right) = \frac{w-1}{wx}.$$

If we now multiply both sides by $(wx)^4$, we obtain

$$\begin{aligned} (1-wx)(1-w^6x)(1-w^{11}x) &= (w-1)w^3x^3, \\ &= w^5 - w^4, \end{aligned}$$

since $x^3 = w$. This is a new relation among elements of the factor base. □

7.5 Elliptic Curves

Elliptic curves are described by the set of solutions to certain equations in two variables. Elliptic curves defined modulo a prime p are of central importance in public-key cryptography. We begin by looking briefly at elliptic curves defined over the real numbers, because some of the basic concepts are easier to motivate in this setting.

7.5.1 Elliptic Curves over the Reals

Definition 7.3: Let $a, b \in \mathbb{R}$ be constants such that $4a^3 + 27b^2 \neq 0$. A ***non-singular elliptic curve*** is the set $\mathcal{E}$ of solutions $(x, y) \in \mathbb{R} \times \mathbb{R}$ to the equation

$$y^2 = x^3 + ax + b, \tag{7.8}$$

together with a special point $\mathcal{O}$ called the ***point at infinity***.

In Figure 7.1, we depict the elliptic curve $y^2 = x^3 - 4x$.

It can be shown that the condition $4a^3 + 27b^2 \neq 0$ is necessary and sufficient to ensure that the equation $x^3 + ax + b = 0$ has three distinct roots (which may be real or complex numbers). If $4a^3 + 27b^2 = 0$, then the corresponding elliptic curve is called a ***singular elliptic curve***.

Suppose $\mathcal{E}$ is a non-singular elliptic curve. We will define a binary operation over $\mathcal{E}$ that makes $\mathcal{E}$ into an abelian group. This operation is usually denoted by addition. The point at infinity, $\mathcal{O}$, will be the identity element, so $P + \mathcal{O} = \mathcal{O} + P = P$ for all $P \in \mathcal{E}$.

Suppose $P, Q \in \mathcal{E}$, where $P = (x_1, y_1)$ and $Q = (x_2, y_2)$. We consider three cases:

1. $x_1 \neq x_2$
2. $x_1 = x_2$ and $y_1 = -y_2$
3. $x_1 = x_2$ and $y_1 = y_2$

In case 1, we define $\mathcal{L}$ to be the line through P and Q. $\mathcal{L}$ intersects $\mathcal{E}$ in the two points P and Q, and it is easy to see that $\mathcal{L}$ will intersect $\mathcal{E}$ in one further point, which we call R'. If we reflect R' in the x-axis, then we get a point that we name R. We define $P + Q = R$.

Let's work out an algebraic formula to compute R. First, the equation of $\mathcal{L}$ is $y = \lambda x + \nu$, where the slope of $\mathcal{L}$ is

$$\lambda = \frac{y_2 - y_1}{x_2 - x_1},$$

and

$$\nu = y_1 - \lambda x_1 = y_2 - \lambda x_2.$$

In order to find the points in $\mathcal{E} \cap \mathcal{L}$, we substitute $y = \lambda x + \nu$ into the equation for $\mathcal{E}$, obtaining the following:

$$(\lambda x + \nu)^2 = x^3 + ax + b,$$

which is the same as

$$x^3 - \lambda^2 x^2 + (a - 2\lambda\nu)x + b - \nu^2 = 0. \qquad (7.9)$$

The roots of equation (7.9) are the x-co-ordinates of the points in $\mathcal{E} \cap \mathcal{L}$. We already know two points in $\mathcal{E} \cap \mathcal{L}$, namely, P and Q. Hence x_1 and x_2 are two roots of equation (7.9).

Since equation (7.9) is a cubic equation over the reals having two real roots, the third root, say x_3, must also be real. The sum of the three roots must be the negative of the coefficient of the quadratic term, or λ^2. Therefore

$$x_3 = \lambda^2 - x_1 - x_2.$$

x_3 is the x-co-ordinate of the point R'. We will denote the y-co-ordinate of R' by

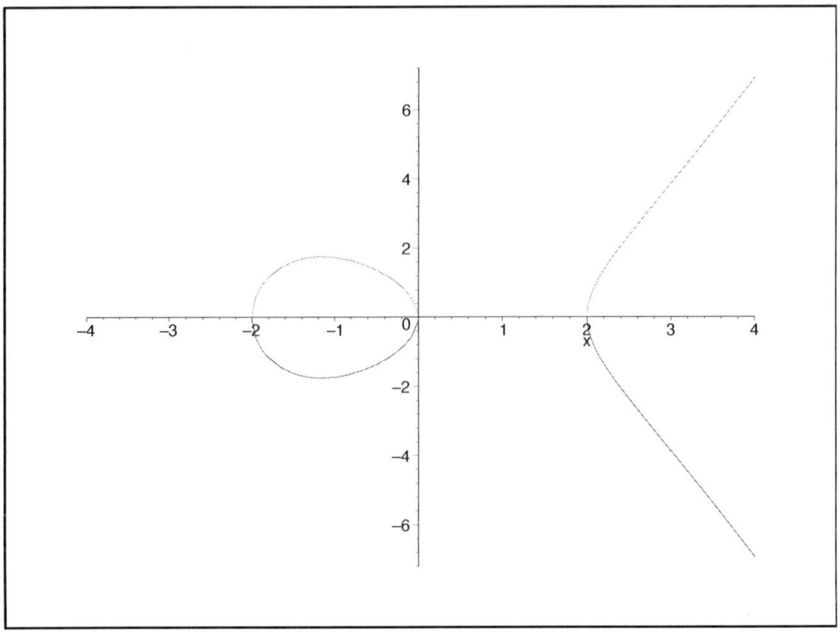

FIGURE 7.1: An elliptic curve over the reals

$-y_3$, so the y-co-ordinate of R will be y_3. An easy way to compute y_3 is to use the fact that the slope of $\mathcal{L}$, namely λ, is determined by any two points on $\mathcal{L}$. If we use the points (x_1, y_1) and $(x_3, -y_3)$ to compute this slope, we get

$$\lambda = \frac{-y_3 - y_1}{x_3 - x_1},$$

or

$$y_3 = \lambda(x_1 - x_3) - y_1.$$

Therefore we have derived a formula for $P + Q$ in case 1: if $x_1 \neq x_2$, then $(x_1, y_1) + (x_2, y_2) = (x_3, y_3)$, where

$$\begin{aligned} x_3 &= \lambda^2 - x_1 - x_2, \\ y_3 &= \lambda(x_1 - x_3) - y_1, \quad \text{and} \\ \lambda &= \frac{y_2 - y_1}{x_2 - x_1}. \end{aligned}$$

Case 2, where $x_1 = x_2$ and $y_1 = -y_2$, is simple: we define $(x, y) + (x, -y) = \mathcal{O}$ for all $(x, y) \in \mathcal{E}$. Therefore (x, y) and $(x, -y)$ are inverses with respect to the elliptic curve addition operation.

Case 3 remains to be considered. Here we are adding a point $P = (x_1, y_1)$ to itself. We can assume that $y_1 \neq 0$, for then we would be in case 2. Case 3 is handled much like case 1, except that we define $\mathcal{L}$ to be the tangent to $\mathcal{E}$ at the point P. A little bit of calculus makes the computation quite simple. The slope of $\mathcal{L}$ can be computed using implicit differentiation of the equation of $\mathcal{E}$:

$$2y\frac{dy}{dx} = 3x^2 + a.$$

Substituting $x = x_1, y = y_1$, we see that the slope of the tangent is

$$\lambda = \frac{3{x_1}^2 + a}{2y_1}.$$

The rest of the analysis in this case is the same as in case 1. The formula obtained is identical, except that λ is computed differently.

At this point the following properties of the addition operation, as defined above, should be clear:

1. addition is closed on the set $\mathcal{E}$,
2. addition is commutative,
3. $\mathcal{O}$ is an identity with respect to addition, and
4. every point on $\mathcal{E}$ has an inverse with respect to addition.

In order to show that $(\mathcal{E}, +)$ is an abelian group, it still must be proven that addition is associative. This is quite messy to prove by algebraic methods. The proof of associativity can be made simpler by using some results from geometry; however, we will not discuss the proof here.

7.5.2 Elliptic Curves Modulo a Prime

Let $p > 3$ be prime. Elliptic curves over $\mathbb{Z}_p$ can be defined exactly as they were over the reals (and the addition operation is also defined in an identical fashion) provided that all operations over $\mathbb{R}$ are replaced by analogous operations in $\mathbb{Z}_p$.

Definition 7.4: Let $p > 3$ be prime. The ***elliptic curve*** $y^2 = x^3 + ax + b$ over $\mathbb{Z}_p$ is the set of solutions $(x, y) \in \mathbb{Z}_p \times \mathbb{Z}_p$ to the congruence

$$y^2 \equiv x^3 + ax + b \pmod{p}, \tag{7.10}$$

where $a, b \in \mathbb{Z}_p$ are constants such that $4a^3 + 27b^2 \not\equiv 0 \pmod{p}$, together with a special point $\mathcal{O}$ called the ***point at infinity***.

The addition operation on $\mathcal{E}$ is defined as follows (where all arithmetic operations are performed in $\mathbb{Z}_p$): Suppose

$$P = (x_1, y_1)$$

and

$$Q = (x_2, y_2)$$

are points on $\mathcal{E}$. If $x_2 = x_1$ and $y_2 = -y_1$, then $P + Q = \mathcal{O}$; otherwise $P + Q = (x_3, y_3)$, where

$$\begin{aligned} x_3 &= \lambda^2 - x_1 - x_2 \\ y_3 &= \lambda(x_1 - x_3) - y_1, \end{aligned}$$

and

$$\lambda = \begin{cases} (y_2 - y_1)(x_2 - x_1)^{-1}, & \text{if } P \neq Q \\ (3{x_1}^2 + a)(2y_1)^{-1}, & \text{if } P = Q. \end{cases}$$

Finally, define

$$P + \mathcal{O} = \mathcal{O} + P = P$$

for all $P \in \mathcal{E}$.

Note that the addition of points on an elliptic curve over $\mathbb{Z}_p$ does not have the nice geometric interpretation that it does on an elliptic curve over the reals. However, the same formulas can be used to define addition, and the resulting pair $(\mathcal{E}, +)$ still forms an abelian group.

Let us look at a small example.

Example 7.9 Let $\mathcal{E}$ be the elliptic curve $y^2 = x^3 + x + 6$ over $\mathbb{Z}_{11}$. Let's first determine the points on $\mathcal{E}$. This can be done by looking at each possible $x \in \mathbb{Z}_{11}$, computing $x^3 + x + 6 \bmod 11$, and then trying to solve equation (7.10) for y. For a given x, we can test to see if $z = x^3 + x + 6 \bmod 11$ is a quadratic residue by applying Euler's criterion. Recall from Section 6.8 that there is an explicit formula to compute square roots of quadratic residues modulo p for primes $p \equiv 3 \pmod 4$. Applying this formula, we have that the square roots of a quadratic residue z are

$$\pm z^{(11+1)/4} \bmod 11 = \pm z^3 \bmod 11.$$

The results of these computations are tabulated in Table 7.1.

$\mathcal{E}$ has 13 points on it. Since any group of prime order is cyclic, it follows that $\mathcal{E}$ is isomorphic to $\mathbb{Z}_{13}$, and any point other than the point at infinity is a generator of $\mathcal{E}$. Suppose we take the generator $\alpha = (2, 7)$. Then we can compute the "powers" of α (which we will write as multiples of α, since the group operation is additive). To compute $2\alpha = (2, 7) + (2, 7)$, we first compute

$$\begin{aligned} \lambda &= (3 \times 2^2 + 1)(2 \times 7)^{-1} \bmod 11 \\ &= 2 \times 3^{-1} \bmod 11 \\ &= 2 \times 4 \bmod 11 \\ &= 8. \end{aligned}$$

Then we have

$$\begin{aligned} x_3 &= 8^2 - 2 - 2 \bmod 11 \\ &= 5 \end{aligned}$$

TABLE 7.1: Points on the elliptic curve $y^2 = x^3 + x + 6$ over $\mathbb{Z}_{11}$

x	$x^3 + x + 6 \bmod 11$	quadratic residue?	y
0	6	no	
1	8	no	
2	5	yes	4,7
3	3	yes	5,6
4	8	no	
5	4	yes	2,9
6	8	no	
7	4	yes	2,9
8	9	yes	3,8
9	7	no	
10	4	yes	2,9

and

$$\begin{aligned} y_3 &= 8(2-5) - 7 \bmod 11 \\ &= 2, \end{aligned}$$

so $2\alpha = (5,2)$.

The next multiple would be $3\alpha = 2\alpha + \alpha = (5,2) + (2,7)$. Again, we begin by computing λ, which in this situation is done as follows:

$$\begin{aligned} \lambda &= (7-2)(2-5)^{-1} \bmod 11 \\ &= 5 \times 8^{-1} \bmod 11 \\ &= 5 \times 7 \bmod 11 \\ &= 2. \end{aligned}$$

Then we have

$$\begin{aligned} x_3 &= 2^2 - 5 - 2 \bmod 11 \\ &= 8 \end{aligned}$$

and

$$\begin{aligned} y_3 &= 2(5-8) - 2 \bmod 11 \\ &= 3, \end{aligned}$$

so $3\alpha = (8,3)$.

Continuing in this fashion, the remaining multiples can be computed to be the following:

$$\begin{array}{lllllllll} \alpha & = & (2,7) & 2\alpha & = & (5,2) & 3\alpha & = & (8,3) \\ 4\alpha & = & (10,2) & 5\alpha & = & (3,6) & 6\alpha & = & (7,9) \\ 7\alpha & = & (7,2) & 8\alpha & = & (3,5) & 9\alpha & = & (10,9) \\ 10\alpha & = & (8,8) & 11\alpha & = & (5,9) & 12\alpha & = & (2,4) \end{array}$$

Hence, as we already knew, $\alpha = (2,7)$ is indeed a primitive element. □

TABLE 7.2: Points on the elliptic curve $y^2 = x^3 + x + 4$ over $\mathbb{F}_{25}$

x	$x^3 + x + 4$	quadratic residue?	y
0	4	yes	2,3
1	1	yes	1,4
2	4	yes	2,3
3	4	yes	2,3
4	2	yes	$4w+3, w+2$
w	2	yes	$4w+3, w+2$
$w+1$	$w+3$	yes	$w, 4w$
$w+2$	$3w$	no	
$w+3$	$w+4$	no	
$w+4$	1	yes	1,4
$2w$	$4w+3$	no	
$2w+1$	$2w+1$	yes	$2w+2, 3w+3$
$2w+2$	$2w$	no	
$2w+3$	$4w+1$	no	
$2w+4$	$3w$	no	
$3w$	w	no	
$3w+1$	$2w+3$	no	
$3w+2$	$w+2$	no	
$3w+3$	$3w+3$	yes	$2w+1, 3w+4$
$3w+4$	$3w+2$	no	
$4w$	1	yes	1,4
$4w+1$	2	yes	$4w+3, w+2$
$4w+2$	$4w+4$	yes	$4w+1, w+4$
$4w+3$	$2w+3$	no	
$4w+4$	$4w$	no	

7.5.3 Elliptic Curves over Finite Fields

We have seen examples of elliptic curves over $\mathbb{Z}_p$. It is also possible to consider elliptic curves over the finite field $\mathbb{F}_{p^n}$, with the addition rule being defined in the same way. (Certain modifications to the relevant formulas are necessary, however, when $p = 2$ or $p = 3$.)

Example 7.10 Let's consider an elliptic curve over $\mathbb{F}_{25}$. We begin by observing that the polynomial $x^2 + 4x + 2$ is irreducible over $\mathbb{F}_5$; hence, we can construct $\mathbb{F}_{25}$ as $\mathbb{Z}_5[x]/(x^2 + 4x + 2)$. The elements of $\mathbb{F}_{25}$ can be written in the form $cw + d$, where $c, d \in \mathbb{F}_5$ and we use the indeterminate w rather than x to avoid confusion with the coordinates used to define our elliptic curve.

Let $\mathcal{E}$ be the elliptic curve $y^2 = x^3 + x + 4$ over $\mathbb{F}_{25}$. The coefficients of $\mathcal{E}$ all happen to belong to the subfield $\mathbb{F}_5$, but we wish nonetheless to consider all points satisfying this equation whose coordinates belong to $\mathbb{F}_{25}$. As in Example 7.9, we can consider each possible x-coordinate and determine any corresponding y-coordinates. The results of these calculations are given in Table 7.2.

Suppose we wish to add the point $(2,3)$ to the point $(w+1,4w)$. Then we begin by computing λ as follows:

$$\begin{aligned}\lambda &= (4w-3)(w+1-2)^{-1},\\ &= (4w+2)(w+4)^{-1},\\ &= (4w+2)(2w),\\ &= 3w^2+4w,\\ &= 2w+4.\end{aligned}$$

Then we have $(2,3)+(w+1,4w)=(x_3,y_3)$, where x_3 is given by

$$\begin{aligned}x_3 &= (2w+4)^2-2-(w-1),\\ &= (4w^2+w+1)+2+4w,\\ &= 4w^2+3,\\ &= 4w,\end{aligned}$$

and y_3 is given by

$$\begin{aligned}y_3 &= (2w+4)(2-4w)-3,\\ &= 2w^2+3w,\\ &= 1.\end{aligned}$$

Thus, $(2,3)+(w+1,4w)=(4w,1)$. □

7.5.4 Properties of Elliptic Curves

An elliptic curve $\mathcal{E}$ defined over $\mathbb{F}_q$ (where $q = p^n$ for p prime,) will have roughly q points on it. More precisely, a well-known theorem due to Hasse asserts that the number of points on $\mathcal{E}$, which we denote by $\#\mathcal{E}$, satisfies the following inequality

$$q+1-2\sqrt{q} \leq \#\mathcal{E} \leq q+1+2\sqrt{q}.$$

Computing the exact value of $\#\mathcal{E}$ is more difficult, but there is an efficient algorithm to do this, due to Schoof. (By "efficient" we mean that it has a running time that is polynomial in $\log q$. Schoof's algorithm has a running time of $O((\log q)^8)$ bit operations and is practical for values of q having several hundred digits.)

Now, given that we can compute $\#\mathcal{E}$, we further want to find a cyclic subgroup of $\mathcal{E}$ in which the **Discrete Logarithm** problem is intractable. So we would like to know something about the structure of the group $\mathcal{E}$. The following theorem gives a considerable amount of information on the group structure of $\mathcal{E}$.

THEOREM 7.1 *Let $\mathcal{E}$ be an elliptic curve defined over $\mathbb{F}_q$, where $q = p^n$ for some prime p. Then there exist positive integers n_1 and n_2 such that $(\mathcal{E},+)$ is isomorphic to $\mathbb{Z}_{n_1} \times \mathbb{Z}_{n_2}$. Further, $n_2 \mid n_1$.*

Note that $n_2 = 1$ is allowed in the above theorem. In fact, $n_2 = 1$ if and only if $\mathcal{E}$ is a cyclic group. Also, if $\#\mathcal{E}$ is a prime, or the product of distinct primes, then $\mathcal{E}$ must be a cyclic group.

In any event, if the integers n_1 and n_2 are computed, then we know that $(\mathcal{E}, +)$ has a cyclic subgroup isomorphic to $\mathbb{Z}_{n_1}$ that can potentially be used as a setting for an *ElGamal Cryptosystem*.

Generic algorithms apply to the elliptic curve **Discrete Logarithm** problem, but there is no known adaptation of the INDEX CALCULUS ALGORITHM to the setting of elliptic curves. However, there is a method of exploiting an explicit isomorphism between subgroups of elliptic curves and finite fields that leads to efficient algorithms for certain classes of elliptic curves. This technique, due to Menezes, Okamoto, and Vanstone, can be applied to some particular examples within a special class of elliptic curves called ***supersingular elliptic curves*** that were suggested for use in cryptosystems. (We describe this technique in Section 7.5.5.)

Another class of elliptic curves for which there are fast techniques for computing discrete logarithms are the so-called curves of ***trace one***. These are elliptic curves defined over $\mathbb{Z}_p$ (where p is prime) having exactly p points on them. The elliptic curve **Discrete Logarithm** problem can easily be solved on these elliptic curves.

If the classes of curves described above are avoided, however, then it appears that an elliptic curve having a cyclic subgroup of size about 2^{224} will provide a secure setting for a cryptosystem, provided that the order of the subgroup is divisible by at least one large prime factor (again, to guard against a Pohlig-Hellman attack).

7.5.5 Pairings on Elliptic Curves

Pairings on elliptic curves were first used in a cryptographic context by Menezes, Okamoto, and Vanstone to assist in solving the **Discrete Logarithm** problem on certain curves. However, despite being introduced initially as a tool for breaking cryptosystems, they have also been shown to have useful applications in constructing a range of cryptosystems; we discuss an application to identity-based encryption in Section 13.1.2. In this section, we introduce the concept of pairings on elliptic curves and discuss their application in attacking the elliptic curve **Discrete Logarithm** problem. We begin with the definition of a pairing; see Definition 7.5.

A function e that maps all pairs of points (P_1, P_2) to the identity in G_3 would technically be a pairing according to Definition 7.5, but it would not be useful for any cryptographic applications.

For our subsequent discussion of pairings, it is useful to define the notion of m-torsion points. Let $\mathcal{E}$ be an elliptic curve defined over the finite field $\mathbb{F}_q$, where $q = p^n$ for some prime p. A point P on $\mathcal{E}$ is an ***m*-torsion point** if $mP = \mathcal{O}$; in other words, if its order is a divisor of m. The set of m-torsion points of $\mathcal{E}$ is finite, and has size at most m^2. If m is coprime to q, then it is always possible to find an extension $\mathbb{F}_{q'}$ of $\mathbb{F}_q$ such that the set of points with coordinates from $\mathbb{F}_{q'}$ that

Definition 7.5: A ***pairing*** is a function e that takes elements P_1 from an abelian group G_1 and P_2 from an abelian group G_2 and returns an element $e(P_1, P_2) = g$ belonging to a group G_3:

$$\begin{aligned} e : G_1 \times G_2 &\rightarrow G_3, \\ (P_1, P_2) &\mapsto g. \end{aligned}$$

We follow the convention of using additive notation for the group operations in G_1 and G_2, but multiplicative notation for G_3.

A pairing e should also satisfy the ***bilinear*** property: for all $P_1, Q_1 \in G_1$ and $P_2, Q_2 \in G_2$, we have

$$e(P_1 + Q_1, P_2) = e(P_1, P_2)e(Q_1, P_2),$$

and

$$e(P_1, P_2 + Q_2) = e(P_1, P_2)e(P_1, Q_2).$$

It is an easy consequence of this definition that

$$e(aP, bQ) = e(P, Q)^{ab}$$

for positive integers a and b.

satisfy the equation of $\mathcal{E}$, together with $\mathcal{O}$, contains precisely m^2 points that are m-torsion points. These points form a subgroup of the group of points of $\mathcal{E}$ defined over $\mathbb{F}_{q'}$, and this subgroup is isomorphic to $\mathbb{Z}_m \times \mathbb{Z}_m$. This subgroup is called the ***m-torsion subgroup*** of $\mathcal{E}$, and it is denoted by $\mathcal{E}[m]$.

In the case where m is a prime that divides the order of $(\mathcal{E}, +)$ but is coprime to q and does not divide $q - 1$, then it is known that the m-torsion subgroup $\mathcal{E}[m]$ is a subgroup of the group of points of $\mathcal{E}$ defined over $\mathbb{F}_{q^k}$, where k is the smallest positive integer such that m divides $q^k - 1$.

Example 7.11 Let $\mathcal{E}$ be the elliptic curve described by the equation $y^2 = x^3 + x + 4$ over $\mathbb{F}_5$. Example 7.10 listed all the points satisfying this equation that have coordinates from $\mathbb{F}_{25}$. There were 27 such points (including $\mathcal{O}$), and nine of these (including $\mathcal{O}$) had coordinates in $\mathbb{F}_5$. Now 3 is a prime that divides 9, is coprime to 5, and does not divide 4. The smallest value of k for which 3 divides $5^k - 1$ is $k = 2$. This implies that the 3-torsion subgroup $\mathcal{E}[3]$ consists of nine points that all have coordinates from $\mathbb{F}_{25}$. By checking the points given in Example 7.10, it is possible to identify that all of the following points are 3-torsion points:

$$E[3] = \{\mathcal{O}, (w+1, 4w), (w+1, w), (4, 4w+3), (4, w+2), (3,2), (3,3), (4w+2, 4w+1), (4w+2, w+4)\}.$$

For example, consider the point $(4, w+2)$. To double this point we calculate

$$\begin{aligned} \lambda &= (3 \times 4^2 + 1)(2 \times (w+2))^{-1}, \\ &= 4(2(w+2))^{-1}, \\ &= 2(3w+1), \\ &= w+2. \end{aligned}$$

Then

$$\begin{aligned} x_3 &= (w+2)^2 - 4 - 4, \\ &= w^2 + 4w + 4, \\ &= 4, \\ y_3 &= (w+2)(4-4) - (w+2), \\ &= 4w+3. \end{aligned}$$

Hence $2(4, w+2) = (4, 4w+3)$, which is equal to $-(4, w+2)$. This implies that $3(4, w+2) = \mathcal{O}$, as claimed. □

Let $\mathcal{E}$ be an elliptic curve and let m be an integer that divides the number of points on $\mathcal{E}$. For our purposes, we will make use of pairings e_m that take as input a point P in a cyclic subgroup G_1 of $\mathcal{E}[m]$ of order m, together with a second point Q in a subgroup G_2 of $\mathcal{E}$, and output a nonzero element of the field $\mathbb{F}_{q^k}$, where k is the smallest positive integer for which m divides $q^k - 1$. Thus,

$$e_m : G_1 \times G_2 \rightarrow \mathbb{F}_{q^k}{}^*.$$

Several different pairings have been proposed for use in cryptography, including the Weil pairing, Tate-Lichtenbaum pairing, Eta pairing, and Ate pairing. We will not discuss the details of algorithms for computing pairings here. However, the most efficient pairings generally recommended for implementation are so-called Type 3 pairings, in which G_2 is also cyclic of order m, yet the isomorphism between G_1 and G_2 cannot be efficiently computed. From now on, we will use the term "pairing" to indicate a pairing of this type. These pairings satisfy the following properties:

THEOREM 7.2 *Let $\mathcal{E}[m]$ be the m-torsion subgroup of an elliptic curve $\mathcal{E}$ over a field $\mathbb{F}_q$, where $q = p^n$ for some prime p and m divides the order of $(\mathcal{E}, +)$, m is coprime to q, and m does not divide $q - 1$. Let k be the smallest positive integer such that m divides $q^k - 1$, and let $e_m : G_1 \times G_2 \rightarrow \mathbb{F}_{q^k}{}^*$ be a pairing. Suppose G_1 and G_2 are as specified above. Then e_m satisfies the following properties:*

Algorithm 7.4: PAIRING-BASED-DL($\mathcal{E}, m, P, R$)

1. Find the smallest integer k for which the points of $\mathcal{E}[m]$ all have coordinates from $\mathbb{F}_{q^k}$.
2. Find $Q \in \mathcal{E}[m]$ for which $\alpha = e_m(P, Q)$ has order m.
3. Compute $\beta = e_m(R, Q)$.
4. Determine the discrete logarithm r of β with respect to the base α.

1. *For all $P \in G_1$ and $Q \in G_2$, the image $e_m(P, Q)$ is an mth* ***root of unity*** *in $\mathbb{F}_{q^k}{}^*$ (i.e., $(e_m(P, Q))^m = 1$).*
2. *The pairing e_m is bilinear, i.e., for all $P_1, Q_1 \in G_1$ and $P_2, Q_2 \in G_2$ we have*
$$e_m(P_1 + Q_1, P_2) = e_m(P_1, P_2)e_m(Q_1, P_2),$$
and
$$e_m(P_1, P_2 + Q_2) = e_m(P_1, P_2)e_m(P_1, Q_2).$$
3. *If $P \in G_1$ has order m, then there exists $Q \in G_2$ such that $e_m(P, Q)$ is a* ***primitive*** *m^{th}* ***root of unity*** *in $\mathbb{F}_{q^k}{}^*$ (i.e., the order of $e_m(P, Q)$ in $\mathbb{F}_{q^k}{}^*$ is equal to m).*

The output of a pairing is an element of the cyclic subgroup μ_m of $\mathbb{F}_{q^k}{}^*$ consisting of the mth roots of unity. Let $P \in G_1$ be an element of order m, and suppose that R belongs to $\langle P \rangle$. Let Q be an element of G_2 for which $e_m(P, Q)$ is a primitive m^{th} root of unity, and hence a generator of μ_m.

The fact that the pairings are bilinear implies that the map σ defined by

$$\begin{array}{rcl} \sigma : \langle P \rangle & \to & \mathbb{F}_{q^k}{}^* \\ R & \mapsto & e_m(R, Q), \end{array}$$

is an isomorphism. The idea behind pairing-based attacks is to use this isomorphism σ to translate the elliptic curve **Discrete Logarithm** problem in $\langle P \rangle$ into the **Discrete Logarithm** problem in μ_m, with the aim of enabling the use of techniques such as the INDEX CALCULUS ALGORITHM that cannot be applied directly to the **Discrete Logarithm** problem on the curve itself.

Suppose that R belongs to $\langle P \rangle$. Algorithm 7.4 gives an outline of the essential steps in the pairing-based approach to determining the discrete logarithm of R with respect to P.

Note that, if $R = rP$, then

$$\begin{array}{rcl} \beta & = & e_m(R, Q), \\ & = & e_m(rP, Q), \\ & = & e_m(P, Q)^r, \\ & = & \alpha^r, \end{array}$$

so the discrete logarithm of β with respect to α, which is the output of Algorithm 7.4, does indeed equal the discrete logarithm of R with respect to P.

In order to apply this approach in practice, we need to have techniques for determining k, for finding a suitable point Q, for computing the pairing, and for solving the **Discrete Logarithm** problem in $\mathbb{F}_{q^k}{}^*$. Pairings can be computed in probabilistic polynomial time using variants of an algorithm due to Miller. The question of whether it is feasible to solve the **Discrete Logarithm** problem in $\mathbb{F}_{q^k}{}^*$ depends on the value of k; if k is sufficiently small, then the INDEX CALCULUS ALGORITHM may be used. For randomly chosen elliptic curves over randomly chosen fields, k is expected to be large in general. However, it was shown by Menezes, Okamoto, and Vanstone that, for the class of supersingular elliptic curves, the value of k is at most 6, and they gave an approach for finding a suitable point Q over such curves. The overall result is a probabilistic subexponential time algorithm for solving the **Discrete Logarithm** problem on supersingular curves.

7.5.6 ElGamal Cryptosystems on Elliptic Curves

Suppose we mimic the operations in an *ElGamal Cryptosystem* on an elliptic curve. We would have two public elliptic curve points P and Q. Q would be a multiple of P, say $Q = mP$, where m is the private key. Encryption would involve choosing a random k and computing kP and kQ. Then kQ would be used to encrypt the plaintext.

It is not advisable for the plaintext space to consist of the points on the curve $\mathcal{E}$, because there is no convenient method known of deterministically generating points on $\mathcal{E}$. So it works better if the plaintext is an arbitrary element in $\mathbb{Z}_p$. Then we can apply a suitable hash function h to kQ and add the result modulo p to x in order to encrypt it. To decrypt, the private key m will allow kQ to be computed from kP. Then the result is hashed and x-ored subtracted modulo p from the ciphertext.

Another standard trick is called ***point compression***. This reduces the storage requirement for points on elliptic curves. A (non-infinite) point on an elliptic curve $\mathcal{E}$ is a pair (x, y), where $y^2 \equiv x^3 + ax + b \pmod{p}$. Given a value for x, there are two possible values for y (unless $x^3 + ax + b \equiv 0 \pmod{p}$). These two possible y-values are negatives of each other modulo p. Since p is odd, one of the two possible values of $y \bmod p$ is even and the other is odd. Therefore we can determine a unique point $P = (x, y)$ on $\mathcal{E}$ by specifying the value of x, together with the single bit $y \bmod 2$. This reduces the storage by (almost) 50%, at the expense of requiring additional computations to reconstruct the y-co-ordinate of P.

The operation of point compression can be expressed as a function

$$\text{POINT-COMPRESS} : \mathcal{E}\backslash\{\mathcal{O}\} \rightarrow \mathbb{Z}_p \times \mathbb{Z}_2,$$

which is defined as follows:

$$\text{POINT-COMPRESS}(P) = (x, y \bmod 2), \text{ where } P = (x, y) \in \mathcal{E}.$$

The inverse operation, POINT-DECOMPRESS, reconstructs the elliptic curve point

Algorithm 7.5: POINT-DECOMPRESS(x, i)

$z \leftarrow x^3 + ax + b \bmod p$
if z is a quadratic non-residue modulo p
  then return ("failure")
  else $\begin{cases} y \leftarrow \sqrt{z} \bmod p \\ \textbf{if } y \equiv i \pmod 2 \\ \quad \textbf{then return } (x, y) \\ \quad \textbf{else return } (x, p - y) \end{cases}$

$P = (x, y)$ from $(x, y \bmod 2)$. It can be implemented as shown in Algorithm 7.5. In this algorithm, as previously mentioned, $\sqrt{z}$ can be computed as $z^{(p+1)/4} \bmod p$ provided that $p \equiv 3 \pmod 4$ and z is a quadratic residue modulo p (or $z = 0$).

Combining the two above-described ideas, we obtain a cryptosystem that we call *Elliptic Curve ElGamal*. This cryptosystem is presented as Cryptosystem 7.2.

Elliptic Curve ElGamal has a message expansion (approximately) equal to two, which is similar to the *ElGamal Cryptosystem* over $\mathbb{Z}_p{}^*$. We illustrate encryption and decryption in *Elliptic Curve ElGamal* using the elliptic curve $y^2 = x^3 + x + 6$ defined over $\mathbb{Z}_{11}$.

Example 7.12 Suppose that $P = (2, 7)$ and Bob's private key is $m = 7$, so

$$Q = 7P = (7, 2).$$

Suppose Alice wants to encrypt the plaintext $x = 9$, and she chooses the random value $k = 6$. First, she computes

$$kP = 6\,(2, 7) = (7, 9)$$

and

$$kQ = 6\,(7, 2) = (8, 3).$$

Then, suppose that $h(8, 3) = 4$ for purposes of illustration. Next, she calculates

$$y_1 = \text{POINT-COMPRESS}(7, 9) = (7, 1)$$

and

$$y_2 = 9 + 4 \bmod 11 = 2.$$

The ciphertext she sends to Bob is

$$y = (y_1, y_2) = ((7, 1), 2).$$

When Bob receives the ciphertext y, he computes

$$\begin{aligned} \text{POINT-DECOMPRESS}(7, 1) &= (7, 9), \\ 7\,(7, 9) &= (8, 3), \\ h(8, 3) &= 4 \quad \text{and} \\ 2 - 4 \bmod 11 &= 9. \end{aligned}$$

Cryptosystem 7.2: *Elliptic Curve ElGamal*

Let $\mathcal{E}$ be an elliptic curve defined over $\mathbb{Z}_p$ (where $p > 3$ is prime) such that $\mathcal{E}$ contains a cyclic subgroup $H = \langle P \rangle$ of prime order n in which the **Discrete Logarithm** problem is infeasible. Let $h : \mathcal{E} \rightarrow \mathbb{Z}_p$ be a secure hash function. Let $\mathcal{P} = \mathbb{Z}_p$ and $\mathcal{C} = (\mathbb{Z}_p \times \mathbb{Z}_2) \times \mathbb{Z}_p$. Define

$$\mathcal{K} = \{(\mathcal{E}, P, m, Q, n, h) : Q = mP\},$$

where P and Q are points on $\mathcal{E}$ and $m \in \mathbb{Z}_n{}^*$. The values $\mathcal{E}, P, Q, n$, and h are the public key and m is the private key.

For $K = (\mathcal{E}, P, m, Q, n, h)$, for a (secret) random number $k \in \mathbb{Z}_n{}^*$, and for a plaintext $x \in \mathbb{Z}_p$, define

$$e_K(x, k) = (\text{POINT-COMPRESS}(kP), x + h(kQ) \bmod p).$$

For a ciphertext $y = (y_1, y_2)$, where $y_1 \in \mathbb{Z}_p \times \mathbb{Z}_2$ and $y_2 \in \mathbb{Z}_p$, define

$$d_K(y) = y_2 - h(R) \bmod p,$$

where

$$R = m\,\text{POINT-DECOMPRESS}(y_1).$$

Hence, the decryption yields the correct plaintext, 9. □

7.5.7 Computing Point Multiples on Elliptic Curves

We can compute powers α^a in a multiplicative group efficiently using the SQUARE-AND-MULTIPLY ALGORITHM (Algorithm 6.5). In an elliptic curve setting, where the group operation is written additively, we would compute a multiple aP of an elliptic curve point P using an analogous DOUBLE-AND-ADD ALGORITHM. (The squaring operation $\alpha \mapsto \alpha^2$ would be replaced by the doubling operation $P \mapsto 2P$, and the multiplication of two group elements would be replaced by the addition of two elliptic curve points.)

The addition operation on an elliptic curve has the property that additive inverses are very easy to compute. This fact can be exploited in a generalization of the DOUBLE-AND-ADD ALGORITHM, which we might call the DOUBLE-AND-(ADD OR SUBTRACT) ALGORITHM. We describe this technique now.

Let c be an integer. A ***signed binary representation*** of c is an equation of the form

$$c = \sum_{i=0}^{\ell-1} c_i 2^i,$$

Algorithm 7.6: DOUBLE-AND-(ADD OR SUBTRACT)$(P, (c_{\ell-1}, \ldots, c_0))$

$Q \leftarrow \mathcal{O}$
for $i \leftarrow \ell - 1$ **downto** 0
do $\begin{cases} Q \leftarrow 2Q \\ \textbf{if } c_i = 1 \\ \quad \textbf{then } Q \leftarrow Q + P \\ \quad \textbf{else if } c_i = -1 \\ \quad \textbf{then } Q \leftarrow Q - P \end{cases}$
return (Q)

where $c_i \in \{-1, 0, 1\}$ for all i. In general, there will be more than one signed binary representation of an integer c. For example, we have that

$$11 = 8 + 2 + 1 = 16 - 4 - 1,$$

so

$$(c_4, c_3, c_2, c_1, c_0) = (0, 1, 0, 1, 1) \quad \text{or} \quad (1, 0, -1, 0, -1)$$

are both signed binary representations of 11.

Let P be a point of order n on an elliptic curve. Given any signed binary representation $(c_{\ell-1}, \ldots, c_0)$ of an integer c, where $0 \leq c \leq n - 1$, it is possible to compute the multiple cP of the elliptic curve point P by a series of doublings, additions, and subtractions, using the following algorithm.

In Algorithm 7.6, the subtraction operation $Q - P$ would be performed by first computing the additive inverse $-P$ of P, and then adding the result to Q.

A signed binary representation $(c_{\ell-1}, \ldots, c_0)$ of an integer c is said to be in ***non-adjacent form*** provided that no two consecutive c_i's are non-zero. Such a representation is denoted as a ***NAF representation***. It is a simple matter to transform a binary representation of a positive integer c into a NAF representation. The basis of this transformation is to replace substrings of the form $(0, 1, \cdots, 1, 1)$ in the binary representation by $(1, 0, \cdots, 0, -1)$. Substitutions of this type do not change the value of c, due to the identity

$$2^i + 2^{i-1} + \cdots + 2^j = 2^{i+1} - 2^j,$$

where $i > j$. This process is repeated as often as needed, starting with the rightmost (i.e., low-order) bits and proceeding to the left.

We illustrate the above-described process with an example:

	1	1	1	1	0	0	1	1	0	1	1	1
	1	1	1	1	0	0	1	1	1	0	0	−1
	1	1	1	1	0	1	0	0	−1	0	0	−1
1	0	0	0	−1	0	1	0	0	−1	0	0	−1

Hence the NAF representation of

$$(1,1,1,1,0,0,1,1,0,1,1,1)$$

is

$$(1,0,0,0,-1,0,1,0,0,-1,0,0,-1).$$

This discussion establishes that every non-negative integer has a NAF representation. It is also possible to prove that the NAF representation of an integer is unique (see the Exercises). Therefore we can speak of the NAF representation of an integer without ambiguity.

In a NAF representation, there do not exist two consecutive non-zero coefficients. We might expect that, on average, a NAF representation contains more zeroes than the traditional binary representation of a positive integer. This is indeed the case: it can be shown that, on average, an ℓ-bit integer contains $\ell/2$ zeroes in its binary representation and $2\ell/3$ zeroes in its NAF representation.

These results make it easy to compare the average efficiency of the DOUBLE-AND-ADD ALGORITHM using a binary representation to the DOUBLE-AND-(ADD OR SUBTRACT) ALGORITHM using the NAF representation. Each algorithm requires ℓ doublings, but the number of additions (or subtractions) is $\ell/2$ in the first case, and $\ell/3$ in the second case. If we assume that a doubling takes roughly the same amount of time as an addition (or subtraction), then the ratio of the average times required by the two algorithms is approximately

$$\frac{\ell + \frac{\ell}{2}}{\ell + \frac{\ell}{3}} = \frac{9}{8}.$$

We have therefore obtained a (roughly) 11% speedup, on average, by this simple technique.

7.6 Discrete Logarithm Algorithms in Practice

Historically, the most important settings (G, α) for the **Discrete Logarithm** problem in cryptographic applications have been the following:

1. $G = (\mathbb{Z}_p{}^*, \cdot)$, p prime, α a primitive element modulo p;

2. $G = (\mathbb{Z}_p{}^*, \cdot)$, p, q prime, $p \equiv 1 \bmod q$, α an element in $\mathbb{Z}_p$ having order q;

3. $G = (\mathbb{F}_{2^n}{}^*, \cdot)$, α a primitive element in $\mathbb{F}_{2^n}{}^*$;

4. $G = (\mathcal{E}, +)$, where $\mathcal{E}$ is an elliptic curve modulo a prime p, $\alpha \in \mathcal{E}$ is a point having prime order $q = \#\mathcal{E}/h$, where (typically) $h = 1, 2$ or 4; and

5. $G = (\mathcal{E}, +)$, where $\mathcal{E}$ is an elliptic curve over a finite field $\mathbb{F}_{2^n}$, $\alpha \in \mathcal{E}$ is a point having prime order $q = \#\mathcal{E}/h$, where (typically) $h = 2$ or 4. (Note that we have defined elliptic curve over finite fields $\mathbb{F}_q$ having characteristic exceeding 3. A different equation is required if the field has characteristic 2 or 3.)

Cases 1, 2, and 3 can be attacked using the appropriate form of the INDEX CALCULUS ALGORITHM in $(\mathbb{Z}_p{}^*, \cdot)$ or $(\mathbb{F}_{2^n}{}^*, \cdot)$. Cases 2, 4, and 5 can be attacked using POLLARD RHO ALGORITHMS in subgroups of order q.

Recommendations published by NIST in 2015 suggest that one should take $p \geq 2^{224}$ in case 4 (or $n \geq 224$ in case 5). In contrast, p needs to be at least 2^{2048} in cases 1 and 2 to achieve the same (predicted) level of security (additionally, in case 2, $q \geq 2^{224}$ is recommended). Case 3 is not recommended.

The reason for the significant differences in the suggested parameter sizes is the lack of a known index calculus attack on elliptic curve discrete logarithms. As a consequence, elliptic curve cryptography has become increasingly popular for practical applications, especially for applications on constrained platforms such as wireless devices and smart cards. On platforms such as these, available memory is very small, and a secure implementation of discrete logarithm based cryptography in $(\mathbb{Z}_p{}^*, \cdot)$, for example, would require too much space to be practical. The smaller space required for elliptic curve based cryptography is therefore very desirable.

The above recommendations take into account the best currently implemented algorithms, as well as reasonable hypotheses concerning possible progress in algorithm development and computing speed in the coming years. It is also of interest to look at the current state-of-the-art of algorithms for the **Discrete Logarithm** problem.

Discrete logarithms in $\mathbb{Z}_p{}^*$ have been computed for a 180-digit (= 596 bits) ***safe prime*** p (i.e., a prime p such that $(p-1)/2$ is also prime) by Bouvier, Gaudry, Imbert, Jejeli, and Thomé in June, 2014. This is the current "record" for $\mathbb{Z}_p{}^*$.

For discrete logarithms in $\mathbb{F}_{2^n}{}^*$, much larger cases have been solved. We already mentioned in Section 7.4 that discrete logarithms in $\mathbb{F}_{2^{9234}}{}^*$ have been computed. For prime values of n, the largest discrete logarithm computation in $\mathbb{F}_{2^n}{}^*$ was accomplished for $n = 1279$. This was announced by Thorsten Kleinjung in October, 2014.

In the case of elliptic curves, *Certicom Corporation* issued a series of "challenges" to encourage the development of efficient implementations of discrete logarithm algorithms. The most recent challenges to be solved were the 109-bit challenges known as ECCp-109, which was solved in November, 2002; and ECC2-109, which was solved in April, 2004. Both of these challenges were solved by a team led by C. Monico.

The largest general instances of elliptic curve discrete logarithm problems to be solved, at the time of writing this book, took place in elliptic curves defined over $\mathbb{F}_{2^{127}}$. The solution was due to Bernstein, Engels, Lange, Niederhagen, Paar, Schwabe, and Zimmermann. It used a parallel version of the POLLARD RHO AL-

GORITHM. The algorithm took approximately six months to run and its successful conclusion was reported on December 2, 2016.

7.7 Security of ElGamal Systems

In this section, we study several aspects of the security of ElGamal-type cryptosystems. First, we look at the bit security of discrete logarithms. Then we consider the semantic security of ElGamal-type cryptosystems, and introduce the **Diffie-Hellman** problems.

7.7.1 Bit Security of Discrete Logarithms

In this section, we consider whether individual bits of a discrete logarithm are easy or hard to compute. To be precise, consider Problem 7.2, which we call the **Discrete Logarithm ith Bit** problem (the setting for the discrete logarithms considered in this section is $(\mathbb{Z}_p{}^*, \cdot)$, where p is prime).

> **Problem 7.2: Discrete Logarithm ith Bit**
>
> **Instance:** $I = (p, \alpha, \beta, i)$, where p is prime, $\alpha \in \mathbb{Z}_p{}^*$ is a primitive element, $\beta \in \mathbb{Z}_p{}^*$, and i is an integer such that $1 \leq i \leq \lceil \log_2(p-1) \rceil$.
> **Question:** Compute $L_i(\beta)$, which (for the specified α and p) denotes the ith least significant bit in the binary representation of $\log_\alpha \beta$.

We will first show that computing the least significant bit of a discrete logarithm is easy. In other words, if $i = 1$, then the **Discrete Logarithm ith Bit** problem can be solved efficiently. This follows from Euler's criterion concerning quadratic residues modulo p, where p is prime.

Consider the mapping $f : \mathbb{Z}_p{}^* \rightarrow \mathbb{Z}_p{}^*$ defined by

$$f(x) = x^2 \bmod p.$$

Denote by $\mathbf{QR}(p)$ the set of quadratic residues modulo p; thus

$$\mathbf{QR}(p) = \{x^2 \bmod p : x \in \mathbb{Z}_p{}^*\}.$$

First, observe that $f(x) = f(p-x)$. Next, note that

$$w^2 \equiv x^2 \pmod{p}$$

if and only if

$$p \mid (w-x)(w+x),$$

which happens if and only if

$$w \equiv \pm x \pmod{p}.$$

It follows that

$$|f^{-1}(y)| = 2$$

for every $y \in \mathbf{QR}(p)$, and hence

$$|\mathbf{QR}(p)| = \frac{p-1}{2}.$$

That is, exactly half the residues in $\mathbb{Z}_p{}^*$ are quadratic residues and half are not.

Now, suppose α is a primitive element of $\mathbb{Z}_p$. Then $\alpha^a \in \mathbf{QR}(p)$ if a is even. Since the $(p-1)/2$ elements $\alpha^0 \bmod p, \alpha^2 \bmod p, \dots, \alpha^{p-3} \bmod p$ are all distinct, it follows that

$$\mathbf{QR}(p) = \{\alpha^{2i} \bmod p : 0 \leq i \leq (p-3)/2\}.$$

Hence, β is a quadratic residue if and only if $\log_\alpha \beta$ is even, that is, if and only if $L_1(\beta) = 0$. But we already know, by Euler's criterion, that β is a quadratic residue if and only if

$$\beta^{(p-1)/2} \equiv 1 \pmod{p}.$$

So we have the following efficient formula to calculate $L_1(\beta)$:

$$L_1(\beta) = \begin{cases} 0 & \text{if } \beta^{(p-1)/2} \equiv 1 \pmod{p} \\ 1 & \text{otherwise.} \end{cases}$$

Let's now consider the computation of $L_i(\beta)$ for values of i exceeding 1. Suppose $p - 1 = 2^s t$, where t is odd. It can be shown that it is easy to compute $L_i(\beta)$ if $i \leq s$. On the other hand, computing $L_{s+1}(\beta)$ is (probably) difficult, in the sense that any hypothetical algorithm (or oracle) to compute $L_{s+1}(\beta)$ could be used to find discrete logarithms in $\mathbb{Z}_p$.

We will prove this result in the case $s = 1$. More precisely, if $p \equiv 3 \pmod{4}$ is prime, then we show how any oracle for computing $L_2(\beta)$ can be used to solve the **Discrete Logarithm** problem in $\mathbb{Z}_p$.

Recall that, if β is a quadratic residue in $\mathbb{Z}_p$ and $p \equiv 3 \pmod{4}$, then the two square roots of β modulo p are $\pm\beta^{(p+1)/4} \bmod p$. It is also important that

$$L_1(\beta) \neq L_1(p - \beta)$$

for any $\beta \neq 0$, if $p \equiv 3 \pmod{4}$. We see this as follows. Suppose

$$\alpha^a \equiv \beta \pmod{p};$$

then

$$\alpha^{a+(p-1)/2} \equiv -\beta \pmod{p}.$$

Since $p \equiv 3 \pmod{4}$, the integer $(p-1)/2$ is odd, and the result follows.

Now, suppose that $\beta = \alpha^a$ for some (unknown) even exponent a. Then either

$$\beta^{(p+1)/4} \equiv \alpha^{a/2} \pmod{p}$$

Algorithm 7.7: L2ORACLE-DISCRETE-LOGARITHM(p, α, β)

```
external L1, ORACLEL2
x0 ← L1(β)
β ← β/α^x0 mod p
i ← 1
while β ≠ 1
     ⎧ xi ← ORACLEL2(β)
     ⎪ γ ← β^((p+1)/4) mod p
     ⎪ if L1(γ) = xi
  do ⎨   then β ← γ
     ⎪   else β ← p − γ
     ⎪ β ← β/α^xi mod p
     ⎩ i ← i + 1
return (x_{i−1}, x_{i−2}, ..., x0)
```

or

$$-\beta^{(p+1)/4} \equiv \alpha^{a/2} \pmod{p}.$$

We can determine which of these two possibilities is correct if we know the value $L_2(\beta)$, since

$$L_2(\beta) = L_1(\alpha^{a/2}).$$

This fact is exploited in our algorithm, which we present as Algorithm 7.7.

At the end of Algorithm 7.7, the x_i's comprise the bits in the binary representation of $\log_\alpha \beta$; that is,

$$\log_\alpha \beta = \sum_{i \geq 0} x_i 2^i.$$

We will work out a small example to illustrate the algorithm.

Example 7.13 Suppose $p = 19$, $\alpha = 2$, and $\beta = 6$. Since the example is so small, we can tabulate the values of $L_1(\gamma)$ and $L_2(\gamma)$ for all $\gamma \in \mathbb{Z}_{19}{}^*$. (In general, L_1 can be computed efficiently using Euler's criterion, and L_2 is is computed using the hypothetical algorithm ORACLEL2.) These values are given in Table 7.3. The reader can then verify that Algorithm 7.7 will compute $\log_2 6 = 1110_2 = 14$. □

It is possible to give formal proof of the algorithm's correctness using mathematical induction. Denote

$$x = \log_\alpha \beta = \sum_{i \geq 0} x_i 2^i.$$

For $i \geq 0$, define

$$Y_i = \left\lfloor \frac{x}{2^{i+1}} \right\rfloor.$$

TABLE 7.3: Values of L_1 and L_2 for $p = 19, \alpha = 2$

γ	$L_1(\gamma)$	$L_2(\gamma)$	γ	$L_1(\gamma)$	$L_2(\gamma)$	γ	$L_1(\gamma)$	$L_2(\gamma)$
1	0	0	7	0	1	13	1	0
2	1	0	8	1	1	14	1	1
3	1	0	9	0	0	15	1	1
4	0	1	10	1	0	16	0	0
5	0	0	11	0	0	17	0	1
6	0	1	12	1	1	18	1	0

Also, define β_0 to be the value of β just before the start of the **while** loop; and, for $i \geq 1$, define β_i to be the value of β at the end of the ith iteration of the **while** loop. It can be proved by induction that

$$\beta_i \equiv \alpha^{2Y_i} \pmod p$$

for all $i \geq 0$. Now, with the observation that

$$2Y_i = Y_{i-1} - x_i,$$

it follows that

$$x_{i+1} = L_2(\beta_i),$$

$i \geq 0$. Since

$$x_0 = L_1(\beta),$$

the algorithm is correct. The details are left to the reader.

7.7.2 Semantic Security of ElGamal Systems

We first observe that the basic *ElGamal Cryptosystem*, as described in Cryptosystem 7.1, is not semantically secure. Recall that $\alpha \in \mathbb{Z}_p{}^*$ is a primitive element and $\beta = \alpha^a \bmod p$ where a is the private key. Given a plaintext element x, a random number k is chosen, and then $e_K(x,k) = (y_1, y_2)$ is computed, where $y_1 = \alpha^k \bmod p$ and $y_2 = x\beta^k \bmod p$.

We make use of the fact that it is easy, using Euler's criterion, to test elements of $\mathbb{Z}_p$ to see if they are quadratic residues modulo p. Recall from Section 7.7.1 that β is a quadratic residue modulo p if and only if a is even. Similarly, y_1 is a quadratic residue modulo p if and only if k is even. We can determine the parity of both a and k, and hence we can compute the parity of ak. Therefore, we can determine if β^k ($= \alpha^{ak}$) is a quadratic residue.

Now, suppose that we wish to distinguish encryptions of x_1 from encryptions of x_2, where x_1 is a quadratic residue and x_2 is a quadratic non-residue modulo p. It is a simple matter to determine the quadratic residuosity of y_2, and we have already discussed how the quadratic residuosity of β^k can be determined. It follows that (y_1, y_2) is an encryption of x_1 if and only if β^k and y_2 are both quadratic residues or both quadratic non-residues.

The above attack does not work if β is a quadratic residue and every plaintext x is required to be a quadratic residue. In fact, if $p = 2q + 1$ where q is prime, then it can be shown that restricting β, y_1, and x to be quadratic residues is equivalent to implementing the *ElGamal Cryptosystem* in the subgroup of quadratic residues modulo p (which is a cyclic subgroup of $\mathbb{Z}_p{}^*$ of order q). This version of the *ElGamal Cryptosystem* is conjectured to be semantically secure if the **Discrete Logarithm** problem in $\mathbb{Z}_p{}^*$ is infeasible.

7.7.3 The Diffie-Hellman Problems

We introduce two variants of the so-called **Diffie-Hellman** problems, a computational version and a decision version. The reason for calling them "**Diffie-Hellman** problems" comes from the origin of these two problems in connection with Diffie-Hellman key agreement protocols, which will be presented in Section 12.2. At this time, we discuss some interesting connections between these problems and security of ElGamal-type cryptosystems.

Here are descriptions of the two problems.

Problem 7.3: Computational Diffie-Hellman

Instance: A multiplicative group $(G, \cdot)$, an element $\alpha \in G$ having order n, and two elements $\beta, \gamma \in \langle\alpha\rangle$.
Question: Find $\delta \in \langle\alpha\rangle$ such that $\log_\alpha \delta \equiv \log_\alpha \beta \times \log_\alpha \gamma \pmod n$. (Equivalently, given α^b and α^c, find α^{bc}.)

Problem 7.4: Decision Diffie-Hellman

Instance: A multiplicative group $(G, \cdot)$, an element $\alpha \in G$ having order n, and three elements $\beta, \gamma, \delta \in \langle\alpha\rangle$.
Question: Is it the case that $\log_\alpha \delta \equiv \log_\alpha \beta \times \log_\alpha \gamma \pmod n$? (Equivalently, given α^b, α^c, and α^d, determine if $d \equiv bc \pmod n$.)

We often denote these two problems by **CDH** and **DDH**, respectively. It is easy to see that there exist Turing reductions

$$\textbf{DDH} \propto_T \textbf{CDH}$$

and

$$\textbf{CDH} \propto_T \textbf{Discrete Logarithm}.$$

The first reduction is proven as follows: Let $\alpha, \beta, \gamma, \delta$ be given. Use an algorithm that solves **CDH** to find the value δ' such that

$$\log_\alpha \delta' \equiv \log_\alpha \beta \times \log_\alpha \gamma \pmod n.$$

Then check to see if $\delta' = \delta$.

The second reduction is also very simple. Let α, β, γ be given. Use an algorithm that solves **Discrete Logarithm** to find $b = \log_\alpha \beta$ and $c = \log_\alpha \gamma$. Then compute $d = bc \bmod n$ and $\delta = \alpha^d$.

These reductions show that the assumption that **DDH** is infeasible is at least as strong as the assumption that **CDH** is infeasible, which in turn is at least as strong as the assumption that **Discrete Logarithm** is infeasible.

It is not hard to show that the semantic security of the *ElGamal Cryptosystem* is equivalent to the infeasibility of **DDH**; and ElGamal decryption (without knowing the private key) is equivalent to solving **CDH**. The assumptions necessary to prove the security of the *ElGamal Cryptosystem* are therefore (potentially) stronger than assuming just that **Discrete Logarithm** is infeasible. Indeed, we already showed that the *ElGamal Cryptosystem* in $\mathbb{Z}_p{}^*$ is not semantically secure, whereas the **Discrete Logarithm** problem is conjectured to be infeasible in $\mathbb{Z}_p{}^*$ for appropriately chosen primes p. This suggests that the security of the three problems may not be equivalent.

Here, we give a proof that any algorithm that solves **CDH** can be used to decrypt ElGamal ciphertexts, and vice versa. Suppose first that ORACLECDH is an algorithm for **CDH**, and let (y_1, y_2) be a ciphertext for the *ElGamal Cryptosystem* with public key α and β. Compute

$$\delta = \text{ORACLECDH}(\alpha, \beta, y_1),$$

and then define

$$x = y_2 \delta^{-1}.$$

It is easy to see that x is the decryption of the ciphertext (y_1, y_2).

Conversely, suppose that ORACLE-ELGAMAL-DECRYPT is an algorithm that decrypts ElGamal ciphertexts. Let α, β, γ be given as in **CDH**. Define α and β to be the public key for the *ElGamal Cryptosystem*. Then define $y_1 = \gamma$ and let $y_2 \in \langle \alpha \rangle$ be chosen randomly. Compute

$$x = \text{ORACLE-ELGAMAL-DECRYPT}(\alpha, \beta, (y_1, y_2)),$$

which is the decryption of the ciphertext (y_1, y_2). Finally, compute

$$\delta = y_2 x^{-1}.$$

Then δ is the solution to the given instance of **CDH**.

7.8 Notes and References

The *ElGamal Cryptosystem* was presented in [80]. For information on the **Discrete Logarithm** problem in general, we recommend the recent survey article by Joux, Odlyzko, and Pierro [102].

The POHLIG-HELLMAN ALGORITHM was published in [163]. The POLLARD RHO ALGORITHM was first described in [165]. Brent [46] described a more efficient method to detect cycles (and, therefore, collisions), which can also be used in the corresponding factoring algorithm. There are many ways of defining the "random walks" used in the algorithm; for a thorough treatment of these topics, see Teske [194].

Different versions of the INDEX CALCULUS ALGORITHM were developed by various researchers, including Western and Miller, Adleman, Merkle and Pollard. One of the earlier papers that presented and analyzed the INDEX CALCULUS ALGORITHM in $\mathbb{Z}_p{}^*$ is Adleman [2].

The lower bound on generic algorithms for the **Discrete Logarithm** problem was proven independently by Nechaev [153] and Shoup [179]. Our discussion is based on the treatment of Chateauneuf, Ling and Stinson [56].

The main reference books for finite fields are Lidl and Niederreiter [122] and Mullen and Panario [143]. Coppersmith [61] describes an INDEX CALCULUS ALGORITHM in $\mathbb{F}_{2^n}{}^*$. Joux's algorithm was presented in [101]. Improvements due to Barbulescu, Gaudry, Joux and Thomé are given in [5].

The idea of using elliptic curves for public-key cryptosystems is due to Koblitz [112] and Miller [139]. For a textbook discussing elliptic curves (including pairings) and elliptic curve cryptography, see Washington [198]. Enge [81] is a useful introduction to pairings.

Galbraith and Gaudry [85] is a recent survey on the elliptic curve discrete logarithm problem. The Menezes-Okamoto-Vanstone reduction of discrete logarithms from elliptic curves to finite fields is given in [133]. The attack on "trace one" curves is due to Smart, Satoh, Araki and Semaev; see, for example, Smart [184].

Recommended secure settings for the discrete logarithm problem (as of 2015) are specified by NIST in [8]. The Wikipedia page [208] is a good source for discrete logarithm "records."

Solinas [187] is an article that presents a thorough treatment of fast arithmetic on elliptic curves, including Algorithm 7.6.

The material we presented concerning the **Discrete Logarithm** ***i*th Bit** problem is based on Peralta [160].

Although it is quite old, Boneh [40] is probably the best survey article on the **Decision Diffie-Hellman** problem.

Exercises

7.1 Implement SHANKS' ALGORITHM for finding discrete logarithms in $\mathbb{Z}_p{}^*$, where p is prime and α is a primitive element modulo p. Use your program to find $\log_{106} 12375$ in $\mathbb{Z}_{24691}{}^*$ and $\log_6 248388$ in $\mathbb{Z}_{458009}{}^*$.

7.2 Describe how to modify SHANKS' ALGORITHM to compute the logarithm of

β to the base α in a group G if it is specified ahead of time that this logarithm lies in the interval $[s,t]$, where s and t are integers such that $0 \leq s < t < n$, where n is the order of α. Prove that your algorithm is correct, and show that its complexity is $O(\sqrt{t-s})$.

7.3 The integer $p = 458009$ is prime and $\alpha = 2$ has order 57251 in $\mathbb{Z}_p{}^*$. Use the POLLARD RHO ALGORITHM to compute the discrete logarithm in $\mathbb{Z}_p{}^*$of $\beta = 56851$ to the base α. Take the initial value $x_0 = 1$, and define the partition (S_1, S_2, S_3) as in Example 7.3. Find the smallest integer i such that $x_i = x_{2i}$, and then compute the desired discrete logarithm.

7.4 Suppose that p is an odd prime and k is a positive integer. The multiplicative group $\mathbb{Z}_{p^k}{}^*$ has order $p^{k-1}(p-1)$, and is known to be cyclic. A generator for this group is called a ***primitive element modulo*** p^k.

(a) Suppose that α is a primitive element modulo p. Prove that at least one of α or $\alpha + p$ is a primitive element modulo p^2.

(b) Describe how to efficiently verify that 3 is a primitive root modulo 29 and modulo 29^2. Note: It can be shown that if α is a primitive root modulo p and modulo p^2, then it is a primitive root modulo p^k for all positive integers k (you do not have to prove this fact). Therefore, it follows that 3 is a primitive root modulo 29^k for all positive integers k.

(c) Find an integer α that is a primitive root modulo 29 but not a primitive root modulo 29^2.

(d) Use the POHLIG-HELLMAN ALGORITHM to compute the discrete logarithm of 3344 to the base 3 in the multiplicative group $\mathbb{Z}_{24389}{}^*$.

7.5 Implement the POHLIG-HELLMAN ALGORITHM for finding discrete logarithms in $\mathbb{Z}_p$, where p is prime and α is a primitive element. Use your program to find $\log_5 8563$ in $\mathbb{Z}_{28703}$ and $\log_{10} 12611$ in $\mathbb{Z}_{31153}$.

7.6 Let $p = 227$. The element $\alpha = 2$ is primitive in $\mathbb{Z}_p{}^*$.

(a) Compute α^{32}, α^{40}, α^{59}, and α^{156} modulo p, and factor them over the factor base $\{2, 3, 5, 7, 11\}$.

(b) Using the fact that $\log 2 = 1$, compute $\log 3$, $\log 5$, $\log 7$, and $\log 11$ from the factorizations obtained above (all logarithms are discrete logarithms in $\mathbb{Z}_p{}^*$ to the base α).

(c) Now suppose we wish to compute $\log 173$. Multiply 173 by the "random" value $2^{177} \bmod p$. Factor the result over the factor base, and proceed to compute $\log 173$ using the previously computed logarithms of the numbers in the factor base.

7.7 Suppose that $n = pq$ is an RSA modulus (i.e., p and q are distinct odd primes), and let $\alpha \in \mathbb{Z}_n{}^*$. For a positive integer m and for any $\alpha \in \mathbb{Z}_m{}^*$, define $\mathbf{ord}_m(\alpha)$ to be the order of α in the group $\mathbb{Z}_m{}^*$.

(a) Prove that
$$\mathbf{ord}_n(\alpha) = \text{lcm}(\mathbf{ord}_p(\alpha), \mathbf{ord}_q(\alpha)).$$

(b) Suppose that $\gcd(p-1, q-1) = d$. Show that there exists an element $\alpha \in \mathbb{Z}_n{}^*$ such that
$$\mathbf{ord}_n(\alpha) = \frac{\phi(n)}{d}.$$

(c) Suppose that $\gcd(p-1, q-1) = 2$, and we have an oracle that solves the **Discrete Logarithm** problem in the subgroup $\langle\alpha\rangle$, where $\alpha \in \mathbb{Z}_n{}^*$ has order $\phi(n)/2$. That is, given any $\beta \in \langle\alpha\rangle$, the oracle will find the discrete logarithm $a = \log_\alpha \beta$, where $0 \leq a \leq \phi(n)/2 - 1$. (The value $\phi(n)/2$ is secret however.) Suppose we compute the value $\beta = \alpha^n \bmod n$ and then we use the oracle to find $a = \log_\alpha \beta$. Assuming that $p > 3$ and $q > 3$, prove that $n - a = \phi(n)$.

(d) Describe how n can easily be factored, given the discrete logarithm $a = \log_\alpha \beta$ from (c).

7.8 In this question, we consider a generic algorithm for the **Discrete Logarithm** problem in $(\mathbb{Z}_{19}, +)$.

(a) Suppose that the set C is defined as follows:
$$C = \{(1 - i^2 \bmod 19, i \bmod 19) : i = 0, 1, 2, 4, 7, 12\}.$$
Compute $\mathbf{Good}(C)$.

(b) Suppose that the output of the group oracle, given the ordered pairs in C, is as follows:
$$\begin{aligned} (0,1) &\mapsto 10111 \\ (1,0) &\mapsto 01100 \\ (16,2) &\mapsto 00110 \\ (4,4) &\mapsto 01010 \\ (9,7) &\mapsto 00100 \\ (9,12) &\mapsto 11001, \end{aligned}$$
where group elements are encoded as (random) binary 5-tuples. What can you say about the value of "a"?

7.9 Decrypt the ElGamal ciphertext presented in Table 7.4. The parameters of the system are $p = 31847$, $\alpha = 5$, $a = 7899$ and $\beta = 18074$. Each element of $\mathbb{Z}_n$ represents three alphabetic characters as in Exercise 6.12.

The plaintext was taken from *The English Patient*, by Michael Ondaatje, Alfred A. Knopf, Inc., New York, 1992.

7.10 Determine which of the following polynomials are irreducible over $\mathbb{Z}_2[x]$: $x^5 + x^4 + 1$, $x^5 + x^3 + 1$, $x^5 + x^4 + x^2 + 1$.

TABLE 7.4: ElGamal Ciphertext

(3781, 14409)	(31552, 3930)	(27214, 15442)	(5809, 30274)
(5400, 31486)	(19936, 721)	(27765, 29284)	(29820, 7710)
(31590, 26470)	(3781, 14409)	(15898, 30844)	(19048, 12914)
(16160, 3129)	(301, 17252)	(24689, 7776)	(28856, 15720)
(30555, 24611)	(20501, 2922)	(13659, 5015)	(5740, 31233)
(1616, 14170)	(4294, 2307)	(2320, 29174)	(3036, 20132)
(14130, 22010)	(25910, 19663)	(19557, 10145)	(18899, 27609)
(26004, 25056)	(5400, 31486)	(9526, 3019)	(12962, 15189)
(29538, 5408)	(3149, 7400)	(9396, 3058)	(27149, 20535)
(1777, 8737)	(26117, 14251)	(7129, 18195)	(25302, 10248)
(23258, 3468)	(26052, 20545)	(21958, 5713)	(346, 31194)
(8836, 25898)	(8794, 17358)	(1777, 8737)	(25038, 12483)
(10422, 5552)	(1777, 8737)	(3780, 16360)	(11685, 133)
(25115, 10840)	(14130, 22010)	(16081, 16414)	(28580, 20845)
(23418, 22058)	(24139, 9580)	(173, 17075)	(2016, 18131)
(19886, 22344)	(21600, 25505)	(27119, 19921)	(23312, 16906)
(21563, 7891)	(28250, 21321)	(28327, 19237)	(15313, 28649)
(24271, 8480)	(26592, 25457)	(9660, 7939)	(10267, 20623)
(30499, 14423)	(5839, 24179)	(12846, 6598)	(9284, 27858)
(24875, 17641)	(1777, 8737)	(18825, 19671)	(31306, 11929)
(3576, 4630)	(26664, 27572)	(27011, 29164)	(22763, 8992)
(3149, 7400)	(8951, 29435)	(2059, 3977)	(16258, 30341)
(21541, 19004)	(5865, 29526)	(10536, 6941)	(1777, 8737)
(17561, 11884)	(2209, 6107)	(10422, 5552)	(19371, 21005)
(26521, 5803)	(14884, 14280)	(4328, 8635)	(28250, 21321)
(28327, 19237)	(15313, 28649)		

7.11 The field $\mathbb{F}_{2^5}$ can be constructed as $\mathbb{Z}_2[x]/(x^5+x^2+1)$. Perform the following computations in this field.

(a) Compute $(x^4+x^2)\times(x^3+x+1)$.

(b) Using the extended Euclidean algorithm, compute $(x^3+x^2)^{-1}$.

(c) Using the square-and-multiply algorithm, compute x^{25}.

7.12 We give an example of the *ElGamal Cryptosystem* implemented in $\mathbb{F}_{3^3}$. The polynomial x^3+2x^2+1 is irreducible over $\mathbb{Z}_3[x]$ and hence $\mathbb{Z}_3[x]/(x^3+2x^2+1)$ is the field $\mathbb{F}_{3^3}$. We can associate the 26 letters of the alphabet with the 26 nonzero field elements, and thus encrypt ordinary text in a convenient way. We will use a lexicographic ordering of the (nonzero) polynomials to set up the correspondence. This correspondence is as follows:

$A \leftrightarrow 1$	$B \leftrightarrow 2$	$C \leftrightarrow x$
$D \leftrightarrow x+1$	$E \leftrightarrow x+2$	$F \leftrightarrow 2x$
$G \leftrightarrow 2x+1$	$H \leftrightarrow 2x+2$	$I \leftrightarrow x^2$
$J \leftrightarrow x^2+1$	$K \leftrightarrow x^2+2$	$L \leftrightarrow x^2+x$
$M \leftrightarrow x^2+x+1$	$N \leftrightarrow x^2+x+2$	$O \leftrightarrow x^2+2x$
$P \leftrightarrow x^2+2x+1$	$Q \leftrightarrow x^2+2x+2$	$R \leftrightarrow 2x^2$
$S \leftrightarrow 2x^2+1$	$T \leftrightarrow 2x^2+2$	$U \leftrightarrow 2x^2+x$
$V \leftrightarrow 2x^2+x+1$	$W \leftrightarrow 2x^2+x+2$	$X \leftrightarrow 2x^2+2x$
$Y \leftrightarrow 2x^2+2x+1$	$Z \leftrightarrow 2x^2+2x+2$	

Suppose Bob uses $\alpha = x$ and $a = 11$ in an *ElGamal Cryptosystem*; then $\beta = x + 2$. Show how Bob will decrypt the following string of ciphertext:

```
(K,H)(P,X)(N,K)(H,R)(T,F)(V,Y)(E,H)(F,A)(T,W)(J,D)(U,J)
```

7.13 Let $\mathcal{E}$ be the elliptic curve $y^2 = x^3 + x + 28$ defined over $\mathbb{Z}_{71}$.

(a) Determine the number of points on $\mathcal{E}$.

(b) Show that $\mathcal{E}$ is not a cyclic group.

(c) What is the maximum order of an element in $\mathcal{E}$? Find an element having this order.

7.14 Suppose that $p > 3$ is an odd prime, and $a, b \in \mathbb{Z}_p$. Further, suppose that the equation $x^3 + ax + b \equiv 0 \pmod{p}$ has three distinct roots in $\mathbb{Z}_p$. Prove that the corresponding elliptic curve group $(\mathcal{E}, +)$ is not cyclic.

HINT Show that the points of order two generate a subgroup of $(\mathcal{E}, +)$ that is isomorphic to $\mathbb{Z}_2 \times \mathbb{Z}_2$.

7.15 Consider an elliptic curve $\mathcal{E}$ described by the formula $y^2 \equiv x^3 + ax + b \pmod{p}$, where $4a^3 + 27b^2 \not\equiv 0 \pmod{p}$ and $p > 3$ is prime.

(a) It is clear that a point $P = (x_1, y_1) \in \mathcal{E}$ has order 3 if and only if $2P = -P$. Use this fact to prove that, if $P = (x_1, y_1) \in \mathcal{E}$ has order 3, then

$$3{x_1}^4 + 6a{x_1}^2 + 12x_1 b - a^2 \equiv 0 \pmod{p}. \tag{7.11}$$

(b) Conclude from equation (7.11) that there are at most 8 points of order 3 on the elliptic curve $\mathcal{E}$.

(c) Using equation (7.11), determine all points of order 3 on the elliptic curve $y^2 \equiv x^3 + 34x \pmod{73}$.

7.16 Suppose that $\mathcal{E}$ is an elliptic curve defined over $\mathbb{Z}_p$, where $p > 3$ is prime. Suppose that $\#\mathcal{E}$ is prime, $P \in \mathcal{E}$, and $P \neq \mathcal{O}$.

(a) Prove that the discrete logarithm $\log_P(-P) = \#\mathcal{E} - 1$.

(b) Describe how to compute $\#\mathcal{E}$ in time $O(p^{1/4})$ by using Hasse's bound on $\#\mathcal{E}$, together with a modification of SHANKS' ALGORITHM. Give a pseudocode description of the algorithm.

7.17 Suppose $e : G_1 \times G_2 \to G_3$ is a bilinear pairing. Prove, for all $P \in G_1$ and $Q \in G_2$, that $e(aP, bQ) = e(P, Q)^{ab}$ for any positive integers a and b.

7.18 Let $\mathcal{E}$ be the elliptic curve described by the equation $y^2 = x^3 + x + 4$ over $\mathbb{F}_5$. Show that the point $(3, 2)$ is a 3-torsion point, and show that the point $(2, 3)$ is not a 3-torsion point.

7.19 (a) Determine the NAF representation of the integer 87.

(b) Using the NAF representation of 87, use Algorithm 7.6 to compute $87P$, where $P = (2, 6)$ is a point on the elliptic curve $y^2 = x^3 + x + 26$ defined over $\mathbb{Z}_{127}$. Show the partial results during each iteration of the algorithm.

7.20 Let $\mathcal{L}_i$ denote the set of positive integers that have exactly i coefficients in their NAF representation, such that the leading coefficient is 1. Denote $k_i = |\mathcal{L}_i|$.

(a) By means of a suitable decomposition of $\mathcal{L}_i$, prove that the k_i's satisfy the following recurrence relation:

$$\begin{aligned} k_1 &= 1 \\ k_2 &= 1 \\ k_{i+1} &= 2(k_1 + k_2 + \ldots + k_{i-1}) + 1 \quad (\text{for } i \geq 2). \end{aligned}$$

(b) Derive a second degree recurrence relation for the k_i's, and obtain an explicit solution of the recurrence relation.

7.21 Find $\log_5 896$ in $\mathbb{Z}_{1103}$ using Algorithm 7.7, given that $L_2(\beta) = 1$ for $\beta = 25$, 219, and 841, and $L_2(\beta) = 0$ for $\beta = 163$, 532, 625, and 656.

7.22 Throughout this question, suppose that $p \equiv 5 \pmod 8$ is prime and suppose that a is a quadratic residue modulo p.

(a) Prove that $a^{(p-1)/4} \equiv \pm 1 \pmod p$.

(b) If $a^{(p-1)/4} \equiv 1 \pmod p$, prove that $a^{(p+3)/8} \bmod p$ is a square root of a modulo p.

(c) If $a^{(p-1)/4} \equiv -1 \pmod p$, prove that $2^{-1}(4a)^{(p+3)/8} \bmod p$ is a square root of a modulo p.

HINT Use the fact that $\left(\frac{2}{p}\right) = -1$ when $p \equiv 5 \pmod 8$ is prime.

(d) Given a primitive element $\alpha \in \mathbb{Z}_p{}^*$, and given any $\beta \in \mathbb{Z}_p{}^*$, show that $L_2(\beta)$ can be computed efficiently.

HINT Use the fact that it is possible to compute square roots modulo p, as well as the fact that $L_1(\beta) = L_1(p - \beta)$ for all $\beta \in \mathbb{Z}_p{}^*$, when $p \equiv 5 \pmod 8$ is prime.

7.23 The *ElGamal Cryptosystem* can be implemented in any subgroup $\langle\alpha\rangle$ of a finite multiplicative group $(G, \cdot)$, as follows: Let $\beta \in \langle\alpha\rangle$ and define (α, β) to be the public key. The plaintext space is $\mathcal{P} = \langle\alpha\rangle$, and the encryption operation is $e_K(x) = (y_1, y_2) = (\alpha^k, x \cdot \beta^k)$, where k is random.

Here we show that distinguishing ElGamal encryptions of two plaintexts can be Turing reduced to **Decision Diffie-Hellman**, and vice versa.

(a) Assume that ORACLEDDH is an oracle that solves **Decision Diffie-Hellman** in $(G, \cdot)$. Prove that ORACLEDDH can be used as a subroutine in an algorithm that distinguishes ElGamal encryptions of two given plaintexts, say x_1 and x_2. (That is, given $x_1, x_2 \in \mathcal{P}$, and given a ciphertext (y_1, y_2) that is an encryption of x_i for some $i \in \{1, 2\}$, the distinguishing algorithm will determine if $i = 1$ or $i = 2$.)

(b) Assume that ORACLE-DISTINGUISH is an oracle that distinguishes ElGamal encryptions of any two given plaintexts x_1 and x_2, for any *ElGamal Cryptosystem* implemented in the group $(G, \cdot)$ as described above. Suppose further that ORACLE-DISTINGUISH will determine if a ciphertext (y_1, y_2) is not a valid encryption of either of x_1 or x_2. Prove that ORACLE-DISTINGUISH can be used as a subroutine in an algorithm that solves **Decision Diffie-Hellman** in $(G, \cdot)$.

Chapter 8

Signature Schemes

In this chapter, we study signature schemes, which are also called digital signatures. We cover various signature schemes based on the **Factoring** and **Discrete Logarithm** problems, including the *Digital Signature Standard*.

8.1 Introduction

A "conventional" handwritten signature attached to a document is used to specify the person responsible for it. A signature is used in everyday situations such as writing a letter, withdrawing money from a bank, signing a contract, etc.

A signature scheme is a method of signing a message stored in electronic form. As such, a signed message can be transmitted over a computer network. In this chapter, we will study several signature schemes, but first we discuss some fundamental differences between conventional and digital signatures.

First is the question of signing a document. With a conventional signature, a signature is part of the physical document being signed. However, a digital signature is not attached physically to the message that is signed, so the algorithm that is used must somehow "bind" the signature to the message.

Second is the question of verification. A conventional signature is verified by comparing it to other, authentic signatures. For example, if someone signs a credit card purchase (which is not so common nowadays, given the prevalence of chip-and-pin technologies), the salesperson is supposed to compare the signature on the sales slip to the signature on the back of the credit card in order to verify the signature. Of course, this is not a very secure method as it is relatively easy to forge someone else's signature. Digital signatures, on the other hand, can be verified using a publicly known verification algorithm. Thus, "anyone" can verify a digital signature. The use of a secure signature scheme prevents the possibility of forgeries.

Another fundamental difference between conventional and digital signatures is that a "copy" of a signed digital message is identical to the original. On the other hand, a copy of a signed paper document can usually be distinguished from an original. This feature means that care must be taken to prevent a signed digital message from being reused. For example, if Alice signs a digital message authorizing Bob to withdraw $100 from her bank account (i.e., a check), she wants Bob to be able to do so only once. So the message itself should contain information, such as a date, that prevents it from being reused.

A signature scheme consists of two components: a signing algorithm and a verification algorithm. Alice can sign a message x using a (private) signing algorithm $\mathbf{sig}_K$ which depends on a private key K. The resulting signature $\mathbf{sig}_K(x)$ can subsequently be verified using a public verification algorithm $\mathbf{ver}_K$. Given a pair (x, y), where x is a message and y is a purported signature on x, the verification algorithm returns an answer *true* or *false* depending on whether or not y is a valid signature for the message x.

Here is a formal definition of a signature scheme.

Definition 8.1: A ***signature scheme*** is a five-tuple $(\mathcal{P}, \mathcal{A}, \mathcal{K}, \mathcal{S}, \mathcal{V})$, where the following conditions are satisfied:

1. $\mathcal{P}$ is a finite set of possible ***messages***
2. $\mathcal{A}$ is a finite set of possible ***signatures***
3. $\mathcal{K}$, the ***keyspace***, is a finite set of possible ***keys***
4. For each $K \in \mathcal{K}$, there is a ***signing algorithm*** $\mathbf{sig}_K \in \mathcal{S}$ and a corresponding ***verification algorithm*** $\mathbf{ver}_K \in \mathcal{V}$. Each $\mathbf{sig}_K : \mathcal{P} \rightarrow \mathcal{A}$ and $\mathbf{ver}_K : \mathcal{P} \times \mathcal{A} \rightarrow \{true, false\}$ are functions[a] such that the following equation is satisfied for every message $x \in \mathcal{P}$ and for every signature $y \in \mathcal{A}$:

$$\mathbf{ver}_K(x, y) = \begin{cases} true & \text{if } y = \mathbf{sig}_K(x) \\ false & \text{if } y \neq \mathbf{sig}_K(x). \end{cases}$$

A pair (x, y) with $x \in \mathcal{P}$ and $y \in \mathcal{A}$ is called a ***signed message***.

[a] In some signature schemes, the signing algorithm is randomized.

For every $K \in \mathcal{K}$, the functions $\mathbf{sig}_K$ and $\mathbf{ver}_K$ should be polynomial-time functions. The verification algorithm, $\mathbf{ver}_K$, will be public and the signing algorithm, $\mathbf{sig}_K$, will be private. Given a message x, it should be computationally infeasible for anyone other than Alice to compute a signature y such that $\mathbf{ver}_K(x, y) = true$ (and note that there might be more than one such y for a given x, depending on how the function **ver** is defined). If Oscar can compute a pair (x, y) such that $\mathbf{ver}_K(x, y) = true$ and x was not previously signed by Alice, then the signature y is called a ***forgery***. Informally, a forged signature is a valid signature produced by someone other than Alice.

8.1.1 RSA Signature Scheme

As our first example of a signature scheme, we observe that the *RSA Cryptosystem* can be used to provide digital signatures; in this context, it is known

Cryptosystem 8.1: *RSA Signature Scheme*

Let $n = pq$, where p and q are primes. Let $\mathcal{P} = \mathcal{A} = \mathbb{Z}_n$, and define

$$\mathcal{K} = \{(n, p, q, a, b) : n = pq, \text{where } p \text{ and } q \text{ are prime}, ab \equiv 1 \pmod{\phi(n)}\}.$$

The values n and b are the public key, and the values p, q, and a are the private key.

For $K = (n, p, q, a, b)$, define

$$\mathbf{sig}_K(x) = x^a \bmod n$$

and

$$\mathbf{ver}_K(x, y) = \mathit{true} \Leftrightarrow x \equiv y^b \pmod{n},$$

for $x, y \in \mathbb{Z}_n$.

as the *RSA Signature Scheme*. See Cryptosystem 8.1 for a "basic" version of the scheme, which will be enhanced a bit later.

Observe that Alice signs a message x using the RSA decryption rule d_K. Alice is the only person who can create the signature because $d_K = \mathbf{sig}_K$ is private. The verification algorithm uses the RSA encryption rule e_K. Anyone can verify a signature because e_K is public.

Note that anyone can forge Alice's RSA signature by choosing a random y and computing $x = e_K(y)$; then $y = \mathbf{sig}_K(x)$ is a valid signature on the message x. (Note, however, that there does not seem to be an obvious way to first choose x and then compute the corresponding signature y; if this could be done, then the *RSA Cryptosystem* would be insecure.) One way to prevent this attack is to require that messages contain sufficient redundancy that a forged signature of this type does not correspond to a "meaningful" message x except with a very small probability. Alternatively, the use of hash functions in conjunction with signature schemes will eliminate this method of forging (cryptographic hash functions were discussed in Chapter 5). We pursue this approach further in the next section.

The rest of this chapter is organized as follows. Section 8.2 introduces the notion of security for signature schemes and how hash functions are used in conjunction with signature schemes. Section 8.3 presents the *ElGamal Signature Scheme* and discusses its security. Section 8.4 deals with three important schemes that evolved from the *ElGamal Signature Scheme*, namely, the *Schnorr Signature Scheme*, the *Digital Signature Algorithm*, and the *Elliptic Curve Digital Signature Algorithm*. A provably secure signature scheme known as *Full Domain Hash* is the topic of Section 8.5, and certificates are discussed in Section 8.6. Finally, some methods of combining signature schemes with encryption schemes are considered in Section 8.7.

8.2 Security Requirements for Signature Schemes

In this section, we discuss what it means for a signature scheme to be "secure." As was the case with a cryptosystem, we need to specify an attack model, the goal of the adversary, and the type of security provided by the scheme.

Recall from Section 2.2 that the attack model defines the information available to the adversary. In the case of signature schemes, the following types of attack models are commonly considered:

key-only attack
: Oscar possesses Alice's public key, i.e., the verification function, $\mathbf{ver}_K$.

known message attack
: Oscar possesses a list of messages previously signed by Alice, say

$$(x_1, y_1), (x_2, y_2), \ldots,$$

where the x_i's are messages and the y_i's are Alice's signatures on these messages (so $y_i = \mathbf{sig}_K(x_i)$, $i = 1, 2, \ldots$).

chosen message attack
: Oscar requests Alice's signatures on a list of messages. Therefore he chooses messages $x_1, x_2, \ldots$, and Alice supplies her signatures on these messages, namely, $y_i = \mathbf{sig}_K(x_i)$, $i = 1, 2, \ldots$.

We consider several possible adversarial goals:

total break
: Oscar is able to determine Alice's private key, i.e., the signing function $\mathbf{sig}_K$. Therefore he can create valid signatures on any message.

selective forgery
: With some non-negligible probability, Oscar is able to create a valid signature on a message chosen by someone else. In other words, if Oscar is given a message x, then he can determine (with some probability) a signature y such that $\mathbf{ver}_K(x, y) = \mathit{true}$. The message x should not be one that has previously been signed by Alice.

existential forgery
: Oscar is able to create a valid signature for at least one message. In other words, Oscar can create a pair (x, y), where x is a message and $\mathbf{ver}_K(x, y) = \mathit{true}$. The message x should not be one that has previously been signed by Alice.

A signature scheme cannot be unconditionally secure, since Oscar can test all possible signatures $y \in \mathcal{A}$ for a given message x, using the public algorithm $\mathbf{ver}_K$, until he finds a valid signature. So, given sufficient time, Oscar can always forge

message	x	$x \in \{0,1\}^*$
	$\downarrow$	
message digest	$z = h(x)$	$z \in \mathcal{Z}$
	$\downarrow$	
signature	$y = sig_K(z)$	$y \in \mathcal{Y}$

FIGURE 8.1: Signing a message digest

Alice's signature on any message. Thus, as was the case with public-key cryptosystems, our goal is to find signature schemes that are computationally or provably secure.

Notice that the above definitions have some similarity to the attacks on MACs that we considered in Section 5.5. In the MAC setting, there is no such thing as a public key, so it does not make sense to speak of a key-only attack (and a MAC does not have separate signing and verifying functions, of course). The attacks in Section 5.5 were existential forgeries using chosen message attacks.

We illustrate the concepts described above with a couple of attacks on the *RSA Signature Scheme*. In Section 8.1, we observed that Oscar can construct a valid signed message by choosing a signature y and then computing x such that $\mathbf{ver}_K(x, y) = true$. This would be an existential forgery using a key-only attack.

Another type of attack is based on the multiplicative property of RSA, which we mentioned in Section 6.9.1. Suppose that $y_1 = \mathbf{sig}_K(x_1)$ and $y_2 = \mathbf{sig}_K(x_2)$ are any two messages previously signed by Alice. Then

$$\mathbf{ver}_K(x_1 x_2 \bmod n, y_1 y_2 \bmod n) = true,$$

and therefore Oscar can create the valid signature $y_1 y_2 \bmod n$ on the message $x_1 x_2 \bmod n$. This is an example of an existential forgery using a known message attack.

Here is one more variation. Suppose Oscar wants to forge a signature on the message x, where x was possibly chosen by someone else. It is a simple matter for him to find $x_1, x_2 \in \mathbb{Z}_n$ such that $x \equiv x_1 x_2 \pmod{n}$. Now suppose he asks Alice for her signatures on messages x_1 and x_2, which we denote by y_1 and y_2 respectively. Then, as in the previous attack, $y_1 y_2 \bmod n$ is the signature for the message $x = x_1 x_2 \bmod n$. This is a selective forgery using a chosen message attack.

8.2.1 Signatures and Hash Functions

Signature schemes are almost always used in conjunction with a secure (public) cryptographic hash function. The hash function $h : \{0,1\}^* \rightarrow \mathcal{Z}$ will take a message of arbitrary length and produce a message digest of a specified size (224 bits is a popular choice). The message digest will then be signed using a signature scheme $(\mathcal{P}, \mathcal{A}, \mathcal{K}, \mathcal{S}, \mathcal{V})$, where $\mathcal{Z} \subseteq \mathcal{P}$. This use of a hash function and signature scheme is depicted diagrammatically in Figure 8.1.

Suppose Alice wants to sign a message x, which is a bitstring of arbitrary length. She first constructs the message digest $z = h(x)$, and then computes the signature on z, namely, $y = sig_K(z)$. Then she transmits the ordered pair (x, y) over the channel. Now the verification can be performed (by anyone) by first reconstructing the message digest $z = h(x)$ using the public hash function h, and then checking that $ver_K(z, y) = true$.

We have to be careful that the use of a hash function h does not weaken the security of the signature scheme, for it is the message digest that is signed, not the message. It will be necessary for h to satisfy certain properties in order to prevent various attacks. The desired properties of hash functions were the ones that were already discussed in Section 5.2.

The most obvious type of attack is for Oscar to start with a valid signed message (x, y), where $y = sig_K(h(x))$. (The pair (x, y) could be any message previously signed by Alice.) Then he computes $z = h(x)$ and attempts to find $x' \neq x$ such that $h(x') = h(x)$. If Oscar can do this, (x', y) would be a valid signed message, so y is a forged signature for the message x'. This is an existential forgery using a known message attack. In order to prevent this type of attack, we require that h be second-preimage resistant.

Another possible attack is the following: Oscar first finds two messages $x \neq x'$ such that $h(x) = h(x')$. Oscar then gives x to Alice and persuades her to sign the message digest $h(x)$, obtaining y. Then (x', y) is a valid signed message and y is a forged signature for the message x'. This is an existential forgery using a chosen message attack; it can be prevented if h is collision resistant.

Here is a third variety of attack. It is often possible with certain signature schemes to forge signatures on random message digests z (we observed already that this could be done with the *RSA Signature Scheme*). That is, we assume that the signature scheme (without the hash function) is subject to existential forgery using a key-only attack. Now, suppose Oscar computes a signature on some message digest z, and then he finds a message x such that $z = h(x)$. If he can do this, then (x, y) is a valid signed message and y is a forged signature for the message x. This is an existential forgery on the signature scheme using a key-only attack. In order to prevent this attack, we desire that h be a preimage resistant hash function.

8.3 The ElGamal Signature Scheme

In this section, we present the *ElGamal Signature Scheme*, which was described in a 1985 paper. A modification of this scheme has been adopted as the *Digital Signature Algorithm* (or *DSA*) by the National Institute of Standards and Technology. The *DSA* also incorporates some ideas used in a scheme known as the *Schnorr Signature Scheme*. All of these schemes are designed specifically for the purpose of signatures, as opposed to the *RSA Cryptosystem*, which can be used both as a public-key cryptosystem and a signature scheme.

Cryptosystem 8.2: *ElGamal Signature Scheme*

Let p be a prime such that the discrete log problem in $\mathbb{Z}_p$ is intractable, and let $\alpha \in \mathbb{Z}_p{}^*$ be a primitive element. Let $\mathcal{P} = \mathbb{Z}_p{}^*$, $\mathcal{A} = \mathbb{Z}_p{}^* \times \mathbb{Z}_{p-1}$, and define

$$\mathcal{K} = \{(p, \alpha, a, \beta) : \beta \equiv \alpha^a \pmod{p}\}.$$

The values p, α, and β are the public key, and a is the private key.

For $K = (p, \alpha, a, \beta)$, and for a (secret) random number $k \in \mathbb{Z}_{p-1}{}^*$, define

$$\mathbf{sig}_K(x, k) = (\gamma, \delta),$$

where

$$\gamma = \alpha^k \bmod p$$

and

$$\delta = (x - a\gamma)k^{-1} \bmod (p-1).$$

For $x, \gamma \in \mathbb{Z}_p{}^*$ and $\delta \in \mathbb{Z}_{p-1}$, define

$$\mathbf{ver}_K(x, (\gamma, \delta)) = true \Leftrightarrow \beta^\gamma \gamma^\delta \equiv \alpha^x \pmod{p}.$$

The *ElGamal Signature Scheme* is randomized (recall that the *ElGamal Public-key Cryptosystem* is also randomized). This means that there are many valid signatures for any given message, and the verification algorithm must be able to accept any of these valid signatures as authentic. The description of the *ElGamal Signature Scheme* is given as Cryptosystem 8.2.

We begin with a couple of preliminary observations. An ElGamal signature consists of two components, which are denoted γ and δ. The first component, γ, is obtained by raising α to a random power modulo p; it does not depend on the message (namely, x) that is being signed. The second component, δ, depends on the message x as well as the private key a. Verifying the signature is accomplished by checking that a certain congruence holds modulo p; this congruence does not involve the private key, of course.

We now show that, if the signature was constructed correctly, then the verification will succeed. This follows easily from the following congruences:

$$\begin{aligned} \beta^\gamma \gamma^\delta &\equiv \alpha^{a\gamma} \alpha^{k\delta} \pmod{p} \\ &\equiv \alpha^x \pmod{p}, \end{aligned}$$

where we use the fact that

$$a\gamma + k\delta \equiv x \pmod{p-1}.$$

Actually, it is probably less mysterious to begin with the verification equation, and then derive the corresponding signing function. Suppose we start with the congruence

$$\alpha^x \equiv \beta^\gamma \gamma^\delta \pmod{p}. \tag{8.1}$$

Then we make the substitutions

$$\gamma \equiv \alpha^k \pmod{p}$$

and

$$\beta \equiv \alpha^a \pmod{p},$$

but we do not substitute for γ in the exponent of (8.1). We obtain the following:

$$\alpha^x \equiv \alpha^{a\gamma + k\delta} \pmod{p}.$$

Now, α is a primitive element modulo p; so this congruence is true if and only if the exponents are congruent modulo $p-1$, i.e., if and only if

$$x \equiv a\gamma + k\delta \pmod{p-1}.$$

Given x, a, γ, and k, this congruence can be solved for δ, yielding the formula used in the signing function of Cryptosystem 8.2.

Alice computes a signature using both the private key, a, and the secret random number, k (which is used to sign one message, x). The verification can be accomplished using only public information.

Let's do a small example to illustrate the arithmetic.

Example 8.1 Suppose we take $p = 467$, $\alpha = 2$, $a = 127$; then

$$\begin{aligned} \beta &= \alpha^a \bmod p \\ &= 2^{127} \bmod 467 \\ &= 132. \end{aligned}$$

Suppose Alice wants to sign the message $x = 100$ and she chooses the random value $k = 213$ (note that $\gcd(213, 466) = 1$ and $213^{-1} \bmod 466 = 431$). Then

$$\gamma = 2^{213} \bmod 467 = 29$$

and

$$\delta = (100 - 127 \times 29)431 \bmod 466 = 51.$$

Anyone can verify the signature $(29, 51)$ by checking that

$$132^{29} 29^{51} \equiv 189 \pmod{467}$$

and

$$2^{100} \equiv 189 \pmod{467}.$$

Hence, the signature is valid. □

8.3.1 Security of the ElGamal Signature Scheme

Let's look at the security of the *ElGamal Signature Scheme*. Suppose Oscar tries to forge a signature for a given message x, without knowing a. If Oscar chooses a value γ and then tries to find the corresponding δ, he must compute the discrete logarithm $\log_\gamma \alpha^x \beta^{-\gamma}$. On the other hand, if he first chooses δ and then tries to find γ, he is trying to "solve" the equation

$$\beta^\gamma \gamma^\delta \equiv \alpha^x \pmod{p}$$

for the "unknown" γ. This is a problem for which no feasible solution is known; however, it does not seem to be related to any well-studied problem such as the **Discrete Logarithm** problem. There also remains the possibility that there might be some way to compute γ and δ simultaneously in such a way that (γ, δ) will be a signature. No one has discovered a way to do this, but conversely, no one has proved that it cannot be done.

If Oscar chooses γ and δ and then tries to solve for x, he is again faced with an instance of the **Discrete Logarithm** problem, namely the computation of $\log_\alpha \beta^\gamma \gamma^\delta$. Hence, Oscar cannot sign a given message x using this approach.

However, there is a method by which Oscar can sign a random message by choosing γ, δ, and x simultaneously. Thus an existential forgery is possible under a key-only attack (assuming a hash function is not used). We describe how to do this now.

Suppose i and j are integers such that $0 \leq i \leq p-2$, $0 \leq j \leq p-2$, and suppose we express γ in the form $\gamma = \alpha^i \beta^j \bmod p$. Then the verification condition is

$$\alpha^x \equiv \beta^\gamma (\alpha^i \beta^j)^\delta \pmod{p}.$$

This is equivalent to

$$\alpha^{x-i\delta} \equiv \beta^{\gamma+j\delta} \pmod{p}.$$

This latter congruence will be satisfied if

$$x - i\delta \equiv 0 \pmod{p-1}$$

and

$$\gamma + j\delta \equiv 0 \pmod{p-1}.$$

Given i and j, we can easily solve these two congruences modulo $p-1$ for δ and x, provided that $\gcd(j, p-1) = 1$. We obtain the following:

$$\begin{aligned} \gamma &= \alpha^i \beta^j \bmod p, \\ \delta &= -\gamma j^{-1} \bmod (p-1), \quad \text{and} \\ x &= -\gamma i j^{-1} \bmod (p-1). \end{aligned}$$

By the way in which we constructed (γ, δ), it is clear that it is a valid signature for the message x.

We illustrate with an example.

Example 8.2 As in the previous example, suppose $p = 467$, $\alpha = 2$, and $\beta = 132$. Suppose Oscar chooses $i = 99$ and $j = 179$; then $j^{-1} \bmod (p-1) = 151$. He would compute the following:

$$\begin{aligned}
\gamma &= 2^{99} 132^{179} \bmod 467 &&= 117\\
\delta &= -117 \times 151 \bmod 466 &&= 41\\
x &= 99 \times 41 \bmod 466 &&= 331.
\end{aligned}$$

Then $(117, 41)$ is a valid signature for the message 331, as may be verified by checking that

$$132^{117} 117^{41} \equiv 303 \pmod{467}$$

and

$$2^{331} \equiv 303 \pmod{467}.$$

□

Here is a second type of forgery, in which Oscar begins with a message previously signed by Alice. This is an existential forgery under a known message attack. Suppose (γ, δ) is a valid signature for a message x. Then it is possible for Oscar to sign various other messages. Suppose h, i, and j are integers, $0 \leq h, i, j \leq p-2$, and $\gcd(h\gamma - j\delta, p-1) = 1$. Compute the following:

$$\begin{aligned}
\lambda &= \gamma^h \alpha^i \beta^j \bmod p\\
\mu &= \delta\lambda(h\gamma - j\delta)^{-1} \bmod (p-1), \quad \text{and}\\
x' &= \lambda(hx + i\delta)(h\gamma - j\delta)^{-1} \bmod (p-1).
\end{aligned}$$

Then, it is tedious but straightforward to check that the verification condition

$$\beta^\lambda \lambda^\mu \equiv \alpha^{x'} \pmod p$$

holds. Hence (λ, μ) is a valid signature for x'.

Both of these methods are existential forgeries, but it does not appear that they can be modified to yield selective forgeries. Hence, they do not seem to represent a threat to the security of the *ElGamal Signature Scheme*, provided that a secure hash function is used as described in Section 8.2.1.

We also mention a couple of ways in which the *ElGamal Signature Scheme* can be broken if it is used carelessly (these are further instances of protocol failures, as introduced in the Exercises of Chapter 6). First, the random value k used in computing a signature should not be revealed. For, if k is known and $gcd(\gamma, p-1) = 1$, then it is a simple matter to compute

$$a = (x - k\delta)\gamma^{-1} \bmod (p-1).$$

Once a is known, then the system is completely broken and Oscar can forge signatures at will.

Another misuse of the system is to use the same value k in signing two different messages. This will result in a repeated γ value, and it also makes it easy for Oscar to compute a and hence break the system. This can be done as follows. Suppose (γ, δ_1) is a signature on x_1 and (γ, δ_2) is a signature on x_2. Then we have

$$\beta^\gamma \gamma^{\delta_1} \equiv \alpha^{x_1} \pmod p$$

and

$$\beta^\gamma \gamma^{\delta_2} \equiv \alpha^{x_2} \pmod p.$$

Thus

$$\alpha^{x_1 - x_2} \equiv \gamma^{\delta_1 - \delta_2} \pmod p.$$

Writing $\gamma = \alpha^k$, we obtain the following equation in the unknown k:

$$\alpha^{x_1 - x_2} \equiv \alpha^{k(\delta_1 - \delta_2)} \pmod p,$$

which is equivalent to

$$x_1 - x_2 \equiv k(\delta_1 - \delta_2) \pmod{p-1}. \tag{8.2}$$

Let $d = \gcd(\delta_1 - \delta_2, p-1)$. If $d = 1$, then we can immediately solve (8.2), obtaining

$$k = (x_1 - x_2)(\delta_1 - \delta_2)^{-1} \pmod{p-1}.$$

However, even if $d > 1$, we still might be able to determine k, provided d is not too large. Since $d \mid (p-1)$ and $d \mid (\delta_1 - \delta_2)$, it follows that $d \mid (x_1 - x_2)$. Define

$$\begin{aligned} x' &= \frac{x_1 - x_2}{d} \\ \delta' &= \frac{\delta_1 - \delta_2}{d} \\ p' &= \frac{p-1}{d}. \end{aligned}$$

Then the congruence (8.2) becomes:

$$x' \equiv k\delta' \pmod{p'}.$$

Since $\gcd(\delta', p') = 1$, we can compute

$$\epsilon = (\delta')^{-1} \bmod p'.$$

The value of k is determined modulo p' to be

$$k = x'\epsilon \bmod p'.$$

This yields d candidate values for k:

$$k = x'\epsilon + ip' \bmod (p-1)$$

for some i, $0 \leq i \leq d-1$. Of these d candidate values, the (unique) correct one can be determined by testing the condition

$$\gamma \equiv \alpha^k \pmod p.$$

8.4 Variants of the ElGamal Signature Scheme

In many situations, a message might be encrypted and decrypted only once, so it suffices to use any cryptosystem that is known to be secure at the time the message is encrypted. On the other hand, a signed message could function as a legal document such as a contract or will; so it is very likely that it would be necessary to verify a signature many years after the message is signed. It is therefore important to take even more precautions regarding the security of a signature scheme as opposed to a cryptosystem. Since the *ElGamal Signature Scheme* is no more secure than the **Discrete Logarithm** problem, this necessitates the use of a large modulus p. Most people would now argue that the length of p should be at least 2048 bits in order to provide present-day security, and even larger to provide security into the foreseeable future (this was already mentioned in Section 7.6).

A 2048 bit modulus leads to an ElGamal signature having 4096 bits. For potential applications, many of which involve the use of smart cards, a shorter signature is desirable. In 1989, Schnorr proposed a signature scheme that can be viewed as a variant of the *ElGamal Signature Scheme* in which the signature size is greatly reduced. The *Digital Signature Algorithm* (or *DSA*) is another modification of the *ElGamal Signature Scheme*, which incorporates some of the ideas used in the *Schnorr Signature Scheme*. The *DSA* was published in the Federal Register on May 19, 1994 and was adopted as a standard on December 1, 1994 (however, it was first proposed in August 1991). We describe the *Schnorr Signature Scheme*, the *DSA*, and a modification of the *DSA* to elliptic curves (called the *Elliptic Curve Digital Signature Algorithm*, or *ECDSA*) in the next subsections.

8.4.1 The Schnorr Signature Scheme

Suppose that p and q are primes such that $p - 1 \equiv 0 \pmod{q}$. Typically we will take $p \approx 2^{2048}$ and $q \approx 2^{224}$. The *Schnorr Signature Scheme* modifies the *ElGamal Signature Scheme* in an ingenious way so that a $\log_2 q$-bit message digest is signed using a $2 \log_2 q$-bit signature, but the computations are done in $\mathbb{Z}_p$. The way that this is accomplished is to work in a subgroup of ${\mathbb{Z}_p}^*$ of size q. The assumed security of the scheme is based on the belief that finding discrete logarithms in this specified subgroup of ${\mathbb{Z}_p}^*$ is secure. (This setting for the **Discrete Logarithm** problem was previously discussed in Section 7.6.)

We will take α to be a qth root of unity modulo p. (It is easy to construct such an α: Let α_0 be a primitive element of $\mathbb{Z}_p$, and define $\alpha = {\alpha_0}^{(p-1)/q} \bmod p$.) The key in the *Schnorr Signature Scheme* is similar to the key in the *ElGamal Signature Scheme* in other respects. However, the *Schnorr Signature Scheme* integrates a hash function directly into the signing algorithm (as opposed to the hash-then-sign method that we discussed in Section 8.2.1). We will assume that $h : \{0,1\}^* \rightarrow \mathbb{Z}_q$ is a secure hash function. A complete description of the *Schnorr Signature Scheme* is given as Cryptosystem 8.3.

Cryptosystem 8.3: *Schnorr Signature Scheme*

Let p be a prime such that the discrete log problem in $\mathbb{Z}_p{}^*$ is intractable, and let q be a prime that divides $p-1$. Let $\alpha \in \mathbb{Z}_p{}^*$ be a qth root of unity modulo p. Let $\mathcal{P} = \{0,1\}^*$, $\mathcal{A} = \mathbb{Z}_q \times \mathbb{Z}_q$, and define

$$\mathcal{K} = \{(p,q,\alpha,a,\beta) : \beta \equiv \alpha^a \pmod p\},$$

where $0 \leq a \leq q-1$. The values p, q, α, and β are the public key, and a is the private key. Finally, let $h : \{0,1\}^* \rightarrow \mathbb{Z}_q$ be a secure hash function.

For $K = (p,q,\alpha,a,\beta)$, and for a (secret) random number k, $1 \leq k \leq q-1$, define

$$\mathbf{sig}_K(x,k) = (\gamma,\delta),$$

where

$$\gamma = h(x \parallel \alpha^k \bmod p)$$

and

$$\delta = k + a\gamma \bmod q.$$

For $x \in \{0,1\}^*$ and $\gamma, \delta \in \mathbb{Z}_q$, verification is done by performing the following computations:

$$\mathbf{ver}_K(x,(\gamma,\delta)) = true \Leftrightarrow h(x \parallel \alpha^\delta \beta^{-\gamma} \bmod p) = \gamma.$$

Observe that each of the two components of a Schnorr signature is an element of $\mathbb{Z}_q$.

It is easy to check that $\alpha^\delta \beta^{-\gamma} \equiv \alpha^k \pmod p$, and hence a Schnorr signature will be verified. Here is a small example to illustrate.

Example 8.3 Suppose we take $q = 101$ and $p = 78q + 1 = 7879$. 3 is a primitive element in $\mathbb{Z}_{7879}{}^*$, so we can take

$$\alpha = 3^{78} \bmod 7879 = 170.$$

α is a qth root of unity modulo p. Suppose $a = 75$; then

$$\beta = \alpha^a \bmod 7879 = 4567.$$

Now, suppose Alice wants to sign the message x, and she chooses the random value $k = 50$. She first computes

$$\alpha^k \bmod p = 170^{50} \bmod 7879 = 2518.$$

The next step is to compute $h(x \parallel 2518)$, where h is a given hash function and 2518 is represented in binary (as a bitstring). Suppose for purposes of illustration that $h(x \parallel 2518) = 96$. Then δ is computed as

$$\delta = 50 + 75 \times 96 \bmod 101 = 79,$$

and the signature is $(96, 79)$.

This signature is verified by computing

$$170^{79} 4567^{-96} \bmod 7879 = 2518,$$

and then checking that $h(x \parallel 2518) = 96$. □

8.4.2 The Digital Signature Algorithm

We will outline the changes that are made to the verification function of the *ElGamal Signature Scheme* in the specification of the *DSA*. The *DSA* uses an order q subgroup of $\mathbb{Z}_p{}^*$, as does the *Schnorr Signature Scheme*. In the *DSA*, one current recommendation is that q is a 224-bit prime and p is a 2048-bit prime. The key in the *DSA* has the same form as in the *Schnorr Signature Scheme*. We will assume that the message will be hashed using *SHA3-224* before it is signed. The result is that a 224-bit message digest is signed with a 448-bit signature, and the computations are done in $\mathbb{Z}_p$ and $\mathbb{Z}_q$.

In the *ElGamal Signature Scheme*, suppose we change the "$-$" to a "$+$" in the definition of δ, so

$$\delta = (x + a\gamma)k^{-1} \bmod (p - 1).$$

It is easy to see that this changes the verification condition to the following:

$$\alpha^x \beta^\gamma \equiv \gamma^\delta \pmod{p}. \tag{8.3}$$

Now, α has order q, and β and γ are powers of α, so they also have order q. This means that all exponents in (8.3) can be reduced modulo q without affecting the validity of the congruence. Since x will be replaced by a 224-bit message digest in the *DSA*, we will assume that $x \in \mathbb{Z}_q$. Further, we will alter the definition of δ, so that $\delta \in \mathbb{Z}_q$, as follows:

$$\delta = (x + a\gamma)k^{-1} \bmod q.$$

It remains to consider $\gamma = \alpha^k \bmod p$. Suppose we temporarily define

$$\gamma' = \gamma \bmod q = (\alpha^k \bmod p) \bmod q.$$

Note that

$$\delta = (x + a\gamma')k^{-1} \bmod q,$$

so δ is unchanged. We can write the verification equation as

$$\alpha^x \beta^{\gamma'} \equiv \gamma^\delta \pmod{p}. \tag{8.4}$$

Cryptosystem 8.4: *Digital Signature Algorithm*

Let p be a 2048-bit prime such that the discrete log problem in $\mathbb{Z}_p$ is intractable, and let q be a 224-bit prime that divides $p-1$. Let $\alpha \in \mathbb{Z}_p{}^*$ be a qth root of unity modulo p. Let $\mathcal{P} = \{0,1\}^*$, $\mathcal{A} = \mathbb{Z}_q{}^* \times \mathbb{Z}_q{}^*$, and define

$$\mathcal{K} = \{(p,q,\alpha,a,\beta) : \beta \equiv \alpha^a \pmod{p}\},$$

where $0 \leq a \leq q-1$. The values p, q, α, and β are the public key, and a is the private key.

For $K = (p,q,\alpha,a,\beta)$, and for a (secret) random number k, $1 \leq k \leq q-1$, define

$$\mathbf{sig}_K(x,k) = (\gamma,\delta),$$

where

$$\begin{aligned} \gamma &= (\alpha^k \bmod p) \bmod q \quad \text{and} \\ \delta &= (\text{SHA3-224}(x) + a\gamma)k^{-1} \bmod q. \end{aligned}$$

(If $\gamma = 0$ or $\delta = 0$, then a new random value of k should be chosen.)

For $x \in \{0,1\}^*$ and $\gamma,\delta \in \mathbb{Z}_q{}^*$, verification is done by performing the following computations:

$$\begin{aligned} e_1 &= \text{SHA3-224}(x)\,\delta^{-1} \bmod q \\ e_2 &= \gamma\,\delta^{-1} \bmod q \\ \mathbf{ver}_K(x,(\gamma,\delta)) = \mathit{true} &\Leftrightarrow (\alpha^{e_1}\beta^{e_2} \bmod p) \bmod q = \gamma. \end{aligned}$$

Notice that we cannot replace the remaining occurrence of γ by γ'.

Now we proceed to rewrite (8.4), by raising both sides to the power $\delta^{-1} \bmod q$ (this requires that $\delta \neq 0$). We obtain the following :

$$\alpha^{x\delta^{-1}}\beta^{\gamma'\delta^{-1}} \bmod p = \gamma. \tag{8.5}$$

Now we can reduce both sides of (8.5) modulo q, which produces the following:

$$(\alpha^{x\delta^{-1}}\beta^{\gamma'\delta^{-1}} \bmod p) \bmod q = \gamma'. \tag{8.6}$$

The complete description of the *DSA* is given as Cryptosystem 8.4, in which we rename γ' as γ and replace x by the message digest SHA3-224(x).

Notice that if Alice computes a value $\delta \equiv 0 \pmod{q}$ in the DSA signing algorithm, she should reject it and construct a new signature with a new random k. We should point out that this is not likely to cause a problem in practice: the probability that $\delta \equiv 0 \pmod{q}$ is likely to be on the order of 2^{-224}; so, for all intents and purposes, it will never happen.

Here is an example (with p and q much smaller than they are required to be in the *DSA*) to illustrate.

Example 8.4 Suppose we take the same values of p, q, α, a, β, and k as in Example 8.3, and suppose Alice wants to sign the message digest SHA3-224$(x) = 22$. Then she computes

$$\begin{aligned} k^{-1} \bmod 101 &= 50^{-1} \bmod 101 \\ &= 99, \end{aligned}$$

$$\begin{aligned} \gamma &= (170^{50} \bmod 7879) \bmod 101 \\ &= 2518 \bmod 101 \\ &= 94, \end{aligned}$$

and

$$\begin{aligned} \delta &= (22 + 75 \times 94)99 \bmod 101 \\ &= 97. \end{aligned}$$

The signature $(94, 97)$ on the message digest 22 is verified by the following computations:

$$\begin{aligned} \delta^{-1} &= 97^{-1} \bmod 101 \\ &= 25, \end{aligned}$$

$$\begin{aligned} e_1 &= 22 \times 25 \bmod 101 \\ &= 45, \end{aligned}$$

$$\begin{aligned} e_2 &= 94 \times 25 \bmod 101 \\ &= 27, \end{aligned}$$

and

$$\begin{aligned} (170^{45} 4567^{27} \bmod 7879) \bmod 101 &= 2518 \bmod 101 \\ &= 94. \end{aligned}$$

□

When the *DSA* was proposed in 1991, there were several criticisms put forward. One complaint was that the selection process by NIST was not public. The standard was developed by the National Security Agency (NSA) without the input of U.S. industry. Regardless of the merits of the resulting scheme, many people resented the "closed-door" approach.

Of the technical criticisms put forward, the most serious was that the size of the modulus p was fixed initially at 512 bits. Many people suggested that the modulus size not be fixed, so that larger modulus sizes could be used if desired. In response to these comments, NIST altered the description of the standard so that a variety of modulus sizes were allowed.

8.4.3 The Elliptic Curve DSA

In 2000, the *Elliptic Curve Digital Signature Algorithm* (*ECDSA*) was approved as FIPS 186-2. This signature scheme can be viewed as a modification of the *DSA* to the setting of elliptic curves. We have two points A and B on an elliptic curve defined over $\mathbb{Z}_p$ for some prime p.[1] The discrete logarithm $m = \log_A B$ is the private key. (This is analogous to the relation $\beta = \alpha^a \bmod p$ in the *DSA*, where a is the private key.) The order of A is a large prime number q. Computing a signature involves first choosing a random value k and computing kA (this is analogous to the computation of α^k in the *DSA*).

Here is the main difference between the *DSA* and the *ECDSA*. In the *DSA*, the value $\alpha^k \bmod p$ is reduced modulo q to yield a value γ that is the first component of the signature (γ, δ). In the *ECDSA*, the analogous value is r, which is the x-coordinate of the elliptic curve point kA, reduced modulo q. This value r is the first component of the signature (r, s).

Finally, in the *ECDSA*, the value s is computed from r, m, k, and the message x in exactly the same way as δ is computed from γ, a, k, and the message x in the *DSA*. We now present the complete description of the *ECDSA* as Cryptosystem 8.5.

We work through a tiny example to illustrate the computations in the *ECDSA*.

Example 8.5 We will base our example on the same elliptic curve that was used in Section 7.5.2, namely, $y^2 = x^3 + x + 6$, defined over $\mathbb{Z}_{11}$. The parameters of the signature scheme are $p = 11$, $q = 13$, $A = (2,7)$, $m = 7$, and $B = (7,2)$.

Suppose we have a message x with $\text{SHA3-224}(x) = 4$, and Alice wants to sign the message x using the random value $k = 3$. She will compute

$$\begin{aligned}
(u,v) &= 3\,(2,7) = (8,3) \\
r &= u \bmod 13 = 8, \quad \text{and} \\
s &= 3^{-1}(4 + 7 \times 8) \bmod 13 = 7.
\end{aligned}$$

Therefore $(8,7)$ is the signature.

Bob verifies the signature by performing the following computations:

$$\begin{aligned}
w &= 7^{-1} \bmod 13 = 2 \\
i &= 2 \times 4 \bmod 13 = 8 \\
j &= 2 \times 8 \bmod 13 = 3 \\
(u,v) &= 8A + 3B = (8,3), \quad \text{and} \\
u \bmod 13 &= 8 = r.
\end{aligned}$$

Hence, the signature is verified. □

[1]We note that the *ECDSA* also permits the use of elliptic curves defined over finite fields $\mathbb{F}_{2^n}$, but we will not describe this variation here.

Cryptosystem 8.5: *Elliptic Curve Digital Signature Algorithm*

Let p be a large prime and let $\mathcal{E}$ be an elliptic curve defined over $\mathbb{Z}_p$. Let A be a point on $\mathcal{E}$ having prime order q, such that the **Discrete Logarithm** problem in $\langle A \rangle$ is infeasible. Let $\mathcal{P} = \{0,1\}^*$, $\mathcal{A} = {\mathbb{Z}_q}^* \times {\mathbb{Z}_q}^*$, and define

$$\mathcal{K} = \{(p, q, \mathcal{E}, A, m, B) : B = mA\},$$

where $0 \leq m \leq q-1$. The values p, q, $\mathcal{E}$, A, and B are the public key, and m is the private key.

For $K = (p, q, \mathcal{E}, A, m, B)$, and for a (secret) random number k, $1 \leq k \leq q-1$, define

$$\mathbf{sig}_K(x, k) = (r, s),$$

where

$$\begin{aligned} kA &= (u, v) \\ r &= u \bmod q, \quad \text{and} \\ s &= k^{-1}(\text{SHA3-224}(x) + mr) \bmod q. \end{aligned}$$

(If either $r = 0$ or $s = 0$, a new random value of k should be chosen.)

For $x \in \{0,1\}^*$ and $r, s \in {\mathbb{Z}_q}^*$, verification is done by performing the following computations:

$$\begin{aligned} & w = s^{-1} \bmod q \\ & i = w \times \text{SHA3-224}(x) \bmod q \\ & j = wr \bmod q \\ & (u, v) = iA + jB \\ & \mathbf{ver}_K(x, (r, s)) = \mathit{true} \Leftrightarrow u \bmod q = r. \end{aligned}$$

8.5 Full Domain Hash

In Section 6.9.2, we showed how to construct provably secure public-key cryptosystems from trapdoor one-way permutations (in the random oracle model). Practical implementations of these systems are based on the *RSA Cryptosystem* and they replace the random oracle by a hash function such as *SHA3-224*. In this section, we use a trapdoor one-way permutation to construct a secure signature scheme in the random oracle model. The scheme we present is called *Full Domain Hash*. The name of this scheme comes from the requirement that the range of the

Cryptosystem 8.6: *Full Domain Hash*

Let k be a positive integer; let $\mathcal{F}$ be a family of trapdoor one-way permutations such that $f : \{0,1\}^k \rightarrow \{0,1\}^k$ for all $f \in \mathcal{F}$; and let $G : \{0,1\}^* \rightarrow \{0,1\}^k$ be a "random" function. Let $\mathcal{P} = \{0,1\}^*$ and $\mathcal{A} = \{0,1\}^k$, and define

$$\mathcal{K} = \{(f, f^{-1}, G) : f \in \mathcal{F}\}.$$

Given a key $K = (f, f^{-1}, G)$, f^{-1} is the private key and (f, G) is the public key.

For $K = (f, f^{-1}, G)$ and $x \in \{0,1\}^*$, define

$$\mathbf{sig}_K(x) = f^{-1}(G(x)).$$

A signature $y = (y_1, \ldots, y_k) \in \{0,1\}^k$ on the message x is verified as follows:

$$\mathbf{ver}_K(x, y) = true \Leftrightarrow f(y) = G(x).$$

random oracle be the same as the domain of the trapdoor one-way permutation used in the scheme. The scheme is presented as Cryptosystem 8.6.

Full Domain Hash uses the familiar hash-then-sign method. $G(x)$ is the message digest produced by the random oracle, G. The function f^{-1} is used to sign the message digest, and f is used to verify it.

Let's briefly consider an RSA-based implementation of this scheme. The function f^{-1} would be the RSA signing (i.e., decryption) function, and f would be the RSA verification (i.e., encryption) function. In order for this to be secure, we would have to take $k = 2048$, say. Now suppose that the random oracle G is replaced by the hash function *SHA3-224*. This hash function constructs a 224-bit message digest, so the range of the hash function, namely $\{0,1\}^{224}$, is a very small subset of $\{0,1\}^k = \{0,1\}^{2048}$. In practice, it is necessary to specify some padding scheme in order to expand a 224-bit message to 2048 bits before applying f^{-1}. This is typically done using a fixed (deterministic) padding scheme.

We now proceed to our security proof, in which we assume that $\mathcal{F}$ is a family of trapdoor one-way permutations and G is a "full domain" random oracle. (Note that the security proofs we will present do not apply when the random oracle is replaced by a fully specified hash function such as *SHA3-224*.) It can be proven that *Full Domain Hash* is secure against existential forgery using a chosen message attack; however, we will only prove the easier result that *Full Domain Hash* is secure against existential forgery using a key-only attack.

As usual, the security proof is a type of reduction. We assume that there is an adversary (i.e., a randomized algorithm, which we denote by FDH-FORGE) that is able to forge signatures (with some specified probability) when it is given the public key and access to the random oracle (recall that it can query the random oracle for values $G(x)$, but there is no algorithm specified to evaluate the function

```
Algorithm 8.1: FDH-INVERT(z_0, q_h)

 external f
 procedure SIMG(x)
  if j > q_h
    then return ("failure")
    else if j = j_0
    then z ← z_0
    else let z ∈ {0,1}^k be chosen at random
  j ← j + 1
  return (z)

 main
  choose j_0 ∈ {1,...,q_h} at random
  j ← 1
  insert the code for FDH-FORGE(f) here
  if FDH-FORGE(f) = (x, y)
           ⎧ if f(y) = z_0
    then   ⎨   then return (y)
           ⎩   else return ("failure")
```

G). FDH-FORGE makes some number of oracle queries, say q_h. Eventually, FDH-FORGE outputs a valid forgery with some probability, denoted by ϵ.

We construct an algorithm, FDH-INVERT, which attempts to invert randomly chosen elements $z_0 \in \{0,1\}^k$. That is, given $z_0 \in \{0,1\}^k$, our hope is that FDH-INVERT$(z_0) = f^{-1}(z_0)$. We now present FDH-INVERT as Algorithm 8.1.

Algorithm 8.1 is fairly simple. It basically consists of running the adversary, FDH-FORGE. Hash queries made by FDH-FORGE are handled by the function SIMG, which is a simulation of a random oracle. We have assumed that FDH-FORGE will make q_h hash queries, say $x_1, \ldots, x_{q_h}$. For simplicity, we assume that the x_i's are distinct. (If they are not, then we need to ensure that SIMG$(x_i) =$ SIMG(x_j) whenever $x_i = x_j$. This is not difficult to do; it just requires some book-keeping, as was done in Algorithm 6.14.) We randomly choose one query, say the j_0th query, and define SIMG$(x_{j_0}) = z_0$ (z_0 is the value we are trying to invert) . For all other queries, the value SIMG(x_j) is chosen to be a random number. Because z_0 is also random, it is easy to see that SIMG is indistinguishable from a true random oracle. It therefore follows that FDH-FORGE outputs a message and a valid forged signature, which we denote by (x, y), with probability ϵ. We then check to see if $f(y) = z_0$; if so, then $y = f^{-1}(z_0)$ and we have succeeded in inverting z_0.

Our main task is to analyze the success probability of the algorithm FDH-INVERT as a function of the success probability, ϵ, of FDH-FORGE. We will assume that $\epsilon > 2^{-k}$, because a random choice of y will be a valid signature for a message x with probability 2^{-k}, and we are only interested in adversaries that have a higher success probability than a random guess. As we did above, we denote the hash queries made by FDH-FORGE by $x_1, \ldots, x_{q_h}$, where x_j is the jth hash query, $1 \leq j \leq q_h$.

We begin by conditioning the success probability, ϵ, on whether or not $x \in \{x_1, \ldots, x_{q_h}\}$:

$$\epsilon = \mathbf{Pr}[\text{FDH-FORGE succeeds } \wedge (x \in \{x_1, \ldots, x_{q_h}\})] \\ + \mathbf{Pr}[\text{FDH-FORGE succeeds } \wedge (x \notin \{x_1, \ldots, x_{q_h}\})]. \quad (8.7)$$

It is not hard to see that

$$\mathbf{Pr}[\text{FDH-FORGE succeeds } \wedge (x \notin \{x_1, \ldots, x_{q_h}\})] = 2^{-k}.$$

This is because the (undetermined) value SIMG(x) is equally likely to take on any given value in $\{0,1\}^k$, and hence the probability that SIMG$(x) = f(y)$ is 2^{-k}. (This is where we use the assumption that the hash function is a "full domain" hash.) Substituting into (8.7), we obtain the following:

$$\mathbf{Pr}[\text{FDH-FORGE succeeds } \wedge (x \in \{x_1, \ldots, x_{q_h}\})] \geq \epsilon - 2^{-k}. \quad (8.8)$$

Now we turn to the success probability of FDH-INVERT. The next inequality is obvious:

$$\mathbf{Pr}[\text{FDH-INVERT succeeds }] \geq \mathbf{Pr}[\text{FDH-FORGE succeeds } \wedge (x = x_{j_0})]. \quad (8.9)$$

Our final observation is that

$$\mathbf{Pr}[\text{FDH-FORGE succeeds } \wedge (x = x_{j_0})] \\ = \frac{1}{q_h} \times \mathbf{Pr}[\text{FDH-FORGE succeeds } \wedge (x \in \{x_1, \ldots, x_{q_h}\})]. \quad (8.10)$$

Note that equation (8.10) is true because there is a $1/q_h$ chance that $x = x_{j_0}$, given that $x \in \{x_1, \ldots, x_{q_h}\}$. Now, if we combine (8.8), (8.9), and (8.10), then we obtain the following bound:

$$\mathbf{Pr}[\text{FDH-INVERT succeeds }] \geq \frac{\epsilon - 2^{-k}}{q_h}. \quad (8.11)$$

Therefore we have obtained a concrete lower bound on the success probability of FDH-INVERT. We have proven the following result.

THEOREM 8.1 *Suppose there exists an algorithm* FDH-FORGE *that will output an existential forgery for Full Domain Hash with probability* $\epsilon > 2^{-k}$ *after making* q_h *queries to the random oracle, using a key-only attack. Then there exists an algorithm* FDH-INVERT *that will find inverses of random elements* $z_0 \in \{0,1\}^k$ *with probability at least* $(\epsilon - 2^{-k})/q_h$.

Observe that the usefulness of the resulting inversion algorithm depends on the ability of FDH-FORGE to find forgeries using as few hash queries as possible.

Protocol 8.1: ISSUING A CERTIFICATE TO ALICE

1. The CA establishes Alice's identity by means of conventional forms of identification such as a birth certificate, passport, etc. Then the CA forms a string, denoted $ID(Alice)$, which contains Alice's identification information.

2. A private signing key for Alice, $\mathbf{sig}_{Alice}$, and a corresponding public verification key, $\mathbf{ver}_{Alice}$, are determined.

3. The CA generates its signature

$$s = \mathbf{sig}_{CA}(ID(Alice) \parallel \mathbf{ver}_{Alice})$$

on Alice's identity string and verification key. The certificate

$$\mathbf{Cert}(Alice) = (ID(Alice) \parallel \mathbf{ver}_{Alice} \parallel s)$$

is given to Alice, along with Alice's private key, $\mathbf{sig}_{Alice}$.

8.6 Certificates

Suppose that Alice and Bob are members of a large network in which every participant has public and private keys for certain prespecified cryptosystems and/or signature schemes. In a setting such as this, it is always necessary to provide a mechanism to authenticate the public keys of other people in the network. This requires some kind of ***public-key infrastructure*** (also denoted as a ***PKI***). In general, we assume that there is trusted ***certification authority***, denoted by CA, who signs the public keys of all people in the network. The (public) verification key of the CA, denoted $\mathbf{ver}_{CA}$, is assumed to be known "by magic" to everyone in the network. This simplified setting is perhaps not completely realistic, but it allows us to concentrate on the design of the schemes.

A ***certificate*** for someone in the network will consist of some identifying information for that person (e.g., their name, email address, etc.), their public key(s), and the signature of the CA on that information. A certificate allows network users to verify the authenticity of each other's keys.

Suppose, for example, that Alice wants to obtain a certificate from the CA that contains a copy of Alice's public verification key for a signature scheme. Then the steps in Protocol 8.1 would be executed.

We are not specifying exactly how Alice identifies herself to the CA, nor do we specify the precise format of $ID(Alice)$, or how the public and private keys of Alice are selected. In general, these implementation details could vary from one PKI to another.

It is possible for anyone who knows the *CA*'s verification key, $\mathbf{ver}_{CA}$, to verify anyone else's certificate. Suppose that Bob wants to be assured that Alice's public key is authentic. Alice can give her certificate to Bob. Bob can then verify the signature of the *CA* by checking that

$$\mathbf{ver}_{CA}(ID(Alice) \parallel \mathbf{ver}_{Alice}, s) = true.$$

The security of a certificate follows immediately from the security of the signature scheme used by the *CA*.

As mentioned above, the purpose of verifying a certificate is to authenticate someone's public key. The certificate itself does not provide any kind of proof of identity, because certificates contain only public information. Certificates can be distributed or redistributed to anyone, and possession of a certificate does not imply ownership of it.

One example where certificates are employed in an essentially transparent fashion is in web browsers. Most web browsers come pre-configured with a set of "independent" *CA*s. There may be on the order of 100 such *CA*s in a typical web browser. The user is implicitly trusting the provider of the web browser to only include valid *CA*s in the browser.

Whenever the user connects to a "secure" website, the user's web browser automatically verifies the website's certificate using the appropriate public key that is loaded into the web browser. This is one of the functions of the *Transport Layer Security* (*TLS*) protocol, without any explicit action required by the user. (*TLS*, which is described in more detail in Section 12.1.1, is used to set up secure keys between a user's web browser and a website.)

8.7 Signing and Encrypting

In this section, we look at how we can securely combine signing and public-key encryption. In some sense, this is the public-key analog of authenticated encryption, a topic that we treated in Section 5.5.3.

Perhaps the most frequently recommended method is called ***sign-then-encrypt***. Suppose Alice wishes to send a signed, encrypted message to Bob. Given a plaintext x, Alice would compute her signature $y = \mathbf{sig}_{\text{Alice}}(x)$, and then encrypt both x and y using Bob's public encryption function e_{Bob}, obtaining $z = e_{\text{Bob}}(x, y)$. This ciphertext z would be transmitted to Bob. When Bob receives z, he first decrypts it with his decryption function d_{Bob} to get (x, y). Then he uses Alice's public verification function to check that $\mathbf{ver}_{\text{Alice}}(x, y) = true$.

However, there is a subtle problem with this approach. Suppose Bob receives a signed, encrypted message from Alice. Bob can decrypt the ciphertext to restore the message signed by Alice, namely, (x, y). Then a malicious Bob can encrypt this message with someone else's public key. For example, Bob might compute $z' = e_{\text{Carol}}(x, y)$ and send z' to Carol. When Carol receives z', she will decrypt it,

Protocol 8.2: SIGN-THEN-ENCRYPT

In what follows, $ID(Alice)$ and $ID(Bob)$ are fixed-length, public ID strings for Alice and Bob, respectively.

Suppose Alice wants to send a signed and encrypted message x to Bob.

1. Alice computes her signature $y = \mathbf{sig}_{\text{Alice}}(x, ID(Bob))$.
2. Alice computes the ciphertext $z = e_{\text{Bob}}(x, y, ID(Alice))$, which she sends to Bob.

When Bob receives z, he carries out the following steps.

1. Bob uses his private key d_{Bob} to decrypt the ciphertext z, obtaining x, y and $ID(Alice))$.
2. Bob obtains Alice's public verification key $\mathbf{ver}_{\text{Alice}}$ and checks that

$$\mathbf{ver}_{\text{Alice}}((x, ID(Bob)), y) = true.$$

obtaining (x, y), which is a valid message that was signed by Alice. The difficulty with this scenario is that Carol might believe that she was the intended recipient of Alice's message, whereas the message was actually intended for Bob. Alice might also assume that no one else has access to the plaintext x. But this is a false assumption, because Bob also knows the plaintext x.

An alternative approach is for Alice to first encrypt x, and then sign the result (this process would be termed ***encrypt-then-sign***). Alice would compute

$$z = e_{\text{Bob}}(x) \text{ and } y = \mathbf{sig}_{\text{Alice}}(z)$$

and then transmit the pair (z, y) to Bob. Bob would first verify the signature y on z using $\mathbf{ver}_{\text{Alice}}$. Provided the signature is valid, Bob would then decrypt z, obtaining x. However, suppose that Oscar intercepts this message and he replaces Alice's signature y by his own signature

$$y' = \mathbf{sig}_{\text{Oscar}}(z).$$

(Note that Oscar can sign the ciphertext $z = e_{\text{Bob}}(x)$ even though he doesn't know the value of the plaintext x.) Then, if Oscar transmits (z, y') to Bob, Oscar's signature will be verified by Bob using $\mathbf{ver}_{\text{Oscar}}$, and Bob might infer that the plaintext x originated with Oscar. Of course the plaintext actually was created by Alice.

We have noted potential attacks against both sign-then-encrypt and encrypt-then-sign. It turns out that it is possible to fix both of these methods, by using the following two rules:

1. before encrypting a message, concatenate identifying information for the sender, and

Protocol 8.3: ENCRYPT-THEN-SIGN

In what follows, $ID(Alice)$ and $ID(Bob)$ are fixed-length, public ID strings for Alice and Bob, respectively.

Suppose Alice wants to send a signed and encrypted message x to Bob.

1. Alice computes the ciphertext $z = e_{\text{Bob}}(x, ID(Alice))$.
2. Alice computes her signature $y = \mathbf{sig}_{\text{Alice}}(z, ID(Bob))$ and she sends $(z, y, ID(Alice)$ to Bob.

When Bob receives $(z, y, ID(Alice)$, he carries out the following steps.

1. Bob obtains Alice's public verification key $\mathbf{ver}_{\text{Alice}}$ and checks that
$$\mathbf{ver}_{\text{Alice}}((z, ID(Bob)), y) = true.$$
2. Bob uses his private key d_{Bob} to decrypt the ciphertext z, obtaining x and $ID(Alice)$.
3. He then checks that $ID(Alice)$, as computed in step 2, matches the initial value that he received in $(z, y, ID(Alice)$.

2. before signing a message, concatenate identifying information for the receiver.

The modified sign-then-encrypt process is detailed in Protocol 8.2. Constructing a modified encrypt-then-sign algorithm is also straightforward; see Protocol 8.3.

Both Protocols 8.2 and 8.3 can be proven to be secure under appropriate assumptions concerning the security of the underlying encryption and signature schemes.

We should also mention that there are examples of specialized schemes, known as ***signcryption schemes***, that combine signature and encryption schemes, but do so in a more computationally efficient manner than sign-then-encrypt or encrypt-then-sign.

8.8 Notes and References

For a nice (but dated) survey of signature schemes, we recommend Mitchell, Piper, and Wild [140]. This paper also contains the two methods of forging ElGamal signatures that we presented in Section 8.3.

The *ElGamal Signature Scheme* was presented by ElGamal [80], and the

Schnorr Signature Scheme is due to Schnorr [174]. A complete description of the *ECDSA* is found in Johnson, Menezes, and Vanstone [100]. The *Digital Signature Algorithm* was first published by NIST in August 1991, and it was adopted as FIPS 186 in December 1994.

The *Digital Signature Standard* incorporates *RSA, DSA,* and *ECDSA*. It has been updated several times. The current version is FIPS 186-4 [147], which was issued in July 2013.

Full Domain Hash is due to Bellare and Rogaway [22, 21]. The paper [21] also includes a more efficient variant, known as the *Probabilistic Signature Scheme* (*PSS*). Provably secure ElGamal-type schemes have also been studied; see, for example, Pointcheval and Stern [164].

Certificates were first suggested as a method of authenticating public keys in a 1978 Bachelor's Thesis by Kohnfelder [116]. For a well-written treatment of public-key infrastructure in general, we recommend Adams and Lloyd [1].

Smith [186] and An, Dodis and Rabin [4] give detailed treatments of sign-then-encrypt and encrypt-then-sign. Signcryption schemes were invented by Zheng [207].

Some of the Exercises point out security problems with ElGamal type signature schemes if the "k" values are reused or generated in a predictable fashion. There are now several works that pursue this theme; see, for example, Bellare, Goldwasser, and Micciancio [14] and Nguyen and Shparlinski [156].

Exercises

8.1 Suppose Alice is using the *ElGamal Signature Scheme* with $p = 31847$, $\alpha = 5$, and $\beta = 25703$. Compute the values of k and a (without solving an instance of the **Discrete Logarithm** problem), given the signature $(23972, 31396)$ for the message $x = 8990$ and the signature $(23972, 20481)$ for the message $x = 31415$.

8.2 Suppose we implement the *ElGamal Signature Scheme* with $p = 31847$, $\alpha = 5$, and $\beta = 26379$. Write a computer program that does the following:

(a) Verify the signature $(20679, 11082)$ on the message $x = 20543$.

(b) Determine the private key, a, by solving an instance of the **Discrete Logarithm** problem.

(c) Then determine the random value k used in signing the message x, without solving an instance of the **Discrete Logarithm** problem.

8.3 Suppose that Alice is using the *ElGamal Signature Scheme*. In order to save time in generating the random numbers k that are used to sign messages, Alice chooses an initial random value k_0, and then signs the ith message using the value $k_i = k_0 + 2i \bmod (p-1)$. Therefore,

$$k_i = k_{i-1} + 2 \bmod (p-1)$$

for all $i \geq 1$. (This is not a recommended method of generating k-values!)

(a) Suppose that Bob observes two consecutive signed messages, say $(x_i, \mathbf{sig}(x_i, k_i))$ and $(x_{i+1}, \mathbf{sig}(x_{i+1}, k_{i+1}))$. Describe how Bob can easily compute Alice's secret key, a, given this information, without solving an instance of the **Discrete Logarithm** problem. (Note that the value of i does not have to be known for the attack to succeed.)

(b) Suppose that the parameters of the scheme are $p = 28703$, $\alpha = 5$, and $\beta = 11339$, and the two messages observed by Bob are

$$\begin{array}{ll} x_i = 12000 & \mathbf{sig}(x_i, k_i) = (26530, 19862) \\ x_{i+1} = 24567 & \mathbf{sig}(x_{i+1}, k_{i+1}) = (3081, 7604). \end{array}$$

Find the value of a using the attack you described in part (a).

8.4 (a) Prove that the second method of forgery on the *ElGamal Signature Scheme*, described in Section 8.3, also yields a signature that satisfies the verification condition.

(b) Suppose Alice is using the *ElGamal Signature Scheme* as implemented in Example 8.1: $p = 467$, $\alpha = 2$, and $\beta = 132$. Suppose Alice has signed the message $x = 100$ with the signature $(29, 51)$. Compute the forged signature that Oscar can then form by using $h = 102$, $i = 45$, and $j = 293$. Check that the resulting signature satisfies the verification condition.

8.5 (a) A signature in the *ElGamal Signature Scheme* or the *DSA* is not allowed to have $\delta = 0$. Show that if a message were signed with a "signature" in which $\delta = 0$, then it would be easy for an adversary to compute the secret key, a.

(b) A signature in the *DSA* is not allowed to have $\gamma = 0$. Show that if a "signature" in which $\gamma = 0$ is known, then the value of k used in that "signature" can be determined. Given that value of k, show that it is now possible to forge a "signature" (with $\gamma = 0$) for any desired message (i.e., a selective forgery can be carried out).

(c) Evaluate the consequences of allowing a signature in the *ECDSA* to have $r = 0$ or $s = 0$.

8.6 Here is a variation of the *ElGamal Signature Scheme*. The key is constructed in a similar manner as before: Alice chooses $\alpha \in \mathbb{Z}_p{}^*$ to be a primitive element, $0 \leq a \leq p-2$ where $\gcd(a, p-1) = 1$, and $\beta = \alpha^a \bmod p$. The key $K = (\alpha, a, \beta)$, where α and β are the public key and a is the private key. Let $x \in \mathbb{Z}_p$ be a message to be signed. Alice computes the signature $\mathbf{sig}(x) = (\gamma, \delta)$, where

$$\gamma = \alpha^k \bmod p$$

and

$$\delta = (x - k\gamma)a^{-1} \bmod (p-1).$$

The only difference from the original *ElGamal Signature Scheme* is in the computation of δ. Answer the following questions concerning this modified scheme.

(a) Describe how a signature (γ, δ) on a message x would be verified using Alice's public key.

(b) Describe a computational advantage of the modified scheme over the original scheme.

(c) Briefly compare the security of the original and modified scheme.

8.7 Suppose Alice uses the *DSA* with $q = 101$, $p = 7879$, $\alpha = 170$, $a = 75$, and $\beta = 4567$, as in Example 8.4. Determine Alice's signature on a message x such that SHA3-224$(x) = 52$, using the random value $k = 49$, and show how the resulting signature is verified.

8.8 We showed that using the same value k to sign two messages in the *ElGamal Signature Scheme* allows the scheme to be broken (i.e., an adversary can determine the secret key without solving an instance of the **Discrete Logarithm** problem). Show how similar attacks can be carried out for the *Schnorr Signature Scheme*, the *DSA*, and the *ECDSA*.

8.9 Suppose that two people (say Alice and Bob) using the *Digital Signature Algorithm* happen to use the same k-value to sign two messages. In addition, suppose that the difference between their secret keys is small. In this situation, the scheme can be broken by an adversary.

More precisely, we assume that Alice and Bob employ the same values of p, q, and α in the *DSA*. Alice has $\beta_1 = \alpha^{a_1}$ and Bob has $\beta_2 = \alpha^{a_2}$ where $|a_1 - a_2| \leq c$, for some small constant $c \leq 1000000$. Additionally, Alice has created a signature (γ, δ_1) on a message x_1 and Bob has created a signature (γ, δ_2) on a message x_2. For simplicity, we assume that the scheme is used without a hash function (though this does not affect the attack).

Then it is almost always possible for an adversary to easily compute Alice's and Bob's secret keys (a_1 and a_2, respectively), without solving the corresponding instances of the **Discrete Logarithm** problem.

(a) Describe how an adversary can first compute $c = a_1 - a_2$ by trial and error, and then compute k, a_1, and a_2. Note that it is only necessary to solve one instance of the **Discrete Logarithm** problem, in order to compute c. Further, since c is small in absolute value, it is feasible to do this by trial and error.

HINT After computing c, consider the equations that are used to define γ and δ.

(b) Suppose that the scheme's parameters are as given below, the two messages are x_1 and x_2, and the signatures are (γ, δ_1) and (γ, δ_2), respectively:

$$\begin{aligned}
p &= 193385032639805360753163840515320974689203045559233170731780025949542944412967019\\
q &= 5636703977890876035746468150425568421101\\
\alpha &= 123656621261045298367389299197720824379615001259891275133080695670324331879335 34\\
\beta_1 &= 101590186979186491501456484061936965361908389572617565733698029960779531023862 82\\
\beta_2 &= 164343217638803565451481602247364908777541341830200409465660453009535461153155 58\\
x_1 &= 3311928858683768754954294150677793289209\\
x_2 &= 2135546260967953305418839258848658265210\\
\gamma &= 3615970285602802148544162492361035626294\\
\delta_1 &= 1166081856389315755292961937802239505182\\
\delta_2 &= 3170739404484160201330661652290161719950
\end{aligned}$$

Verify that Alice's and Bob's signatures are both valid.

(c) Illustrate the attack by breaking the instance of *DSA* given above, computing Alice's and Bob's secret keys.

8.10 Suppose that $x_0 \in \{0,1\}^*$ is a bitstring such that SHA3-224$(x_0) = 00 \cdots 0$. Therefore, when used in *DSA* or *ECDSA*, we have that SHA3-224$(x_0) \equiv 0 \pmod{q}$.

(a) Show how it is possible to forge a *DSA* signature for the message x_0.

HINT Let $\delta = \gamma$, where γ is chosen appropriately.

(b) Show how it is possible to forge an *ECDSA* signature for the message x_0.

8.11 (a) We describe a potential attack against the *DSA*. Suppose that x is given, let $z = (\text{SHA3-224}(x))^{-1} \bmod q$, and let $\epsilon = \beta^z \bmod p$. Now suppose it is possible to find $\gamma, \lambda \in \mathbb{Z}_q{}^*$ such that

$$\left((\alpha\, \epsilon^{\gamma})^{\lambda^{-1} \bmod q}\right) \bmod p \bmod q = \gamma.$$

Define $\delta = \lambda\, \text{SHA3-224}(x) \bmod q$. Prove that (γ, δ) is a valid signature for x.

(b) Describe a similar (possible) attack against the *ECDSA*.

8.12 In a verification of a signature constructed using the *ElGamal Signature Scheme* (or many of its variants), it is necessary to compute a value of the form $\alpha^c \beta^d$. If c and d are random ℓ-bit exponents, then a straightforward use of the SQUARE-AND-MULTIPLY algorithm would require (on average) $\ell/2$ multiplications and ℓ squarings to compute each of α^c and β^d. The purpose of this exercise is to show that the product $\alpha^c \beta^d$ can be computed much more efficiently.

(a) Suppose that c and d are represented in binary, as in Algorithm 6.5. Suppose also that the product $\alpha\beta$ is precomputed. Describe a modification of Algorithm 6.5, in which at most one multiplication is performed in each iteration of the algorithm.

(b) Suppose that $c = 26$ and $d = 17$. Show how your algorithm would compute $\alpha^c \beta^d$, i.e., what are the values of the exponents i and j at the end of each iteration of your algorithm (where $z = \alpha^i \beta^j$).

(c) Explain why, on average, this algorithm requires ℓ squarings and $3\ell/4$ multiplications to compute $\alpha^c \beta^d$, if c and d are randomly chosen ℓ-bit integers.

(d) Estimate the average speedup achieved, as compared to using the original SQUARE-AND-MULTIPLY algorithm to compute α^c and β^d separately, assuming that a squaring operation takes roughly the same time as a multiplication operation.

8.13 Prove that a correctly constructed signature in the *ECDSA* will satisfy the verification condition.

8.14 Let $\mathcal{E}$ denote the elliptic curve $y^2 \equiv x^3 + x + 26 \bmod 127$. It can be shown that $\#\mathcal{E} = 131$, which is a prime number. Therefore any non-identity element in $\mathcal{E}$ is a generator for $(\mathcal{E}, +)$. Suppose the *ECDSA* is implemented in $\mathcal{E}$, with $A = (2, 6)$ and $m = 54$.

(a) Compute the public key $B = mA$.

(b) Compute the signature on a message x if SHA3-224$(x) = 10$, when $k = 75$.

(c) Show the computations used to verify the signature constructed in part (b).

8.15 This exercise looks at an RSA-type signature scheme due to Gennero, Halevi, and Rabin. Suppose $n = pq$, where p and q are distinct large safe primes. Let $h : \{0,1\}^* \rightarrow \mathbb{Z}_n$ be a public hash function with the property that h only takes on odd values. Let $s \in {\mathbb{Z}_n}^*$ be a random value. The public key consists of the hash function h, n, and s, and the private key is p, q. To sign a message x, perform the following computations:

1. Compute $e = h(s)$

2. Compute $f = e^{-1} \bmod \phi(n)$

3. Compute $y = s^f \bmod n$.

The signature is the value y.

(a) Explain why it is necessary that h only takes on odd values.

(b) Derive a formula to verify a signature.

Chapter 9

Post-Quantum Cryptography

In this chapter, we discuss several techniques for creating public-key cryptosystems and signature schemes in the setting of post-quantum cryptography. We include lattice-based cryptography (specifically *NTRU* and cryptography based on the **Learning With Errors** problem), code-based cryptography, multivariate cryptography, and hash-based signature schemes.

9.1 Introduction

The two previous chapters have dealt with public-key cryptography based on the presumed difficulty of the **Factoring** and **Discrete Logarithm** problems, respectively. However, there has been increased interest, especially in recent years, in developing public-key cryptosystems based on other underlying computational problems. One specific motivation for this interest is the ongoing research in ***quantum computing*** and the possible impact it might have on existing cryptographic schemes, in particular, public-key cryptography based on the **Factoring** and **Discrete Logarithm** problems.

Here is a useful high-level explanation of the basics of quantum computing:

> A traditional computer uses long strings of "bits," which encode either a zero or a one. A quantum computer, on the other hand, uses ***quantum bits***, or ***qubits***. What's the difference? Well a qubit is a quantum system that encodes the zero and the one into two distinguishable quantum states. But, because qubits behave quantumly, we can capitalize on the phenomena of ***superposition*** and ***entanglement***. Superposition is essentially the ability of a quantum system to be in multiple states at the same time—that is, something can be "here" and "there," or "up" and "down" at the same time. Entanglement is an extremely strong correlation that exists between quantum particles—so strong, in fact, that two or more quantum particles can be inextricably linked in perfect unison, even if separated by great distances. Thanks to superposition and entanglement, a quantum computer can process a vast number of calculations simultaneously. Think of it this way: whereas a classical computer works with ones and zeros, a quantum computer will have

the advantage of using ones, zeros and "superpositions" of ones and zeros."[1]

Explaining these ideas in detail would require considerable background, so we are not going to attempt to discuss quantum computing except in the most broad terms. Historically, the basic idea of quantum computing dates back to at least 1980, and the relevance of quantum computing became evident with the publication of SHOR'S ALGORITHM in 1994.

Before we delve into the implications of SHOR'S ALGORITHM, we should mention that the development of a practical quantum computer appears to be some years in the future. Despite intense research during the last twenty years, construction of a scalable, fault-tolerant quantum computer has not been achieved yet. However, some experts have expressed the opinion that such a computer has a reasonable chance of being constructed by 2030 or thereabouts. More precisely, Mike Mosca, a leading researcher in quantum computing, predicted in 2016 that there is a one-in-seven chance that a quantum computer would be able to factor a 2048-bit RSA modulus by 2026, and a 50% probability that this would be achieved by 2031.

SHOR'S ALGORITHM shows that integers could be factored quickly using a quantum computer.[2] More precisely, SHOR'S ALGORITHM has complexity $O((\log n)^2(\log \log n)(\log \log \log n))$ to factor a positive integer n. This is a polynomial-time algorithm as a function of the "size" of n, which is $\log n$. SHOR'S ALGORITHM can also be used to solve the **Discrete Logarithm** problem efficiently (i.e., in polynomial time).

So, the consequence of SHOR'S ALGORITHM is the following: if a scalable, fault-tolerant quantum computer can be built, then public-key cryptography based on **Factoring** and **Discrete Logarithm** problems is irretrievably broken. Based on this possible scenario, researchers have been studying potential ways of constructing public-key cryptosystems based on different computational problems, which hopefully would not be susceptible to attacks carried out by quantum computers. The term ***post-quantum cryptography*** is used to describe such cryptographic schemes. More generally, the phrase "post-quantum cryptography" can also apply to other types of public-key primitives (such as signature schemes, for example, which we introduced in Chapter 8).

We should also take a moment to clarify the distinction between quantum cryptography and post-quantum cryptography. ***Quantum cryptography*** refers to cryptographic algorithms or primitives that rely on quantum mechanical techniques for their implementation. Quantum cryptography includes algorithms for ***quantum key distribution*** and ***quantum bit commitment***, among other things. The basic idea of quantum cryptography dates back to Stephen Wiesner's work in the early

[1] `https://uwaterloo.ca/institute-for-quantum-computing/quantum-computing-101#What-is-quantum-computing`

[2] The idea of this algorithm is to compute the order of the element 3 (or some other small integer) in $\mathbb{Z}_n{}^*$. This order will be a divisor of $\phi(n)$ and "usually" it will lead to the determination of a nontrivial factor of n.

1970s. The advantage of quantum cryptography is that it allows the construction of unconditionally secure schemes that cannot exist in the setting of ***classical cryptography***. However, quantum cryptography is a topic that we do not cover in this book.

The potential impact of quantum computers on secret-key cryptography appears to be much less drastic than on public-key cryptography. The main attack method that could be carried out by a quantum computer is based on ***Grover's Algorithm***. Roughly speaking, this permits certain types of exhaustive searches that would require time $O(m)$ on a "classical" (i.e., nonquantum) computer to be carried out in time $O(\sqrt{m})$ on a quantum computer. What this means is that a secure secret-key cryptosystem having key length ℓ should be replaced by one having key length 2ℓ in order to remain secure against a quantum computer. This is because an exhaustive search of an ℓ-bit key on a classical computer takes time $O(2^\ell)$, and an exhaustive search of a 2ℓ-bit key on a quantum computer takes time $O(\sqrt{2^{2\ell}})$, which is the same as $O(2^\ell)$ because $\sqrt{2^{2\ell}} = (2^{2\ell})^{1/2} = 2^\ell$.

There have been several interesting approaches to post-quantum cryptography that have been investigated in recent years. These include the following:

lattice-based cryptography
We discuss *NTRUEncrypt* in Section 9.2.1; this system is defined using arithmetic in certain polynomial rings. Other examples of lattice-based cryptography are based on the **Learning With Errors** problem, which originated in the field of machine learning. We present a simple cryptosystem based on this problem, the *Regev Cryptosystem*, in Section 9.2.3.

code-based cryptography
Section 9.3 gives a short description of the *McEliece Cryptosystem*, which involves error-correcting codes.

multivariate cryptography
Techniques of multivariate cryptography have been considered in the context of cryptosystems (*Hidden Field Equations*; see Section 9.4.1) as well as signature schemes (*Oil and Vinegar*; see Section 9.4.2) .

hash-based cryptography
Hash-based cryptography is used primarily for signature schemes; see Section 9.5.

isogeny-based cryptography
The idea of isogeny-based cryptography is based on certain morphisms between different elliptic curves. However, we do not discuss these techniques in this book.

In any discussion of post-quantum cryptography, it should be pointed out that the proposed techniques are not proven to be immune to attacks by quantum computers. Rather, the approach is to utilize problems that, at present, are not susceptible to quantum attacks based on currently known algorithms.

It should be emphasized that even 15 years of lead time to develop post-quantum cryptographic algorithms leaves little margin for delay. The development, standardization, and deployment of new cryptography technologies can easily take 20 years or more. Thus, post-quantum cryptography is viewed by many as a serious problem of immediate and pressing concern. In particular, NIST is giving a high priority to quantum cryptography, sponsoring the first standardization conference for post-quantum cryptography in 2018. It is also worth noting that, in 2015, the NSA announced its intention to transition to post-quantum cryptography.

9.2 Lattice-based Cryptography

Lattice-based cryptography has been of interest for over twenty years. We first describe the *NTRU* public-key cryptosystem. Then, after a discussion of the basic theory of lattices, we give a brief introduction to cryptography whose security rests on the presumed difficulty of the **Learning With Errors** problem.

9.2.1 NTRU

NTRU is a public-key cryptosystem, due to Hoffstein, Pipher, and Silverman, that was introduced at the CRYPTO '96 rump session. The current version of *NTRU* is known as *NTRUEncrypt*. It is a very fast cryptosystem that is easy to implement. It is also of interest because its security is based on certain lattice problems and thus it is considered to be a practical example of post-quantum cryptography. (We will discuss these lattice problems in the next section.)

NTRUEncrypt is defined in terms of three parameters, N, p, and q, which are fixed integers. Computations are performed in the ring $\mathcal{R} = \mathbb{Z}[x]/(x^N - 1)$. Multiplication of two polynomials is easy in $\mathcal{R}$; it suffices to compute the product of two polynomials in $\mathbb{Z}[x]$ and then reduce all exponents modulo N.

For example, suppose $N = 3$ and we want to compute the product $(x^2 + 3x + 1)(2x^2 + x - 4)$ in $\mathcal{R}$. We compute as follows:

$$\begin{aligned}(x^2+3x+1)(2x^2+x-4) &= 2x^4+7x^3+x^2-11x-4\\ &= 2x+7+x^2-11x-4\\ &= x^2-9x+3.\end{aligned}$$

It is often convenient to represent a polynomial in $\mathcal{R}$ by its vector of coefficients:

$$a(x) = \sum_{i=0}^{N-1} a_i x^i \text{ corresponds to } \mathbf{a} = (a_0, a_1, \ldots, a_{N-1}).$$

Suppose we have

$$a(x) = \sum_{i=0}^{N-1} a_i x^i,$$

$$b(x) = \sum_{i=0}^{N-1} b_i x^i,$$

and

$$c(x) = a(x)b(x) = \sum_{i=0}^{N-1} c_i x^i.$$

The corresponding coefficient vectors have the relation

$$\mathbf{c} = \mathbf{a} \star \mathbf{b},$$

where "$\star$" is a ***convolution operation***. Specifically, for $0 \leq i \leq N-1$, we have that

$$c_i = \sum_{j=0}^{N-1} a_j b_{i-j}, \tag{9.1}$$

where all subscripts are reduced modulo N.

In our description of *NTRUEncrypt*, we will sometimes use the notation of coefficient vectors and the convolution operation. But of course this is exactly the same thing as multiplication in $\mathcal{R}$.

At various points in the *NTRUEncrypt* encryption and decryption process, coefficients will be reduced modulo p or modulo q. These parameters have the following properties: q will be quite a bit larger than p, and q and p should be relatively prime. Also, p should be odd. The values $p = 3$ and $q = 2048$ are popular choices. Finally, N is usually taken to be a prime; $N = 401$ is a currently recommended value.

Various operations in *NTRUEncrypt* require certain "centered" modular reductions, which we define now.

Definition 9.1: For an odd integer n and integers a and b, define

$$a \text{ mods } n = b \text{ if } a \equiv b \pmod{n} \text{ and } -\tfrac{n-1}{2} \leq b \leq \tfrac{n}{2},$$

For example a mods $5 \in \{-2,-1,0,1,2\}$, whereas a mod $5 \in \{0,1,2,3,4\}$.

We now describe the public and private keys used in *NTRUEncrypt*. First, $F(x)$ and $G(x)$ are secret polynomials chosen from $\mathcal{R}$. All coefficients of $F(x)$ and $G(x)$ will be in the set $\{-1,0,1\}$. Next, define $f(x) = 1 + pF(x)$ and $g(x) = pG(x)$. Finally, compute $f^{-1}(x)$ in the ring $\mathcal{R}$ mods q, and then compute $h(x) = f^{-1}(x)g(x)$ mods q. After this is done, F and G can be discarded.

The public key is the coefficient vector $\mathbf{h}$ and the private key is the coefficient vector $\mathbf{f}$. The polynomial $g(x)$ is used in the construction of the public key $h(x)$; $g(x)$ is not part of the public or private key, but it should be kept secret and then discarded after $h(x)$ is formed.

The polynomial $f^{-1}(x)$ can be computed using the EXTENDED EUCLIDEAN ALGORITHM for polynomials. Let $c(x) = \gcd(f(x), x^N - 1)$, which is computed in $\mathbb{Z}_q[x]$. The extended Euclidean algorithm computes polynomials $a(x), b(x) \in \mathbb{Z}_q[x]$ such that

$$a(x)f(x) + b(x)(x^N - 1) = c(x).$$

Then $f^{-1}(x)$ exists if and only if $c(x) = 1$. Further, if $c(x) = 1$, then $f^{-1}(x) = a(x) \bmod\mathsf{s}\ q$.

A plaintext $\mathbf{m}$ is an N-tuple in the set $\{-1, 0, 1\}^N$. The encryption operation in *NTRUEncrypt* is randomized. First, $\mathbf{r} \in \{-1, 0, 1\}^N$ is chosen uniformly at random from a specified subset of $\mathcal{R}$. The ciphertext $\mathbf{y}$ is computed as

$$\mathbf{y} = \mathbf{r} \star \mathbf{h} + \mathbf{m} \ \mathsf{mods}\ q.$$

To decrypt a ciphertext $\mathbf{y}$, perform the following operations:

1. Compute $\mathbf{a} = \mathbf{f} \star \mathbf{y} \ \mathsf{mods}\ q$.
2. Compute $\mathbf{m}' = \mathbf{a} \ \mathsf{mods}\ p$.

If all goes well, it will be the case that $\mathbf{m}' = \mathbf{m}$.

First, it is easy to verify that $\mathbf{a} \equiv \mathbf{r} \star \mathbf{g} + \mathbf{f} \star \mathbf{m} \pmod{q}$. This is done as follows:

$$\begin{aligned}
\mathbf{a} &\equiv \mathbf{f} \star \mathbf{y} \pmod{q} \\
&\equiv \mathbf{f} \star (\mathbf{r} \star \mathbf{h} + \mathbf{m}) \pmod{q} \\
&\equiv \mathbf{f} \star (\mathbf{r} \star \mathbf{f}^{-1} \star \mathbf{g} + \mathbf{m}) \pmod{q} \\
&\equiv \mathbf{r} \star \mathbf{g} + \mathbf{f} \star \mathbf{m} \pmod{q}.
\end{aligned}$$

Now, suppose that this congruence is actually an equality in $\mathcal{R}$, i.e.,

$$\mathbf{a} = \mathbf{r} \star \mathbf{g} + \mathbf{f} \star \mathbf{m}. \tag{9.2}$$

This happens if and only if every coefficient of $\mathbf{r} \star \mathbf{g} + \mathbf{f} \star \mathbf{m}$ lies in the interval

$$\left[-\frac{q-1}{2}, \frac{q}{2} \right],$$

which will hold with high probability if the parameters of the system are chosen in a suitable way.

Now, assuming that (9.2) holds and reducing modulo p, we have

$$\begin{aligned}
\mathbf{a} &\equiv \mathbf{r} \star \mathbf{g} + \mathbf{f} \star \mathbf{m} \pmod{p} \\
&\equiv \mathbf{r} \star p\mathbf{G} + (1 + p\mathbf{F}) \star \mathbf{m} \pmod{p} \\
&\equiv \mathbf{m} \pmod{p}.
\end{aligned}$$

From this relation, we see that

$$\mathbf{m} = \mathbf{a} \ \mathsf{mods}\ p,$$

because all of the coefficients of $\mathbf{m}$ are in the set $\{-1, 0, 1\}$. Therefore the ciphertext is decrypted correctly.

A concise description of *NTRUEncrypt* is presented as Cryptosystem 9.1. We illustrate the encryption and decryption processes with an example.

Cryptosystem 9.1: *NTRUEncrypt*

Suppose p, q, and N are integers, where $q \gg p$, q and p are relatively prime, p is odd, and N is prime. Typical values for these parameters are $p = 3$, $q = 2048$, and $N = 401$.

Let $\mathcal{P} = \{-1,0,1\}^N$ and $\mathcal{C} = (\mathbb{Z}_q)^N$. Choose $\mathbf{F}, \mathbf{G} \in \{-1,0,1\}^N$, let $\mathbf{f} = 1 + p\,\mathbf{F}$, let $\mathbf{g} = p\,\mathbf{G}$, and define $\mathbf{h} = \mathbf{f}^{-1}\mathbf{g} \text{ mods } q$. The associated key is $K = (\mathbf{f}, \mathbf{h})$, where $\mathbf{f}$ is private and $\mathbf{h}$ is public.

Now define

$$e_K(\mathbf{m}) = \mathbf{y} = \mathbf{r} \star \mathbf{h} + \mathbf{m} \text{ mods } q$$

for a randomly chosen $\mathbf{r} \in \{-1,0,1\}^N$, and

$$d_K(\mathbf{y}) = (\mathbf{f} \star \mathbf{y} \text{ mods } q) \text{ mods } p.$$

Example 9.1 Suppose we take $N = 23$, $p = 3$ and $q = 31$. Let

$$F(x) = x^{18} - x^9 + x^8 - x^4 - x^2, \text{ so } f(x) = 3x^{18} - 3x^9 + 3x^8 - 3x^4 - 3x^2 + 1$$

and

$$G(x) = x^{17} + x^{12} + x^9 + x^3 - x, \text{ so } g(x) = 3x^{17} + 3x^{12} + 3x^9 + 3x^3 - 3x.$$

Next we compute

$$\begin{aligned} h(x) = {} & -13x^{22} - 15x^{21} + 12x^{19} - 14x^{18} + 8x^{16} - 14x^{15} - 6x^{14} + 14x^{13} \\ & - 3x^{12} + 7x^{11} - 5x^{10} - 14x^9 + 3x^8 + 10x^7 + 5x^6 - 8x^5 + 4x^2 + x + 8. \end{aligned}$$

Suppose we wish to encrypt the plaintext

$$m(x) = x^{15} - x^{12} + x^7 - 1,$$

and we choose

$$r(x) = x^{19} + x^{10} + x^6 - x^2.$$

The ciphertext is

$$\begin{aligned} y(x) = {} & 5x^{22} - 15x^{21} + 4x^{20} + 8x^{19} + 10x^{18} - 15x^{17} + 6x^{16} + 8x^{15} - 8x^{14} \\ & + 3x^{13} - 10x^{12} - 7x^{11} - x^{10} - 9x^9 + 12x^8 - 14x^7 + 15x^6 - 10x^5 \\ & + 15x^4 - 14x^3 - 5x^2 - 15x - 3. \end{aligned}$$

The decryption process will begin by computing

$$\begin{aligned} a(x) = {} & 6x^{22} + 3x^{21} - 6x^{20} - 3x^{19} - 3x^{17} + 7x^{15} + 6x^{13} - x^{12} - 9x^{11} + 3x^{10} \\ & + 3x^9 - 5x^7 + 6x^4 + 3x^3 + 6x^2 - 3x + 5. \end{aligned}$$

Reducing the coefficients of $a(x)$ modulo 3 yields

$$x^{15} - x^{12} + x^7 - 1,$$

which is the plaintext.

In this example, decryption yielded the correct plaintext because (9.2) is satisfied, as can easily be verified. □

There are a couple of additional conditions on the parameters that we should mention. First, it is usually recommended that each of **F**, **G**, **r**, and **m** have (roughly) one third of their coefficients equal to each of 0, −1, and 1. These requirements are related to the security of the scheme. The second condition is that q should be large compared to N, so the decryption condition (9.2) holds with certainty, or at least with very high probability. The parameter choices mentioned above ensure that this will be the case.

We briefly discuss the decryption operation in a bit more detail now. Suppose we focus on a specific co-ordinate of $\mathbf{r} \star \mathbf{g} + \mathbf{f} \star \mathbf{m}$, say the ith co-ordinate. This would be computed as the sum of the ith co-ordinates of $\mathbf{r} \star \mathbf{g}$ and $\mathbf{f} \star \mathbf{m}$, each of which are obtained from the convolution formula (9.1). First, let's focus on a co-ordinate of $\mathbf{r} \star \mathbf{g}$. The formula (9.1) is the sum of N terms. The N-tuple $\mathbf{r}$ has (approximately) $N/3$ co-ordinates equal to each of 1, 0, and −1, and $\mathbf{g}$ has (approximately) $N/3$ co-ordinates equal to each of p, 0, and $-p$. The maximum value taken on by a particular co-ordinate of $\mathbf{r} \star \mathbf{g}$ would therefore be

$$\frac{N}{3}(p \times 1 + (-p) \times (-1)) = \frac{2Np}{3}.$$

The maximum value of a co-ordinate of $\mathbf{f} \star \mathbf{m}$ is also $2Np/3$. So the maximum value of a co-ordinate of $\mathbf{r} \star \mathbf{g} + \mathbf{f} \star \mathbf{m}$ is $4Np/3 = 4N = 1604$, using the values $p = 3$ and $N = 401$. Similarly, the minimum value is −1604. So the maximum and minimum values are outside the interval

$$\left[-\frac{q-1}{2}, \frac{q}{2}\right] = [-1023, 1024],$$

which means that a decryption error is possible. However, it is very unlikely that all the co-ordinates "line up" so the maximum or minimum is actually achieved. A more detailed analysis shows that the probability of a decryption error is very small.

9.2.2 Lattices and the Security of NTRU

We mentioned that the security of *NTRUEncrypt* is related to certain lattice problems. However, before discussing security, we need to develop some of the basic theory of lattices.

A lattice is very similar to a vector space. A ***real vector space*** can be defined by starting with a ***basis***, which is a set of linearly independent vectors in $\mathbb{R}^n$ for some integer n. The vector space generated by the given basis consists of all linear combinations of basis vectors. If there are r vectors in the basis, then we have an ***r-dimensional vector space***. Restating this using mathematical notation, suppose the r basis vectors are $\mathbf{b}^1, \ldots, \mathbf{b}^r$. The vector space generated by this basis consists of all the vectors of the form

$$\alpha_1 \mathbf{b}^1 + \cdots + \alpha_r \mathbf{b}^r,$$

where $\alpha_1, \ldots, \alpha_r$ are arbitrary real numbers.

Now, a lattice is very similar, except that the vectors in the lattice are ***integer linear combinations*** of basis vectors. That is, the lattice generated by the basis consists of all the vectors of the form

$$\alpha_1 \mathbf{b}^1 + \cdots + \alpha_r \mathbf{b}^r,$$

where $\alpha_1, \ldots, \alpha_r$ are arbitrary integers.

For a vector $\mathbf{v} = (v_1, \ldots, v_n) \in \mathbb{R}^n$, we define the ***norm*** of $\mathbf{v}$, which is denoted by $\|\mathbf{v}\|$, as follows:

$$\|\mathbf{v}\| = \sqrt{\sum_{i=1}^{n} v_i{}^2}.$$

Two fundamental problems in the setting of lattices are the **Shortest Vector** problem and the **Closest Vector** problem. We define these as Problems 9.1 and 9.2. In the specification of these problems, an "instance" consists of a lattice. However, we will always consider a lattice to be specified or represented by giving a basis for the lattice; this is the phraseology used in these problems.

Problem 9.1: Shortest Vector

Instance: A basis for a lattice $\mathcal{L}$ in $\mathbb{R}^n$.
Find: A vector $\mathbf{v} \in \mathcal{L}$, $\mathbf{v} \neq (0, \ldots, 0)$, such that $\|\mathbf{v}\|$ is minimized. Such a vector $\mathbf{v}$ is called a ***shortest vector*** in $\mathcal{L}$.

Problem 9.2: Closest Vector

Instance: A basis for a lattice $\mathcal{L}$ in $\mathbb{R}^n$ and a vector $\mathbf{w} \in \mathbb{R}^n$ that is not in $\mathcal{L}$.
Find: A vector $\mathbf{v} \in \mathcal{L}$ such that $\|\mathbf{v} - \mathbf{w}\|$ is minimized. Such a vector $\mathbf{v}$ is called a ***closest vector*** to $\mathbf{w}$ in $\mathcal{L}$.

One way in which an adversary could break *NTRUEncrypt* would be to compute the polynomials $f(x)$ and $g(x)$ that were used to construct the public key $\mathbf{h}$. Denote $\mathbf{h} = (h_0, \ldots, h_{N-1})$ and consider the lattice $\mathcal{L}_\mathbf{h}$ whose basis consists of the rows of the following $2N$ by $2N$ matrix:

$$M = \left(\begin{array}{cccc|cccc} 1 & 0 & \cdots & 0 & h_0 & h_1 & \cdots & h_{N-1} \\ 0 & 1 & \cdots & 0 & h_{N-1} & h_0 & \cdots & h_{N-2} \\ \vdots & \vdots & \ddots & \vdots & \vdots & \vdots & \ddots & \vdots \\ 0 & 0 & \cdots & 1 & h_1 & h_2 & \cdots & h_0 \\ \hline 0 & 0 & \cdots & 0 & q & 0 & \cdots & 0 \\ 0 & 0 & \cdots & 0 & 0 & q & \cdots & 0 \\ \vdots & \vdots & \ddots & \vdots & \vdots & \vdots & \ddots & \vdots \\ 0 & 0 & \cdots & 0 & 0 & 0 & \cdots & q \end{array}\right)$$

The lattice $\mathcal{L}_{\mathbf{h}}$ consists of the following vectors:

$$\mathcal{L}_{\mathbf{h}} = \{(\mathbf{a}, \mathbf{b}) \in \mathbb{Z}^{2N} : \mathbf{a} \star \mathbf{h} = \mathbf{b}\}.$$

From the way in which $\mathbf{h}$ is constructed, we have that

$$\mathbf{f} \star \mathbf{h} \equiv \mathbf{g} \bmod q,$$

where $\mathbf{f}$ and $\mathbf{g}$ are the coefficient vectors of $f(x)$ and $g(x)$, respectively. This means that

$$\mathbf{f} \star \mathbf{h} - \mathbf{g} = q\,\mathbf{t}$$

for some integer vector $\mathbf{t}$. It is then straightforward to compute

$$(\mathbf{f}, -\mathbf{t})M = (\mathbf{f}, \mathbf{g}),$$

so $(\mathbf{f}, \mathbf{g}) \in \mathcal{L}_{\mathbf{h}}$.

Further, the vector $(\mathbf{f}, \mathbf{g})$ has a small norm, since all of its coefficients are in the set $\{-p, -1, 0, 1, p\}$. More precisely, $(\mathbf{f}, \mathbf{g})$ has roughly $N/3$ coefficients equal to each of $-p$, -1, 1, and p, and the remaining $2N/3$ coefficients are equal to 0. So the norm of $(\mathbf{f}, \mathbf{g})$ is approximately

$$\sqrt{\frac{2N(1+p^2)}{3}} = 2\sqrt{\frac{5N}{3}}.$$

However, a vector of length $2N$ whose co-ordinates take on random values in $[-q/2, q/2]$ would (on average) have norm approximately equal to

$$q\sqrt{\frac{N}{6}},$$

which is much larger (recall that we are assuming $p = 3$ and $q = 2048$).

It therefore seems plausible that $(\mathbf{f}, \mathbf{g})$ is the shortest vector in the lattice $\mathcal{L}_{\mathbf{h}}$. Since solving the **Shortest Vector** problem is believed to be difficult, it should not be possible for the adversary to find the private key $\mathbf{f}$.

An adversary might also attempt to decrypt a specific ciphertext. It turns out that this can be modeled as an instance of the **Closest Vector** problem. The vector

$(0, \mathbf{y})$ is in fact quite close to the vector $(\mathbf{r}, \mathbf{r} \star \mathbf{h} \text{ mods } q)$, which is in the lattice $\mathcal{L}_{\mathbf{h}}$. More precisely,

$$(0, \mathbf{y}) = (\mathbf{r}, \mathbf{r} \star \mathbf{h} \text{ mods } q) + (-\mathbf{r}, \mathbf{m}),$$

and $(-\mathbf{r}, \mathbf{m})$ has small norm. Therefore, it seems reasonable that the closest vector to $(0, \mathbf{y})$ is $(\mathbf{r}, \mathbf{r} \star \mathbf{h} \text{ mods } q)$. Solving the **Closest Vector** problem reveals the vector $\mathbf{r}$, which allows $\mathbf{y}$ to be decrypted.

It is important to emphasize that there is no proof that breaking *NTRUEncrypt* is as hard as solving the **Shortest Vector** problem or the **Closest Vector** problem. Thus, *NTRUEncrypt* cannot (currently, at least) be regarded as a provably secure cryptosystem. We give an example of a provably secure lattice-based public-key cryptosystem in the next subsection.

9.2.3 Learning With Errors

Given a prime q, it is possible to find solutions to a system of linear equations in n variables over $\mathbb{Z}_q$ efficiently. However, by carefully introducing randomness into the system we obtain a new problem, known as the **Learning With Errors** (or **LWE**) problem, that is believed to be difficult to solve.

Problem 9.3: Learning With Errors

Instance: A prime q, an integer n, a discrete random variable $\mathbf{E}$ with probability distribution χ defined on the set $\mathbb{Z}_q$ and m samples $(\mathbf{a^i}, b^i) \in (\mathbb{Z}_q)^{n+1}$. The m samples are all constructed from a secret $\mathbf{s} = (s_1, s_2, \ldots, s_n) \in (\mathbb{Z}_q)^n$. For $1 \leq i \leq m$, $\mathbf{a^i} = (a^i_1, \ldots, a^i_n)$ is chosen uniformly at random from $(\mathbb{Z}_q)^n$, e^i is chosen using the probability distribution χ and

$$b^i = e^i + \sum_{j=1}^{n} a^i_j s_j \bmod q.$$

Find: The secret $(s_1, s_2, \ldots, s_n)$.

Informally, **LWE** can be regarded as the problem of finding a solution modulo q to the approximate system of linear equations

$$\begin{aligned}
a^1_1 x_1 + a^1_2 x_2 + \cdots + a^1_n x_n &\approx b^1,\\
a^2_1 x_1 + a^2_2 x_2 + \cdots + a^2_n x_n &\approx b^2,\\
&\vdots\\
a^m_1 x_1 + a^m_2 x_2 + \cdots + a^m_n x_n &\approx b^m.
\end{aligned}$$

The solution is unique if there are enough equations in the system. Constructing an **LWE** system that is difficult to solve requires a value of q that is significantly larger than n as well as a careful choice of probability distribution χ for the random variable **E**. Most proposals use a distribution in which the probability of a given error e increases the closer e is to 0 (where closeness is defined by treating e as an integer in the range $-\lfloor q/2 \rfloor$ to $\lfloor q/2 \rfloor$), with 0 being the most likely error. If the variance of the distribution is too high then the errors risk obscuring the rest of the information in the system. However, if it is too small (for example, if **E** is uniformly zero) then the corresponding **LWE** problem will not be secure.

Existing literature has used very sophisticated techniques to show that particular families of distributions are suitable for use in cryptographic constructions based on **LWE**. The **LWE** problem is of interest to cryptographers because the average-case difficulty of solving instances of **LWE** can be shown to be based on the worst-case difficulty of solving a certain lattice problem that is believed to be hard. The particular lattice problem used in the reduction is a decision problem known as the **Gap Shortest Vector** problem, which (naturally) is closely related to the **Shortest Vector** problem. There is no known efficient quantum algorithm to solve this problem, so cryptographic systems based on **LWE** are regarded as post-quantum and fall under the general category of lattice-based cryptography.

Cryptosystem 9.2 (the *Regev Cryptosystem*) is an example of a cryptosystem based on **LWE** that can be used to encrypt a single bit. We note that the ciphertext involves a sum of samples, and that the "errors" in the samples are also summed. It follows that in order for decryption to be possible, it is necessary to choose the distribution χ in such a way that the overall error in the ciphertext is not so great as to obscure the distinction between values close to 0 and values close to $q/2$. It is possible to show that Cryptosystem 9.2 is secure against chosen plaintext attacks provided that

- samples constructed from a secret $\mathbf{s}$ and errors following the probability distribution χ cannot be distinguished from uniformly distributed elements of $(\mathbb{Z}_q)^{n+1}$, and
- an algorithm for distinguishing these LWE samples from uniform elements can be used to break **LWE**.

Example 9.2 Let $n = 3$ and $q = 11$. Suppose that **E** is the discrete random variable that takes on each of the values 0, 1, or -1 with probability $1/3$. Suppose the secret key is $(1,2,3)$, and the public key consists of the three samples $((5,8,10),7)$, $((4,9,1),4)$, and $((3,6,0),3)$.

To encrypt the message $t = 1$ we choose a random subset of $\{1,2,3\}$, say $S = \{1,3\}$. Then the ciphertext is $(\mathbf{a}, b)$ where

$$\mathbf{a} = (5,8,10) + (3,6,0) = (8,3,10)$$

and

$$b = 7 + 3 + \lfloor 11/2 \rfloor = 4.$$

To decrypt the ciphertext $((8,3,10),4)$ we compute

$$4 - 8 \times 1 - 3 \times 2 - 10 \times 3 = 4.$$

Because 4 is closer to 5 than to 0, the decrypted message is 1, as expected. □

Cryptosystem 9.2: *Regev Cryptosystem*

Let n and m be integers and let q be a prime. Let $\mathbf{E}$ be a discrete random variable defined on $\mathbb{Z}_q$.

The private key is an element $\mathbf{s} \in (\mathbb{Z}_q)^n$.

The public key consists of m samples $(\mathbf{a^i}, b^i)$ where $\mathbf{a^i}$ is drawn uniformly from $\mathbb{Z}_q$ and b^i is taken to be $b^i = \mathbf{E} + \sum_{j=1}^{n} a_j^i s_j$.

To encrypt a one-bit message x, choose a random subset $S \subseteq \{1, 2, \ldots, m\}$. The ciphertext y is given by

$$y = \begin{cases} (\sum_{i \in S} \mathbf{a^i}, \sum_{i \in S} b^i) & \text{if } x = 0, \\ (\sum_{i \in S} \mathbf{a^i}, \lfloor \frac{q}{2} \rfloor + \sum_{i \in S} b^i) & \text{if } x = 1. \end{cases}$$

To decrypt a ciphertext $(\mathbf{a}, b)$, compute the quantity $b - \sum_{j=1}^{m} a_j s_j$. The decrypted message is 0 if the result is closer to 0 than $\lfloor q/2 \rfloor$ and 1 otherwise.

Cryptosystem 9.2 is not practical due to the large overhead required to encrypt a single bit. There exist more efficient cryptosystems based on **LWE** (including ones that are secure against chosen ciphertext attacks) as well as many other cryptographic primitives. However, these schemes tend to require large keys. One way to reduce the size of the public key in Cryptosystem 9.2 is to replace the uniformly generated $\mathbf{a^i}$ with a more structured set of elements of $(\mathbb{Z}_q)^n$. This idea has led to the development of cryptosystems based on a variant of **LWE** known as **ring-LWE**. This approach permits the construction of more efficient schemes, but addressing the question of how to determine suitable error distributions is even more subtle than in the case of **LWE**.

9.3 Code-based Cryptography and the McEliece Cryptosystem

The ***NP-complete problems*** comprise a large class of decision problems (i.e., problems that have a yes/no answer) that are believed to be impossible to solve in polynomial time. The ***NP-hard problems*** are a class of problems, which may or may not be decision problems, that are at least as difficult to solve as the NP-complete problems.

In the *McEliece Cryptosystem*, decryption is an easy special case of an NP-hard problem, disguised so that it looks like a (presumably difficult) general instance of the problem. In this system, the NP-hard problem that is employed is related to decoding a general linear (binary) error-correcting code. However, for many special classes of codes, polynomial-time algorithms are known to exist. One such class of codes is used as the basis of the *McEliece Cryptosystem*.

We begin with some essential definitions. First we define the notion of a linear code and a generating matrix.

Definition 9.2: Let k, n be positive integers, $k \leq n$. A ***linear code*** is a k-dimensional subspace of $(\mathbb{Z}_2)^n$, the vector space of all binary n-tuples. A ***linear*** $[n,k]$ ***code***, $\mathbf{C}$, is a k-dimensional subspace of $(\mathbb{Z}_2)^n$.

A ***generating matrix*** for a linear $[n,k]$ code, $\mathbf{C}$, is a $k \times n$ binary matrix whose rows form a basis for $\mathbf{C}$.

Next, we define the distance of a (linear) code.

Definition 9.3: Let $\mathbf{x}, \mathbf{y} \in (\mathbb{Z}_2)^n$, where $\mathbf{x} = (x_1, \dots, x_n)$ and $\mathbf{y} = (y_1, \dots, y_n)$. Define the ***Hamming distance***

$$\mathbf{dist}(\mathbf{x}, \mathbf{y}) = |\{i : 1 \leq i \leq n, x_i \neq y_i\}|,$$

i.e., the number of co-ordinates in which $\mathbf{x}$ and $\mathbf{y}$ differ.

Let $\mathbf{C}$ be a linear $[n,k]$ code. Define the ***distance*** of $\mathbf{C}$ to be the quantity

$$\mathbf{dist}(\mathbf{C}) = \min\{\mathbf{dist}(\mathbf{x}, \mathbf{y}) : \mathbf{x}, \mathbf{y} \in \mathbf{C}, \mathbf{x} \neq \mathbf{y}\}.$$

A ***linear*** $[n,k,d]$ ***code*** is a linear $[n,k]$ code, say $\mathbf{C}$, in which $\mathbf{dist}(\mathbf{C}) \geq d$.

Finally, we define the dual code (of a linear code) and the parity-check matrix.

Definition 9.4: Two vectors $\mathbf{x}, \mathbf{y} \in (\mathbb{Z}_2)^n$, say $\mathbf{x} = (x_1, \dots, x_n)$ and $\mathbf{y} = (y_1, \dots, y_n)$, are ***orthogonal*** if

$$\sum_{i=1}^{n} x_i y_i \equiv 0 \pmod{2}.$$

The ***orthogonal complement*** of a linear $[n,k,d]$ code, $\mathbf{C}$, consists of all the vectors that are orthogonal to all the vectors in $\mathbf{C}$. This set of vectors is denoted by $\mathbf{C}^{\perp}$ and it is called the ***dual code*** to $\mathbf{C}$.

A ***parity-check matrix*** for a linear $[n,k,d]$ code $\mathbf{C}$ having generating matrix G is a generating matrix H for $\mathbf{C}^{\perp}$. This matrix H is an $(n-k)$ by n matrix. (Stated another way, the rows of H are linearly independent vectors, and GH^T is a k by $n-k$ matrix of zeroes.)

The purpose of an error-correcting code is to correct random errors that occur in the transmission of (binary) data through a noisy channel. Briefly, this is done as follows. Let G be a generating matrix for a linear $[n,k,d]$ code. Suppose $\mathbf{x}$ is the binary k-tuple we wish to transmit. Then we encode $\mathbf{x}$ as the n-tuple $\mathbf{y} = \mathbf{x}G$ and we transmit $\mathbf{y}$ through the channel.

Now, suppose Bob receives the n-tuple $\mathbf{r}$, which may not be the same as $\mathbf{y}$. He will decode $\mathbf{r}$ using the strategy of "nearest neighbor decoding." The idea is that Bob finds a codeword $\mathbf{y}' \neq \mathbf{r}$ that has minimum distance to $\mathbf{r}$. Such a codeword will be called a ***nearest neighbor*** to $\mathbf{r}$ and it will be denoted as by $\mathbf{nn}(\mathbf{r})$ (note that it is possible that there might be more than one nearest neighbor). The process of computing $\mathbf{nn}(\mathbf{r})$ is called ***nearest neighbor decoding***.

After decoding $\mathbf{r}$ to $\mathbf{y}' = \mathbf{nn}(\mathbf{r})$, Bob would determine the k-tuple $\mathbf{x}'$ such that $\mathbf{y}' = \mathbf{x}'G$. Bob is hoping that $\mathbf{y}' = \mathbf{y}$, so $\mathbf{x}' = \mathbf{x}$ (i.e., he is hoping that any transmission errors have been corrected).

It is fairly easy to show that if at most $(d-1)/2$ errors occurred during transmission, then nearest neighbor decoding does in fact correct all the errors. In this case, any received vector $\mathbf{r}$ will have a unique nearest neighbor, and $\mathbf{nn}(\mathbf{r}) = \mathbf{y}$.

Let us think about how nearest neighbor decoding would be done in practice. The number of possible codewords is equal to 2^k. If Bob compares $\mathbf{r}$ to every codeword, then he will have to examine 2^k vectors, which is an exponentially large number compared to k. In other words, this obvious decoding algorithm is not a polynomial-time algorithm.

Another approach, which forms the basis for many practical decoding algorithms, is based on the idea of a syndrome. Suppose $\mathbf{C}$ is a linear $[n,k]$ code having parity-check matrix H. Given a vector $\mathbf{r} \in (\mathbb{Z}_2)^n$, we define the ***syndrome*** of $\mathbf{r}$ to be Hr^T. A syndrome is a column vector with $n-k$ components.

The following basic result can be proven using straightforward techniques from linear algebra.

THEOREM 9.1 *Suppose* $\mathbf{C}$ *is a linear* $[n,k]$ *code with parity-check matrix* H. *Then* $\mathbf{x} \in (\mathbb{Z}_2)^n$ *is a codeword if and only if*

$$H\mathbf{x}^T = \begin{pmatrix} 0 \\ 0 \\ \vdots \\ 0 \end{pmatrix}.$$

Further, if $\mathbf{x} \in \mathbf{C}$, $\mathbf{e} \in (\mathbb{Z}_2)^n$ *and we define* $\mathbf{r} = \mathbf{x} + \mathbf{e}$, *then* $H\mathbf{r}^T = H\mathbf{e}^T$.

Think of $\mathbf{e}$ as being the vector of errors that occur during transmission of a codeword $\mathbf{x}$. Then $\mathbf{r}$ represents the vector that is received. The above theorem is saying that the syndrome depends only on the errors, and not on the particular codeword that was transmitted.

This suggests the following approach to decoding, known as ***syndrome decoding***: First, compute $\mathbf{s} = H\mathbf{r}^T$. If $\mathbf{s}$ is a vector of zeroes, then decode $\mathbf{r}$ as $\mathbf{r}$. If not, then generate all possible error vectors of weight 1 in turn, where the ***weight*** of a vector is the number of nonzero components it contains. For each such error vector $\mathbf{e}$, compute $H\mathbf{e}^T$. If, for any of these vectors $\mathbf{e}$, it holds that $H\mathbf{e}^T = \mathbf{s}$, then decode $\mathbf{r}$ to $\mathbf{r} - \mathbf{e}$. Otherwise, continue on to generate all error vectors of weight

$2, \ldots, \lfloor (d-1)/2 \rfloor$. If, at any time, we have $H\mathbf{e}^T = \mathbf{s}$ for a candidate error vector $\mathbf{e}$, then we decode $\mathbf{r}$ to $\mathbf{r} - \mathbf{e}$ and quit. If this equation is never satisfied, then we conclude that more than $\lfloor (d-1)/2 \rfloor$ errors have occurred during transmission.

Using this approach, we can decode a received vector in at most

$$1 + \binom{n}{1} + \cdots + \binom{n}{\lfloor (d-1)/2 \rfloor}$$

steps.

This method works on any linear code. Further, for certain specific types of codes, the decoding procedure can be speeded up. However, nearest neighbor decoding is in fact an NP-hard problem. Thus, no polynomial-time algorithm is known for the general problem of nearest neighbor decoding.

It turns out that it is possible to identify an "easy" special case of the decoding problem and then disguise it so that it looks like a "difficult" general case of the problem. It would take us too long to go into the theory here, so we will just summarize the results. The "easy" special case that was suggested by McEliece is to use a code from a class of codes known as the ***Goppa codes***. These codes do in fact have efficient decoding algorithms. Also, they are easy to generate and there are a large number of inequivalent Goppa codes with the same parameters.

The parameters of the Goppa codes have the form $n = 2^m$, $d = 2t + 1$, and $k = n - mt$ for an integer t. For a practical implementation of the public-key cryptosystem, McEliece originally suggested taking $m = 10$ and $t = 50$. This gives rise to a Goppa code that is a linear $[1024, 524, 101]$ code. Each plaintext is a binary 524-tuple, and each ciphertext is a binary 1024-tuple. The public key is a 524×1024 binary matrix. However, current recommended parameter sizes are considerably larger than these. For example, a 2008 study by Bernstein, Lange, and Peters recommended taking $m = 11$ and $t = 27$, which utilizes a linear $[2048, 1751, 55]$ Goppa code, for a minimum acceptable level of security.

A description of the *McEliece Cryptosystem* is given in Cryptosystem 9.3. We present a toy example to illustrate the encoding and decoding procedures.

Example 9.3 The matrix

$$G = \begin{pmatrix} 1 & 0 & 0 & 0 & 1 & 1 & 0 \\ 0 & 1 & 0 & 0 & 1 & 0 & 1 \\ 0 & 0 & 1 & 0 & 0 & 1 & 1 \\ 0 & 0 & 0 & 1 & 1 & 1 & 1 \end{pmatrix}$$

is a generating matrix for a linear $[7, 4, 3]$ code, known as a ***Hamming code***. Suppose Bob chooses the matrices

$$S = \begin{pmatrix} 1 & 1 & 0 & 1 \\ 1 & 0 & 0 & 1 \\ 0 & 1 & 1 & 1 \\ 1 & 1 & 0 & 0 \end{pmatrix}$$

Cryptosystem 9.3: *McEliece Cryptosystem*

Let G be a generating matrix for a linear $[n,k,d]$ Goppa code $\mathbf{C}$, where $n = 2^m$, $d = 2t+1$, and $k = n - mt$. Let S be a $k \times k$ matrix that is invertible over $\mathbb{Z}_2$, let P be an $n \times n$ permutation matrix, and let $G' = SGP$. Let $\mathcal{P} = (\mathbb{Z}_2)^k, \mathcal{C} = (\mathbb{Z}_2)^n$, and let

$$\mathcal{K} = \{(G, S, P, G')\},$$

where G, S, P, and G' are constructed as described above. The matrix G' is the public key and G, S, and P comprise the private key.

For a public key G', a plaintext $\mathbf{x} \in (\mathbb{Z}_2)^k$ is encrypted by computing

$$\mathbf{y} = \mathbf{x}G' + \mathbf{e},$$

where $\mathbf{e} \in (\mathbb{Z}_2)^n$ is a random error vector of weight t.

A ciphertext $\mathbf{y} \in (\mathbb{Z}_2)^n$ is decrypted by means of the following operations:

1. Compute $\mathbf{y}_1 = \mathbf{y}P^{-1}$.
2. Decode $\mathbf{y}_1$, obtaining $\mathbf{y}_1 = \mathbf{x}_1 + \mathbf{e}_1$, where $\mathbf{x}_1 \in \mathbf{C}$.
3. Compute $\mathbf{x}_0 \in (\mathbb{Z}_2)^k$ such that $\mathbf{x}_0 G = \mathbf{x}_1$.
4. Compute $\mathbf{x} = \mathbf{x}_0 S^{-1}$.

and

$$P = \begin{pmatrix} 0 & 1 & 0 & 0 & 0 & 0 & 0 \\ 0 & 0 & 0 & 1 & 0 & 0 & 0 \\ 0 & 0 & 0 & 0 & 0 & 0 & 1 \\ 1 & 0 & 0 & 0 & 0 & 0 & 0 \\ 0 & 0 & 1 & 0 & 0 & 0 & 0 \\ 0 & 0 & 0 & 0 & 0 & 1 & 0 \\ 0 & 0 & 0 & 0 & 1 & 0 & 0 \end{pmatrix}.$$

Then, the public generating matrix is

$$G' = \begin{pmatrix} 1 & 1 & 1 & 1 & 0 & 0 & 0 \\ 1 & 1 & 0 & 0 & 1 & 0 & 0 \\ 1 & 0 & 0 & 1 & 1 & 0 & 1 \\ 0 & 1 & 0 & 1 & 1 & 1 & 0 \end{pmatrix}.$$

Now, suppose Alice encrypts the plaintext $\mathbf{x} = (1,1,0,1)$ using the vector $\mathbf{e} = (0,0,0,0,1,0,0)$ as the random error vector of weight 1. The ciphertext is computed to be

$$\begin{aligned}
\mathbf{y} &= \mathbf{x}G' + \mathbf{e} \\
&= (1,1,0,1)\begin{pmatrix} 1 & 1 & 1 & 1 & 0 & 0 & 0 \\ 1 & 1 & 0 & 0 & 1 & 0 & 0 \\ 1 & 0 & 0 & 1 & 1 & 0 & 1 \\ 0 & 1 & 0 & 1 & 1 & 1 & 0 \end{pmatrix} + (0,0,0,0,1,0,0) \\
&= (0,1,1,0,0,1,0) + (0,0,0,0,1,0,0) \\
&= (0,1,1,0,1,1,0).
\end{aligned}$$

When Bob receives the ciphertext $\mathbf{y}$, he first computes

$$\begin{aligned}
\mathbf{y}_1 &= \mathbf{y}P^{-1} \\
&= (0,1,1,0,1,1,0)\begin{pmatrix} 0 & 0 & 0 & 1 & 0 & 0 & 0 \\ 1 & 0 & 0 & 0 & 0 & 0 & 0 \\ 0 & 0 & 0 & 0 & 1 & 0 & 0 \\ 0 & 1 & 0 & 0 & 0 & 0 & 0 \\ 0 & 0 & 0 & 0 & 0 & 0 & 1 \\ 0 & 0 & 0 & 0 & 0 & 1 & 0 \\ 0 & 0 & 1 & 0 & 0 & 0 & 0 \end{pmatrix} \\
&= (1,0,0,0,1,1,1).
\end{aligned}$$

Next, he decodes $\mathbf{y}_1$ to get $\mathbf{x}_1 = (1,0,0,0,1,1,0)$. It is worth noting that $\mathbf{e}_1 \neq \mathbf{e}$ due to the multiplication by P^{-1}. However, since P is a permutation matrix, the multiplication only changes the position of the error(s).

Next, Bob forms $\mathbf{x}_0 = (1,0,0,0)$ (the first four components of $\mathbf{x}_1$).

Finally, Bob calculates

$$\begin{aligned}
\mathbf{x} &= \mathbf{x}_0 S^{-1} \\
&= (1,0,0,0)\begin{pmatrix} 1 & 1 & 0 & 1 \\ 1 & 1 & 0 & 0 \\ 0 & 1 & 1 & 1 \\ 1 & 0 & 0 & 1 \end{pmatrix} \\
&= (1,1,0,1).
\end{aligned}$$

This is indeed the plaintext that Alice encrypted. □

9.4 Multivariate Cryptography

Another example of a problem suggested for use in the design of post-quantum cryptosystems is that of finding solutions to large systems of quadratic equations in many variables over a finite field. This is known as the **Multivariate Quadratic**

Problem 9.4: Multivariate Quadratic Equations

Instance: A finite field $\mathbb{F}_q$ and a system of m quadratic equations in n variables:

$$\begin{array}{rcl} f_1(x_1, x_2, \ldots, x_n) & = & d_1, \\ f_2(x_1, x_2, \ldots, x_n) & = & d_2, \\ \vdots & \vdots & \vdots \\ f_m(x_1, x_2, \ldots, x_n) & = & d_m, \end{array} \tag{9.3}$$

where, for all k with $1 \leq k \leq m$, the polynomial f_k has the form

$$f_k(x_1, x_2, \ldots, x_n) = \sum_{i=1}^{n} \sum_{j=i}^{n} a_{ij} x_i x_j + \sum_{i=1}^{n} b_i x_i + c,$$

with a_{ij}, b_i, and c chosen uniformly at random from $\mathbb{F}_q$, for all i, j with $1 \leq i, j \leq n$.

Question: Find a vector $(s_1, s_2, \ldots, s_n) \in (\mathbb{F}_q)^n$ that satisfies the equations $f_i(s_1, s_2, \ldots, s_n) = d_i$ for all i with $1 \leq i \leq m$.

Equations problem, and is usually abbreviated as the **MQ** problem. We present it as Problem 9.4.

The **MQ** problem is NP-hard over any finite field. It is used as the basis of both public-key cryptosystems and signature scheme, which we discuss in the next two subsections.

9.4.1 Hidden Field Equations

The design of cryptosystems based on the **MQ** problem follows a similar strategy to that of the *McEliece Cryptosystem*. Namely, it involves starting with a special case of the problem that is easy to solve, and then disguising it with the aim of making it appear like a general instance of the problem. In order to explore an example of this approach, we consider the cryptosystem known as *Hidden Field Equations* (*HFE*), proposed in 1996 by Jacques Patarin. In *HFE*, the special system of multivariate quadratic equations is constructed from a univariate polynomial over an extension $\mathbb{F}_{q^n}$ of the field $\mathbb{F}_q$. The following example shows how this can be done.

Example 9.4 In Example 7.7, we saw that the field $\mathbb{F}_8 = \mathbb{F}_{2^3}$ is an extension of the field $\mathbb{F}_2$. All of its elements can be written in the form $a\theta^2 + b\theta + c$ with $a, b, c \in \mathbb{F}_2$, where θ satisfies $\theta^3 + \theta + 1 = 0$ (note that we have switched to using θ in place of x here to avoid confusion with the other notation used in this example). As discussed in Example 7.7, there is a one-to-one correspondence between elements of $(\mathbb{F}_2)^3$ and $\mathbb{F}_8$ where (a, b, c) corresponds to $a\theta^2 + b\theta + c$.

Consider the polynomial $f(X) = X^5 + X^2 + 1$ with coefficients from $\mathbb{F}_8$. Any solution to the equation

$$f(X) = 0 \tag{9.4}$$

in $\mathbb{F}_8$ can be written in the form $x_1\theta^2 + x_2\theta + x_3$. Substituting this expression in place of X gives

$$\begin{aligned} f(X) &= X^5 + X^2 + 1, \\ &= (x_1\theta^2 + x_2\theta + x_3)^5 + (x_1\theta^2 + x_2\theta + x_3)^2 + 1. \end{aligned}$$

Now, if we expand the terms in parentheses, we will obtain an expression involving powers of θ up to θ^{10}. However, as in Example 7.7, each of these powers of θ can be expressed in the form $a\theta^2 + b\theta + c$ for suitable $a, b, c \in \mathbb{F}_2$. Once we express the powers of θ in this form and then collect together like powers of θ, we obtain

$$f(x_1\theta^2 + x_2\theta + x_3) = \theta^2(x_3x_1 + x_3x_2 + x_1) + \theta(x_3x_1 + x_2) + x_2x_1 + x_2 + x_1 + 1.$$

(We will discuss how this simplification can be carried out conveniently in Examples 9.5 and 9.6.) If we compare the coefficients of the various powers of θ, we can observe that a solution $a\theta^2 + b\theta + c \in \mathbb{F}_8$ to (9.4) corresponds to a solution $(a, b, c) \in (\mathbb{F}_2)^3$ to the following system of multivariate quadratic equations:

$$\begin{aligned} g_1(\mathbf{x}) &= x_3x_1 + x_3x_2 + x_1 &= 0 \\ g_2(\mathbf{x}) &= x_3x_1 + x_2 &= 0 \\ g_3(\mathbf{x}) &= x_2x_1 + x_2 + x_1 + 1 &= 0. \end{aligned} \tag{9.5}$$

There exist efficient algorithms for finding solutions of systems of univariate polynomial equations over finite fields. Hence, applying these techniques to (9.4) over $\mathbb{F}_8$ gives an efficient way to find solutions to the system of multivariate quadratic equations (9.5) over $\mathbb{F}_2$. □

Any univariate polynomial over an extension field $\mathbb{F}_{q^n}$ can be turned into a system of multivariate polynomial equations over the field $\mathbb{F}_q$ by following the approach illustrated in Example 9.4. However, in general, these equations will have degree greater than two. To ensure we obtain a system of quadratic equations, we have to choose our univariate polynomial carefully. The following examples show how this can be done.

Example 9.5 Consider the term X^2 that appears in the polynomial $f(X)$ of Example 9.4. We observe that, if a and b are elements of $\mathbb{F}_8$, then

$$(a + b)^2 = a^2 + ab + ab + b^2 = a^2 + b^2,$$

since $\mathbb{F}_8$ has characteristic 2. This means that, when we substitute $x_1\theta^2 + x_2\theta + x_3$ in place of X, we see that

$$\begin{aligned} X^2 &= (x_1\theta^2 + x_2\theta + x_3)^2, \\ &= (x_1\theta^2)^2 + (x_2\theta)^2 + x_3{}^2, \\ &= x_1{}^2\theta^4 + x_2{}^2\theta^2 + x_3{}^2. \end{aligned}$$

However, we observe that x_3, x_2, and x_1 all represent elements of $\mathbb{F}_2$ and hence they can take on only the values 0 or 1. Now $0^2 = 0$ and $1^2 = 1$, so we can conclude that $x_1{}^2 = x_1$, $x_2{}^2 = x_2$, and $x_3{}^2 = x_3$, so the above expression becomes

$$\begin{aligned} x_1\theta^4 + x_2\theta^2 + x_3{}^2 &= x_1(\theta^2 + \theta) + x_2\theta^2 + x_3, \\ &= \theta^2(x_1 + x_2) + \theta x_1 + x_3. \end{aligned}$$

Thus we see that X^2 translates into linear equations in the variables x_3, x_2 and x_1. Similarly, we can deduce that X^{2^i} will also give rise to linear equations for any $i \geq 0$. □

Example 9.6 Now consider the term X^5 that appears in $f(X)$. We note that

$$X^5 = (X^4)X = X^{2^2}X^{2^0}.$$

Hence we see that X^5 can be written as the product of two terms that are linear in the variables x_3, x_2 and x_1, and so the resulting expression will contain terms that are quadratic in these variables:

$$\begin{aligned} X^5 &= X^4X \\ &= (x_1\theta^2 + x_2\theta + x_3)^4(x_1\theta^2 + x_2\theta + x_3), \\ &= (x_1\theta^8 + x_2\theta^4 + x_3)(x_1\theta^2 + x_2\theta + x_3), \\ &= x_1\theta^{10} + x_2x_1\theta^9 + x_3x_1\theta^8 + x_2x_1\theta^6 + x_2\theta^5 + x_3x_2\theta^4 + x_3x_1\theta^2 + x_3x_2\theta + x_3, \\ &= x_1(\theta + 1) + x_2x_1\theta^2 + x_3x_1\theta + x_2x_1(\theta^2 + 1) + x_2(\theta^2 + \theta + 1) \\ &\quad + x_3x_2(\theta^2 + \theta) + x_3x_1\theta^2 + x_3x_2\theta + x_3, \\ &= (x_2 + x_3x_2 + x_3x_1)\theta^2 + (x_1 + x_3x_1 + x_2)\theta + x_1 + x_2x_1 + x_2 + x_3. \end{aligned}$$

Similarly, any term of the form $X^{2^i+2^j}$ with $i, j \geq 0$ will give rise to terms that have degree at most two in the variables x_i. □

More generally, we can extend the arguments used in Examples 9.5 and 9.6 to the case of the field $\mathbb{F}_q$ to show that choosing a univariate polynomial over $\mathbb{F}_{q^n}$ of the form

$$\sum_{i=0}^{n-1}\sum_{j=0}^{n-1} a_{ij}X^{q^i+q^j} + \sum_{i=0}^{n-1} b_iX^{q^i} + c$$

ensures that, when we translate this polynomial into a system of multivariate polynomials over $\mathbb{F}_q$, the degree of these polynomials is at most two.

Cryptosystem 9.4: *Hidden Field Equations*

Let $\mathbb{F}_q$ be a finite field and let $n > 0$ be an integer. The private key consists of invertible affine transformations R and S, together with a univariate polynomial $f(X)$ with coefficients from $\mathbb{F}_{q^n}$ that has the form

$$f(X) = \sum_{i=0}^{n-1}\sum_{j=0}^{n-1} a_{ij}X^{q^i+q^j} + \sum_{i=0}^{n-1} b_i X^{q^i} + c.$$

Let $\mathbf{x} = (x_1, x_2, \ldots, x_n)$ and, for i from 1 to n, let $g_i(\mathbf{x})$ be the polynomials obtained by representing $f(X)$ as a system of n quadratic polynomials in n variables over $\mathbb{F}_q$. The public key consists of the system of n quadratic polynomials in n variables over $\mathbb{F}_q$ that are given by

$$\begin{pmatrix} g_1^{\mathsf{pub}}(\mathbf{x}) \\ g_2^{\mathsf{pub}}(\mathbf{x}) \\ \vdots \\ g_n^{\mathsf{pub}}(\mathbf{x}) \end{pmatrix} = R \begin{pmatrix} g_1(S(\mathbf{x})) \\ g_2(S(\mathbf{x})) \\ \vdots \\ g_n(S(\mathbf{x})) \end{pmatrix}.$$

The plaintexts are elements of $(\mathbb{F}_q)^n$, and a plaintext $\mathbf{a} = (a_1, a_2, \ldots, a_n)$ is encrypted by computing $\mathbf{y} = (g_1^{\mathsf{pub}}(\mathbf{a}), g_2^{\mathsf{pub}}(\mathbf{a}), \ldots, g_n^{\mathsf{pub}}(\mathbf{a}))$.

A ciphertext $\mathbf{y}$ is decrypted through the following steps:

1. Compute $\mathbf{y}' = R^{-1}(\mathbf{y})$.

2. Find all solutions $\mathbf{z} \in \mathbb{F}_{q^n}$ to the equation $f(X) = \mathbf{y}'$ (here $\mathbf{y}'$ is interpreted to be an element of $\mathbb{F}_{q^n}$).

3. Calculate $S^{-1}(\mathbf{z})$ for all solutions found in the previous step. One of these solutions is the desired plaintext, $\mathbf{a}$. (By using some redundancy when representing plaintexts as elements of $(\mathbb{F}_q)^n$, it is possible to ensure that the correct solution can be identified at this point.)

Now that we have a way to construct a system of equations for which we know how to find solutions, it is necessary to "hide" the fact that they were obtained from a polynomial over an extension field in this way. This is done by changing variables through the use of affine transformations that map an n-tuple $\mathbf{x} = (x_1, x_2, \ldots, x_n)$ to the element $M\mathbf{x}^T + \mathbf{v}^T$, where M is an invertible $n \times n$ matrix with entries from $\mathbb{F}_q$, the vector $\mathbf{v} \in (\mathbb{F}_q)^n$ and the superscript "T" denotes "transpose." A full description of *HFE* is presented as Cryptosystem 9.4. We now give a toy example to show how encryption and decryption work.

Example 9.7 Suppose we continue to work with the extension $\mathbb{F}_8$ of the field $\mathbb{F}_2$ constructed in Example 7.7. Take $f(X) = X^3 \in \mathbb{F}_8[X]$ and define S and R to be the following linear transformations:

$$S : \begin{pmatrix} x_1 \\ x_2 \\ x_3 \end{pmatrix} \mapsto \begin{pmatrix} 0 & 1 & 0 \\ 1 & 0 & 0 \\ 0 & 0 & 1 \end{pmatrix} \begin{pmatrix} x_1 \\ x_2 \\ x_3 \end{pmatrix} + \begin{pmatrix} 1 \\ 0 \\ 1 \end{pmatrix} = \begin{pmatrix} x_2 + 1 \\ x_1 \\ x_3 + 1 \end{pmatrix}$$

$$R : \begin{pmatrix} x_1 \\ x_2 \\ x_3 \end{pmatrix} \mapsto \begin{pmatrix} 0 & 1 & 0 \\ 0 & 0 & 1 \\ 1 & 0 & 0 \end{pmatrix} \begin{pmatrix} x_1 \\ x_2 \\ x_3 \end{pmatrix} + \begin{pmatrix} 1 \\ 0 \\ 0 \end{pmatrix} = \begin{pmatrix} x_2 + 1 \\ x_3 \\ x_1 \end{pmatrix}.$$

We then have

$$\begin{aligned} g_1(\mathbf{x}) &= x_2x_3 + x_1, \\ g_2(\mathbf{x}) &= x_1x_3 + x_2x_3 + x_2, \\ g_3(\mathbf{x}) &= x_1x_2 + x_1 + x_2 + x_3. \end{aligned}$$

It then follows that

$$g_1(S(\mathbf{x})) = x_1(x_3 + 1) + (x_2 + 1) = x_1x_3 + x_1 + x_2 + 1,$$

and, similarly, we can determine that

$$g_2(S(\mathbf{x})) = x_1x_3 + x_2x_3 + x_1 + x_3 + 1$$

and

$$g_3(S(\mathbf{x})) = x_1x_2 + x_2 + x_3.$$

Thus we have

$$\begin{aligned} g_1^{\mathsf{pub}}(\mathbf{x}) &= x_1x_3 + x_2x_3 + x_1 + x_3, \\ g_2^{\mathsf{pub}}(\mathbf{x}) &= x_1x_2 + x_2 + x_3, \\ g_3^{\mathsf{pub}}(\mathbf{x}) &= x_1x_3 + x_1 + x_2 + 1. \end{aligned}$$

We can encrypt the plaintext $\mathbf{a} = (1, 1, 0)$ by evaluating g_1^{pub}, g_2^{pub}, and g_3^{pub} at $\mathbf{a}$, which results in the ciphertext $\mathbf{y} = (1, 0, 1)$.

Now the inverse of the transformation R sends $(x_1, x_2, x_3)^T$ to the value $(x_3, x_1 + 1, x_2)^T$. Hence $R^{-1}(\mathbf{y}^T) = (1, 0, 0)^T$, which corresponds to the element θ^2 in $\mathbb{F}_8$.

The next step in decryption is to find the solutions to the equation $X^3 = \theta^2$; over $\mathbb{F}_8$ there is a unique solution $X = \theta + 1$, which corresponds to $(0, 1, 1)$. The inverse of S sends $(x_1, x_2, x_3)^T$ to $(x_2, x_1 + 1, x_3 + 1)^T$. Hence $S^{-1}((0, 1, 1)^T) = (1, 1, 0)^T$, and we have recovered our original plaintext. □

Experiments involving ***Gröbner basis algorithms*** (a class of algorithms that include some of the fastest known techniques for solving general instances of the

MQ problem) indicate that solving systems of equations arising from *HFE* may be significantly easier than solving randomly generated systems of equations. This suggests that the use of affine transformations is not entirely effective in hiding the specialized structure of these systems of equations. Many variations on *HFE* have been proposed in order to strengthen the scheme, including omitting some of the multivariate equations from the system or adding random quadratic equations to the equations in the system. Although *HFE* itself is no longer regarded as secure for any practical parameter sizes, the underlying ideas continue to inspire the design of new multivariate cryptosystems.

9.4.2 The Oil and Vinegar Signature Scheme

The *Oil and Vinegar Signature Scheme*, proposed by Jacques Patarin in 1997, is an example of a multivariate signature scheme. As in the case of *HFE* (which was discussed in Section 9.4.1), it involves a system of multivariate quadratic equations that is easy to solve, disguised by the use of an affine transformation. In this case, the initial system consists of n polynomial equations in $2n$ variables $x_1, x_2, \dots, x_{2n}$ over a finite field $\mathbb{F}_q$. The first n variables $x_1, x_2, \dots, x_n$ are referred to as the ***vinegar variables*** and the remaining variables $x_{n+1}, x_{n+2}, \dots, x_{2n}$ are the ***oil variables***. These names reflect the fact that, when oil and vinegar are combined to make a salad dressing, they are initially separated into distinct layers, and then they are shaken up to mix them. For this scheme, the oil and vinegar variables are "separated" in the quadratic polynomials used in the signing key, but are "mixed" by the application of an affine transformation in order to construct the verification key.

Specifically, the n quadratic polynomials that make up the signing key for the *Oil and Vinegar Signature Scheme* have the form

$$f_k(x_1, x_2, \dots, x_{2n}) = \sum_{i=1}^{2n} \sum_{j=1}^{n} a_{ij}^k x_i x_j + \sum_{i=1}^{2n} b_i^k x_i + c^k \tag{9.6}$$

for all k with $1 \leq k \leq n$. What is special about these equations is that there are terms involving the product of two vinegar variables, or one vinegar variable and one oil variable, but there are no terms involving the product of two oil variables. Given a vector $(m_1, m_2, \dots, m_n) \in (\mathbb{F}_q)^n$, we can easily exploit this structure to find a solution to the following system of multivariate quadratic equations:

$$\begin{aligned} f_1(x_1, x_2, \dots, x_{2n}) &= m_1, \\ f_2(x_1, x_2, \dots, x_{2n}) &= m_2, \\ &\vdots \\ f_n(x_1, x_2, \dots, x_{2n}) &= m_n. \end{aligned} \tag{9.7}$$

To do this, we choose random values $v_1, v_2, \dots, v_n \in \mathbb{F}_q$ for the vinegar variables. When these values are substituted in (9.7), we are left with a system of n linear equations in the n oil variables, which we can then solve to find a solution

$(v_1, v_2, \ldots, v_n, o_1, o_2, \ldots, o_n)$ to (9.7). In the case where the system of linear equations has no solutions, we try new values for the vinegar variables until we find a system that does have solutions.

In order to disguise the special nature of the polynomial (9.6), we "mix" the oil and vinegar variables with the use of an affine transformation $S : (\mathbb{F}_q)^{2n} \rightarrow (\mathbb{F}_q)^{2n}$ defined by

$$S(x_1, x_2, \ldots, x_{2n}) = (x_1, x_2, \ldots, x_{2n})M + (r_1, r_2, \ldots, r_{2n}), \tag{9.8}$$

where M is a $2n \times 2n$ invertible matrix over $\mathbb{F}_q$ and $(r_1, r_2, \ldots, r_{2n}) \in (\mathbb{F}_q)^{2n}$ is a random vector. This will allow us to define public verification polynomials that appear to be more complicated than the private signing polynomials that are used to compute the signature in the first place. Note that the inverse transformation to (9.8) is simply

$$S^{-1}(y_1, y_2, \ldots, y_{2n}) = ((y_1, y_2, \ldots, y_{2n}) - (r_1, r_2, \ldots, r_{2n}))M^{-1}. \tag{9.9}$$

The *Oil and Vinegar Signature Scheme* is outlined as Cryptosystem 9.5. A signature on a message $(m_1, m_2, \ldots, m_n) \in (\mathbb{F}_q)^n$ is a vector $(s_1, s_2, \ldots, s_{2n}) \in (\mathbb{F}_q)^{2n}$ such that

$$f_k^{\mathsf{pub}}(s_1, s_2, \ldots, s_{2n}) = m_k \tag{9.10}$$

where

$$f_k^{\mathsf{pub}}(x_1, x_2, \ldots, x_{2n}) = f_k(S(x_1, x_2, \ldots, x_{2n})) \tag{9.11}$$

for all k with $1 \leq k \leq n$.[3] Verifying a signature $(s_1, s_2, \ldots, s_{2n})$ on a message $(m_1, m_2, \ldots, m_n)$ simply requires evaluating the public polynomials at the signature value to determine whether (9.10) holds as required. However, forging a signature on a message $(m_1, m_2, \ldots, m_n)$ without knowledge of the public key requires solving the system (9.10) of n multivariate quadratic equations in $2n$ variables. The security of this scheme relies on the hope that the affine transformation S can disguise the structure of the signing equations. The system (9.10) should look like a general system of multivariate quadratic equations that is presumably difficult to solve.

Signing a message $(m_1, m_2, \ldots, m_n)$ can be carried out efficiently as follows:

1. Find a solution $(v_1, v_2, \ldots, v_n, o_1, o_2, \ldots, o_n)$ to the system of equations (9.7).

2. Apply the inverse of the transformation S to this solution (as specified in (9.9)), giving $(s_1, s_2, \ldots, s_{2n}) = S^{-1}(v_1, v_2, \ldots, v_n, o_1, o_2, \ldots, o_n)$.

Note that the chosen structure of the signing polynomials means that both of these steps can be carried out efficiently using just linear algebra. This makes signing fast, which is a nice feature of this scheme.

[3] As usual, we would probably sign a message digest, rather than a message. But this is not important for the discussion of this scheme, as well as other schemes described in the following sections.

Cryptosystem 9.5: *Oil and Vinegar Signature Scheme*

Let $\mathbb{F}_q$ be a finite field and let $n > 0$ be an integer. The signing key consists of an invertible affine transformation $S : (\mathbb{F}_q)^{2n} \rightarrow (\mathbb{F}_q)^{2n}$, together with a system of n quadratic polynomials in $2n$ variables over $\mathbb{F}_q$, each of the form

$$f_k(x_1, x_2, \ldots, x_{2n}) = \sum_{i=1}^{2n}\sum_{j=1}^{n} a_{ij}^k x_i x_j + \sum_{i=1}^{2n} b_i^k x_i + c^k,$$

for all k with $1 \leq k \leq n$, where a_{ij}^k, b_i^k, c^k are drawn randomly from $\mathbb{F}_q$ for all i with $1 \leq i \leq 2n$ and all j with $1 \leq j \leq n$.

The public verification key is the system of n quadratic functions in $2n$ variables, namely $f_k^{\mathsf{pub}}(x_1, x_2, \ldots, x_{2n})$ for $1 \leq k \leq n$, which are defined by the formulas

$$f_k^{\mathsf{pub}}(x_1, x_2, \ldots, x_{2n}) = f_k(S(x_1, x_2, \ldots, x_{2n})),$$

for all k with $1 \leq k \leq n$.

Messages are elements of $(\mathbb{F}_q)^n$. A message $(m_1, \ldots, m_n)$ is signed by first finding a solution $(v_1, v_2, \ldots, v_n, o_1, o_2, \ldots, o_n)$ to the system of equations $f_k(x_1, x_2, \ldots, x_{2n}) = m_k$ for all k with $1 \leq k \leq n$. The inverse transformation S^{-1} is then applied to obtain the signature

$$(s_1, s_2, \ldots, s_{2n}) = S^{-1}(v_1, v_2, \ldots, v_n, o_1, o_2, \ldots, o_n).$$

Verification of a signature $(s_1, s_2, \ldots, s_{2n})$ on a message $(m_1, m_2, \ldots, m_n)$ consists of checking that the relationship $f_k^{\mathsf{pub}}(s_1, s_2, \ldots, s_{2n}) = m_k$ holds for all k with $1 \leq k \leq n$.

We can check that the resulting signature is valid by observing that

$$\begin{aligned}
f_k^{\mathsf{pub}}(s_1, s_2, \ldots, s_{2n}) &= f_k^{\mathsf{pub}}(S^{-1}(v_1, v_2, \ldots, v_n, o_1, o_2, \ldots, o_n)) \\
&= f_k(S(S^{-1}(v_1, v_2, \ldots, v_n, o_1, o_2, \ldots, o_n))) \\
&= f_k(v_1, v_2, \ldots, v_n, o_1, o_2, \ldots, o_n), \\
&= m_k,
\end{aligned}$$

as required, for all k with $1 \leq k \leq n$.

Example 9.8 Let f_1 and f_2 be polynomials in four variables over $\mathbb{F}_2$ given by

$$\begin{aligned}
f_1(x_1, x_2, x_3, x_4) &= x_1x_2 + x_2x_3 + x_4, \\
f_2(x_1, x_2, x_3, x_4) &= x_1x_3 + x_2x_4 + x_2.
\end{aligned}$$

Let S be the affine transformation that maps

$$(x_1, x_2, x_3, x_4) \mapsto (x_2 + x_4 + 1, x_1 + x_4 + 1, x_2 + x_3 + x_4, x_1 + x_2 + x_3 + x_4).$$

Using (9.9). it can be shown that S^{-1} is the transformation that maps

$$(y_1, y_2, y_3, y_4) \mapsto (y_3 + y_4, y_1 + y_2 + y_3 + y_4, y_1 + y_3 + 1, y_2 + y_3 + y_4 + 1).$$

Applying (9.11), the polynomials f_1^{pub} and f_2^{pub} are given by

$$\begin{aligned} f_1^{\text{pub}}(x_1, x_2, x_3, x_4) &= x_1x_3 + x_3x_4 + x_2 + 1 \quad \text{and} \\ f_2^{\text{pub}}(x_1, x_2, x_3, x_4) &= x_1x_2 + x_1x_3 + x_2x_3 + x_2x_4 + x_1 + x_2 + x_4 + 1. \end{aligned}$$

Suppose we wish to sign the message $(0, 1)$. This requires solving the system of equations $f_1(x_1, x_2, x_3, x_4) = 0$ and $f_2(x_1, x_2, x_3, x_4) = 1$. To do this, we guess values for the vinegar variables, say $x_1 = 0$ and $x_2 = 1$. Then the equations we need to solve become

$$\begin{aligned} f_1(0, 1, x_3, x_4) &= x_3 + x_4 = 0, \\ f_2(0, 1, x_3, x_4) &= x_4 + 1 = 1. \end{aligned}$$

This system has the unique solution $x_3 = x_4 = 0$, and hence we conclude that $(0, 1, 0, 0)$ is a solution to our original system of equations. The required signature is then given by $S^{-1}(0, 1, 0, 0) = (0, 1, 1, 0)$.

To verify that $(0, 1, 1, 0)$ is a valid signature for the message $(0, 1)$, we simply evaluate $f_1^{\text{pub}}(0, 1, 1, 0)$ and $f_2^{\text{pub}}(0, 1, 1, 0)$, which gives the results 0 and 1 respectively. □

As in the case of *HFE*, the *Oil and Vinegar Signature Scheme* has been broken, as the affine transformation does not adequately hide the very structured nature of the original system of equations. Suggested approaches to improving the security of this scheme have included increasing the number of vinegar variables (the so-called *Unbalanced Oil and Vinegar Signature Scheme*). One set of parameters, proposed by Kipnis, Patarin, and Goubin in 2009, uses the field $\mathbb{F}_2$ with 64 oil variables and 128 vinegar variables.

9.5 Hash-based Signature Schemes

In this section, we describe some nice techniques to construct signature schemes based only on hash functions (or possibly even one-way functions). Thus these signature schemes are of considerable interest in the setting of post-quantum cryptography.

Cryptosystem 9.6: *Lamport Signature Scheme*

Let k be a positive integer and let $\mathcal{P} = \{0,1\}^k$. Suppose $f : Y \to Z$ is a one-way function (in practice, the function f would probably be a secure hash function). Let $\mathcal{A} = Y^k$. Let $y_{i,j} \in Y$ be chosen at random, $1 \leq i \leq k$, $j = 0, 1$, and let $z_{i,j} = f(y_{i,j})$, $1 \leq i \leq k$, $j = 0, 1$. The key K consists of the $2k$ y's and the $2k$ z's. The y's are the private key while the z's are the public key.

For $K = (y_{i,j}, z_{i,j} : 1 \leq i \leq k, j = 0, 1)$, define

$$\mathbf{sig}_K(x_1, \ldots, x_k) = (y_{1,x_1}, \ldots, y_{k,x_k}).$$

A signature $(a_1, \ldots, a_k)$ on the message $(x_1, \ldots, x_k)$ is verified as follows:

$$\mathbf{ver}_K((x_1, \ldots, x_k), (a_1, \ldots, a_k)) = \text{true} \Leftrightarrow f(a_i) = z_{i,x_i}, 1 \leq i \leq k.$$

9.5.1 Lamport Signature Scheme

First, we discuss a conceptually simple way to construct a provably secure one-time signature scheme from a one-way function. (A signature scheme is a ***one-time signature scheme*** if it is secure when only one message is signed. The signature can be verified an arbitrary number of times, of course.) The description of the scheme, which is known as the *Lamport Signature Scheme*, is given in Cryptosystem 9.6. This scheme was published in 1979, so it is one of the earliest examples of a signature scheme.

Informally, this is how the system works. A message to be signed is a binary k-tuple. In order to not have to worry about the length of the message, we assume that the value of k is fixed ahead of time.

Each bit of the message is signed individually. If the ith bit of the message equals j (where $j \in \{0,1\}$), then the ith element of the signature is the value $y_{i,j}$, which is a preimage of the public key value $z_{i,j}$. The verification consists simply of checking that each element in the signature is a preimage of the public key element $z_{i,j}$ that corresponds to the ith bit of the message. This can be done using the public function f.

We illustrate the scheme by considering one possible implementation using the exponentiation function $f(x) = \alpha^x \bmod p$, where α is a primitive element modulo p. Here $f : \{0, \ldots, p-2\} \to \mathbb{Z}_p{}^*$. We present a toy example to demonstrate the computations that take place in the scheme.

Example 9.9 7879 is prime and 3 is a primitive element in $\mathbb{Z}_{7879}{}^*$. Define

$$f(x) = 3^x \bmod 7879.$$

Suppose $k = 3$, and Alice chooses the six (secret) random numbers

$$\begin{aligned} y_{1,0} &= 5831 \\ y_{1,1} &= 735 \\ y_{2,0} &= 803 \\ y_{2,1} &= 2467 \\ y_{3,0} &= 4285 \\ y_{3,1} &= 6449. \end{aligned}$$

Then Alice computes the images of these six y's under the function f:

$$\begin{aligned} z_{1,0} &= 2009 \\ z_{1,1} &= 3810 \\ z_{2,0} &= 4672 \\ z_{2,1} &= 4721 \\ z_{3,0} &= 268 \\ z_{3,1} &= 5731. \end{aligned}$$

These z's are published. Now, suppose Alice wants to sign the message

$$x = (1,1,0).$$

The signature for x is

$$(y_{1,1}, y_{2,1}, y_{3,0}) = (735, 2467, 4285).$$

To verify this signature, it suffices to compute the following:

$$\begin{aligned} 3^{735} \bmod 7879 &= 3810 \\ 3^{2467} \bmod 7879 &= 4721 \\ 3^{4285} \bmod 7879 &= 268. \end{aligned}$$

Hence, the signature is verified. □

We argue that, if Oscar sees one message and its signature, then he will be unable to forge a signature on a second message. Suppose that $(x_1, \dots, x_k)$ is a message and $(y_{1,x_1}, \dots, y_{k,x_k})$ is its signature. Now suppose Oscar tries to sign the new message $(x'_1, \dots, x'_k)$. Since this message is different from the first message, there is at least one co-ordinate i such that $x'_i \neq x_i$. Signing the new message requires computing a value a such that $f(a) = z_{i,x'_i}$. Since Oscar has not seen a preimage of z_{i,x'_i}, and f is a one-way function, he is unable to find a value a which would could be used in a valid signature for $(x'_1, \dots, x'_k)$.

However, this signature scheme can be used to sign only one message securely. Given signatures on two different messages, it is an easy matter for Oscar to construct signatures for another message different from the first two (unless the first two messages differ in exactly one bit).

For example, suppose the messages $(0,1,1)$ and $(1,0,1)$ are both signed using the same key. The message $(0,1,1)$ has as its signature the triple $(y_{1,0}, y_{2,1}, y_{3,1})$, and the message $(1,0,1)$ is signed with $(y_{1,1}, y_{2,0}, y_{3,1})$. Given these two signatures, Oscar can manufacture signatures for the messages $(1,1,1)$ (namely, $(y_{1,1}, y_{2,1}, y_{3,1})$) and $(0,0,1)$ (namely, $(y_{1,0}, y_{2,0}, y_{3,1})$).

9.5.2 Winternitz Signature Scheme

The *Lamport Signature Scheme*, as described in Section 9.5.1, has a very large key size. To sign a k-bit message, we require a public key consisting of $2k$ values $z_{i,j}$ from the set $\mathcal{Z}$. Since these z-values are probably outputs of a secure hash function, they would each be least 224 bits in length (for example, if we used the hash function *SHA3-224*). The *Winternitz Signature Scheme* provides a significant reduction in key size, by allowing multiple bits to be signed by each application of the one-way function f.

We first present the basic idea, which, however, is not secure. Then we describe how to fix the security problem.

We will sign w bits at a time, where w is a pre-specified parameter. Suppose f is a secure hash function. To illustrate the basic idea, let's fix $w = 3$ for the time being. Suppose a random value y_0 is chosen and we compute the ***hash chain***

$$y^0 \to y^1 \to y^2 \to y^3 \to y^4 \to y^5 \to y^6 \to y^7 \to z$$

according to the rules $y^j = f(y^{j-1})$ for $1 \leq j \leq 7$, and $z = f(y_7)$. We can equivalently define $y^j = f^j(y_0)$ for $1 \leq j \leq 7$, and $z = f^8(y_0)$, where f^j denotes j applications of the function f. The value z would be the public key for this hash chain. In general, the hash chain would consist of $2^w + 1$ values, namely, $y^0, \ldots, y^{2^w-1}, z$.

For a k-bit message, we would construct $\ell = k/w$ hash chains (let's assume that k is a multiple of w, for convenience). Denote the initial values in these hash chains by $y_1^0, y_2^0, \ldots, y_\ell^0$. These initial values comprise the private key.

Now consider a message $(x_1, \ldots, x_\ell)$, where each x_i is a binary w-tuple. Thus we can view each x_i as an integer between 0 and $2^w - 1$ (inclusive). As a first attempt at a signature, consider releasing the values $a_i = y_i^{x_i} = f^{x_i}(y_i)$ for $i = 1, \ldots, \ell$ as a signature. (Note that we do not need to store the entire hash chains; we can compute the a_i's, as needed, from the initial values.) Then, to verify a given a_i, it suffices to check that $f^{2^w - x_i}(a_i) = z_i$.

Example 9.10 Suppose that $k = 9$ (and hence $\ell = 3$). There are three hash chains:

$$y_1^0 \to y_1^1 \to y_1^2 \to y_1^3 \to y_1^4 \to y_1^5 \to y_1^6 \to y_1^7 \to z_1$$

$$y_2^0 \to y_2^1 \to y_2^2 \to y_2^3 \to y_2^4 \to y_2^5 \to y_2^6 \to y_2^7 \to z_2$$

$$y_3^0 \to y_3^1 \to y_3^2 \to y_3^3 \to y_3^4 \to y_3^5 \to y_3^6 \to y_3^7 \to z_3.$$

Therefore, the public key is (z_1, z_2, z_3). Now suppose we want to sign the message 011101001. We have $x_1 = 011 = 3$, $x_2 = 101 = 5$, and $x_3 = 001 = 1$. So we release the values $a_1 = y_1^3$, $a_2 = y_2^5$, and $a_3 = y_3^1$:

$$y_1^0 \to y_1^1 \to y_1^2 \to \boxed{y_1^3} \to y_1^4 \to y_1^5 \to y_1^6 \to y_1^7 \to z_1$$

$$y_2^0 \to y_2^1 \to y_2^2 \to y_2^3 \to y_2^4 \to \boxed{y_2^5} \to y_2^6 \to y_2^7 \to z_2$$

$$y_3^0 \to \boxed{y_3^1} \to y_3^2 \to y_3^3 \to y_3^4 \to y_3^5 \to y_3^6 \to y_3^7 \to z_3.$$

The verification requires checking that

$$\begin{aligned} f^5(a_1) &= z_1, \\ f^3(a_2) &= z_2, \quad \text{and} \\ f^7(a_3) &= z_3. \end{aligned}$$

□

The above process is quite ingenious, but it is not secure. Let us look at Example 9.10 to see what the problem is. Consider the signature (a_1, a_2, a_3) given in this example. The released values are just elements in the three hash chains, and once an element in a hash chain is known, anyone can compute any later values in the hash chains, as desired. So, for example, Oscar could compute

$$\begin{aligned} y_1^5 &= f^2(a_1), \\ y_2^6 &= f(a_2), \quad \text{and} \\ y_3^4 &= f^3(a_3). \end{aligned}$$

Therefore, Oscar can now create the signature (y_1^5, y_2^6, y_3^4) for the message 101110100.

Fortunately, a small tweak will yield a secure signature scheme. The fix is to include a ***checksum*** in the message, and also sign the checksum. The checksum is defined to be

$$C = \sum_{i=1}^{\ell} (2^w - 1 - x_i).$$

In Example 9.10, we would have

$$C = (7-3) + (7-5) + (7-1) = 4 + 2 + 6 = 12.$$

In binary, we have $C = 1100$. After padding on the left with two zeroes, we can break C into two chunks of three bits: $x_4 = 001$ and $x_5 = 100$. Now we create two additional hash chains and use them to sign x_4 and x_5, releasing the values $a_4 = y_4^1 = f(y_4)$ and $a_5 = y_5^4 = f^4(y_5)$. These two hash chains have public keys z_4 and z_5, respectively.

Pictorially, we have

$$y_4^0 \to \boxed{y_4^1} \to y_4^2 \to y_4^3 \to y_4^4 \to y_4^5 \to y_4^6 \to y_4^7 \to z_4$$

$$y_5^0 \to y_5^1 \to y_5^2 \to y_5^3 \to \boxed{y_5^4} \to y_5^5 \to y_5^6 \to y_5^7 \to z_5.$$

So the entire signature on the message (x_1, x_2, x_3) is $(a_1, a_2, a_3, a_4, a_5)$. To verify this signature, the following steps are performed:

1. Verify that (a_1, a_2, a_3) is the correct signature for (x_1, x_2, x_3).
2. Form the checksum and create (x_4, x_5).
3. Verify that (a_4, a_5) is the correct signature for (x_4, x_5).

We now argue informally that the signature scheme is secure when we include a checksum as described above. Suppose that Oscar sees a message (x_1, x_2, x_3) and its signature $(a_1, a_2, a_3, a_4, a_5)$ (where a_4 and a_5 comprise the signature on the checksum). Oscar then wants to create a signature on a second message (x_1', x_2', x_3'). Since Oscar can only move "forward" in the hash chains, it must be the case that $x_i' \geq x_i$ for $i = 1, 2, 3$. Also, because $(x_1', x_2', x_3') \neq (x_1, x_2, x_3)$, it follows that $x_{i_0}' > x_{i_0}$ for some i_0. From this, we see that $C' < C$, where C' is the checksum for (x_1', x_2', x_3'). This means that $x_4' < x_4$ or $x_5' < x_5$ (or both). For purposes of illustration, suppose that $x_4' < x_4$. Then Oscar cannot sign x_4' because he would need to move "backwards" in the hash chain, which is not possible due to the one-way property of f. A similar contradiction arises if $x_5' < x_5$.

In general, the checksum C will satisfy the inequality $0 \leq C \leq \ell(2^w - 1)$. The number of bits in the binary representation of C is at most $w + \log_2 \ell$. Let B denote the number of w-bit blocks that are required to store C. Then

$$B \leq 1 + \left\lceil \frac{\log_2 \ell}{w} \right\rceil .$$

So we will have ℓ message blocks and B checksum blocks, giving rise to a total of $\ell + B$ hash chains.

Note that the value of w should be chosen carefully. As w increases, the number of hash chains decreases. However, the time to create and verify signatures increases, because we have to traverse longer hash chains.

There is one other improvement that can be made to shrink the size of the public key. Instead of using the tuple $(z_1, z_2, z_3, z_4, z_5)$ as the public key, we concatenate these values and pass them through as secure cryptographic hash function, say h. Thus, the public key is defined to be $\mathbf{z} = h(z_1 \parallel z_2 \parallel z_3 \parallel z_4 \parallel z_5)$. The verification of the signature would comprise the following steps:

1. compute the ends of all the hash chains,
2. concatenate the results,
3. apply the hash function h,
4. compare the output of h to the public key $\mathbf{z}$.

We summarize by giving a description of the *Winternitz Signature Scheme* for arbitrary values of w in Cryptosystem 9.7.

Cryptosystem 9.7: *Winternitz Signature Scheme*

Let k and w be positive integers, where $\ell = k/w$ is an integer, and define

$$B = 1 + \left\lceil \frac{\log_2 \ell}{w} \right\rceil.$$

Suppose $f : Y \to Y$ is a one-way function and suppose $h : Y \to Z$ is a secure hash function.

Construct $k + B$ hash chains using f, each of length 2^w. The starting points of the hash chains (which comprise the private key) are y_i^0 and the ending points are z_i, for $1 \leq i \leq k + B$. The public key is

$$\mathbf{z} = h(z_1 \parallel z_2 \parallel \cdots \parallel z_{\ell+B}).$$

Let $\mathcal{P} = (\{0,1\}^w)^\ell$ and let $\mathcal{A} = Z$.

For a message $(x_1, \dots, x_k) \in \mathcal{P}$, the signature $\mathbf{sig}_K(x_1, \dots, x_k)$ is computed as follows:

1. Compute the checksum $C = (x_{k+1}, \dots, x_{k+B})$.
2. For $1 \leq i \leq \ell + B$, compute $a_i = f^{x_i}(y_i)$.
3. The signature $\mathbf{sig}_K(x_1, \dots, x_k) = (a_1, \dots, a_{\ell+B})$.

A signature $(a_1, \dots, a_{\ell+B})$ on the message $(x_1, \dots, x_\ell)$ is verified as follows:

1. Compute the checksum $C = (x_{k+1}, \dots, x_{k+B})$.
2. For $1 \leq i \leq \ell + B$, compute $z_i = f^{2^w - x_i}(a_i)$.
3. Check to see if $\mathbf{z} = h(z_1 \parallel z_2 \parallel \cdots \parallel z_{\ell+B})$.

9.5.3 Merkle Signature Scheme

The signature schemes in Sections 9.5.1 and 9.5.2 are one-time schemes. Merkle invented a useful method of extending a one-time scheme so it could be used for a large (but fixed) number of signatures, without increasing the size of the public key. We describe Merkle's technique in this section.

The basic idea is to create a binary tree (which is now called a ***Merkle tree***) by hashing combinations of various public keys (i.e., verification keys) of one-time signature schemes.[4] The particular one-time scheme that is used is not important.

[4]The Merkle tree will be used only to authenticate public keys; it is not used to create signatures in the component one-time signature schemes.

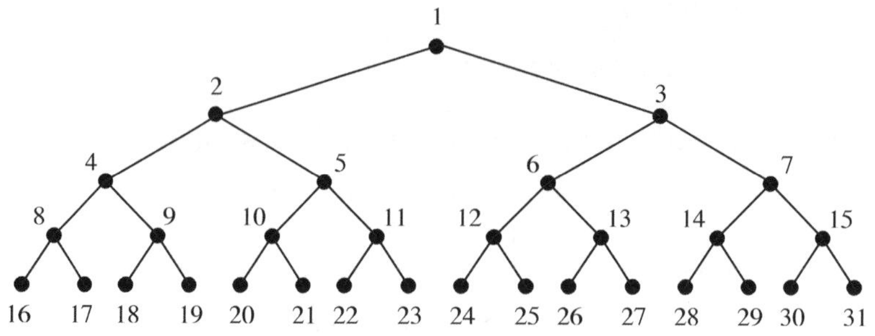

FIGURE 9.1: A binary tree with 16 leaf nodes

Let d be a prespecified positive integer, and suppose we have 2^d instances of a one-time signature scheme, with verification keys denoted by $K_1, \ldots, K_{2^d}$, respectively. It is then possible to sign a series of 2^d messages, where the signature on the ith message will be verified using K_i, for $1 \leq i \leq 2^d$.

The Merkle tree is a complete binary tree, say $\mathcal{T}$, of depth d. We will assume that the nodes of $\mathcal{T}$ are labeled as shown in Figure 9.1, so they satisfy the following properties:

1. For $0 \leq \ell \leq d$, the 2^ℓ nodes at depth ℓ are labeled (in order) $2^\ell, 2^\ell + 1, \ldots, 2^{\ell+1} - 1$.

2. For $j \neq 1$, the parent of node j is node $\lfloor \frac{j}{2} \rfloor$.

3. The left child of node j is node $2j$ and the right child of node j is node $2j + 1$, assuming that one or both of these children exist.

4. For $j \neq 1$, the sibling of node j is node $j + 1$, if j is even; or node $j - 1$, if j is odd.

Let h be a secure hash function. Each node j in $\mathcal{T}$ is assigned a value $V(j)$, according to the following rules.

1. For $2^d \leq j \leq 2^{d+1} - 1$, let $V(j) = h(K_{j-2^d+1})$.

2. For $1 \leq j \leq 2^d - 1$, let $V(j) = h(V(2j) \parallel V(2j+1))$.

Observe that all the values $V(j)$ are strings of a fixed length, namely, the length of a message digest for the hash function h.

The values stored in the 2^d leaf nodes are obtained by hashing the 2^d public keys. The value stored in a nonleaf node is computed by hashing the concatenation of the values stored in its two children. The value stored in the root node, which is $V(1)$, is the public key K for the scheme.

We now discuss how to create a signature for the ith message, say m_i. First the ith private (signing) key is used to create a signature for m_i, which we denote by s_i. This signature can be verified using the public key K_i, which must also be supplied as part of the signature. In addition, the public key must be authenticated, which is done using the Merkle tree $\mathcal{T}$. This is done by supplying enough information for the verifier to be able to recompute the value in the root, $V(1)$, and compare it to the stored value K. This necessary information consists of $V(i + 2^d - 1)$, along with the values of the siblings of all the nodes in the path in $\mathcal{T}$ from node $i + 2^d - 1$ to the root node (node 1).

Example 9.11 Suppose $d = 4$ and suppose we want to create a signature for message m_{11}. The relevant path contains nodes 26, 13, 6, 3, and 1. The siblings of the nodes on this path are nodes 27, 12, 7, and 2, so $V(27)$, $V(12)$, $V(7)$, and $V(2)$ are supplied as part of the signature. The key K_{11} would then be authenticated by performing the following computations:

1. compute $V(26) = h(K_{11})$
2. compute $V(13) = h(V(26) \parallel V(27))$
3. compute $V(6) = h(V(12) \parallel V(13))$
4. compute $V(3) = h(V(6) \parallel V(7))$
5. compute $V(1) = h(V(2) \parallel V(3))$
6. verify that $V(1) = K$.

Therefore, the entire signature consists of the list $K_{11}, s_{11}, V(27), V(12), V(7), V(2)$.
□

We now argue informally that this method of authenticating public keys is secure. The situation we need to consider is where an adversary tries to authenticate a false public key. That is, Oscar may try to convince the recipient of a signature that $K_i' \neq K_i$ is a valid key in the scheme (this would take place before K_i is actually used). For purposes of illustration, let's take $i = 11$. The adversary must also supply values $V'(27)$, $V'(12)$, $V'(7)$, and $V'(2)$. These values are all required to have the same (fixed) length as the values they are replacing.

Now, suppose we consider the "validation chain" resulting from K_{11}', namely,

$$V'(26), V'(13), V'(6), V'(3), V'(1) = K.$$

We have

$$h(V'(2) \parallel V'(3)) = V'(1) = K = V(1) = h(V(2) \parallel V(3)).$$

Since h is collision resistant, it must be the case that $V'(2) = V(2)$ and $V'(3) = V(3)$. Working backwards, we have

$$h(V'(6) \parallel V'(7)) = V'(3) = V(3) = h(V(6) \parallel V(7)),$$

so $V'(6) = V(6)$ and $V'(7) = V(7)$. Continuing in this fashion, we eventually see that

$$V'(26) = h(K_{11}') = h(K_{11}) = V(26).$$

Finally, $h(K_{11}') = h(K_{11})$ yields $K_{11}' = K_{11}$, which is a contradiction.

9.6 Notes and References

We recommend the book edited by Bernstein, Buchmann, and Dahmen [24] for an introduction to post-quantum cryptography, including many of the specific systems we discuss in this chapter. For a recent survey, see Bernstein and Lange [25].

Mosca's predictions regarding practical quantum computing are from [142].

For a good survey on the **Learning With Errors** problem, we recommend Regev [169]. Cryptosystem 9.2 is from [169]. The **Learning With Errors** problem has also been used as the basis for key agreement schemes; see Ding, Xie, and Lin [74].

McEliece proposed his code-based cryptosystem in [131] in 1978. This was in fact one of the very first public-key cryptosystems. It did not receive as much attention as cryptosystems based on the **Factoring** and **Discrete Logarithm** problems due to the large key lengths required. However, the *McEliece Cryptosystem* has received much more attention since the advent of post-quantum cryptography. Recommended parameters (as of 2008) for the *McEliece Cryptosystem* can be found in [23].

Hidden Field Equations was presented by Patarin in [158]. It is actually a modified version of an earlier cryptosystem due to Matsumoto and Imai [130]. Currently, the most promising cryptosystem based on these techniques is *SimpleMatrix*; see [193].

Oil and Vinegar Signature Scheme is from [159] and *Unbalanced Oil and Vinegar* was presented in [105]. A newer, more secure signature scheme using similar ideas, due to Ding and Schmidt, is known as *Rainbow* [73].

The *Lamport Signature Scheme* is described in the 1976 paper by Diffie and Hellman [71]. Merkle's tree-based scheme was published in [136], though it dates back to 1979. The *Winternitz Signature Scheme* is also of the same vintage; for a good description of it, see [75]. *XMSS* is an updated and improved variation of the *Merkle Signature Scheme,* due to Buchmann, Dahmen, and Hülsing [50].

Exercises

9.1 Compute

$$g(x) = (x^4 + 3x^3 + x^2 + 2x + 3)(2x^4 + 5x^2 + 6x + 2)$$

in $\mathbb{Z}[x]/(x^5 - 1)$. Compute $(1,3,1,2,3) \star (2,0,5,6,2)$ and confirm that the entries in the resulting vector correspond to the coefficients of g.

9.2 Let $f(x) = 3x^5 + 3x + 1 \in \mathbb{Z}_{11}[x]$.

(a) Find polynomials $a(x)$ and $b(x)$ in $\mathbb{Z}_{11}[x]$ such that

$$a(x)f(x) + b(x)(x^7 - 1) = 1.$$

(b) Determine the inverse of f in $\mathbb{Z}_{11}[x]/(x^7 - 1)$ mods 11.

9.3 Let $\{\mathbf{u}, \mathbf{v}\}$ be a basis for a lattice $\mathcal{L}$ in $\mathbb{R}^2$ where $\mathbf{u} = (3,7)$ and $\mathbf{v} = (5,10)$.

(a) Find the norm of $\mathbf{u}$ and the norm of $\mathbf{v}$.

(b) Determine the shortest vectors in $\mathcal{L}$.

9.4 Let C be the $[7,4,3]$ Hamming code with generating matrix

$$G = \begin{pmatrix} 1 & 0 & 0 & 0 & 1 & 1 & 0 \\ 0 & 1 & 0 & 0 & 1 & 0 & 1 \\ 0 & 0 & 1 & 0 & 0 & 1 & 1 \\ 0 & 0 & 0 & 1 & 1 & 1 & 1 \end{pmatrix}.$$

Find a parity-check matrix for C having the form $(Y \mid I_3)$, and use syndrome decoding to decode the received vector $(1,0,1,0,0,0,0)$.

9.5 Consider Cryptosystem 9.2 with the parameters $n = 3$ and $q = 17$, and public key consisting of the samples

$$((6,4,3),0), ((1,5,8),12) \text{ and } ((7,11,2),4).$$

(a) Encrypt the plaintext bit 1 using each of the seven possible nonempty choices of a random subset S.

(b) Given that the private key is $(5,0,1)$, determine whether the decryption algorithm succeeds for each of the seven ciphertexts computed above.

9.6 Suppose the elements of $\mathbb{F}_8$ are written in the form $a\theta^2 + b\theta + c$, where $a, b, c \in \mathbb{F}_2$ and where θ satisfies $\theta^3 + \theta + 1 = 0$. Express the equation $X^6 + 1 = 0$ over $\mathbb{F}_8$ as a system of multivariate polynomial equations over $\mathbb{F}_2$.

9.7 The ciphertext $(0,1,0)$ was obtained by performing HFE encryption with the parameters and public keys given in Example 9.7. Determine the corresponding plaintext.

9.8 Using exhaustive search, determine all solutions to the following system of equations over $\mathbb{F}_2$:

$$\begin{aligned} x_1x_2 + x_2x_3 + x_3 &= 0 \\ x_1x_3 + x_1 + x_2 &= 1 \\ x_2x_3 + x_1 &= 1. \end{aligned}$$

9.9 Exploit the particular structure of the following system of equations over $\mathbb{F}_2$ in order to find a solution:

$$\begin{aligned} x_1x_2 + x_3x_4 + x_2 + x_5 &= 1 \\ x_2x_3 + x_2x_5 + x_1 &= 1 \\ x_1x_4 + x_2x_5 + x_6 &= 0. \end{aligned}$$

9.10 In the *Lamport Signature Scheme*, suppose that two k-tuples, x and x', were signed by Alice using the same key. Let ℓ denote the number of co-ordinates in which x and x' differ, i.e.,

$$\ell = |\{i : x_i \neq x'_i\}|.$$

Show that Oscar can now sign $2^\ell - 2$ new messages.

Chapter 10

Identification Schemes and Entity Authentication

In this chapter, we discuss various mechanisms that allow one user to "prove" their identity to another user. Among the techniques we describe are passwords and challenge-and-response protocols, some of which involve "zero-knowledge" techniques.

10.1 Introduction

The topic of this chapter is ***identification***, which is also known as ***entity authentication***. Roughly speaking, the goal of an ***identification scheme*** is to allow someone's identity to be confirmed. Normally this is done in "real time." In contrast, cryptographic tools such as signature schemes allow the authentication of data, which can be performed any time after the relevant message has been signed.

Suppose you want to prove your identity to someone else. It is sometimes said that this can be done in one of three ways, namely, based on what you are, what you have, or what you know. These are often referred to as ***factors***. "What you are" refers to behavioral and physical attributes; "what you have" refers to documents or credentials; and "what you know" encompasses passwords, personal information, etc.

Some examples of typical identification scenarios are described here in more detail.

physical attributes

People often identify other people already known to them by their appearance. This could include family and friends as well as famous celebrities. Specific features used for this purpose include sex, height, weight, racial origin, eye color, hair color, etc. Attributes that are unique to an individual are often more useful; these include fingerprints or retina scans. Sometimes automated identification schemes are based on biometrics such as these, and it seems likely that biometrics might frequently be used in the future.

credentials

A *credential* is defined, in the diplomatic usage of the word, as a letter of introduction. Trusted documents or cards such as driver's licences and passports function as credentials in many situations. Note that credentials often include photographs, which enable physical identification of the bearer of the credential.

knowledge

Knowledge is often used for identification when the person being identified is not in the same physical location as the person or entity performing the identification. In the context of identification, ***knowledge*** could be a password or ***PIN*** (***personal identification number***), or "your mother's maiden name" (a favorite of credit card companies). The difficulty with using knowledge for identification is that such knowledge may not be secret in the first place, and, moreover, it is usually revealed as part of the identification process. This allows for possible future impersonation of the person being identified, which is not a good thing! However, suitable cryptographic protocols will enable the construction of secure identification schemes, which will prevent these kinds of impersonation attacks.

Let's consider some everyday situations where it is common to "prove" one's identity, either in person or electronically. Some typical scenarios are as follows:

remote login

To remotely login to a computer or a website over the internet, it suffices to know a valid user name and the corresponding password. The user name is often just an email address. This is an example of "what you know."

in-store chip credit card purchases

When a purchase is made at a store using a chip-enabled credit card, the owner of the card is required to enter a personal identification number (or, PIN), which is verified by the card terminal. This is a combination of "what you have" (the card) and "what you know" (the PIN).

contactless credit card payments

For these types of payments, possession of the credit card is sufficient to allow it to be used; there is nothing at all to prevent the use of a stolen card. Here the card just needs to be in close physical proximity to the terminal reader. Communication is done using RFID. So this method is based on "what you have."

card-not-present credit card purchases

In many situations, possession of the actual credit card is not required in order to use it. It suffices to have knowledge of information that is written on the card. For example, to charge purchases made over the internet to a credit card, all that usually is necessary is a valid credit card number, the expiry date, and the CCV2 (card verification value), which is a three-digit number on the back of the card. This is another example of "what you know."

bank machine withdrawals

To withdraw money from an automated teller machine (or ATM), we use a bank card together with a four- or six-digit PIN. The card contains the owner's name and information about his or her bank accounts. The purpose of the PIN is to protect against fraudulent use of the card by someone else. The assumption is that the only person who knows the correct PIN is the owner of the card. This again is a combination of "what you have" (the card) and "what you know" (the PIN).

Some of the above-described techniques employ more than one of the three factors. However, remote login traditionally is handled by password-based methods ("what you know"). In recent years, ***two-factor authentication*** has become a popular technique to improve security of remote login. Two-factor authentication typically augments a password with a second factor, such a fob that displays a dynamically changing random number. The user who is attempting to gain access to a system must supply their password along with the current random number displayed on the fob. Thus the second factor is "what you have," since possession of the fob is required in order to know the value of the current random number. Another frequently-used second factor involves sending an access code by SMS to a mobile device belonging to the user.[1]

10.1.1 Passwords

Passwords are, by and large, the most common technique used for identification over the internet. Although, strictly speaking, they are not a cryptographic tool, it might be useful to discuss some of the methods to improve the security of password-based identification.

One of the weaknesses of passwords is that people often choose weak passwords that are easy to guess, such as *password*, *123456* and *abc123*. For this reason, many websites have requirements that are intended to force the user to choose a password that would be harder to guess. Typical rules relate to password length, inclusion of upper-case as well as lower-case letters, inclusion of numbers and/or special symbols, etc. Such measures may help to prevent ***online attacks*** from succeeding, where an online attack involves an attacker trying to guess a password in real time.

It is more challenging to protect against ***offline attacks***. An offline attack can be carried out after a data breach has occurred. In a typical data breach, the adversary gains access to a file of user ids and associated passwords. If the passwords are stored as plaintext, then the adversary can gain access to any user's account, of course. Therefore, some additional precautions must be taken if this kind of outcome is to be prevented.

Perhaps the most obvious safeguard would be to encrypt the password file. However, this approach is not usually followed due to the possibility that, if a data breach occurs, then the attacker may gain access to the decryption key as

[1]However, there are various ways for an attacker to intercept text messages and send them to a third party, rather than the intended recipient. So this might not be a secure method of two-factor authentication in every situation.

well as the password file. It is therefore recommended to hash the passwords, and only store the resulting message digests. A hashed password is often referred to as a ***fingerprint***, and we will follow this terminology for the rest of this section. Of course the hash function that is used should be a secure cryptographic hash function, which provides collision resistance, preimage resistance, and second preimage resistance (these concepts were introduced in Section 5.2).

Thus, the password file will contain user ids and corresponding fingerprints. When a user supplies a password, the system will verify it by hashing it and comparing it to the fingerprint that is stored in the password file. If this approach is adopted, then an attacker, after obtaining a copy of the password file, can guess passwords, hashing them and comparing the results with the stored fingerprints. This is the basic idea of an offline attack.

Two common types of offline attacks are ***dictionary attacks*** and ***brute force attacks***. In a dictionary attack, the adversary tries various commonly used weak passwords, as compiled in a "dictionary." In a brute force attack, all passwords of a specified length might be tested, until the correct password is found. A more sophisticated approach is to construct a ***rainbow table***, which is a type of time-memory tradeoff.

It is important to observe that the attacker can spend a large amount of time and resources carrying out an offline attack, if they wish to do so. Also, a table of common passwords and corresponding fingerprints can be precomputed by the adversary, if desired, before the data breach occurs. Then the attacker can just look for fingerprints in the password file that match precomputed fingerprints. This would allow the adversary to quickly identify many weak passwords. One final comment is that two users who have identical fingerprints almost surely have identical passwords. A weak password might be used by several different users, and all occurrences of the same weak password can be detected at once.

There is one additional technique that is frequently used to provide additional security against offline attacks. The idea is to include some randomness, called a ***salt***, before the hash function is computed. This method has been used in Unix systems since the 1970's. So, instead of computing

$$\mathit{fingerprint} = h(\mathit{userid}),$$

we instead compute

$$\mathit{fingerprint} = h(\mathit{userid} \parallel \mathit{salt}),$$

where h is a hash function. Each entry in the password file will now consist of a userid, a salt, and a fingerprint. Note that a new random salt should be used for every user. The salt may or may not be secret.[2]

For the purposes of our discussion, suppose the salt is a random bitstring of length 128. Then, even if every user has the same password, the fingerprints would be different (from the birthday paradox, we would not expect two uses of the same salt value until approximately 2^{64} salt values are generated). Since the hash

[2]Actually, Unix does not reveal salt values to users in the Unix `/etc/passwd` file. The actual salt values are in the Unix `/etc/shadow` file, which is normally accessibly only to root.

function is assumed to be collision resistant, we will not encounter two identical fingerprints, even if the corresponding passwords are identical.

The other advantage provided by the salt is that it makes it impractical to precompute a useful table of fingerprints. This is because, for each possible password, there are 2^{128} possible fingerprints, and it is infeasible to compute all of them.

On the other hand, the salt does not provide any additional security against an attacker that is trying to determine the password for a particular user. The attacker is assumed to know the salt for the user, and therefore the attacker can attempt to find the password, using a dictionary attack or a brute force attack, exactly as might be done if the passwords were not salted.

Another commonly used technique to make exhaustive password searches less efficient is that of ***key stretching***. Instead of hashing a password and salt once to create a fingerprint, a slower, more complicated process is used. For example, the password (and perhaps the salt) could be iteratively hashed 10000 times instead of just once. This slows down the computation of the fingerprint, and hence it also makes it more difficult to carry out exhaustive searches. *Argon2* is a recently proposed key stretching algorithm.

We have one final comment relating to hashing passwords as opposed to encrypting passwords. If passwords are encrypted, then it is possible to recover a lost password, because the system can decrypt the encrypted password that is stored in the password file. However, if the password file just contains fingerprints of passwords, then password recovery is not practical. Instead, the user would be required to reset their password in the event that their password is lost or compromised.

10.1.2 Secure Identification Schemes

The objective of a secure identification scheme would be that someone "listening in" as Alice identifies herself to Bob, say, should not subsequently be able to misrepresent herself as Alice. At the very least, the attack model allows the adversary to observe all the information being transmitted between Alice and Bob. The adversarial goal is to be able to impersonate Alice. Furthermore, we may even try to guard against the possibility that Bob himself might try to impersonate Alice after she has identified herself to him. Ultimately, we would like to devise "zero-knowledge" schemes whereby Alice can prove her identity electronically, without "giving away" the knowledge (or partial information about the knowledge) that is used as her identifying information.

Several practical and secure identification schemes have been discovered. One objective is to find a scheme that is simple enough that it can be implemented on a smart card, which is essentially a credit card equipped with a chip that can perform arithmetic computations. However, it is important to note that the "extra" security pertains to someone monitoring the communication line. Since it is the card that is "proving" its identity, we have no extra protection against a lost card. It would still be necessary to include a PIN in order to establish that it is the real owner of the card who is initiating the identification scheme.

Protocol 10.1: INSECURE CHALLENGE-AND-RESPONSE

1. Bob chooses a random challenge, denoted by r, which he sends to Alice.
2. Alice computes her response
$$y = MAC_K(r)$$
and she sends y to Bob.
3. Bob computes
$$y' = MAC_K(r).$$
If $y' = y$, then Bob "accepts"; otherwise, Bob "rejects."

A first observation is that any identification scheme should involve randomization in some way. If the information that Alice transmits to Bob to identify herself never changes, then the scheme is insecure in the model we introduced above. This is because the identifying information can be stored and reused by anyone observing a run of the protocol (including Bob)—this is known as a ***replay attack***. Therefore, secure identification schemes usually include "random challenges" in them. This concept is explored more deeply in the next section.

We will take two approaches to the design of identification schemes. First, we explore the idea of building secure identification schemes from simpler cryptographic primitives, namely, message authentication codes or signature schemes. Schemes of this type are developed and analyzed in Sections 10.2 and 10.3. Then, in the remaining sections of this chapter, we discuss two identification schemes that are built "from scratch." These schemes are due to Schnorr and Feige-Fiat-Shamir.

10.2 Challenge-and-Response in the Secret-key Setting

In later sections, we will describe some of the more popular zero-knowledge identification schemes. First we look at identification in the secret-key setting, where Alice and Bob both have the same secret key. We begin by examining a very simple (but insecure) scheme that can be based on any message authentication code, e.g., the MACs discussed in Chapter 5. The scheme, which is described as Protocol 10.1, is called a ***challenge-and-response protocol***. In it, we assume that Alice is identifying herself to Bob, and their common secret key is denoted by K. (Bob can also identify himself to Alice, by interchanging the roles of Alice and Bob

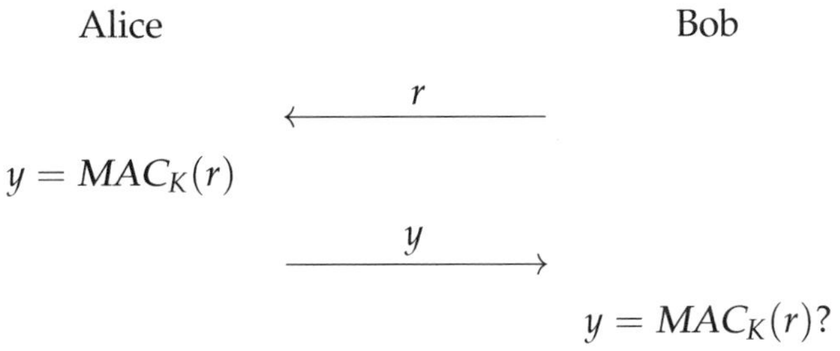

FIGURE 10.1: Information flows in Protocol 10.1

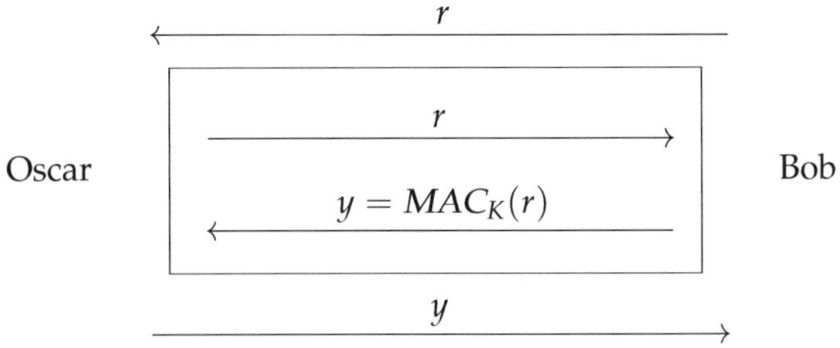

FIGURE 10.2: Attack on Protocol 10.1

in the scheme.) In this scheme, Bob sends a ***challenge*** to Alice, and then Alice sends Bob her ***response***.

We will often depict interactive protocols in diagrammatic fashion. Protocol 10.1 could be presented as shown in Figure 10.1.

Before analyzing the weaknesses of this scheme, let us define some basic terminology related to interactive protocols. In general, an ***interactive protocol*** will involve two or more parties that are communicating with each other. Each party is modeled by an algorithm that alternately sends and receives information. Each run of a protocol will be called a ***session***. Each step within a session of the protocol is called a ***flow***; a flow consists of information transmitted from one party to another party. (Protocol 10.1 consists of two flows, the first one being from Bob to Alice, and the second one being from Alice to Bob.) At the end of a session, Bob (the initiator of the session) "accepts" or "rejects" (this is Bob's ***internal state*** at the end of the session). It may not be known to Alice whether Bob accepts or rejects.

It is not hard to see that Protocol 10.1 is insecure, even if the message authentication code used in it is secure. It is susceptible to a fairly standard type of attack known as a ***parallel session attack***, wherein Oscar impersonates Alice. The attack is depicted in Figure 10.2.

Within the first session (in which it is supposed that Oscar is impersonating Alice to Bob), Oscar initiates a second session in which he asks Bob to identify himself. This second session is boxed in Figure 10.2. In this second session, Oscar gives Bob the same challenge that he received from Bob in the first session. Once he

Protocol 10.2: (SECURE) CHALLENGE-AND-RESPONSE

1. Bob chooses a random challenge, r, which he sends to Alice.
2. Alice computes
$$y = MAC_K(ID(Alice) \parallel r)$$
and sends y to Bob.
3. Bob computes
$$y' = MAC_K(ID(Alice) \parallel r).$$
If $y' = y$, then Bob "accepts"; otherwise, Bob "rejects."

receives Bob's response, Oscar resumes the first session, in which he relays Bob's response back to him. Thus Oscar is able to successfully complete the first session!

The reader might object to the premise that parallel sessions constitute a realistic threat. However, there are scenarios in which parallel sessions might be reasonable, or even desirable, and it would seem to be prudent to design an identification scheme to withstand such attacks. We present one easy way to rectify the problem in Protocol 10.2. The only change that is made is to include the identity of the person creating the MAC tag into the computation of the tag.

In Protocol 10.2, we will assume that the random challenge is a bitstring of a specified, predetermined length, say k bits (in practice, $k = 100$ will be a suitable choice). We also assume that the identity string (*ID*(*Alice*) or *ID*(*Bob*), depending on whose identity is being authenticated) is also a bitstring of a specified length, formatted in some standard, fixed manner. We assume that an identity string contains enough information to specify a unique individual in the network (so Bob does not have to worry about which "Alice" he is talking to).

We claim that a parallel session attack cannot be carried out against Protocol 10.2. If Oscar attempted to mount the same attack as before, he would receive the value $MAC_K(ID(Bob) \parallel r)$ from Bob in the second session. This is of no help in computing the value $MAC_K(ID(Alice) \parallel r)$ that is required in the first session to successfully respond to Bob's challenge.

The preceding discussion may convince the reader that the parallel session attack cannot be mounted against Protocol 10.2, but it does not present a proof of security against all possible attacks. We shortly will give a proof of security. First, however, we explicitly list all the assumptions we make regarding the cryptographic components used in the scheme. These assumptions are as follows:

secret key

We assume that the secret key, K, is known only to Alice and Bob.

random challenges

We assume that Alice and Bob both have perfect random number generators which they use to determine their challenges. Therefore, there is only a very small probability that the same challenge occurs by chance in two different sessions.

MAC security

We assume that the message authentication code is secure. More precisely, we assume there does not exist an (ϵ, Q)-forger for the MAC, for appropriate values of ϵ and Q. That is, the probability that Oscar can correctly compute $MAC_K(x)$ is at most ϵ, even when he is given Q other tags, say $MAC_K(x_i)$, $i = 1, 2, \ldots, Q$, provided that $x \neq x_i$ for any i. Reasonable choices for Q might be 10000 or 100000, depending on the application.

Oscar may observe several sessions between Alice and Bob. Oscar's goal is to deceive Alice or Bob, i.e., to cause Bob to "accept" in a session in which Alice is not taking part, or to cause Alice to "accept" in a session in which Bob is not taking part. We show that Oscar will not succeed in deceiving Alice or Bob in this way, except with small probability, when the above assumptions are valid. This is done fairly easily by analyzing the structure of the identification scheme.

Suppose that Bob "accepts." Then $y = MAC_K(ID(Alice) \parallel r)$, where y is the value he receives in the second flow and r was his challenge from the first flow of the scheme. We claim that, with high probability, this value y must have been constructed by Alice in response to the challenge r from the first flow of the scheme. To justify this claim, let's consider the possible sources of a response if it did not come directly from Alice. First, because the key K is assumed to be known only to Alice and Bob, we do not have to consider the possibility that $y = MAC_K(ID(Alice) \parallel r)$ was computed by some other party that knows the key K. So either Oscar (or someone else) computed y without knowing the key K, or the value y was computed by Alice or Bob in some previous session, copied, and then reused by Oscar in the current session.

We now consider these possible cases in turn:

1. Suppose the value $y = MAC_K(ID(Alice) \parallel r)$ was previously constructed by Bob himself in some previous session. However, Bob only computes tags of the form $MAC_K(ID(Bob) \parallel r)$, so he would not have created y himself. Therefore this case does not arise.

2. Suppose the value y was previously constructed by Alice in some earlier session. This can happen only if the challenge r is reused. However, the challenge r is assumed to be a challenge that is newly created by Bob using a perfect random number generator, so Bob would not have issued the same challenge in some other session, except with a very small probability.

3. Suppose the value y is a new tag that is constructed by Oscar. Assuming that the message authentication code is secure and Oscar does not know the key K, Oscar cannot do this, except with a very small probability.

The informal proof given above can be made more precise. If we can prove an explicit, precise statement of the security of the underlying MAC, then we can give a precise security guarantee for the identification scheme. This is possible if the MAC is unconditionally secure. Alternatively, if we make an assumption about the MAC's security, then we can provide a security result for the identification scheme that depends on this assumption (this is the usual model of provable security). The security guarantees for the identification scheme quantify the probability that the adversary can fool Bob into accepting when the adversary is an active participant in the scheme.

A MAC is said to be ***unconditionally*** (ϵ, Q)***-secure*** if the adversary cannot construct a valid tag for any new message with probability greater than ϵ, given that the adversary has previously seen valid tags for at most Q messages (i.e., there does not exist an (ϵ, Q)-forger). As usual, we assume a fixed key, K, whose value is not known to the adversary, is used to construct all Q of the tags. An identification scheme is defined to be ***unconditionally*** (ϵ, Q)***-secure*** if the adversary cannot fool Alice or Bob into accepting with probability greater then ϵ, given that the adversary has observed at most Q previous sessions between Alice and Bob.

Unconditionally secure (ϵ, Q)-secure MACs exist for any desired values of Q and ϵ (e.g., using almost strongly universal hash families, which are considered in Exercise 20 of Chapter 5). However, unconditionally secure MACs typically require fairly large keys (especially if Q is large). As a consequence, computationally secure MACs, such as *CBC-MAC*, are more often used in practice. In this situation, an assumption about the security of the MAC is necessary. This assumption would take a similar form, but could include time as an explicit parameter. A MAC would be said to be (ϵ, Q, T)***-secure*** if the adversary cannot construct a valid tag for any new message with probability greater than ϵ, given that his computation time is at most T and given that he has previously seen valid tags for at most Q messages. An identification scheme is defined to be (ϵ, Q, T)***-secure*** if the adversary cannot fool Alice or Bob into accepting with probability greater then ϵ, given that the adversary has observed at most Q previous sessions between Alice and Bob, and given that the adversary's computation time is at most T.

For simplicity of notation, we will usually omit an explicit specification of the time parameter. This allows us to use similar notations in both the computationally secure and unconditionally secure settings. Whether we are talking about unconditional or computational security should be clear from the context.

Suppose first that we base the identification scheme on an unconditionally secure MAC. Then the resulting identification scheme will also be unconditionally secure, provided that the adversary has access to at most Q valid tags during some collection of sessions that all use the same MAC key. We need to recall one additional parameter, namely, the size (in bits) of the random challenge used in the scheme, which is denoted by k. Under these conditions, we can easily give an upper bound on the adversary's probability of deceiving Bob. We consider the same three cases as before:

1. As argued before, the value $y = MAC_K(ID(Alice) \parallel r)$ would not have been

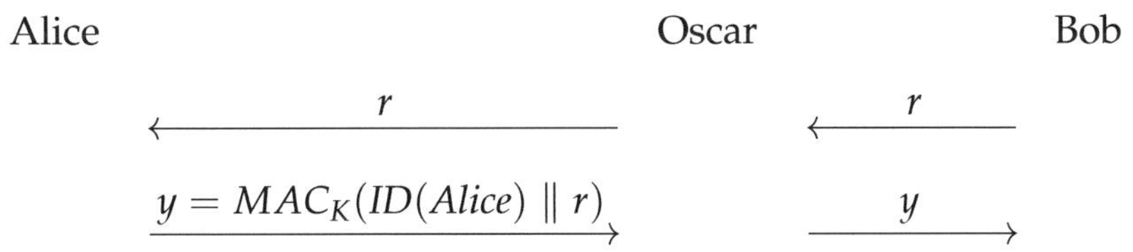

FIGURE 10.3: An intruder-in-the-middle?

previously constructed by Bob himself in some other session. (So this case does not occur.)

2. Suppose the value y was previously constructed by Alice in some other session. The challenge r is assumed to be a random challenge newly created by Bob. The probability that Bob already used the challenge r in a specific previous session is $1/2^k$. There are at most Q previous sessions under consideration, so the probability that r was used as a challenge in one of these previous sessions is at most $Q/2^k$. If this happens, then the adversary can re-use a MAC from a previous session.[3]

3. Suppose the value y is a new tag that is constructed by Oscar. Then, Oscar will be successful in his deception with probability at most ϵ; this follows from the security of the message authentication code being used.

Summing up, Oscar's probability of deceiving Bob is at most $Q/2^k + \epsilon$. We therefore have established the security of the identification scheme as a function of the security of the underlying primitives.

The analysis is essentially identical if a computationally secure MAC is used. We summarize the results of this section in the following theorem.

THEOREM 10.1 *Suppose that MAC is an (ϵ, Q)-secure message authentication code, and suppose that random challenges are k bits in length. Then Protocol 10.2 is a $(Q/2^k + \epsilon, Q)$-secure identification scheme.*

10.2.1 Attack Model and Adversarial Goals

There are several subtleties associated with the attack model and the adversarial goals in an identification scheme. To illustrate, we depict a possible ***intruder-in-the-middle*** scenario in Figure 10.3.

At first glance, this might appear to be a parallel session attack. It could be argued that Oscar impersonates Alice to Bob in one session, and he impersonates Bob to Alice in a parallel session. When Oscar receives Bob's challenge, r, he sends it to Alice. Then Alice's response (namely, y) is sent by Oscar to Bob, and Bob will "accept."

[3]The exact probability that a given challenge is repeated from a previous session is $1 - (1 - 2^{-k})^Q$, which is less than $Q/2^k$.

However, we do not consider this to be a real attack, because the "union" of the two "sessions" is a single session in which Alice has successfully identified herself to Bob. The overall result is that Bob issued a challenge r and Alice computed the correct response y to the challenge. Oscar simply forwarded messages to their intended recipients without modifying the messages, so Oscar was not an active participant in the scheme. The session executed exactly as it would have if Oscar had not been present.

A clear formulation of the adversarial goal should allow us to demonstrate that this is not an attack. We adopt the following approach. We will define the adversary (Oscar) to be an ***active adversary*** in a particular session if one of the following conditions holds:

1. Oscar creates a new message and places it in the channel,
2. Oscar changes a message in the channel, or
3. Oscar diverts a message in the channel so it is sent to someone other than the intended receiver.

An adversary who simply forwards messages is not considered to be an active adversary.

The goal of the adversary is to have the initiator of the scheme (e.g., Bob, who is assumed to be honest) "accept" in some session in which the adversary is active. According to this definition as well, Oscar is not active in the intruder-in-the-middle scenario considered above, and therefore the adversarial goal is not realized.

Another, essentially equivalent, way to decide if an adversary is active is to consider Alice and Bob's view of a particular session. Both Alice and Bob are interacting with an ***intended peer***: Alice's intended peer is Bob and vice versa. Further, if there is no active adversary, then Alice and Bob will have compatible views of the session: every message sent by Alice is received by Bob, and vice versa. Moreover, no message will be received out of order. The term ***matching conversations*** is often used to describe this situation; Alice and Bob will have matching conversations if and only if there is no active adversary.

The above discussion of the model assumes that the legitimate participants in a session are honest. To be precise, a participant in a session of the scheme (e.g., Alice or Bob) is said to be an ***honest participant*** if she/he follows the scheme, performs correct computations, and does not reveal information to the adversary (Oscar). If a participant is not honest, then the scheme is completely broken, so statements of security generally require that participants are honest.

Let's now turn to a consideration of attack models. Before he actually tries to deceive Bob, say, Oscar carries out an ***information-gathering phase***. Oscar is a ***passive adversary*** during this phase if he simply observes sessions between Alice and Bob. Alternatively, we might consider an attack model in which Oscar is active during the information-gathering phase. For example, Oscar might be given temporary access to an oracle that computes authentication tags $MAC_K(\cdot)$ for the (unknown) key K being used by Alice and Bob. During this time period, Oscar can successfully deceive Alice and Bob, of course, by using the oracle to respond to challenges. However, after the information-gathering phase, the MAC oracle is confiscated, and then Oscar carries out his attack, trying to get Alice or Bob to "accept" in a new session in which Oscar does not have a MAC oracle.

The security analysis that was performed in Section 10.2 applies to both of these attack models. The identification scheme is provably secure (more precisely, the adversary's success probability is at most $Q/2^k + \epsilon$) in the passive information-gathering model provided that the MAC is (ϵ, Q)-secure against a known message attack. Furthermore, the identification scheme is secure in the active information-gathering model provided that the MAC is (ϵ, Q)-secure against a chosen message attack.[4]

10.2.2 Mutual Authentication

A scheme in which Alice and Bob are both proving their identities to each other is called ***mutual authentication*** or ***mutual identification***. Both participants are required to "accept" if a session of the scheme is to be considered a successfully completed session. The adversary could be trying to fool Alice, Bob, or both of them into accepting. The adversarial goal is to cause an honest participant to "accept" after a flow in which the adversary is active.

The following conditions specify what the outcome of a mutual identification scheme should be, if the scheme is to be considered secure:

1. Suppose Alice and Bob are the two participants in a session of the scheme and they are both honest. Suppose also that the adversary is passive. Then Alice and Bob will both "accept."

2. If the adversary is active during a given flow of the scheme, then no honest participant will "accept" after that flow.

Note that the adversary might be inactive in a particular session until after one participant accepts, and then become active. Therefore it is possible that one honest participant "accepts" and then the other honest participant "rejects." The adversary does not achieve his goal in this scenario, even though the session did not successfully complete, because the adversary was inactive before the first participant accepted. The outcome of the session is that Alice successfully identifies herself to Bob (say), but Bob does not successfully identify himself to Alice. This could be considered a ***disruption*** of the scheme, but it is not a successful attack.

There are several ways in which the adversary could be active in a session of a scheme. We list some of these now:

1. The adversary impersonates Alice, hoping to cause Bob to accept.

[4]The attack model for MACs that we described in Chapter 5 is basically a chosen message attack. The notion of a known message attack for MACs is analogous to the corresponding notion for signature schemes that was introduced in Chapter 8.

Protocol 10.3: INSECURE MUTUAL CHALLENGE-AND-RESPONSE

1. Bob chooses a random challenge, r_1, which he sends to Alice.
2. Alice chooses a random challenge, r_2. She also computes
$$y_1 = MAC_K(ID(Alice) \parallel r_1)$$
and she sends r_2 and y_1 to Bob.
3. Bob computes
$$y_1' = MAC_K(ID(Alice) \parallel r_1).$$
If $y_1' = y_1$, then Bob "accepts"; otherwise, Bob "rejects." Bob also computes
$$y_2 = MAC_K(ID(Bob) \parallel r_2)$$
and he sends y_2 to Alice.
4. Alice computes
$$y_2' = MAC_K(ID(Bob) \parallel r_2).$$
If $y_2' = y_2$, then Alice "accepts"; otherwise, Alice "rejects."

2. The adversary impersonates Bob, hoping to cause Alice to accept.
3. The adversary is active in some session involving Alice and Bob, and he is trying to cause both Alice and Bob to accept.

We might try to achieve mutual authentication by running Protocol 10.2 twice (i.e., Alice verifies Bob's identity, and then Bob verifies Alice's identity in a separate session). However, it is generally more efficient to design a single scheme to accomplish both identifications at once.

What if Alice and Bob were to combine two sessions of one-way identification into a single scheme, in the obvious way? This is what is done in Protocol 10.3, and it reduces the number of flows required (compared to running the original one-way scheme twice) from four to three. However, it turns out that the resulting mutual identification scheme is flawed and can be attacked.

Protocol 10.3 is insecure because Oscar can fool Alice in a parallel session attack. Oscar, pretending to be Bob, initiates a session with Alice. When Oscar receives Alice's challenge, r_2, in the second flow, he "accepts," and then he initiates a second session (pretending to be Alice) with Bob. In this second session, Oscar sends r_2 to Bob as his challenge in the first flow. When Oscar receives Bob's response (in the second flow in the second session), he forwards it to Alice as the third flow in the first session. Alice will "accept," and therefore Oscar has successfully impersonated Bob in this session. (The second session is dropped, i.e., it is

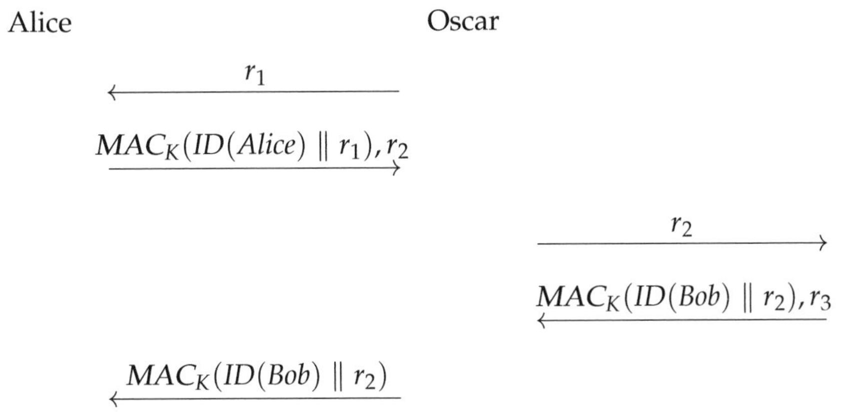

FIGURE 10.4: Attack on Protocol 10.3

never completed.) This constitutes a successful attack, because the honest participant in the first session (namely, Alice) accepted after a flow in which Oscar was active (namely, the initial flow of the session). An illustration of the attack is given in Figure 10.4.

Clearly, the attack is based on re-using a flow from one session in a different flow in another session. It is not difficult to rectify the problem, and there are in fact several ways to modify the scheme so that it is secure. Basically, what is required is to design the flows so that each flow contains information that is computed in a different manner. One solution along these lines is shown in Protocol 10.4.

The only change that was made in Protocol 10.4 is in the definition of y_1 in step 2. Now this tag depends on two challenges, r_1 and r_2. This serves to distinguish the second flow from the third flow (in which the tag depends only on the challenge r_2).

Protocol 10.4 can be analyzed in a similar fashion as Protocol 10.2. The analysis is a bit more complicated, however, because the adversary could try to play the role of Bob (fooling Alice) or Alice (fooling Bob). The probability that a value y_1 or y_2 can be "reused" from a previous session can be computed, as can the probability that the adversary can compute a new tag from scratch.

First, because any value y_1 is computed differently than any value y_2, it is impossible that a y_1 value from one session can be re-used as a y_2 value from another session (or vice versa). Oscar could try to play the role of Bob (fooling Alice) or Alice (fooling Bob), by determining y_2 or y_1, respectively. The probability that y_1 or y_2 can be reused from a previous session is at most $Q/2^k$, under the assumption that Oscar has seen at most Q tags from previous sessions (this limits the number of previous sessions to $Q/2$, because there are two tags per session). The probability that Oscar can compute a new y_1 is at most ϵ, and the probability that he can compute a new y_2 is at most ϵ. Therefore Oscar's probability of deceiving one of Alice or Bob is at most $Q/2^k + 2\epsilon$. Summarizing, we have the following theorem.

Protocol 10.4: (SECURE) MUTUAL CHALLENGE-AND-RESPONSE

1. Bob chooses a random challenge, r_1, which he sends to Alice.
2. Alice chooses a random challenge, r_2. She also computes
$$y_1 = MAC_K(ID(Alice) \parallel r_1 \parallel r_2)$$
and she sends r_2 and y_1 to Bob.
3. Bob computes
$$y_1' = MAC_K(ID(Alice) \parallel r_1 \parallel r_2).$$
If $y_1' = y_1$, then Bob "accepts"; otherwise, Bob "rejects." Bob also computes
$$y_2 = MAC_K(ID(Bob) \parallel r_2)$$
and he sends y_2 to Alice.
4. Alice computes
$$y_2' = MAC_K(ID(Bob) \parallel r_2).$$
If $y_2' = y_2$, then Alice "accepts"; otherwise, Alice "rejects."

THEOREM 10.2 *Suppose that MAC is an (ϵ, Q)-secure message authentication code, and suppose that random challenges are k bits in length. Then Protocol 10.4 is a $(Q/2^k + 2\epsilon, Q/2)$-secure mutual identification scheme.*

10.3 Challenge-and-Response in the Public-key Setting

10.3.1 Public-key Identification Schemes

We now look at mutual identification schemes in the public-key setting. Our strategy is to modify Protocol 10.4 by replacing MAC tags by signatures. Another difference is that, in the secret-key setting, we included the name of the person who produced the tag in each tag (this was important because a secret key K, being known to two parties, allows either party to create tags). In the public-key setting, only one person can create signatures using a specified private signing key, namely, the person possessing that key. Therefore we do not need to explicitly designate who created a particular signature.

As in the secret-key setting, at the beginning of a session, each participant has an intended peer (the person with whom each of them thinks they are communicating). Each participant will use the intended peer's verification key to verify

Protocol 10.5: PUBLIC-KEY MUTUAL AUTHENTICATION (VERSION 1)

1. Bob chooses a random challenge, r_1. He sends $\mathbf{Cert}(Bob)$ and r_1 to Alice.
2. Alice chooses a random challenge, r_2. She also computes $y_1 = \mathbf{sig}_{Alice}(ID(Bob) \parallel r_1 \parallel r_2)$ and sends $\mathbf{Cert}(Alice)$, r_2 and y_1 to Bob.
3. Bob verifies Alice's public key, $\mathbf{ver}_{Alice}$, on the certificate $\mathbf{Cert}(Alice)$. Then he checks that $\mathbf{ver}_{Alice}(ID(Bob) \parallel r_1 \parallel r_2, y_1) = true$. If so, then Bob "accepts"; otherwise, Bob "rejects." Bob also computes $y_2 = \mathbf{sig}_{Bob}(ID(Alice) \parallel r_2)$ and sends y_2 to Alice.
4. Alice verifies Bob's public key, $\mathbf{ver}_{Bob}$, on the certificate $\mathbf{Cert}(Bob)$. Then she checks that $\mathbf{ver}_{Bob}(ID(Alice) \parallel r_2, y_2) = true$. If so, then Alice "accepts"; otherwise, Alice "rejects."

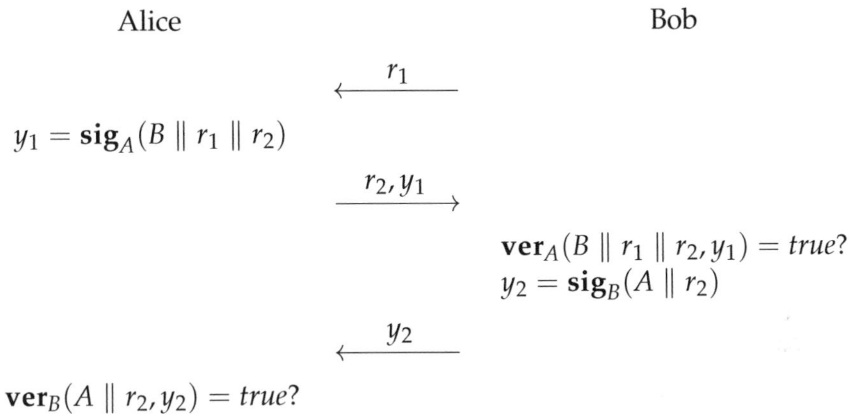

FIGURE 10.5: Information flows in Protocol 10.5

signatures received in the forthcoming session. They will also include the name of the intended peer in all signatures that they create during the scheme.

Protocol 10.5 is a typical mutual identification scheme in the public-key setting. It can be proven to be secure if the signature scheme is secure and challenges are generated randomly. Figure 10.5 illustrates the scheme, omitting the transmission of the certificates of Alice and Bob. In this figure and elsewhere, "A" denotes "$ID(Alice)$" and "B" denotes "$ID(Bob)$."

Here is a theorem stating the security of Protocol 10.5 as a function of the security of the underlying signature scheme (where security of signature schemes is described using notation similar to that of MACs). The proof is left as an Exercise.

THEOREM 10.3 *Suppose that* **sig** *is an (ϵ, Q)-secure signature scheme, and suppose that random challenges are k bits in length. Then Protocol 10.5 is a $(Q/2^{k-1} + 2\epsilon, Q)$-secure mutual identification scheme.*

Protocol 10.6: (INSECURE) PUBLIC-KEY MUTUAL AUTHENTICATION

1. Bob chooses a random challenge, r_1. He sends **Cert**(Bob) and r_1 to Alice.

2. Alice chooses a random challenge, r_2. She also computes $y_1 = \mathbf{sig}_{Alice}(ID(Bob) \parallel r_1 \parallel r_2)$ and sends **Cert**$(Alice)$, r_2 and y_1 to Bob.

3. Bob verifies Alice's public key, $\mathbf{ver}_{Alice}$, on the certificate **Cert**$(Alice)$. Then he checks that $\mathbf{ver}_{Alice}(ID(Bob) \parallel r_1 \parallel r_2, y_1) = true$. If so, then Bob "accepts"; otherwise, Bob "rejects." Bob also chooses a random number r_3, computes $y_2 = \mathbf{sig}_{Bob}(ID(Alice) \parallel r_2 \parallel r_3)$, and sends r_3 and y_2 to Alice.

4. Alice verifies Bob's public key, $\mathbf{ver}_{Bob}$, on the certificate **Cert**(Bob). Then she checks that $\mathbf{ver}_{Bob}(ID(Alice) \parallel r_2 \parallel r_3, y_2) = true$. If so, then Alice "accepts"; otherwise, Alice "rejects."

REMARK In Theorem 10.3, the number of previous sessions is Q, whereas in Theorem 10.2, the number of previous sessions was limited to $Q/2$. This is because the signatures created by Alice and Bob in Protocol 10.5 use different keys. The adversary is allowed to view Q signatures created by each of Alice and Bob. In contrast, in Protocol 10.4, both Alice and Bob use the same key to create tags. Since we want to limit the adversary to seeing Q tags created with any given key, this forces us to require that the adversary be allowed to eavesdrop in at most $Q/2$ previous sessions. ∎

It is instructive to consider various modifications of this scheme. Some modifications turn out to be insecure, while others are secure. An example of an insecure (modified) scheme includes a third random number r_3 that is signed by Bob; with this modification, the scheme becomes vulnerable to a parallel session attack. The scheme is presented as Protocol 10.6.

In Protocol 10.6, the random value, r_3, is chosen by Bob and is signed by him (along with r_2) in the third flow of the scheme. Including this extra piece of information in the signature makes the scheme insecure because the signature in the third flow is now constructed in a similar fashion as the signature in the second flow. This allows the parallel session attack depicted in Figure 10.6 to be carried out. In this attack, Oscar initiates a session with Alice, pretending to be Bob. Then he initiates a second session, with Bob, pretending to Alice. Bob's response in the second flow of the second session is forwarded to Alice in the third flow of the first session.

Finally, we note that another variation of Protocol 10.5 that is secure is discussed in the Exercises.

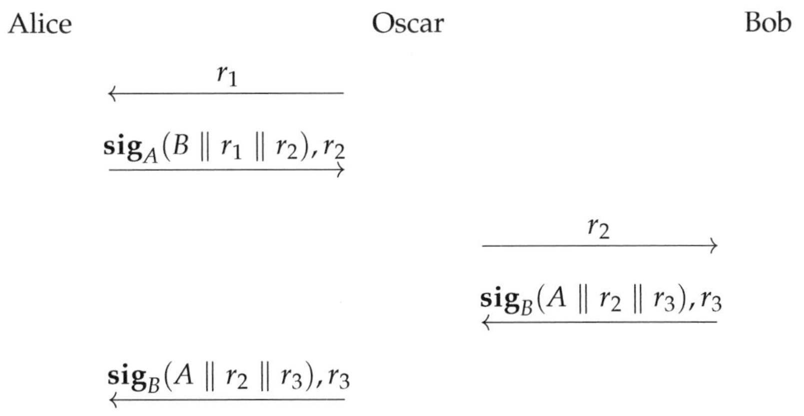

Note that "A" denotes "$ID(Alice)$" and "B" denotes "$ID(Bob)$."

FIGURE 10.6: Attack on Protocol 10.6

10.4 The Schnorr Identification Scheme

Another approach to identification schemes is to design schemes "from scratch," without using any other cryptographic tools as building blocks. A potential advantage of schemes of this type is that they might be more efficient and have a lower communication complexity than the schemes considered in the previous sections. Such schemes typically involve having someone identify themselves by proving that they know the value of some secret quantity (i.e., a private key) without having to reveal its value.

The *Schnorr Identification Scheme* (Protocol 10.7) is an example of such a scheme. This scheme is based on the **Discrete Logarithm** problem, which we introduced as Problem 7.1. Here, we will take α to be an element having prime order q in the group $\mathbb{Z}_p{}^*$ (where p is prime and $p - 1 \equiv 0 \pmod{q}$). Then $\log_\alpha \beta$ is defined for any element $\beta \in \langle\alpha\rangle$, and $0 \leq \log_\alpha \beta \leq q - 1$. This is the same setting of the **Discrete Logarithm** problem that was used in the *Schnorr Signature Scheme* and the *Digital Signature Algorithm* (see Section 8.4). In order for this setting to be considered secure, we will specify that $p \approx 2^{2048}$ and $q \approx 2^{224}$.

The scheme requires a trusted authority, or *TA*, who chooses some common system parameters (***domain parameters***) for the scheme, as follows:

1. p is a large prime (i.e., $p \approx 2^{2048}$).
2. q is a large prime divisor of $p - 1$ (i.e., $q \approx 2^{224}$).
3. $\alpha \in \mathbb{Z}_p{}^*$ has order q.
4. t is a security parameter such that $q > 2^t$. (A security parameter is a parameter whose value can be chosen in order to provide a desired level of security

Protocol 10.7: SCHNORR IDENTIFICATION SCHEME

1. Alice chooses a random number, k, where $0 \leq k \leq q-1$, and she computes $\gamma = \alpha^k \bmod p$. She sends **Cert**$(Alice)$ and the commitment γ to Bob.
2. Bob verifies Alice's public key, v, on the certificate **Cert**$(Alice)$. Bob chooses a random challenge r, $1 \leq r \leq 2^t$, and he sends r to Alice.
3. Alice computes $y = k + ar \bmod q$ and she sends the response y to Bob.
4. Bob verifies that $\gamma \equiv \alpha^y v^r \pmod{p}$. If so, then Bob "accepts"; otherwise, Bob "rejects."

in a given scheme. Here, the adversary's probability of deceiving Alice or Bob will be 2^{-t}, so $t = 80$ will provide adequate security for most practical applications.)

The domain parameters p, q, α, and t are all public, and they will be used by everyone in the network.

Every user in the network chooses their own private key, a, where $0 \leq a \leq q-1$, and constructs a corresponding public key $v = \alpha^{-a} \bmod p$. Observe that v can be computed as $(\alpha^a)^{-1} \bmod p$, or (more efficiently) as $\alpha^{q-a} \bmod p$. The TA issues certificates for everyone in the network. Each user's certificate will contain their public key (and, perhaps, the public domain parameters). This information, as well as the user's identifying information, is signed by the TA, of course.

Observe that Protocol 10.7 incorporates a ***commitment*** by Alice, followed by a challenge (issued by Bob) and finally a response by Alice.

The following congruences demonstrate that Alice will be able to prove her identity to Bob using Protocol 10.7, assuming that both parties are honest and perform correct computations:

$$\begin{aligned} \alpha^y v^r &\equiv \alpha^{k+ar} v^r \pmod{p} \\ &\equiv \alpha^{k+ar} \alpha^{-ar} \pmod{p} \\ &\equiv \alpha^k \pmod{p} \\ &\equiv \gamma \pmod{p}. \end{aligned}$$

The fact that Bob will accept Alice's proof of identity (assuming that he and Alice are honest) is sometimes called the ***completeness*** property of the scheme.

Let's work out a small, toy example. The following example omits the authentication of Alice's public key by Bob.

Example 10.1 Suppose $p = 88667$, $q = 1031$, and $t = 10$. The element $\alpha = 70322$

has order q in $\mathbb{Z}_p{}^*$. Suppose Alice's private key is $a = 755$; then

$$\begin{aligned} v &= \alpha^{-a} \bmod p \\ &= 70322^{1031-755} \bmod 88667 \\ &= 13136. \end{aligned}$$

Now suppose Alice chooses the random number $k = 543$. Then she computes

$$\begin{aligned} \gamma &= \alpha^k \bmod p \\ &= 70322^{543} \bmod 88667 \\ &= 84109, \end{aligned}$$

and she sends γ to Bob. Suppose Bob issues the challenge $r = 1000$. Then Alice computes

$$\begin{aligned} y &= k + ar \bmod q \\ &= 543 + 755 \times 1000 \bmod 1031 \\ &= 851, \end{aligned}$$

and she sends y to Bob as her response. Bob then verifies that

$$84109 \equiv 70322^{851} 13136^{1000} \pmod{88667}.$$

Finally, Bob "accepts." □

The *Schnorr Identification Scheme* was designed to be very fast and efficient, both from a computational point of view and in the amount of information that needs to be exchanged in the scheme. It is also designed to minimize the amount of computation performed by Alice, in particular. This is desirable because, in many practical applications, Alice's computations will be performed by a smart card with low computing power, while Bob's computations will be performed by a more powerful computer.

Let us consider Alice's computations. Step 1 requires an exponentiation (modulo p) to be performed; step 3 comprises one addition and one multiplication (modulo q). It is the modular exponentiation that is computationally intensive, but this can be precomputed offline, before the scheme is executed, if desired. The online computations to be performed by Alice are very modest.

It is also a simple matter to calculate the number of bits that are communicated during the scheme. We depict the information that is communicated (excluding Alice's certificate) in Figure 10.7. In that diagram, the notation $\in_R$ is used to denote a random choice made from a specified set.

Alice gives Bob 2048 bits of information (excluding her certificate) in the first flow; Bob sends Alice 80 bits in the second flow; and Alice transmits 224 bits to Bob in the third flow. So the communication requirements are quite modest, as well.

The information transmitted in the second and third flows of the scheme have been reduced by the way in which the scheme is designed. In the second flow,

Alice		Bob
$k \in_R \{0, \dots, q-1\}$ $\gamma = \alpha^k \bmod p$		
	γ (2048 bits) $\longrightarrow$	
		$r \in_R \{1, \dots, 2^t\}$
	$\longleftarrow$ r (80 bits)	
$y = k + ar \bmod q$		
	y (224 bits) $\longrightarrow$	
		$\gamma \equiv \alpha^y v^r \pmod{p}$?

FIGURE 10.7: Information flows in Protocol 10.7

a challenge could be taken to be any integer between 0 and $q-1$; however, this would yield a 224-bit challenge. An 80-bit challenge provides sufficient security for many applications.

In the third flow, the value y is an exponent. This value is only 224 bits in length because the scheme is working inside a subgroup of $\mathbb{Z}_p{}^*$ of order $q \approx 2^{224}$. This permits the information transmitted in the third flow to be reduced significantly, as compared to an implementation of the scheme in the "whole group," $\mathbb{Z}_p{}^*$, in which an exponent would be 2048 bits in length.

The first flow clearly requires the most information to be transmitted. One possible way to reduce the amount of information is to replace the 2048-bit value γ by a 224-bit message digest, e.g., $\gamma' = \mathit{SHA3\text{-}224}(\gamma)$. Then, in the last step of the scheme, Bob would verify that (the message digest) $\gamma' = \mathit{SHA3\text{-}224}(\alpha^y v^r \pmod{p}))$.

10.4.1 Security of the Schnorr Identification Scheme

We now study the security of the *Schnorr Identification Scheme*. As mentioned previously, t is a security parameter. It is sufficiently large to prevent an impostor posing as Alice, say Olga, from guessing Bob's challenge, r. (If Olga guessed the correct value of r, she could choose any value for y and precompute

$$\gamma = \alpha^y v^r \bmod p.$$

She would give Bob the value γ in the first flow of the scheme, and when she receives the challenge r, she would respond with the value y she has already chosen. Then the congruence involving γ would be verified by Bob, and he would "accept.") The probability that Olga will guess the value of r correctly is 2^{-t} if r is chosen uniformly at random by Bob.

Notice that Bob should choose a new, random challenge, r, every time Alice identifies herself to him. If Bob always used the same challenge r, then Olga could impersonate Alice by the method we just described.

Alice's computations in the scheme involve the use of her private key, a. The value a functions somewhat like a PIN, in that it convinces Bob that the person (or entity) carrying out the identification scheme is, indeed, Alice. But there is an important difference from a PIN: in this identification scheme, the value of a is not revealed. Instead, Alice (or more accurately, Alice's smart card) "proves" that she/it knows the value of a in the third flow of the scheme, by computing the correct response, y, to the challenge, r, issued by Bob. An adversary could attempt to compute a, because a is just a discrete logarithm of a known quantity: $a = -\log_\alpha v$ in $\mathbb{Z}_p{}^*$. However, we are assuming that this computation is infeasible.

We have argued that Olga can guess Bob's challenge, r, and thereby impersonate Alice, with probability 2^{-t}. Suppose that Olga can do better than this. In particular, suppose that Olga knows some γ (a value of her choosing), and two possible challenges, r_1 and r_2, such that she can compute responses y_1 and y_2, respectively, which would cause Bob to accept. (If Olga could only compute a correct response for one challenge for each γ, then her probability of success would be only 2^{-t}.)

So we assume that Olga knows (or she can compute) values r_1, r_2, y_1, and y_2 such that

$$\gamma \equiv \alpha^{y_1} v^{r_1} \equiv \alpha^{y_2} v^{r_2} \pmod{p}.$$

It follows that

$$\alpha^{y_1 - y_2} \equiv v^{r_2 - r_1} \pmod{p}.$$

It holds that $v \equiv \alpha^{-a} \pmod{p}$, where a is Alice's private key. Hence,

$$\alpha^{y_1 - y_2} \equiv \alpha^{-a(r_2 - r_1)} \pmod{p}.$$

The element α has order q, so it must be the case that

$$y_1 - y_2 \equiv a(r_1 - r_2) \pmod{q}.$$

Now, $0 < |r_2 - r_1| < 2^t$ and $q > 2^t$ is prime. Hence $\gcd(r_2 - r_1, q) = 1$, and therefore $(r_1 - r_2)^{-1} \bmod q$ exists. Hence, Olga can compute Alice's private key, a, as follows:

$$a = (y_1 - y_2)(r_1 - r_2)^{-1} \bmod q.$$

It may not be clear what conditions would be sufficient for Olga to be able to find this particular γ in the first place. However, we will present a method by which γ can be computed, along with correct responses to two different challenges, by any attacker with a sufficiently high success probability. We will prove the following theorem due to Schnorr:

THEOREM 10.4 *Suppose that* IMPERSONATEALICE *is an algorithm that succeeds in completing the Schnorr Identification Scheme (impersonating Alice) with success probability $\epsilon \geq 2^{-t+1}$. Then* IMPERSONATEALICE *can be used as an oracle in an algorithm* COMPUTEPRIVATEKEY *that computes Alice's private key, where* COMPUTEPRIVATEKEY *is a Las Vegas algorithm having expected complexity $O(1/\epsilon)$.*

Algorithm 10.1: COMPUTEPRIVATEKEY(G, n, α, β)

1. Choose random pairs (k, r) and run IMPERSONATEALICE(k, r), until $m_{k,r} = 1$. Define (k_1, r_1) to be the current pair (k, r) and proceed to step 2.
2. Denote by u the number of trials that were required in step 1. Choose random pairs (k_1, r) and run IMPERSONATEALICE(k_1, r), until $m_{k_1,r} = 1$ or until $4u$ trials have finished unsuccessfully.
3. If step 2 terminates with a successful pair (k_1, r_2) where $r_2 \neq r_1$, then we already showed that it is easy to compute Alice's private key and the algorithm COMPUTEPRIVATEKEY terminates successfully. Otherwise, go back to step 1 and start again.

The algorithm COMPUTEPRIVATEKEY is presented as Algorithm 10.1. In the analysis of this algorithm, we will use the following technical lemma, which we state without proof.

LEMMA 10.5 *For $\delta \approx 0$, it holds that $(1 - \delta)^{c/\delta} \approx e^{-c}$.*

The first step is to define a $q \times 2^t$ matrix $M = (m_{k,r})$, having entries equal to 0 or 1, where

$$m_{k,r} = 1 \text{ if and only if IMPERSONATEALICE}(k, r) \text{ executes successfully,}$$

where $\gamma = \alpha^k$. Since IMPERSONATEALICE has success probability ϵ, it follows that M contains exactly $\epsilon q 2^t$ 1's. Therefore, the average number of 1's in a row k of M is $\epsilon 2^t$. Now, define a row k of M to be a ***heavy row*** if it contains more than $\epsilon 2^{t-1}$ 1's. Observe that a heavy row contains at least two 1's, because $\epsilon \geq 2^{-t+1}$.

LEMMA 10.6 *Given a random entry $m_{k,r}$ having the value 1, the probability that k is a heavy row is at least 1/2.*

PROOF The total number of 1's in light rows is at most

$$q \times \epsilon 2^{t-1} = \epsilon q 2^{t-1}.$$

Hence, the total number of 1's in heavy rows is at least

$$\epsilon q 2^t - \epsilon q 2^{t-1} = \epsilon q 2^{t-1}.$$

∎

We now begin the analysis of the various steps in Algorithm 10.1. Since ϵ is the success probability of IMPERSONATEALICE, it follows that $E[u] = 1/\epsilon$, where $E[u]$ denotes the expected number of trials in step 1.

LEMMA 10.7 $\mathbf{Pr}[u > 1/(2\epsilon)] \approx 0.6$.

PROOF It is clear that $u > 1/(2\epsilon)$ if and only if the first $1/(2\epsilon)$ random trials are unsuccessful. This happens with probability

$$(1-\epsilon)^{1/(2\epsilon)} \approx e^{-1/2} \approx 0.6,$$

from Lemma 10.6. ∎

LEMMA 10.8 *Suppose that $u > 1/(2\epsilon)$ and suppose also that k_1 is a heavy row. Then the probability that step 2 of* COMPUTEPRIVATEKEY *succeeds in finding a value r_2 such that* IMPERSONATEALICE$(k_1, r_2) = 1$ *is at least .63.*

PROOF Under the given assumptions, $4u > 2/\epsilon$. The probability that a random entry in row k_1 is a 1 is at least $\epsilon/2$. Therefore the probability that step 2 of COMPUTEPRIVATEKEY succeeds is at least

$$1 - \left(1 - \frac{\epsilon}{2}\right)^{2/\epsilon} \approx 1 - e^{-1} \approx 0.63,$$

where the estimate is obtained from Lemma 10.6. ∎

LEMMA 10.9 *Given that k_1 is a heavy row and step 2 of* COMPUTEPRIVATEKEY *succeeds in finding a pair (k_1, r_2) such that* IMPERSONATEALICE$(k_1, r_2) = 1$, *the probability that $r_2 \neq r_1$ is at least $1/2$.*

PROOF This follows immediately from the fact that a heavy row contains at least two 1's. ∎

Now we can analyze the expected complexity of algorithm COMPUTEPRIVATEKEY. We have shown that:

1. The probability that step 1 terminates with k being a heavy row is at least $1/2$.

2. The probability that $u > 1/(2\epsilon)$ is 0.6.

3. Assuming 1 and 2 hold, the probability that step 2 of COMPUTEPRIVATEKEY succeeds is at least .63.

4. Given that 1, 2 and 3, all hold, the probability that $r_2 \neq r_1$ in step 3 of COMPUTEPRIVATEKEY is at least $1/2$.

The probability that 1–4 all hold (and hence algorithm COMPUTEPRIVATEKEY succeeds) is at least

$$\frac{1}{2} \times 0.6 \times 0.63 \times \frac{1}{2} \approx 0.095.$$

Therefore, the expected number of iterations of steps 1 and 2 in COMPUTEPRIVATEKEY is at most $1/0.095 \approx 10.6$.

Finally, we determine the complexity of executing the operations in a single iteration of steps 1 and 2 of COMPUTEPRIVATEKEY. The expected complexity of step 1 of COMPUTEPRIVATEKEY (i.e., the expected number of trials in step 1) is $E[u] = 1/\epsilon$. Also, the complexity of step 2 is at most $4u$. Therefore, the expected complexity of steps 1 and 2 of COMPUTEPRIVATEKEY is at most $5/\epsilon$. Finally, the expected complexity of COMPUTEPRIVATEKEY is at most

$$10.6 \times \frac{5}{\epsilon} = \frac{53}{\epsilon} \in O\left(\frac{1}{\epsilon}\right).$$

The above analysis shows that anyone who is able to successfully impersonate Alice with a probability exceeding 2^{-t} must know (or be able to easily compute) Alice's private key, a. (We proved above that a "successful" impersonator can compute a. Conversely, it is obvious that anyone who knows the value of a can impersonate Alice, with probability equal to 1.) It therefore follows, roughly speaking, that being able to impersonate Alice is equivalent to knowing Alice's private key. This property is sometimes termed ***soundness***.

An identification scheme that is both sound and complete is called a ***proof of knowledge***. Our analysis so far has established that the *Schnorr Identification Scheme* is a proof of knowledge. We provide an example to illustrate the above discussion.

Example 10.2 Suppose we have the same parameters as in Example 10.1: $p = 88667$, $q = 1031$, $t = 10$, $\alpha = 70322$, and $v = 13136$. Suppose, for $\gamma = 84109$, that Olga is able somehow to determine two correct responses: $y_1 = 851$ is the correct response for the challenge $r_1 = 1000$; and $y_1 = 454$ is the correct response for the challenge $r_1 = 19$. In other words,

$$84109 \equiv \alpha^{851} v^{1000} \equiv \alpha^{454} v^{19} \pmod{p}.$$

Then Olga can compute

$$a = (851 - 454)(1000 - 19)^{-1} \bmod 1031 = 755,$$

and thus discover Alice's private key. □

We have proved that the scheme is a proof of knowledge. But this is not sufficient to ensure that the scheme is "secure." We still need to consider the possibility that secret information (namely, Alice's private key) might be leaked to a verifier who takes part in the scheme, or to an observer. (This could be thought of as the information gathering phase of an attack.) Our hope is that no information about a will be gained by Olga when Alice proves her identity. If this is true, then Olga will not be able subsequently to masquerade as Alice (assuming that the computation of the discrete logarithm a is infeasible).

In general, we could envision a situation whereby Alice proves her identity to Olga, say, on several different occasions. After several sessions of the scheme, Olga will try to determine the value of a so she can subsequently impersonate Alice. If Olga can determine no information about the value of a by taking part in a "reasonable" number of sessions of the scheme as the verifier, and then performing a "reasonable" amount of computation, then the scheme is termed a ***zero-knowledge identification scheme***. This would prove that the scheme is secure, under the assumption that a is infeasible to compute. (Of course, it would be necessary to define, in a precise way, the term "reasonable," in order to have a meaningful statement of security.)

We will show that the *Schnorr Identification Scheme* is zero-knowledge for an honest verifier, where an ***honest verifier*** is defined to be one who chooses his or her challenges r at random, as specified by the scheme.

We require the notion of a ***transcript*** of a session, which consists of a triple $T = (\gamma, r, y)$ in which $\gamma \equiv \alpha^y v^r \pmod p$. The verifier (or an observer) can obtain a transcript $T(S)$ of each session S. The set of possible transcripts is

$$\mathcal{T} = \{(\gamma, r, y) : 1 \leq r \leq 2^t, 0 \leq y \leq q-1, \gamma \equiv \alpha^y v^r \pmod p\}.$$

It is easy to see that $|\mathcal{T}| = q\,2^t$. Further, it is not difficult to prove that the probability that any particular transcript occurs in any given session is $1/(q\,2^t)$, assuming that the challenges r are generated at random. We argue this as follows: for any fixed value of r, there is a one-to-one correspondence between the value of $\gamma \in \langle\alpha\rangle$ and the value of $y \in \{0, \ldots, q-1\}$ on a particular transcript. We are assuming that Alice chooses γ at random (namely, by choosing a random k and computing $\gamma = \alpha^k \bmod p$), and we also assume that Bob chooses r at random (because he is an honest verifier). These two values determine the value of y. Since there are q possible choices for γ and 2^t possible choices for r, it follows that every possible transcript occurs with the same probability, $1/(q\,2^t)$, in sessions involving an honest verifier.

The key point of the zero-knowledge aspect of the scheme is a property called ***simulatability***. It turns out that Olga (or anyone else, for that matter) can generate simulated transcripts, having exactly the same probability distribution as real transcripts, without taking part in the scheme. This is done by the following three simple steps:

1. choose r uniformly at random from the set $\{1, \ldots, 2^t\}$
2. choose y uniformly at random from the set $\{0, \ldots, q-1\}$
3. compute $\gamma = \alpha^y v^r \bmod p$.

It is easy to see that the probability that any $T \in \mathcal{T}$ is generated by the above procedure is $1/(q\,2^t)$. Therefore, it holds that

$$\mathbf{Pr}_{\text{real}}[T] = \mathbf{Pr}_{\text{sim}}[T] = \frac{1}{q\,2^t}$$

for all $T \in \mathcal{T}$, where $\mathbf{Pr}_{\text{real}}[T]$ is the probability of generating the transcript T during a real session, and $\mathbf{Pr}_{\text{sim}}[T]$ is the probability of generating T as a simulated transcript.

What is the significance of the fact that transcripts can be simulated? We claim that anything an honest verifier can compute after taking part in several sessions of the scheme can also be computed without taking part in any sessions of the scheme. In particular, computing Alice's private key, a, which is necessary for Olga to be able to impersonate Alice, is not made easier for Olga if she plays the role of the verifier in one or more sessions in which challenges are chosen randomly.

The above statements can be justified further, as follows. Suppose there exists an algorithm EXTRACT which, when given a set of transcripts, say $T_1, \dots, T_\ell$, computes a private key, say a, with some probability, say ϵ. We assume that the transcripts are actual transcripts of sessions, in which the participants follow the scheme. Suppose that $T'_1, \dots, T'_\ell$ are simulated transcripts. We have noted that the probability distribution on simulated transcripts is identical to the probability distribution on real transcripts. Therefore EXTRACT$(T'_1, \dots, T'_\ell)$ will also compute a with probability ϵ. This establishes that executing the scheme does not make computing a easier, so the scheme is zero-knowledge.

Let's consider the possibility that Olga (a ***dishonest verifier***) might obtain some useful information by choosing her challenges r in a non-uniform way. To be specific, suppose that Olga chooses her challenge r using some function that depends, in a complicated way, on Alice's choice of γ. There does not seem to be any way to perfectly simulate the resulting probability distribution on transcripts, and therefore we cannot prove that the scheme is zero-knowledge in the way that we did for an honest verifier.

We should emphasize that there is no known attack on the scheme based on making non-random challenges; we are just saying that the proof technique we used previously does not seem to apply in this case. The only known security proofs of the scheme for arbitrary verifiers require additional assumptions.

To summarize, an interactive scheme is a proof of knowledge if it is impossible (except with a very small probability) to impersonate Alice without knowing the value of Alice's key. This means that the only way to "break" the scheme is to actually compute a. A scheme is termed zero-knowledge if it reveals no information about Alice's private key. Stated another way, computing Alice's private key is not made easier by taking part in the scheme (in Bob's role as the verifier) in some specified number of sessions. If a scheme is a zero-knowledge proof of knowledge, then it is "secure."

10.5 The Feige-Fiat-Shamir Identification Scheme

In this section, we study another identification scheme that follows the "zero-knowledge proof of knowledge" methodology, namely, the *Feige-Fiat-Shamir*

Identification Scheme. Many of the concepts and terminologies we use in this section follow those that are employed in the discussion of the *Schnorr Identification Scheme* in the previous section.

The trusted authority (which, as usual, is denoted by *TA*) chooses system parameters p and q, which are large primes that are both congruent to 3 modulo 4. The product $n = pq$ is public. For such a value of n, the Jacobi symbol $\left(\frac{-1}{n}\right) = 1$ but -1 is a quadratic non-residue modulo n (these facts follow from basic properties of Legendre and Jacobi symbols; see Section 6.4.1). The security of the *Feige-Fiat-Shamir Identification Scheme* is based on the assumed difficulty of the **Computational Composite Quadratic Residues** problem, which we present as Problem 10.1.

Problem 10.1: Computational Composite Quadratic Residues

Instance: A positive integer n that is the product of two unknown distinct primes p and q where $p, q \equiv 3 \pmod 4$, and an integer $x \in \mathbb{Z}_n{}^*$ such that the Jacobi symbol $\left(\frac{x}{n}\right) = 1$.
Question: Find $y \in \mathbb{Z}_n{}^*$ such that $y^2 \equiv \pm x \pmod n$.

Observe that, in Problem 10.1, we are required to actually find a square root of x or $-x$. Note that, if $\left(\frac{x}{n}\right) = 1$, then exactly one of x and $-x$ (modulo n) is a quadratic residue.

We also observe that the **Computational Composite Quadratic Residues** problem is essentially equivalent in difficulty to the factoring problem. Showing that an algorithm that solves Problem 10.1 can be used to factor n is basically the same as the proof that a decryption oracle for the *Rabin Cryptosystem* yields the factorization of the modulus (see Section 6.12). Conversely, if n can be factored, then it is easy to solve Problem 10.1.

A network user, say Alice, chooses private and public keys for the *Feige-Fiat-Shamir Identification Scheme* as follows:

1. Choose random values $S_1, \dots, S_k \in \mathbb{Z}_n$, for an integer value $k \approx \log_2 \log_2 n$.
2. For $1 \leq j \leq k$, compute $I_j = \pm 1/S_j{}^2 \bmod n$, where the sign ($\pm$) is chosen randomly.
3. $\mathbf{I} = (I_1, \dots, I_k)$ is Alice's public key and $\mathbf{S} = (S_1, \dots, S_k)$ is Alice's private key.

Alice's public key, $\mathbf{I}$, would be stored on a certificate that is signed by the *TA*.

The *Feige-Fiat-Shamir Identification Scheme* is presented as Protocol 10.8. This protocol follows the same structure as the *Schnorr Identification Scheme*, namely, a commitment by Alice, a challenge by Bob, and a response by Alice.

Proving completeness of the *Feige-Fiat-Shamir Identification Scheme* is straightforward; it is just a matter of proving that the verification condition holds:

$$Y^2 \left(\prod_{\{j:E_j=1\}} I_j \right) \equiv \left(R \prod_{\{j:E_j=1\}} S_j \right)^2 \left(\prod_{\{j:E_j=1\}} I_j \right) \pmod n$$

Protocol 10.8: FEIGE-FIAT-SHAMIR IDENTIFICATION SCHEME

Repeat the following four steps $t = \log_2 n$ times:

1. Alice chooses a random value $R \in \mathbb{Z}_n$, she computes $X = \pm R^2 \bmod n$ with the sign chosen randomly, and she sends the commitment X to Bob.

2. Bob sends a random boolean vector (the challenge) $\mathbf{E} = (E_1, \ldots, E_k) \in \{0,1\}^k$ to Alice.

3. Alice computes
$$Y = R \prod_{\{j:E_j=1\}} S_j \bmod n$$
and sends the response Y to Bob.

4. Bob verifies that
$$X = \pm Y^2 \prod_{\{j:E_j=1\}} I_j \bmod n.$$
If so, then Bob "accepts"; otherwise, Bob "rejects."

$$\begin{aligned} &\equiv R^2 \left(\prod_{\{j:E_j=1\}} {S_j}^2 I_j \right) \pmod{n} \\ &\equiv \pm R^2 \pmod{n} \\ &\equiv \pm X \pmod{n}. \end{aligned}$$

Let's now think a bit about soundness. A dishonest Alice (who does not possess the private key) can fool Bob in one of the t main iterations of the protocol with probability $1/2^k$ by correctly guessing Bob's challenge $\mathbf{E}$ ahead of time. Given a particular $\mathbf{E}$, it is straightforward to compute X and Y such that the verification condition in step 4 of protocol 10.8 will be satisfied. So the cheating prover provides this X in step 1. If the guessed challenge happens to be issued in step 2, then the correct response Y can by provided in step 3. However, because steps 1–4 are repeated t times, this strategy will succeed only with probability $1/2^{tk}$.

Now suppose that a cheating prover can fool Bob with probability exceeding $1/2^t$. Since the proof consists of t identical iterations, this means that the prover can fool Bob with probability $\rho > 1/2$ in any one of the t iterations of the protocol. Thus there must be a commitment X for which the prover can find a valid Y for more than 2^{k-1} of the 2^k possible challenges that Bob might issue. Assuming this is the case, we proceed as follows. Choose any integer j such that $1 \leq j \leq k$. Then there must be two challenges, $\mathbf{E}$ and $\mathbf{E}'$, such that

1. the prover can compute valid responses Y and Y' respectively, for the same X, and

2. $\mathbf{E}$ and $\mathbf{E}'$ differ only in the jth coordinate, say $E_j = 0$ and $E'_j = 1$.

We provide a bit of detail to justify these two statements. Consider the set of all 2^k possible challenges. This set can be partitioned into 2^{k-1} pairs, where each pair consists of two challenges that differ only in the jth co-ordinate. Then it is obvious that any subset of more than 2^{k-1} of these challenges must contain two challenges from the same pair.

It is then straightforward to see that

$$\begin{aligned} X &\equiv \pm Y^2 \prod_{\{j:E_j=1\}} I_j \pmod{n} \\ &\equiv \pm (Y')^2 \prod_{\{j:E'_j=1\}} I_j \pmod{n}, \end{aligned}$$

so

$$Y^2 \equiv \pm (Y')^2 I_j \pmod{n},$$

and hence

$$(Y/Y')^2 \equiv \pm I_j \pmod{n}.$$

Hence, it is then possible to compute a square root of I_j or $-I_j$ modulo n (exactly one of I_j or $-I_j$ is a quadratic residue). However, this implies that n can be factored, since solving the **Composite Quadratic Residues** problem in $\mathbb{Z}_n$ is equivalent to factoring n. So this line of reasoning shows that a cheating prover who can succeed in fooling Bob with a certain probability is able to actually factor n. That is, the scheme is a proof of knowledge.

Now, in order make this discussion rigorous, we would need to show how the desired X can be found (in polynomial time). This is somewhat similar to the process used to show that the *Schnorr Identification Scheme* is a proof of knowledge. We do not provide the formal soundness proof here; however, we provide a reference for the proof at the end of the chapter.

We next want to discuss the "zero-knowledge" aspect of this scheme in a bit more detail. Proving zero-knowledge for an honest verifier is fairly easy. We need to describe how to construct a simulated transcript that looks identical to a real transcript. The trick is to start by choosing the vector $\mathbf{E} = (E_1, \dots, E_k)$ first, then using $\mathbf{E}$ to construct X, and then constructing Y. In more detail, here are the steps to be followed for each of the t main iterations of the algorithm, letting i range from 1 to t:

1. Choose a random boolean vector $\mathbf{E} = (E_1, \dots, E_k) \in \{0,1\}^k$.

2. Choose a random value $R \in \mathbb{Z}_n$ and compute

$$X = \pm R^2 \left(\prod_{\{j:E_j=1\}} I_j \right) \bmod n.$$

3. Define $Y = R$.

4. Let $T_i = (X, \mathbf{E}, Y)$.

The final simulated transcript is $T = (T_1, T_2, \ldots, T_t)$. We leave as an exercise for the reader to show that a simulated transcript is identical to a real transcript.

The *Feige-Fiat-Shamir Identification Scheme* is also zero-knowledge against a cheating verifier, where a cheating verifier is one who chooses the challenge vector E in some nonuniform or nonrandom way (recall we were not able to show this in the case of the *Schnorr Identification Scheme*). For example, suppose that Bob applies a hash function to the X value he receives in step 1, and then takes the k low-order bits of $h(X)$ to determine E. This prevents us from simulating transcripts as we did for an honest verifier, because the previous simulation chose E before X is chosen. So the strategy has to be modified.

Roughly speaking, we can simulate transcripts produced by a dishonest verifier by the following process. As before, we let i range from 1 to t.

1. Choose a random boolean vector $\mathbf{E} = (E_1, \ldots, E_k) \in \{0,1\}^k$.

2. Choose a random value $R \in \mathbb{Z}_n$ and compute

$$X = \pm R^2 \left(\prod_{\{j : E_j = 1\}} I_j \right) \bmod n.$$

3. Now compute the challenge $\mathbf{E}'$ according to the strategy employed by the dishonest verifier. If $\mathbf{E} = \mathbf{E}'$, then move to step 4; otherwise return to step 1.

4. Define $Y = R$.

5. Let $T_i = (X, \mathbf{E}, Y)$.

Observe that, in step 3, the challenge may depend on X as well as on previous parts of the transcript, $T_1, \ldots, T_{i-1}$. Steps 1 and 2 are identical to the corresponding steps in the simulation process for an honest verifier. Step 3 throws away the $\mathbf{E}$ and R if the challenge $\mathbf{E}'$ produced by the dishonest verifier is different from the guessed random challenge $\mathbf{E}$. When we return to step 1, we "reset" the algorithm generating the dishonest verifier's challenges to its state before the previous (discarded) commitment R was chosen.

It is important to note that the algorithm generating the dishonest verifier's challenges is a randomized algorithm, so when we reset this algorithm, it will probably generate a different challenge R than it did previously. However, the probability of success in step 2 (namely, that $\mathbf{E} = \mathbf{E}'$) is $1/2^k$. To see this, we compute

$$\begin{aligned} &\sum_{(E_1,\ldots,E_k) \in \{0,1\}^k} \left(\mathbf{Pr}[\mathbf{E} = (E_1, \ldots, E_k)] \times \mathbf{Pr}[\mathbf{E}' = (E_1, \ldots, E_k)] \right) \\ = &\sum_{(E_1,\ldots,E_k) \in \{0,1\}^k} \frac{1}{2^k} \times \mathbf{Pr}[\mathbf{E}' = (E_1, \ldots, E_k)] \end{aligned}$$

$$= \frac{1}{2^k} \times \sum_{(E_1,\dots,E_k)\in\{0,1\}^k} \mathbf{Pr}[\mathbf{E}' = (E_1,\dots,E_k)]$$
$$= \frac{1}{2^k}.$$

Next, we observe that

$$\frac{1}{2^k} = \frac{1}{2^{\log_2 \log_2 n}} = \frac{1}{\log_2 n}.$$

So the expected number of repetitions of steps 1–3, until $\mathbf{E} = \mathbf{E}'$, is $\log_2 n$, which is polynomial (linear, actually) in the size of n (the size of n is the number of bits in the binary representation of n). Since $t = \log n$, it follows that a transcript can be simulated in expected time $\Theta((\log_2 n)^2)$, which is expected polynomial time.

It is interesting to note that this strategy to simulate transcripts of a dishonest verifier cannot be employed in the case of the *Schnorr Identification Scheme*. The problem is that the number of possible challenges is too large, so a guessed value for the challenge will be correct with probability $1/2^t$, so 2^t challenges would be required (on average) until a guessed challenge matches one produced by the dishonest verifier.

10.6 Notes and References

Two-factor authentication was patented in the U.S. by AT&T in 1995. Strangely, another patent for two-factor authentication was also issued in the U.S., to Kim Dotcom, in 2000. See [48] for a discussion. The *Argon2* key-stretching technique is presented in [30].

The security model we use for identification schemes is adapted from the models described in Diffie, van Oorschot, and Wiener [72], Bellare and Rogaway [18] and Blake-Wilson and Menezes [33]. Diffie [69] notes that cryptographic challenge-and-response protocols for identification date from the early 1950s. Research into identification schemes in the context of computer networks was initiated by Needham and Schroeder [154] in the late 1970s.

Protocol 10.5 is from [33]. Protocol 10.10 is very similar to one version of the mutual authentication scheme described in FIPS publication 196 [148]. The attack on Protocol 10.6 is from [69].

The *Schnorr Identification Scheme* is from [174]. A proof of security under certain reasonable computational assumptions was provided by Bellare and Palacio [17]. The *Feige-Fiat-Shamir Identification Scheme* is from [82, 83].

Exercises

10.1 Prove that it is impossible to design a secure two-flow mutual identification scheme based on random challenges. (A two-flow scheme consists of one flow from Bob to Alice (say) followed by a flow from Alice to Bob. In a mutual identification scheme, both parties are required to "accept" in order for a session to terminate successfully.)

10.2 Consider the mutual identification scheme presented in Protocol 10.9. Prove that this scheme is insecure. (In particular, show that Olga can impersonate Bob by means of a certain type of parallel session attack, assuming that Olga has observed a previous session of the scheme between Alice and Bob.)

Protocol 10.9: INSECURE PUBLIC-KEY MUTUAL AUTHENTICATION

1. Bob chooses a random challenge, r_1. He also computes $y_1 = \mathbf{sig}_{Bob}(r_1)$ and he sends $\mathbf{Cert}(Bob)$, r_1 and y_1 to Alice.

2. Alice verifies Bob's public key, $\mathbf{ver}_{Bob}$, on the certificate $\mathbf{Cert}(Bob)$. Then she checks that $\mathbf{ver}_{Bob}(r_1, y_1) = true$. If not, then Alice "rejects" and quits. Otherwise, Alice chooses a random challenge, r_2. She also computes $y_2 = \mathbf{sig}_{Alice}(r_1)$ and $y_3 = \mathbf{sig}_{Alice}(r_2)$ and she sends $\mathbf{Cert}(Alice)$, r_2, y_2, and y_3 to Bob.

3. Bob verifies Alice's public key, $\mathbf{ver}_{Alice}$, on the certificate $\mathbf{Cert}(Alice)$. Then he checks that $\mathbf{ver}_{Alice}(r_1, y_2) = true$ and $\mathbf{ver}_{Alice}(r_2, y_3) = true$. If so, then Bob "accepts"; otherwise, Bob "rejects." Bob also computes $y_4 = \mathbf{sig}_{Bob}(r_2)$ and he sends y_4 to Alice.

4. Alice checks that $\mathbf{ver}_{Bob}(r_2, y_4) = true$. If so, then Alice "accepts"; otherwise, Alice "rejects."

10.3 Give a complete proof that Protocol 10.10 is secure. (This scheme is essentially identical to one of the schemes standardized in FIPS publication 196.)

10.4 Discuss whether Protocol 10.11 is secure. (Certificates are omitted from its description, but they are assumed to be inlcuded in the scheme in the usual way.)

10.5 Prove that Protocol 10.5 and Protocol 10.10 are both insecure if the identity of Alice (Bob, resp.) is omitted from the signature computed by Bob (Alice, resp.).

10.6 Consider the following possible identification scheme. Alice possesses a secret key $n = pq$, where p and q are prime and $p \equiv q \equiv 3 \pmod 4$. The value

Protocol 10.10: PUBLIC-KEY MUTUAL AUTHENTICATION (VERSION 2)

1. Bob chooses a random challenge, r_1. He sends **Cert**(*Bob*) and r_1 to Alice.
2. Alice chooses a random challenge, r_2. She also computes $y_1 = \mathbf{sig}_{Alice}(ID(Bob) \parallel r_1 \parallel r_2)$ and she sends **Cert**(*Alice*), r_2 and y_1 to Bob.
3. Bob verifies Alice's public key, $\mathbf{ver}_{Alice}$, on the certificate **Cert**(*Alice*). Then he checks that $\mathbf{ver}_{Alice}(ID(Bob) \parallel r_1 \parallel r_2, y_1) = true$. If so, then Bob "accepts"; otherwise, Bob "rejects." Bob also computes $y_2 = \mathbf{sig}_{Bob}(ID(Alice) \parallel r_2 \parallel r_1)$ and he sends y_2 to Alice.
4. Alice verifies Bob's public key, $\mathbf{ver}_{Bob}$, on the certificate **Cert**(*Bob*). Then she checks that $\mathbf{ver}_{Bob}(ID(Alice) \parallel r_2 \parallel r_1, y_2) = true$. If so, then Alice "accepts"; otherwise, Alice "rejects."

Protocol 10.11: UNKNOWN PROTOCOL

1. Bob chooses a random challenge, r_1, and he sends it to Alice.
2. Alice chooses a random challenge r_2, she computes $y_1 = \mathbf{sig}_{Alice}(r_1)$, and she sends r_2 and y_1 to Bob.
3. Bob checks that $\mathbf{ver}_{Alice}(r_1, y_1) = true$; if so, then Bob "accepts"; otherwise, Bob "rejects." Bob also computes $y_2 = \mathbf{sig}_{Bob}(r_2)$ and he sends y_2 to Alice.
4. Alice checks that $\mathbf{ver}_{Bob}(r_2, y_2) = true$. If so, then Alice "accepts"; otherwise, Alice "rejects."

of n will be stored on Alice's certificate. When Alice wants to identify herself to Bob, say, Bob will present Alice with a random quadratic residue modulo n, say x. Then Alice will compute a square root y of x and give it to Bob. Bob then verifies that $y^2 \equiv x \pmod{n}$. Explain why this scheme is insecure.

10.7 Suppose Alice is using the *Schnorr Identification Scheme* where $q = 1201$, $p = 122503$, $t = 10$, and $\alpha = 11538$.

(a) Verify that α has order q in $\mathbb{Z}_p{}^*$.

(b) Suppose that Alice's secret exponent is $a = 357$. Compute v.

(c) Suppose that $k = 868$. Compute γ.

(d) Suppose that Bob issues the challenge $r = 501$. Compute Alice's response y.

(e) Perform Bob's calculations to verify y.

10.8 Suppose that Alice uses the *Schnorr Identification Scheme* with p, q, t, and α as in the previous exercise. Now suppose that $v = 51131$, and Olga has learned that

$$\alpha^3 v^{148} \equiv \alpha^{151} v^{1077} \pmod{p}.$$

Show how Olga can compute Alice's secret exponent a.

10.9 Show that the *Schnorr Identification Scheme* is not secure against an active adversary who changes the messages that are sent from Alice to Bob.

10.10 Consider the following identification scheme. Alice has a public key $v = g^a$ and a private key a. Assume that g is a primitive element in a $\mathbb{Z}_p{}^*$ where p is prime, and assume that the **Discrete Logarithm** problem in $\mathbb{Z}_p{}^*$ is infeasible. Bob chooses a random b, computes $w = g^b$, and sends w to Alice. Alice computes $K = w^a$ and sends it to Bob. Bob accepts if and only if $K = v^b$. Prove that the above-described scheme zero-knowledge against an honest verifier (i.e., a verifier Bob who chooses the challenges w as described above). That is, show that it is possible to simulate transcripts of Bob's view of the protocol.

10.11 Prove that Protocol 10.2 is not secure if ID strings and random challenges are not required to have a prespecified, fixed length.

HINT Recall that, before the protocol is executed, each user claims an identity to the other user, so each user has an "intended peer." The protocol will proceed only if these two users have a shared secret key. You can assume that the users' IDs are represented in ASCII. Consider the situation where the two users who share a key K are Ali and Alice.

Chapter 11

Key Distribution

In this chapter, we describe several methods to manage cryptographic keys, including various techniques to distribute as well as update keys. All the methods discussed in this chapter involve a trusted authority who is ultimately responsible for choosing keys and keying information, and then distributing this information to network users who require it.

11.1 Introduction

We have observed that public-key cryptosystems have the advantage over secret-key cryptosystems that a secure channel is not needed to exchange a key. But, unfortunately, most public-key cryptosystems (e.g., *RSA*) are much slower than secret-key systems (e.g., *AES*). So, in practice, secret-key systems are usually used to encrypt "long" messages. We already discussed hybrid cryptography, which is one commonly used method, in Section 6.1. But there are many other techniques that allow Alice and Bob to determine a secret key in a secure manner. This is called ***key establishment***, which is the topic of this and the next chapter.

We will discuss several approaches to the problem of establishing secret keys. As our setting, we have an insecure network $\mathcal{U}$ of n users. In all the schemes discussed in this chapter, we will have a trusted authority (denoted by *TA*) who is responsible for such things as verifying the identities of users, issuing certificates, choosing and transmitting keys to users, etc. There are many possible scenarios, including the following:

key predistribution

In a ***key predistribution scheme*** (or ***KPS***), a *TA* distributes keying information "ahead of time" in a secure fashion to everyone in the network. Note that a secure channel is required at the time that keys are distributed. Later, network users can use these secret keys to encrypt messages they transmit over the network. Secret keys can also be used for the purposes of message authentication, by employing a suitable MAC. Typically, every pair of users in the network will be able to determine a key (or keys), known only to them, as a result of the keying information they hold.

session key distribution

In session key distribution, an online *TA* chooses session keys and distributes them to network users, when requested to do so, via an interactive protocol. Such a protocol is called a ***session key distribution scheme*** and denoted ***SKDS***. Session keys are used to encrypt information for a specified, fairly short, period of time. The session keys will be encrypted by the *TA* using previously distributed secret keys (under the assumption that every network user possesses a secret key whose value is known to the *TA*).

key agreement

Key agreement refers to the situation where network users employ an interactive protocol to construct a session key. Such a protocol is called a ***key agreement scheme***, and it is denoted by ***KAS***. These may be secret-key based or public-key based schemes, and they do not require an online *TA*.

Key predistribution schemes and session key distribution schemes are considered in this chapter, while key agreement schemes will be studied in Chapter 12. As in Chapter 10, we consider aspects such as active and/or passive adversaries, various adversarial goals, attack models, and security levels.

We now compare and contrast the above-mentioned methods of key establishment in more detail. First, we can distinguish between key distribution and key agreement as follows. Key distribution is a mechanism whereby a *TA* chooses a secret key or keys and then transmits them to another party or parties in encrypted form. Key agreement denotes a protocol whereby two (or more) network users jointly establish a secret key (usually a session key) by communicating over a public channel. In a key agreement scheme, the value of the key is most often determined as a function of inputs provided by both parties and secret information of the two users. However, there are protocols in which one user chooses the key (or keying information) and sends it to another user in encrypted form (similar to what a *TA* does in a session key distribution scheme). This particular scenario can be termed ***key transport*** if we wish to distinguish it from a KAS in which the key depends on input from both users.

It is important to distinguish between long-lived keys and session keys. We summarize the main features of these two types of keys now.

long-lived keys

Users (or pairs of users) may have ***long-lived keys*** (***LL-keys***) that are precomputed and then stored securely. Alternatively, LL-keys might be computed non-interactively, as needed, from securely stored secret information. LL-keys could be secret keys known to a pair of users, or, alternatively, to a user and the *TA*. On the other hand, they could be private keys corresponding to public keys that are stored on users' certificates.

session keys

Pairs of users will often employ secret short-lived ***session keys*** in a particular session, and then throw them away when the session has ended. Session keys are usually secret keys, for use in a secret-key cryptosystem or MAC. LL-keys are often used in protocols to transmit encrypted session keys (e.g., they may be used as "key-encrypting keys" in an SKDS). LL-keys are also used to authenticate data—using a message authentication code or signature scheme—that is sent in a session of a scheme.

A key predistribution scheme provides one method to distribute secret LL-keys ahead of time. It requires a secure channel between the *TA* and each network user at the time that the keys are distributed. At a later time, a KAS might be used by pairs of network users to generate session keys, as needed. One main consideration in the study of KPS is the amount of secret information that must be stored by each user in the network.

A session key distribution scheme is a three-party protocol, involving two users U and V (say) and the *TA*. SKDS are usually based on long-lived secret keys held by individual users and the *TA*. That is, U holds a secret key whose value is known to the *TA*, and V holds a (different) secret key whose value is known to the *TA*.

A key agreement scheme can be a secret-key based or a public/private-key based scheme. A KAS usually involves two users, say U and V, but it does not require an online *TA*. However, an offline *TA* might have distributed secret LL-keys in the past, say in the case of a secret-key based scheme. If the KAS is based on public keys, then a *TA* is implicitly required to issue certificates and (perhaps) to maintain a suitable public-key infrastructure. However, the *TA* does not take an active role in a session of the KAS.

There are several reasons why session keys are useful. First, they limit the amount of ciphertext (that is encrypted with one particular key) available to an attacker, because session keys are changed on a regular basis. Another advantage of session keys is that they limit exposure in the event of session key compromise, provided that the scheme is designed well (e.g., it is desirable that the compromise of a session key should not reveal information about the LL-key, or about other session keys). Session keys can therefore be used in "risky" environments where there is a higher possibility of exposure. Finally, the use of session keys often reduces the amount of long-term information that needs to be securely stored by each user, because keys for pairs of users are generated only when they are needed.

Long-lived keys should satisfy several requirements. The "type" of scheme used to construct session keys (i.e., a public-key based scheme as opposed to a secret-key based scheme) dictates the type of LL-keys required. As well, users' storage requirements depend on the type of keys used. We consider these requirements now, assuming that we have a network of n users.

First, as mentioned above, if an SKDS is to be used for session key distribution, then each network user must have a secret LL-key in common with the *TA*. This entails a low storage requirement for network users, but the *TA* has a very high storage requirement.

A secret-key based KAS requires that every pair of the n network users has a secret LL-key known only to them. In a "naive" implementation, each user stores $n-1$ long-lived keys, which necessitates a high storage requirement if n is large. The total number of secret keys in the network is $\binom{n}{2}$, which grows quadratically as a function of n. So this is sometimes called the ***n^2 problem***. Appropriate key predistribution schemes can reduce this storage requirement significantly, however.

Finally, in a public-key based KAS, we require that all network users have their own public/private LL-key pair. This yields a low storage requirement, because users only store their own private key and a certificate containing their public key.

11.1.1 Attack Models and Adversarial Goals

Since the network is insecure, we need to protect against potential adversaries. Our opponent, Oscar, may be one of the users in the network. He may be active or passive during an information-gathering phase. Later, when he carries out his attack, he might be a passive adversary, which means that his actions are restricted to eavesdropping on messages that are transmitted over the channel. On the other hand, we might want to guard against the possibility that Oscar is an active adversary. Recall that an active adversary can carry out various types of malicious actions, such as:

1. alter messages that he observes being transmitted over the network,
2. save messages for reuse at a later time, or
3. attempt to masquerade as various users in the network.

The objective of an adversary might be:

1. to fool U and V into accepting an "invalid" key as valid (an invalid key could be an old key that has expired, or a key chosen by the adversary, to mention two possibilities),
2. to make U or V believe that they have exchanged a key with each other, when they have not done so, or
3. to determine some (partial) information about the key exchanged by U and V.

The first two of these adversarial goals involve active attacks, while the third goal could perhaps be accomplished within the context of a passive attack.

Summarizing, the objective of a session key distribution scheme or a key agreement scheme is that, at the end of a session of the scheme, the two parties involved in the session both have possession of the same key K, and the value of K is not known to any other party (except possibly the TA).

We sometimes desire ***authenticated key agreement schemes***, which include (mutual) identification of U and V. Therefore, the schemes should be secure identification schemes (as defined in Chapter 10), and, in addition, U and V should possess a new secret key at the end of a session, whose value is not known to the adversary.

Extended attack models can also be considered. Suppose that the adversary learns the value of a particular session key (this is called a ***known session key attack***). In this attack model, we would still want other session keys (as well as the LL-keys) to remain secure. As another possibility, suppose that the adversary learns the LL-keys of the participants (this is a ***known LL-key attack***). This is a catastrophic attack, and consequently, a new scheme must be set up. However, can we limit the damage that is done in this type of attack? If the adversary cannot learn the values of previous session keys (or partial information about those keys), then the scheme is said to possess the property of ***perfect forward secrecy***. This is clearly a desirable attribute of a session key distribution scheme or a key agreement scheme.

In this chapter, we concentrate on key predistribution and session key distribution. In Section 11.2, for the problem of key predistribution, we study the classical *Diffie-Hellman Scheme*, as well as an unconditionally secure scheme (the *Blom Scheme*) that uses algebraic techniques. We also consider a specialized scheme that is suitable for the setting of a sensor network. For session key distribution, we analyze some insecure schemes, and then we present a secure scheme due to Bellare and Rogaway, in Section 11.3. Sections 11.4 and 11.5 discuss two additional techniques for key predistribution in different scenarios. One setting is a dynamic network, where users may join or leave the network. This necessitates a mechanism for ***re-keying***, which updates users' keys appropriately so as to maintain the security of keys in a dynamic network. We describe the *Logical Key Hierarchy*, which is a nice solution to this problem. The other setting permits parts of a key (called "shares") to be distributed to network users in such a way that certain subsets of users can reconstruct the key at a later time; this is called a ***threshold scheme***.

11.2 Key Predistribution

11.2.1 Diffie-Hellman Key Predistribution

In this section, we describe a key predistribution scheme that is a modification of the well-known *Diffie-Hellman Key Agreement Scheme*, which we will discuss in the next chapter. The scheme we describe now is the *Diffie-Hellman Key Predistribution Scheme*. This scheme is computationally secure provided that the **Decision Diffie-Hellman** problem (Problem 7.4) is intractable. Suppose that $(G, \cdot)$ is a group and suppose that $\alpha \in G$ is an element of order n such that the **Decision Diffie-Hellman** problem is intractable in the subgroup of G generated by α.

Every user U in the network has a private LL-key a_U (where $0 \leq a_U \leq n-1$)

Protocol 11.1: DIFFIE-HELLMAN KPS

1. The public domain parameters consist of a group $(G, \cdot)$ and an element $\alpha \in G$ having order n.

2. V computes
$$K_{U,V} = \alpha^{a_U a_V} = b_U{}^{a_V},$$
using the public key b_U from U's certificate, together with her own private key a_V.

3. U computes
$$K_{U,V} = \alpha^{a_U a_V} = b_V{}^{a_U},$$
using the public key b_V from V's certificate, together with his own private key a_U.

and a corresponding public key
$$b_U = \alpha^{a_U}.$$

The users' public keys are signed by the *TA* and stored on certificates, as usual. The common LL-key for any two users, say U and V, is defined to be
$$K_{U,V} = \alpha^{a_U a_V}.$$

The *Diffie-Hellman KPS* is summarized in Protocol 11.1.

We illustrate Protocol 11.1 with a toy example.

Example 11.1 Suppose $p = 12987461$, $q = 1291$, and $\alpha = 3606738$ are the public domain parameters. Here, p and q are prime, $p - 1 \equiv 0 \pmod q$, and α has order q. We implement Protocol 11.1 in the subgroup of $(\mathbb{Z}_p{}^*, \cdot)$ generated by α. This subgroup has order q.

Suppose U chooses $a_U = 357$. Then he computes
$$\begin{aligned} b_U &= \alpha^{a_U} \bmod p \\ &= 3606738^{357} \bmod 12987461 \\ &= 7317197, \end{aligned}$$
which is placed on his certificate. Suppose V chooses $a_V = 199$. Then she computes
$$\begin{aligned} b_V &= \alpha^{a_V} \bmod p \\ &= 3606738^{199} \bmod 12987461 \\ &= 138432, \end{aligned}$$
which is placed on her certificate.

Now U can compute the key

$$\begin{aligned} K_{U,V} &= {b_V}^{a_U} \bmod p \\ &= 138432^{357} \bmod 12987461 \\ &= 11829605, \end{aligned}$$

and V can compute the same key

$$\begin{aligned} K_{U,V} &= {b_U}^{a_V} \bmod p \\ &= 7317197^{199} \bmod 12987461 \\ &= 11829605. \end{aligned}$$

□

Let us think about the security of the *Diffie-Hellman KPS* in the presence of an adversary. Since there is no interaction in the scheme[1] and we assume that users' private keys are secure, we do not need to consider the possibility of an active adversary. Therefore, we need only to consider whether a user (say W) can compute $K_{U,V}$ if $W \neq U, V$. In other words, given public keys α^{a_U} and α^{a_V} (but not a_U or a_V), is it feasible to compute the secret key $K_{U,V} = \alpha^{a_U a_V}$? This is precisely the **Computational Diffie-Hellman** problem, which was defined as Problem 7.3. Therefore, the *Diffie-Hellman KPS* is secure against an adversary if and only if the **Computational Diffie-Hellman** problem in the subgroup $\langle \alpha \rangle$ is intractable.

Even if the adversary is unable to compute a Diffie-Hellman key, perhaps there is the possibility that he could determine some partial information about the key in polynomial time. Therefore, we desire semantic security of the keys, which means that an adversary cannot compute any partial information about the key in polynomial time. In other words, distinguishing Diffie-Hellman keys from random elements of the subgroup $\langle \alpha \rangle$ should be intractable. The semantic security of Diffie-Hellman keys is easily seen to be equivalent to the intractability of the **Decision Diffie-Hellman** problem (which was presented as Problem 7.4).

11.2.2 The Blom Scheme

In this section, we consider unconditionally secure key predistribution schemes. We begin by describing a "trivial" solution. For every pair of users $\{U, V\}$, the *TA* chooses a random key $K_{U,V} = K_{V,U}$ and transmits it "off-band" to U and V over a secure channel. (That is, the transmission of keys does not take place over the network, because the network is not secure.) Unfortunately, each user must store $n - 1$ keys, and the *TA* needs to transmit a total of $\binom{n}{2}$ keys securely (as mentioned in the introduction to this chapter, this is sometimes called the "n^2 problem"). Even for relatively small networks, this approach can become prohibitively expensive, and so it is not really a practical solution.

[1]It might happen that two users exchange their IDs and/or their certificates, but this information is regarded as fixed, public information. Therefore we do not view the scheme as an interactive scheme.

Thus, it is of interest to try to reduce the amount of information that needs to be transmitted and stored, while still allowing each pair of users U and V to be able to (independently) compute a secret key $K_{U,V}$. A particularly elegant scheme to accomplish this, called the *Blom Key Predistribution Scheme*, is discussed next.

We begin by briefly discussing the security model used in the study of unconditionally secure KPS. We assume that the *TA* distributes secret information securely to the n network users. The adversary can corrupt a subset of at most k users, and obtain all their secret information, where k is a pre-specified ***security parameter***. The adversary's goal is to determine the secret LL-key of a pair of uncorrupted users. The *Blom Key Predistribution Scheme* is a KPS that is unconditionally secure against adversaries of this type.

It is desired that each pair of users U and V will be able to compute a key $K_{U,V} = K_{V,U}$. Therefore, the security condition is as follows: any set of at most k users disjoint from $\{U, V\}$ must be unable to determine any information about $K_{U,V}$ (where we are speaking here about unconditional security).

In the *Blom Key Predistribution Scheme*, keys are chosen from a finite field $\mathbb{Z}_p$, where $p \geq n$ is prime. The *TA* will transmit $k+1$ elements of $\mathbb{Z}_p$ to each user over a secure channel (as opposed to $n-1$ elements in the trivial key predistribution scheme). Note that the amount of information transmitted by the *TA* is independent of n.

We first present the special case of the *Blom Key Predistribution Scheme* where $k = 1$. Here, the *TA* will transmit two elements of $\mathbb{Z}_p$ to each user over a secure channel. The security achieved is that no individual user, W, say, will be able to determine any information about $K_{U,V}$ if $W \neq U, V$. The *Blom KPS* with $k = 1$ is presented as Protocol 11.2.

In Protocol 11.2, the *TA* generates a random bivariate polynomial (in x and y) and evaluates it at different y-values. These evaluations are given to the network users. A network user can then evaluate the polynomial they receive from the *TA* at different x-values in order to compute keys. An important feature of Protocol 11.2 is that the polynomial f is symmetric: $f(x,y) = f(y,x)$ for all x, y. This property ensures that $g_U(r_V) = g_V(r_U)$, so U and V compute the same key in step 4 of the scheme.

We illustrate the *Blom KPS* with $k = 1$ in the following example.

Example 11.2 Suppose the three users are U, V, and W, $p = 17$, and their public elements are $r_U = 12$, $r_V = 7$, and $r_W = 1$. Suppose that the *TA* chooses $a = 8$, $b = 7$, and $c = 2$, so the polynomial f is

$$f(x,y) = 8 + 7(x+y) + 2xy.$$

The g polynomials are as follows:

$$\begin{aligned} g_U(x) &= 7 + 14x \\ g_V(x) &= 6 + 4x \\ g_W(x) &= 15 + 9x. \end{aligned}$$

Protocol 11.2: BLOM KPS ($k = 1$)

1. A prime number p is made public, and for each user U, an element $r_U \in \mathbb{Z}_p$ is made public. The elements r_U must be distinct.

2. The *TA* chooses three random elements $a, b, c \in \mathbb{Z}_p$ (not necessarily distinct), and forms the polynomial

$$f(x,y) = a + b(x+y) + cxy \bmod p.$$

3. For each user U, the *TA* computes the polynomial

$$g_U(x) = f(x, r_U) \bmod p$$

and transmits $g_U(x)$ to U over a secure channel. Note that $g_U(x)$ is a linear polynomial in x, so it can be written as

$$g_U(x) = a_U + b_U x,$$

where

$$a_U = a + br_U \bmod p \quad \text{and} \quad b_U = b + cr_U \bmod p.$$

4. If U and V want to communicate, then they use the common key

$$K_{U,V} = K_{V,U} = f(r_U, r_V) = a + b(r_U + r_V) + cr_U r_V \bmod p,$$

where U computes

$$K_{U,V} = g_U(r_V)$$

and V computes

$$K_{V,U} = g_V(r_U).$$

The three keys are thus

$$\begin{aligned} K_{U,V} &= 3 \\ K_{U,W} &= 4 \\ K_{V,W} &= 10. \end{aligned}$$

U would compute

$$K_{U,V} = g_U(r_V) = 7 + 14 \times 7 \bmod 17 = 3.$$

while V would compute

$$K_{V,U} = g_V(r_U) = 6 + 4 \times 12 \bmod 17 = 3.$$

We leave the computation of the other keys as an exercise for the reader. □

We now prove that no one user can determine the key of two other users.[2]

THEOREM 11.1 *The Blom Key Predistribution Scheme with $k = 1$ is unconditionally secure against any individual user.*

PROOF Let's suppose that user W wants to try to compute the key

$$K_{U,V} = a + b(r_U + r_V) + c\, r_U r_V \bmod p,$$

where $W \neq U, V$. The values r_U and r_V are public, but a, b, and c are unknown. W knows the values

$$a_W = a + b\, r_W \bmod p$$

and

$$b_W = b + c\, r_W \bmod p,$$

because these are the coefficients of the polynomial $g_W(x)$ that was sent to W by the TA.

We will show that the information known by W is consistent with any possible value $K^* \in \mathbb{Z}_p$ of the key $K_{U,V}$, and therefore, W cannot rule out any values for $K_{U,V}$. Consider the following matrix equation (in $\mathbb{Z}_p$):

$$\begin{pmatrix} 1 & r_U + r_V & r_U r_V \\ 1 & r_W & 0 \\ 0 & 1 & r_W \end{pmatrix} \begin{pmatrix} a \\ b \\ c \end{pmatrix} = \begin{pmatrix} K^* \\ a_W \\ b_W \end{pmatrix}.$$

The first equation represents the hypothesis that $K_{U,V} = K^*$; the second and third equations contain the information that W knows about a, b, and c from $g_W(x)$. The determinant of the coefficient matrix is

$$r_W{}^2 + r_U r_V - (r_U + r_V) r_W = (r_W - r_U)(r_W - r_V),$$

where all arithmetic is done in $\mathbb{Z}_p$. Because $r_W \neq r_U$, $r_W \neq r_V$, and p is prime, it follows that the coefficient matrix has non-zero determinant modulo p, and hence the matrix equation has a unique solution in $\mathbb{Z}_p$ for a, b, and c. Therefore, we have shown that any possible value K^* of $K_{U,V}$ is consistent with the information known to W. Hence, W cannot compute $K_{U,V}$. ∎

On the other hand, a coalition of two users, say $\{W, X\}$, will be able to determine any key $K_{U,V}$ where $\{W, X\} \cap \{U, V\} = \emptyset$. W and X together know that

$$\begin{aligned} a_W &= a + b\, r_W, \\ b_W &= b + c\, r_W, \\ a_X &= a + b\, r_X, \quad \text{and} \\ b_X &= b + c\, r_X. \end{aligned}$$

[2]Here, we just show that no user W can rule out any possible value of a key $K_{U,V}$. It is in fact possible to prove a stronger result, analogous to a perfect secrecy type condition (Section 3.3). Such a result would have the form $\mathbf{Pr}[K_{U,V} = K^* | g_W(x)] = \mathbf{Pr}[K_{U,V} = K^*]$ for all $K^* \in \mathbb{Z}_p$.

Thus they have four equations in three unknowns, and they can easily compute the unique solution for a, b, and c. Once they know a, b, and c, they can form the polynomial $f(x, y)$ and compute any key they desire. Hence, we have shown the following:

THEOREM 11.2 *The Blom Key Predistribution Scheme with $k = 1$ can be broken by any coalition of two users.*

It is straightforward to generalize the *Blom Key Predistribution Scheme* to be secure against coalitions of size $k \geq 1$. The only thing that changes is that the polynomial $f(x, y)$ has degree equal to k (in x and y). Therefore, the *TA* uses a polynomial $f(x, y)$ having the form

$$f(x,y) = \sum_{i=0}^{k} \sum_{j=0}^{k} a_{i,j}\, x^i y^j \bmod p,$$

where $a_{i,j} \in \mathbb{Z}_p$ $(0 \leq i \leq k, 0 \leq j \leq k)$, and $a_{i,j} = a_{j,i}$ for all i, j. Note that the polynomial $f(x, y)$ is symmetric, as before. The remainder of the scheme is unchanged; see Protocol 11.3.

We will show that the *Blom KPS* satisfies the following security properties:

1. No set of k users, say $W_1, \ldots, W_k$, can determine any information about a key for two other users, say $K_{U,V}$.
2. Any set of $k + 1$ users, say $W_1, \ldots, W_{k+1}$, can break the scheme.

First, we consider how $k + 1$ users can break the scheme. A set of users $W_1, \ldots, W_{k+1}$ collectively know the polynomials

$$g_{W_i}(x) = f(x, r_{W_i}) \bmod p,$$

for $1 \leq i \leq k + 1$. Rather than attempt to modify the attack we presented in the case $k = 1$, we will present a more general and elegant approach. This attack makes use of certain formulas for ***polynomial interpolation***, which are presented in the next two theorems.

THEOREM 11.3 (Lagrange interpolation formula) *Suppose p is prime, suppose $x_1, x_2, \ldots, x_{m+1}$ are distinct elements in $\mathbb{Z}_p$, and suppose $a_1, a_2, \ldots, a_{m+1}$ are (not necessarily distinct) elements in $\mathbb{Z}_p$. Then there is a unique polynomial $A(x) \in \mathbb{Z}_p[x]$ having degree at most m, such that $A(x_i) = a_i$, $1 \leq i \leq m + 1$. The polynomial $A(x)$ is as follows:*

$$A(x) = \sum_{j=1}^{m+1} a_j \prod_{1 \leq h \leq m+1, h \neq j} \frac{x - x_h}{x_j - x_h}. \tag{11.1}$$

This formula might appear to be rather mysterious! But it actually is not too difficult to prove that it works. If we set $x = x_i$ in (11.1), we see that every term in the summation evaluates to 0, except for the term $j = i$. For this term, the product evaluates to 1 and hence $A(x_i) = a_i$.

The Lagrange interpolation formula also has a bivariate form, which we state now.

Protocol 11.3: BLOM KPS (GENERAL VERSION)

1. A prime number p is made public, and for each user U, an element $r_U \in \mathbb{Z}_p$ is made public. The elements r_U must be distinct.

2. For $0 \leq i,j \leq k$, the TA chooses random elements $a_{i,j} \in \mathbb{Z}_p$, such that $a_{i,j} = a_{j,i}$ for all i,j. Then the TA forms the polynomial

$$f(x,y) = \sum_{i=0}^{k} \sum_{j=0}^{k} a_{i,j}\, x^i y^j \bmod p.$$

3. For each user U, the TA computes the polynomial

$$g_U(x) = f(x, r_U) \bmod p = \sum_{i=0}^{k} a_{U,i}\, x^i$$

and transmits the coefficient vector $(a_{U,0}, \ldots, a_{U,k})$ to U over a secure channel.

4. For any two users U and V, the key $K_{U,V} = f(r_U, r_V)$, where U computes

$$K_{U,V} = g_U(r_V)$$

and V computes

$$K_{V,U} = g_V(r_U).$$

THEOREM 11.4 (Bivariate Lagrange interpolation formula) *Suppose p is prime, suppose that $y_1, y_2, \ldots, y_{m+1}$ are distinct elements in $\mathbb{Z}_p$, and suppose that $a_1(x), a_2(x), \ldots, a_{m+1}(x) \in \mathbb{Z}_p[x]$ are polynomials of degree at most m. Then there is a unique polynomial $A(x,y) \in \mathbb{Z}_p[x,y]$ having degree at most m (in x and y), such that $A(x, y_i) = a_i(x)$, $1 \leq i \leq m+1$. The polynomial $A(x,y)$ is as follows:*

$$A(x,y) = \sum_{j=1}^{m+1} a_j(x) \prod_{1 \leq h \leq m+1, h \neq j} \frac{y - y_h}{y_j - y_h}.$$

We provide an example of bivariate Lagrange interpolation.

Example 11.3 Suppose that $p = 13$, $m = 2$, $y_1 = 1$, $y_2 = 2$, $y_3 = 3$,

$$\begin{aligned} a_1(x) &= 1 + x + x^2, \\ a_2(x) &= 7 + 4x^2, \quad \text{and} \\ a_3(x) &= 2 + 9x. \end{aligned}$$

Then

$$\begin{aligned}\frac{(y-2)(y-3)}{(1-2)(1-3)} &= 7y^2+4y+3,\\ \frac{(y-1)(y-3)}{(2-1)(2-3)} &= 12y^2+4y+10, \quad \text{and}\\ \frac{(y-1)(y-2)}{(3-1)(3-2)} &= 7y^2+5y+1.\end{aligned}$$

Therefore,

$$\begin{aligned}A(x,y) &= (1+x+x^2)(7y^2+4y+3)+(7+4x^2)(12y^2+4y+10)\\ &\qquad +(2+9x)(7y^2+5y+1) \bmod 13\\ &= y^2+3y+10+5xy^2+10xy+12x+3x^2y^2+7x^2y+4x^2.\end{aligned}$$

It can easily be verified that $A(x,i) = a_i(x)$, $i = 1,2,3$. For example, when $i = 1$, we have

$$\begin{aligned}A(x,1) &= 1+3+10+5x+10x+12x+3x^2+7x^2+4x^2 \bmod 13\\ &= 14+27x+14x^2 \bmod 13\\ &= 1+x+x^2.\end{aligned}$$

□

It is straightforward to show that the *Blom Key Predistribution Scheme* is insecure against a coalition of size $k+1$. A coalition of size $k+1$, say $W_1, \dots, W_{k+1}$, collectively know $k+1$ polynomials of degree k, namely,

$$g_{W_i}(x) = f(x, r_{W_i}) \bmod p,$$

for $1 \leq i \leq k+1$. Using the bivariate interpolation formula, they can compute $f(x,y)$. This is done exactly as in Example 11.3. After having computed $f(x,y)$, they can compute any key $K_{U,V}$ that they desire.

We can also show that the *Blom Key Predistribution Scheme* is secure against a coalition of size k by a modification of the preceding argument. A coalition of size k, say $W_1, \dots, W_k$, collectively know k polynomials of degree k, namely,

$$g_{W_i}(x) = f(x, r_{W_i}) \bmod p,$$

for $1 \leq i \leq k$. We show that this information is consistent with any possible value of the key. Let K be the real key (whose value is unknown to the coalition), and let K^* be arbitrary. We will show that there is a symmetric polynomial $f^*(x,y)$ that is consistent with the information known to the coalition, and such that the secret key associated with the polynomial $f^*(x,y)$ is K^*. Therefore, the coalition cannot rule out any possible values of the key.

We define the polynomial $f^*(x,y)$ as follows:

$$f^*(x,y) = f(x,y) + (K^* - K) \prod_{1 \leq i \leq k} \frac{(x - r_{W_i})(y - r_{W_i})}{(r_U - r_{W_i})(r_V - r_{W_i})}. \tag{11.2}$$

We list some properties of $f^*(x,y)$:

1. First, it is easy to see that f^* is a symmetric polynomial (i.e., $f(x,y) = f(y,x)$), because $f(x,y)$ is symmetric and the product in (11.2) is also symmetric in x and y.

2. Next, for $1 \leq i \leq k$, it holds that
$$f^*(x, r_{W_i}) = f(x, r_{W_i}) = g_{W_i}(x).$$
This is because every product in (11.2) contains a term equal to 0 when $y = r_{W_i}$, and hence the product is 0.

3. Finally,
$$f^*(r_U, r_V) = f(r_U, r_V) + K^* - K = K^*,$$
because the product in (11.2) is equal to 1.

These three properties establish that, for any possible value K^* of the key, there is a symmetric polynomial $f^*(x,y)$ such that the key $f^*(U,V) = K^*$ and such that the secret information held by the coalition of size k is unchanged.

Summarizing, we have proven the following theorem.

THEOREM 11.5 *The* Blom Key Predistribution Scheme *is unconditionally secure against any coalition of k users. However, any coalition of size $k+1$ can break the scheme.*

One drawback of the *Blom Key Predistribution Scheme* is that there is a sharp security threshold (namely, the value of k) that must be prespecified. Once more than k users decide to collaborate, the whole scheme can be broken. On the other hand, the *Blom Key Predistribution Scheme* is optimal with respect to its storage requirements: It has been proven that any unconditionally secure key predistribution that is secure against coalitions of size k requires each user's storage to be at least $k+1$ times the length of a key.

11.2.3 Key Predistribution in Sensor Networks

A ***wireless sensor network*** (or WSN) consists of a large number, say m, of identical sensor nodes that are randomly deployed over a target area. After deployment, each node communicates in a wireless manner with other nodes that are within communication range, thus forming an ***ad hoc network***. Because it is easy to eavesdrop on wireless communication, it is desirable for appropriate cryptographic tools to be used to provide confidentiality and message authentication. Sensor nodes typically are restricted in their computational ability and power. Hence, in many situations, it is preferable to use secret-key cryptography rather than relying on more computationally-intensive public-key techniques. This of course requires nodes to share secret keys; one standard approach to providing such keys is the use of a KPS, in which a set of keys is stored in each node's ***keyring*** prior to deployment.

For reasons of security (which we will discuss in detail a bit later), it is not advisable to use a single, common key throughout the network. On the other hand, it may not be feasible, due to storage limitations, for each node to store $m-1$ different secret keys (one for every other node in the network). A keyring will typically contain a set of fewer than $m-1$ keys, enabling a node to communicate directly with some subset of the other nodes in the network.

After the nodes have been deployed, nodes that are within communication range execute a ***shared key discovery*** protocol to determine which keys they have in common. Two nodes that share at least one key will be able to derive a new key that is used to secure communication between them using an appropriate key derivation function. This is referred to as a secure ***link*** between these nodes.

Key predistribution schemes for WSNs can be evaluated using certain metrics that measure the performance of the resulting networks. As we have mentioned, we wish to restrict the total amount of memory each node must use for storing keys. Second, after the nodes have been deployed, it is desirable for there to be as many secure links as possible between neighbouring nodes, so as to increase the (secure) ***connectivity*** of the resulting network. The extent to which a KPS facilitates achieving this objective is frequently measured by computing the probability that two random nodes in the network can establish a secure link, provided that they are within wireless communication range.

In addition, we might wish to measure the scheme's ability to withstand adversarial attack. A widely studied attack model is called ***random node capture***. We assume that the adversary can eavesdrop on all communication in the network, and it can also compromise a certain number of randomly chosen nodes and extract any keys that they contain.

Note that it is possible that a link $\{A, B\}$ is "broken" after the capture of another node C. This happens when $A \cap B \subseteq C$, where A, B, and C denote the sets of keys held by the three corresponding nodes. For example, suppose A holds keys K_1, K_3, K_6, and K_9; suppose B holds keys K_2, K_3, K_7, and K_8; and suppose C holds keys K_1, K_2, K_3, and K_9. Then the capture of node C breaks the link $\{A, B\}$, because $A \cap B = \{K_3\}$ and $K_3 \in C$.

One measure of the ***resilience*** of a KPS against an attacker is the probability that a randomly chosen link is broken when an attacker compromises a single node (not in the link) that is chosen uniformly at random, and then extracts the keys in this node. There is an inherent tradeoff between the need to provide good connectivity and the need to maintain a high level of resilience, without requiring an excessive number of keys to be stored.

We provide a small example to illustrate the concepts introduced above.

Example 11.4 Suppose we have a network $\mathcal{U}$ with $m = 12$ nodes, denoted $U_1, \dots, U_{12}$. Suppose there are a total of 9 keys, denoted $K_1, \dots, K_9$, and each node has exactly three nodes in its keyring, as follows:

node	keyring	node	keyring	node	keyring
U_1	K_1, K_2, K_3	U_2	K_4, K_5, K_6	U_3	K_7, K_8, K_9
U_4	K_1, K_4, K_7	U_5	K_2, K_5, K_8	U_6	K_3, K_6, K_9
U_7	K_1, K_5, K_9	U_8	K_2, K_6, K_7	U_9	K_3, K_4, K_8
U_{10}	K_1, K_6, K_8	U_{11}	K_2, K_4, K_9	U_{12}	K_3, K_5, K_8

Protocol 11.4: LEE-STINSON LINEAR KPS

1. Let p be a prime number and let $k \leq p$.
2. There are kp keys in the scheme, denoted $K_{i,j}$, $0 \leq i \leq k-1, 0 \leq j \leq p-1$.
3. There are p^2 nodes in the network, denoted as $U_{a,b}$, $0 \leq a, b \leq p-1$.
4. There are k keys to given to each node by the TA. The keys given to $U_{a,b}$ are

$$K_{i,ai+b \bmod p},$$

$0 \leq i \leq k-1$.

This scheme satisfies two important properties:

1. Each key is contained in four nodes, and
2. two nodes have exactly zero or one common key.

From these two properties, we see that every node is contained in exactly nine links. The total number of links is $12 \times 9/2 = 54$ and the probability that a random pair of nodes forms a link is

$$\frac{54}{\binom{12}{2}} = \frac{54}{66} = \frac{9}{11}.$$

Now we consider resilience against an adversary who captures one node. Consider any link $\{U_i, U_j\}$. There is exactly one key held by U_i and U_j, say K_h. The key K_h is held by two of the other ten nodes in the network, so the probability this link is broken in the described attack model is $2/10 = 1/5$. □

We now present in Protocol 11.4 a useful and flexible scheme known as the *Lee-Stinson Linear KPS*. In this scheme, each node is labeled by an ordered pair (a, b), which defines a "line" $y = ax + b$. The points on this line identify the keys given to that node.

LEMMA 11.6 *Each key $K_{i,j}$ in the Lee-Stinson Linear Scheme is held by exactly p nodes.*

PROOF Node $U_{a,b}$ holds key $K_{i,j}$ if and only if

$$ai + b \equiv j \pmod{p}.$$

For any $a \in \mathbb{Z}_p$, this congruence has a unique solution for b, namely,

$$b = j - ai \bmod p.$$

Therefore, the p nodes that hold $K_{i,j}$ are

$$U_{a,j-ai \bmod p},$$

$a \in \mathbb{Z}_p$. ∎

LEMMA 11.7 *Suppose $U_{a,b}$ and $U_{a',b'}$ are two distinct nodes. Then the following hold:*

1. *If $a = a'$ (and hence $b \neq b'$), then $U_{a,b}$ and $U_{a',b'}$ do not share a common key.*
2. *Otherwise, compute $i = (b' - b)(a - a')^{-1} \bmod p$. If $0 \leq i \leq k-1$, then $U_{a,b}$ and $U_{a',b'}$ share the common key $K_{(i,ai+b \bmod p)}$. If $i \geq k$, then $U_{a,b}$ and $U_{a',b'}$ do not share a common key.*

PROOF A key $K_{i,j}$ is held by $U_{a,b}$ and $U_{a',b'}$ if and only if

$$ai + b \equiv j \pmod p$$

and

$$a'i + b' \equiv j \pmod p.$$

Suppose first that $a = a'$. Then, subtracting the two congruences, we see that $b = b'$, which contradicts the requirement that $U_{a,b}$ and $U_{a',b'}$ are two distinct nodes.

Now suppose that $a \neq a'$. When we subtract the second congruence from the first one, we get

$$(a - a')i + b - b' \equiv 0 \pmod p.$$

Thus we can solve for i modulo p, obtaining

$$i = (b' - b)(a - a')^{-1} \bmod p.$$

Then j is determined uniquely modulo p as $j = ai + b \bmod p$. This pair (i,j) specifies a (unique) key held by $U_{a,b}$ and $U_{a',b'}$, provided that $K_{i,j}$ is in fact a key in the scheme. This requires that $0 \leq i \leq k-1$, which determines the two possible subcases that arise when $a \neq a'$. ∎

The formulas derived in Lemma 11.7 allow any two nodes $U_{a,b}$ and $U_{a',b'}$ to easily determine if they hold a common key, based only on their "identifiers" (a,b) and (a',b').

THEOREM 11.8 *The probability that a random pair of nodes in the Lee-Stinson Linear KPS forms a link is $k/(p+1)$.*

PROOF Consider any node $U_{a,b}$. Each of the k keys held by $U_{a,b}$ occurs in $p-1$ other nodes. No pair of nodes has two keys in common, so $U_{a,b}$ has a shared key with exactly $k(p-1)$ nodes. There are $p^2 - 1$ nodes other than p, so the probability that $U_{a,b}$ forms a link with another node $U_{a',b'}$ is exactly

$$\frac{k(p-1)}{p^2-1} = \frac{k}{p+1}.$$

∎

THEOREM 11.9 *The probability that a link* $\{U_{a,b}, U_{a',b'}\}$*in the Lee-Stinson Linear KPS is broken by the capture of a random node (distinct from* $U_{a,b}$ *and* $U_{a',b'}$*) is* $(p-2)/(p^2-2)$.

PROOF The nodes $U_{a,b}$ and $U_{a',b'}$ contain exactly one common key. This key occurs in $p-2$ additional nodes. There are p^2-2 nodes distinct from $U_{a,b}$ and $U_{a',b'}$, so the probability that a node chosen randomly from these p^2-2 nodes breaks the given link is $(p-2)/(p^2-2)$. ∎

11.3 Session Key Distribution Schemes

Recall from the introduction of this chapter that the *TA* is assumed to have a shared secret key with every network user in a session key distribution scheme. We will use K_{Alice} to denote Alice's secret key, K_{Bob} is Bob's secret key, etc. In a session key distribution scheme, the *TA* chooses session keys and distributes them on-line in encrypted form, upon the request of network users.

Eventually, we will define attack models and adversarial goals for session key distribution. However, it is not easy to formulate precise definitions because session key distribution schemes sometimes do not include mutual identification of the users in a session of the scheme. Therefore, we begin by giving a historical tour of some important SKDSs and describing some attacks on them, before we proceed to a more formal treatment of the subject.

11.3.1 The Needham-Schroeder Scheme

One of the first session key distribution schemes is the *Needham-Schroeder SKDS*, which was proposed in 1978; this scheme is presented in Protocol 11.5. The diagram in Figure 11.1 depicts the five flows in the *Needham-Schroeder SKDS*.

Here is a summary of the main steps in the scheme. In flow 1, Alice asks the *TA* for a session key to communicate with Bob. At this point, Bob might not even be aware of Alice's request. The *TA* transmits the encrypted session key to Alice in flow 2, and Alice sends an encrypted session key to Bob in flow 3. Thus flows 1–3 of *Needham-Schroeder* comprise the session key distribution: the session key K is encrypted using the secret keys of Alice and Bob and it is distributed to both of them. The purpose of flows 4 and 5 is to convince Bob that Alice actually possesses the session key K. This is accomplished by having Alice use the new session key to encrypt the challenge $r_B - 1$; the process is called ***key confirmation*** (from Alice to Bob).

There are some validity checks required in the *Needham-Schroeder SKDS*, where the term ***validity check*** refers to verifying that decrypted data has the correct format and contains expected information. (Note that there are no message

Protocol 11.5: NEEDHAM-SCHROEDER SKDS

1. Alice chooses a random number, r_A. Alice sends $ID(Alice)$, $ID(Bob)$, and r_A to the TA.

2. The TA chooses a random session key, K. Then it computes

$$t_{Bob} = e_{K_{Bob}}(K \parallel ID(Alice))$$

(which is called a ***ticket to Bob***) and

$$y_1 = e_{K_{Alice}}(r_A \parallel ID(Bob) \parallel K \parallel t_{Bob}),$$

and it sends y_1 to Alice.

3. Alice decrypts y_1 using her key K_{Alice}, obtaining K and t_{Bob}. Then Alice sends t_{Bob} to Bob.

4. Bob decrypts t_{Bob} using his key K_{Bob}, obtaining K. Then, Bob chooses a random number r_B and computes $y_2 = e_K(r_B)$. Bob sends y_2 to Alice.

5. Alice decrypts y_2 using the session key K, obtaining r_B. Then Alice computes $y_3 = e_K(r_B - 1)$ and she sends y_3 to Bob.

authentication codes being used in the *Needham-Schroeder SKDS*.) These validity checks are as follows:

1. When Alice decrypts y_1, she checks to see that the plaintext $d_{K_{Alice}}(y_1)$ has the form
$$d_{K_{Alice}}(y_1) = r_A \parallel ID(Bob) \parallel K \parallel t_{Bob}$$
for some K and t_{Bob}. If this above condition holds, then Alice "accepts"; otherwise, Alice "rejects" and aborts the session.

2. When Bob decrypts y_3, he checks to see that the plaintext
$$d_K(y_3) = r_B - 1.$$
If this condition holds, then Bob "accepts"; otherwise, Bob "rejects."

11.3.2 The Denning-Sacco Attack on the NS Scheme

In 1981, Denning and Sacco discovered a replay attack on the *Needham-Schroeder SKDS*. We present this attack now. Suppose Oscar records a session, say $\mathcal{S}$, of the *Needham-Schroeder SKDS* scheme between Alice and Bob, and somehow he obtains the session key, K, for the session $\mathcal{S}$. (Recall that this attack model is called a "known session key attack.") Then Oscar can initiate a new session, say

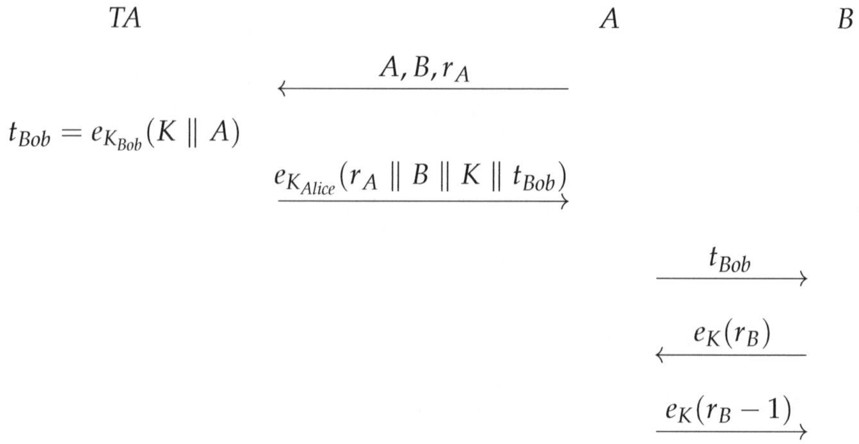

TA		A		B
	$\xleftarrow{A,B,r_A}$			
$t_{Bob} = e_{K_{Bob}}(K \parallel A)$				
	$\xrightarrow{e_{K_{Alice}}(r_A \parallel B \parallel K \parallel t_{Bob})}$			
			$\xrightarrow{t_{Bob}}$	
			$\xleftarrow{e_K(r_B)}$	
			$\xrightarrow{e_K(r_B - 1)}$	

Note that "A" denotes "$ID(Alice)$" and "B" denotes "$ID(Bob)$."

FIGURE 11.1: Information flows in the *Needham-Schroeder SKDS*

$\mathcal{S}'$, of the *Needham-Schroeder SKDS* with Bob, starting with the third flow of the session $\mathcal{S}'$, by sending the previously used ticket, t_{Bob}, to Bob:

Oscar		Bob
	$\xrightarrow{t_{Bob} = e_{K_{Bob}}(K \parallel A)}$	
	$\xleftarrow{e_K(r'_B)}$	
	$\xrightarrow{e_K(r'_B - 1)}$	

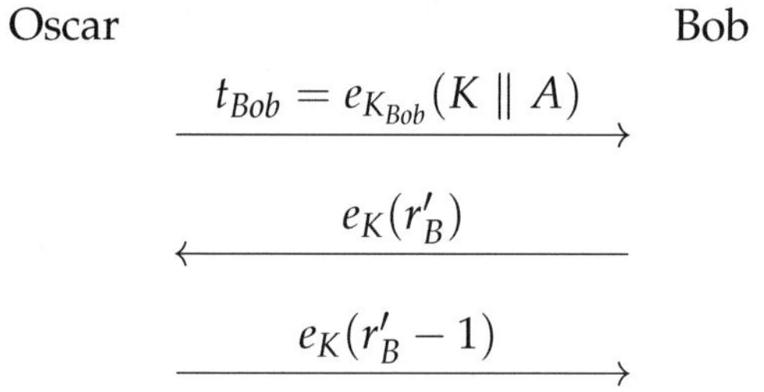

Notice that when Bob replies with $e_K(r'_B)$, Oscar can decrypt this using the known key K, subtract 1, and then encrypt the result. The value $e_K(r'_B - 1)$ is sent to Bob in the last flow of the session $\mathcal{S}'$. Bob will decrypt this and "accept."

Let's consider the consequences of this attack. At the end of the session $\mathcal{S}'$ between Oscar and Bob, Bob thinks he has a "new" session key, K, shared with Alice (this is because $ID(Alice)$ occurs in the ticket t_{Bob}). This key K is known to Oscar, but it may not be known to Alice, because Alice might have thrown away the key K after the previous session with Bob, namely $\mathcal{S}$, terminated. Hence, there are two ways in which Bob is deceived by this attack:

1. The key K that is distributed in the session $\mathcal{S}'$ is not known to Bob's intended peer, Alice.

2. The key K for the session $\mathcal{S}'$ is known to someone other than Bob's intended peer (namely, it is known to Oscar).

Protocol 11.6: SIMPLIFIED KERBEROS V5

1. Alice chooses a random number, r_A. Alice sends $ID(Alice)$, $ID(Bob)$, and r_A to the TA.

2. The TA chooses a random session key K and a validity period (or lifetime), L. Then it computes a ticket to Bob,

$$t_{Bob} = e_{K_{Bob}}(K \parallel ID(Alice) \parallel L),$$

and

$$y_1 = e_{K_{Alice}}(r_A \parallel ID(Bob) \parallel K \parallel L).$$

The TA sends t_{Bob} and y_1 to Alice.

3. Alice decrypts y_1 using her key K_{Alice}, obtaining K. Then Alice determines the current time, *time*, and she computes

$$y_2 = e_K(ID(Alice) \parallel time).$$

Finally, Alice sends t_{Bob} and y_2 to Bob.

4. Bob decrypts t_{Bob} using his key K_{Bob}, obtaining K. He also decrypts y_2 using the key K, obtaining *time*. Then, Bob computes

$$y_3 = e_K(time + 1).$$

Finally, Bob sends y_3 to Alice.

11.3.3 Kerberos

Kerberos comprises a popular series of schemes for session key distribution that were developed at MIT in the late 1980s and early 1990s. We provide a simplified treatment of version five of the scheme. This is presented as Protocol 11.6. A diagram depicting the four flows in a session of the scheme is given in Figure 11.2.

As was the case with *Needham-Schroeder*, there are certain validity checks required in *Kerberos*. These are as follows:

1. When Alice decrypts y_1, she checks to see that the plaintext $d_{K_{Alice}}(y_1)$ has the form
$$d_{K_{Alice}}(y_1) = r_A \parallel ID(Bob) \parallel K \parallel L,$$
for some K and L. If this condition does not hold, then Alice "rejects" and aborts the current session.

2. When Bob decrypts y_2 and t_{Bob}, he checks to see that the plaintext $d_K(y_2)$ has

$$TA \qquad\qquad\qquad A \qquad\qquad\qquad B$$

$$\xleftarrow{\quad A, B, r_A \quad}$$

$$\xrightarrow{\quad t_{Bob}, y_1 \quad}$$

$$\xrightarrow{\quad t_{Bob}, y_2 \quad}$$

$$\xleftarrow{\quad y_3 \quad}$$

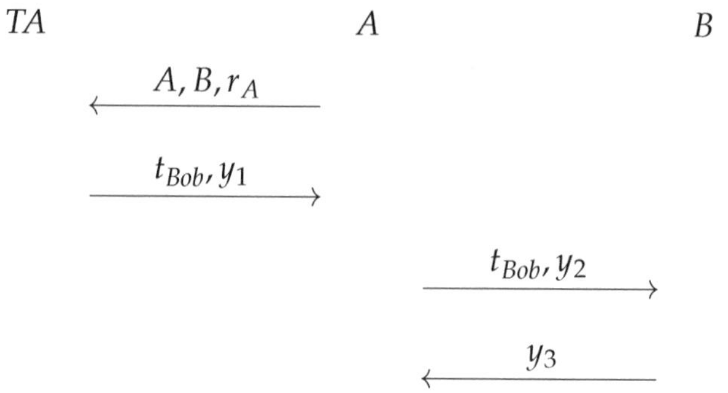

where

$$
\begin{aligned}
A &= ID(Alice), \\
B &= ID(Bob), \\
t_{Bob} &= e_{K_{Bob}}(K \parallel A \parallel L), \\
y_1 &= e_{K_{Alice}}(r_A \parallel B \parallel K \parallel L) \\
y_2 &= e_K(A \parallel time), \quad \text{and} \\
y_3 &= e_K(time + 1).
\end{aligned}
$$

FIGURE 11.2: The flows in *Kerberos V5*

the form

$$d_K(y_2) = ID(Alice) \parallel time$$

and the plaintext $d_{K_{Bob}}(t_{Bob})$ has the form

$$d_{K_{Bob}}(t_{Bob}) = K \parallel ID(Alice) \parallel L,$$

where $ID(Alice)$ is the same in both plaintexts and $time \leq L$. If these conditions hold, then Bob "accepts"; otherwise, Bob "rejects."

3. When Alice decrypts y_3, she checks that $d_K(y_3) = time + 1$. If this condition holds, then Alice "accepts"; otherwise, Alice "rejects."

Here is a summary of the rationale behind some of the features in *Kerberos*. When a request for a session key is sent by Alice to the *TA*, the *TA* will generate a new random session key K. As well, the *TA* will specify the **lifetime**, L, during which K will be valid. That is, the session key K is to be regarded as a valid key until time L. All this information is encrypted before it is transmitted to Alice.

Alice can use her secret key to decrypt y_1, and thus obtain K and L. She will verify that the current time is within the lifetime of the key and that y_1 contains Alice's random challenge, r_A. She can also verify that y_1 contains $ID(Bob)$, where Bob is Alice's intended peer. These checks prevent Oscar from replaying an "old" y_1, which might have been transmitted by the *TA* in a previous session.

Next, Alice will relay t_{Bob} to Bob. As well, Alice will use the new session key K to encrypt the current time (denoted by *time*) and $ID(Alice)$. Then she sends the resulting ciphertext y_2 to Bob.

When Bob receives t_{Bob} and y_2 from Alice, he decrypts t_{Bob} to obtain K, L, and $ID(Alice)$. Then he uses the new session key K to decrypt y_2 and he verifies that $ID(Alice)$, as decrypted from t_{Bob} and y_2, are the same. This assures Bob that the session key encrypted within t_{Bob} is the same key that was used to encrypt y_2. He should also check that $time \leq L$ to verify that the key K has not expired.

Finally, Bob encrypts the value $time + 1$ using the new session key K and sends the result back to Alice. When Alice receives this message, y_3, she decrypts it using K and verifies that the result is $time + 1$. This assures Alice that the session key K has been successfully transmitted to Bob, since K is needed in order to produce the message y_3.

The purpose of the lifetime L is to prevent an active adversary from storing "old" messages for retransmission at a later time, as was done in the Denning-Sacco attack on the *Needham-Schroeder SKDS*. One of the drawbacks of *Kerberos* is that all the users in the network should have synchronized clocks, since the current time is used to determine if a given session key K is valid. In practice, it is very difficult to provide perfect synchronization, so some amount of variation in times must be allowed.

We make a few comments comparing *Needham-Schroeder* to *Kerberos*.

1. In *Kerberos*, ***mutual key confirmation*** is accomplished in flows 3 and 4. By using the new session key K to encrypt $ID(Alice)$, Alice is trying to convince Bob that she knows the value of K. Similarly, when Bob encrypts $time + 1$ using K, he is demonstrating to Alice that he knows the value of K.

2. In *Needham-Schroeder*, information intended for Bob is doubly encrypted: the ticket t_{Bob}, which is already encrypted, is re-encrypted using Alice's secret key. This seems to serve no useful purpose and it adds unnecessary complexity to the scheme. In *Kerberos*, this double encryption was removed.

3. Partial protection against the Denning-Sacco attack is provided in *Kerberos* by verifying that the current time (namely, the value *time*, which is often referred to as a ***timestamp***) lies within the lifetime L. Basically, this limits the time period during which a Denning-Sacco type attack can be carried out.

Needham-Schroeder and *Kerberos* have some features that are not generally regarded as useful in present day SKDSs. We discuss these briefly before proceeding to the development of a secure SKDS.

1. Timestamps require reliable, synchronized clocks. Schemes using timestamps are hard to analyze and it is difficult to give convincing security proofs for them. For this reason, it is generally preferred to use random challenges rather than timestamps, if possible.

2. Key confirmation is not necessarily an important attribute of a session key distribution scheme. For example, possession of a key during a session of the SKDS does not imply possession of the key at a later time, when it is actually going to be used. For this reason, it is now often recommended that key confirmation be omitted from SKDSs.

3. In *Needham-Schroeder* and *Kerberos*, encryption is used to provide both secrecy and authenticity. However, it is preferable to use encryption for secrecy and a message authentication code to provide authenticity. For example, in the second flow of *Needham-Schroeder*, we could remove the double encryption and use MACs for authentication, as follows:

   The *TA* chooses a random session key K. Then it computes

$$y_1 = (e_{K_{Bob}}(K), MAC_{Bob}(ID(Alice) \parallel e_{K_{Bob}}(K))),$$

   and

$$y_1' = (e_{K_{Alice}}(K), MAC_{Alice}(ID(Bob) \parallel r_A \parallel e_{K_{Alice}}(K))).$$

   The *TA* would send y_1 and y_1' to Alice, who would then relay y_1 to Bob.

   The revised second flow does not fix the flaw found by Denning and Sacco, however.

4. In order to prevent the Denning-Sacco attack, the flow structure of the scheme must be modified. Any "secure" scheme should involve Bob as an active participant prior to his receiving the session key, in order to prevent Denning-Sacco type replay attacks. The solution requires Alice to contact Bob (or vice versa) before sending a request for a session key to the *TA*.

11.3.4 The Bellare-Rogaway Scheme

Bellare and Rogaway proposed an SKDS in 1995 and provided a proof of security for their scheme, under certain assumptions. We begin by describing the *Bellare-Rogaway SKDS* in Protocol 11.7. Then we will proceed to a more formal analysis of the scheme, which will require developing rigorous definitions of the attack model and adversarial goals.

Protocol 11.7 has a different flow structure than the schemes we have considered so far. Alice and Bob both choose random challenges, which are sent to the *TA*. Thus Bob is involved in the scheme before the *TA* issues the session key. The information that the *TA* sends to Alice consists of the following components.

1. A session key (encrypted using Alice's secret key), and
2. a MAC tag computed on the encrypted session key, the identities of Alice and Bob, and Alice's challenge.

The information sent to Bob is analogous.

Alice and Bob will "accept" if their respective tags are valid (where these tags are computed using secret MAC keys that are known to the *TA*). For example,

Protocol 11.7: BELLARE-ROGAWAY SKDS

1. Alice chooses a random number, r_A, and she sends $ID(Alice)$, $ID(Bob)$, and r_A to Bob.
2. Bob chooses a random number, r_B, and he sends $ID(Alice)$, $ID(Bob)$, r_A and r_B to the *TA*.
3. The *TA* chooses a random session key K. Then it computes

$$y_B = (e_{K_{Bob}}(K), MAC_{Bob}(ID(Alice) \parallel ID(Bob) \parallel r_B \parallel e_{K_{Bob}}(K)))$$

and

$$y_A = (e_{K_{Alice}}(K), MAC_{Alice}(ID(Bob) \parallel ID(Alice) \parallel r_A \parallel e_{K_{Alice}}(K))).$$

The *TA* sends y_B to Bob and y_A to Alice.

when Bob receives the encrypted session key, say $y_{B,1}$, and the tag, say $y_{B,2}$, he verifies that

$$y_{B,2} = MAC_{Bob}(ID(Alice) \parallel ID(Bob) \parallel r_B \parallel y_{B,1}).$$

Note the use of "encrypt-then-MAC" in this construction.

Observe that no key confirmation is provided in this scheme. When Alice accepts, for example, she does not know if Bob has accepted, or even if Bob has received the message sent by the *TA*. When Alice accepts, it just means that she has received the information she expected, and this information is valid (or, more precisely, the tag is valid). From Alice's point of view, when she accepts, she believes that she has received a new session key from the *TA*. Moreover, because this session key was encrypted using Alice's secret key, Alice is confident that no one else can compute the session key K from the information that she just received. Of course, Bob should also have received an encryption of the same session key. Alice does not know if this, in fact, transpired, but we will argue that Alice can be confident that no one other than Bob can compute the new session key. The analysis will be similar when the session is examined from Bob's point of view. In other words, we have removed the objective of key confirmation (one-way or two-way) from the SKDS. This is replaced with the somewhat weaker (but still useful) objective that, from the point of view of a participant in the scheme who "accepts," no one other than their intended peer should be able to compute the new session key.

The objective of an adversary will be to cause an honest participant to "accept" in a situation where someone other than the intended peer of that participant knows the value of the session key K. For example, suppose that an honest Alice "accepts" and her intended peer is Bob. The adversary, Oscar, achieves his goal if he (Oscar) can compute the session key, or if some other network user (say Charlie) can compute the session key. On the other hand, Oscar's attack is not considered to be successful if Alice is the only network user who can compute the session key. In this situation, Bob can't compute the session key, but neither can anyone else (except Alice).

Summarizing the above discussion, we will define a ***secure session key distribution scheme*** to be an SKDS in which the following property holds: if a participant in a session "accepts," then the probability that someone other than that participant's intended peer knows the session key is negligible.

We now consider how to go about proving that the *Bellare-Rogaway SKDS* is secure. As usual, we make several reasonable assumptions:

1. Alice, Bob, and the *TA* are honest,
2. the encryption scheme and MAC are secure,
3. secret keys are known only to their intended owners,
4. random challenges are generated using secure random number generators, and
5. the *TA* generates session keys using a secure random number generator.

These assumptions are similar to those made in Chapter 10 in the study of identification schemes.

Let us consider various ways in which Oscar can carry out an attack. For each of these possibilities, we argue that Oscar will not be successful, except with a small probability. These possibilities are not all mutually exclusive.

1. Oscar is a passive adversary.
2. Oscar is an active adversary and Alice is a legitimate participant in the scheme. Oscar may impersonate Bob or the *TA*, and Oscar may intercept and change messages sent during the scheme.
3. Oscar is an active adversary and Bob is a legitimate participant in the scheme. Oscar may impersonate Alice or the *TA*, and Oscar may intercept and change messages sent during the scheme.

We now go on to analyze the possible attacks enumerated above. In each situation, we discuss what the outcome of the scheme will be, subject to our underlying "reasonable" assumptions.

1. If the adversary is passive, then Alice and Bob will both output "accept" in any session in which they are the two participants. Further, they will both be able to decrypt the same session key, K. No one else (including Oscar) is able to compute K, because the encryption scheme is secure.

2. Suppose Alice is a legitimate participant in the scheme. She wishes to obtain a new session key that will be known only to Bob and to herself. However, Alice does not know if she really is communicating with Bob, because Oscar may be impersonating Bob.

   When Alice receives the message y_A, she checks to see that the tag is valid. This tag incorporates Alice's random challenge, r_A, as well as the identities of Alice and Bob, and the encrypted session key $e_{K_{Alice}}(K)$. This convinces Alice that the tag was newly computed by the *TA*, because the *TA* is the only party other than Alice who knows the key MAC_{Alice}. Furthermore, the random challenge r_A prevents replay of a tag from a previous session. Finally, including $e_{K_{Alice}}(K)$ in the tag prevents an adversary from replacing the session key chosen by the *TA* with something else. Therefore, Alice can be confident that Bob (her intended peer) is the only other user who is able to decrypt the session key K, even if Oscar has impersonated Bob in the current session of the scheme.

3. This analysis is essentially the same as in the previous case. Suppose Bob is a legitimate participant in the scheme. He believes he will obtain a new session key that will be known only to Alice and to himself. However, Bob does not know if he really is communicating with Alice, because Oscar may be impersonating Alice.

   When Bob receives the message y_B, he checks to see that the tag is valid. This tag incorporates Bob's random challenge, r_B, as well as the identities of Alice and Bob, and the encrypted session key $e_{K_{Bob}}(K)$. This convinces Bob that the tag was newly computed by the *TA*, because the *TA* is the only party other than Bob who knows the key MAC_{Bob}. Furthermore, the random challenge r_B prevents replay of a tag from a previous session. Finally, including $e_{K_{Bob}}(K)$ in the tag prevents an adversary from replacing the session key chosen by the *TA* with something else. Therefore, Bob can be confident that Alice (his intended peer) is the only other user who is able to decrypt the session key K, even if Oscar has impersonated Alice in the current session of the scheme.

11.4 Re-keying and the Logical Key Hierarchy

In this section, we consider the setting of a long-lived dynamic group of network users, say $\mathcal{U}$, with an online *TA*. The *TA* might want to broadcast messages to every user in the group, but members may join or leave the group over time. Communications to the group are encrypted with a single ***group key***, and every user has a copy of the group key. Users may also have additional long-lived keys (or LL-keys), which are used to update the system as the group evolves over time. The system is initialized in a ***key predistribution phase***, during which the *TA* gives LL-keys and an initial group key to the users in the network.

When a new user joins the group, that user is given a copy of the current group key, as well as appropriate long-lived keys; this is called a ***user join operation***. When a user U leaves the group, a ***user revocation operation*** is necessary in order to remove the user from the group. The user revocation operation will establish a new group key for the remaining users, namely, all the users in $\mathcal{U}\backslash\{U\}$; this is an example of re-keying. In addition, updating of LL-keys may be required as part of the user revocation operation.

Criteria used to evaluate multicast re-keying schemes include the following:

communication and storage complexity
: This includes the size of broadcasts required for key updating and the size (and number) of secret LL-keys that have to be stored by users.

security
: Here, we mainly consider security against revoked users and coalitions of revoked users. Note that a revoked user has more information than someone who never belonged to the group in the first place. Hence, if we achieve security against revoked users, then this automatically implies security against "outsiders."

flexibility of user revocation
: Flexibility and efficiency of user revocation operations is an important consideration. For example, it might be the case that users must be revoked one at a time. In some schemes, however, ***multiple user revocation*** may be possible (up to some specified number of revoked users). This would be more convenient, because users would need to update their keys less frequently.

flexibility of user join
: There are several possible variations. In some systems, it may be that any number of new users may be added easily to the system. In other systems, it might be the case that the entire system has to be re-initialized in order to add new users (this would be thought of as a one-time system). Obviously, a flexible and efficient user join operation is desirable in the situation where it is expected that new users will want to join the group.

efficiency of updating LL-keys
: Here there are also many possibilities. Perhaps no updating is required (i.e., the LL-keys are static). On the other hand, LL-keys might require updating by an efficient update operation (e.g., via a broadcast). In the worst case, the entire system would have to be reinitialized after a user revocation (basically, this would mean that the system does not accommodate revocation).

We will now present the *Logical Key Hierarchy*, which is a tree-based re-keying scheme. It was suggested (independently) by Wallner, Harder, and Agee, and by Wong and Lam.

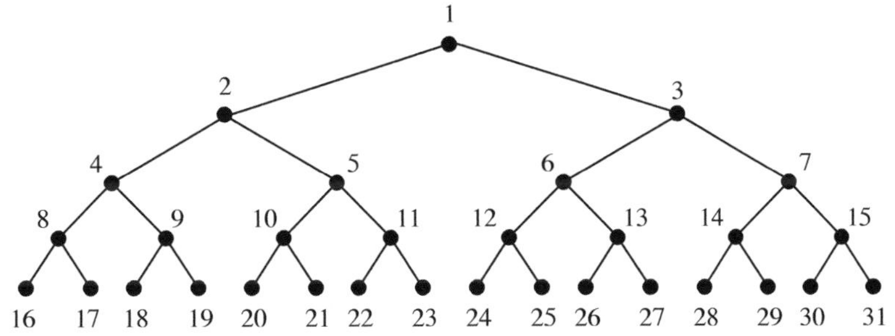

FIGURE 11.3: A binary tree with 16 leaf nodes

We will use a binary tree with nodes labeled in the same fashion as Merkle trees (Section 9.5.3). The only difference is that we do not require that the tree is complete. Suppose the number of users, n, satisfies $2^{d-1} < n \leq 2^d$. We will initially construct a binary tree, say $\mathcal{T}$, of depth d, having exactly n leaf nodes. All the levels of the tree will be filled, except (possibly) for the last level. The n leaf nodes of $\mathcal{T}$ correspond to the n users. For every user U, let U also denote the (leaf) node corresponding to the user U.

There is a key associated with every node in $\mathcal{T}$ (i.e., there is a different key for every leaf node and every internal node). For every node X, let $k(X)$ denote the key for node X. Then $k(R)$ is the group key, where R is the root node of $\mathcal{T}$. Every user U is given the $d+1$ keys corresponding to the nodes of $\mathcal{T}$ that lie on the unique path from U to R in $\mathcal{T}$. Therefore every user has $O(\log n)$ keys.

Example 11.5 A binary tree with $d = 4$ and $n = 16$, having nodes labeled $1, 2 \dots, 2^{d+1} - 1 = 31$, is depicted in Figure 11.3. The 16 users are named $16, \dots, 31$. The group key is $k(1)$ and the keys given to user 25 are $k(1)$, $k(3)$, $k(6)$, $k(12)$, and $k(25)$. □

Now we can describe the basic user revocation operation in the *Logical Key Hierarchy*. Suppose that we wish to remove user U. Let $\mathcal{P}(U)$ denote the set of nodes in the unique path from a leaf node U to the root node R (recall that R has the label 1). It is necessary to change the keys corresponding to the d nodes in $\mathcal{P}(U) \backslash \{U\}$. For each node $X \in \mathcal{P}(U) \backslash \{U\}$, let $k'(X)$ denote the new key for node X. Let $\mathbf{sib}(\cdot)$ denote the sibling of a given node, and let $\mathbf{par}(\cdot)$ denote the parent of a given node. Then, the following $2d - 1$ items are broadcasted by the *TA*:

1. $e_{k(\mathbf{sib}(U))}(k'(\mathbf{par}(U)))$
2. $e_{k(\mathbf{sib}(X))}(k'(\mathbf{par}(X)))$ and $e_{k'(X)}(k'(\mathbf{par}(X)))$, for all nodes $X \in \mathcal{P}(U)$, $X \neq U, R$.

We claim that this broadcast allows any non-revoked user V to update all the keys in the intersection $\mathcal{P}(U) \cap \mathcal{P}(V)$. Perhaps the most convincing way to demonstrate this is to consider an example.

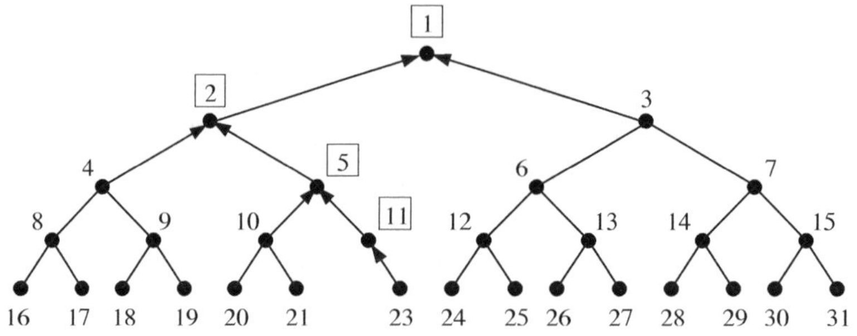

FIGURE 11.4: A broadcast updating a binary tree

Example 11.6 Suppose the *TA* wants to revoke user $U = 22$. The path $\mathcal{P}(U) = \{22, 11, 5, 2, 1\}$. The *TA* creates new keys k'_{11}, k'_5, k'_2, and k'_1. The siblings of the nodes in $\mathcal{P}(U)$ are $\{23, 10, 4, 3\}$. The broadcast consists of

$$\begin{array}{llll} e_{k(23)}(k'(11)) & e_{k(10)}(k'(5)) & e_{k(4)}(k'(2)) & e_{k(3)}(k'(1)) \\ & e_{k'(11)}(k'(5)) & e_{k'(5)}(k'(2)) & e_{k'(2)}(k'(1)). \end{array}$$

This example is depicted in Figure 11.4, where labels of the nodes receiving new keys are boxed and encryptions of new keys are indicated by arrows.

Let's consider how user 23 would update her keys. First she can use her key $k(23)$ to decrypt $e_{k(23)}(k'(11))$; in this way, she computes $k'(11)$. Next, she uses $k'(11)$ to compute $k'(5)$. Then, she uses $k'(5)$ to compute $k'(2)$; and, finally, she uses $k'(2)$ to compute $k'(1)$. □

The depth of the tree used in the *Logical Key Hierarchy* is d. Since d is $\Theta(\log n)$, it follows that every user stores $O(\log n)$ keys and the broadcast has size $O(\log n)$. These quantities are larger than the comparable values for the schemes considered previously. However, because the LL-keys are updated each time a user is revoked, there is no limit on the number of users that can be revoked over time. That is, any number of users can be revoked, without affecting the security of the system.

Simultaneous revocation of more than one user can be done, but it is somewhat complicated (see the Exercises). New users can be added to the *Logical Key Hierarchy*, whenever the current number of users is less than 2^d, by assigning the new user to the leftmost "unoccupied" leaf node of the tree. When the number of users exceeds 2^d, one more level of nodes in the tree must be created. This increases the depth of the tree by one, and it allows the number of users to be doubled.

11.5 Threshold Schemes

In a bank, there is a vault that must be opened every day. The bank employs three senior tellers, but they do not trust the combination to any individual teller.

Hence, we would like to design a system whereby any two of the three senior tellers can gain access to the vault, but no individual teller can do so. This problem can be solved by means of a threshold scheme.

Here is an interesting "real-world" example of this situation: According to *Time Magazine*,[3] control of nuclear weapons in Russia in the early 1990s depended upon a similar "two-out-of-three" access mechanism. The three parties involved were the President, the Defense Minister, and the Defense Ministry.

Here is an informal definition of a threshold scheme.

> **Definition 11.1:** Let t, w be positive integers, $t \leq w$. A (t, w)-***threshold scheme*** is a method of sharing a ***key*** K among a set of w participants (denoted by $\mathcal{P}$), in such a way that any t participants can compute the value of K, but no group of $t-1$ participants can do so.

We will study the unconditional security of threshold schemes. That is, we do not place any limit on the amount of computation that can be performed by any subset of participants.

Note that the examples described above are $(2,3)$-threshold schemes.

The value of K is chosen by a special participant called the ***dealer***. The dealer, who is nothing more than a *TA*, is denoted by D and we assume $D \notin \mathcal{P}$. When D wants to share the key K among the participants in $\mathcal{P}$, he gives each participant some partial information called a ***share***. The shares should be distributed secretly, so no participant knows the share given to another participant.

At a later time, a subset of participants $B \subseteq \mathcal{P}$ will pool their shares in an attempt to compute the key K. (Alternatively, they could give their shares to a trusted entity that will perform the computation for them.) If $|B| \geq t$, then they should be able to compute the value of K as a function of the shares they collectively hold; if $|B| < t$, then they should not be able to compute K. Thus, a threshold scheme can be viewed as a method of ***distributed key predistribution***.

We will use the following notation. Let

$$\mathcal{P} = \{P_i : 1 \leq i \leq w\}$$

be the set of w participants. $\mathcal{K}$ is the ***key set*** (i.e., the set of all possible keys); and $\mathcal{S}$ is the ***share set*** (i.e., the set of all possible shares).

11.5.1 The Shamir Scheme

In this section, we present a method of constructing a (t, w)-threshold scheme, called the *Shamir Threshold Scheme*, which was invented by Shamir in 1979. Let $\mathcal{K} = \mathbb{Z}_p$, where $p \geq w+1$ is prime. Also, let $\mathcal{S} = \mathbb{Z}_p$. Hence, the key will be an element of $\mathbb{Z}_p$, as will be each share given to a participant. The *Shamir Threshold Scheme* is presented as Cryptosystem 11.1.

In this scheme, the dealer constructs a random polynomial $a(x)$ of degree at

[3] *Time Magazine*, May 4, 1992, p. 13

Cryptosystem 11.1: *Shamir (t, w)-Threshold Scheme*

Initialization Phase

1. D chooses w distinct, non-zero elements of $\mathbb{Z}_p$, denoted x_i, $1 \leq i \leq w$ (this is where we require $p \geq w + 1$). For $1 \leq i \leq w$, D gives the value x_i to P_i. The values x_i are public.

Share Distribution

2. Suppose D wants to share a key $K \in \mathbb{Z}_p$. D secretly chooses (independently at random) $t - 1$ elements of $\mathbb{Z}_p$, which are denoted $a_1, \ldots, a_{t-1}$.

3. For $1 \leq i \leq w$, D computes $y_i = a(x_i)$, where

$$a(x) = K + \sum_{j=1}^{t-1} a_j x^j \bmod p.$$

4. For $1 \leq i \leq w$, D gives the share y_i to P_i.

most $t - 1$ in which the constant term is the key, K. Every participant P_i obtains a point (x_i, y_i) on this polynomial.

Let's look at how a subset B of t participants can reconstruct the key. This is basically accomplished by means of polynomial interpolation. We will describe a couple of methods of doing this.

Suppose that participants $P_{i_1}, \ldots, P_{i_t}$ want to determine K. They know that

$$y_{i_j} = a(x_{i_j}),$$

for $1 \leq j \leq t$, where $a(x) \in \mathbb{Z}_p[x]$ is the (secret) polynomial chosen by D. Since $a(x)$ has degree at most $t - 1$, $a(x)$ can be written as

$$a(x) = a_0 + a_1 x + \cdots + a_{t-1} x^{t-1},$$

where the coefficients $a_0, \ldots, a_{t-1}$ are unknown elements of $\mathbb{Z}_p$, and $a_0 = K$ is the key. Since $y_{i_j} = a(x_{i_j})$, $1 \leq j \leq t$, the subset B can obtain t linear equations in the t unknowns $a_0, \ldots, a_{t-1}$, where all arithmetic is done in $\mathbb{Z}_p$. If the equations are linearly independent, there will be a unique solution, and a_0 will be revealed as the key.

Here is a small example to illustrate.

Example 11.7 Suppose that $p = 17$, $t = 3$, and $w = 5$; and the public x-co-ordinates are $x_i = i$, $1 \leq i \leq 5$. Suppose that $B = \{P_1, P_3, P_5\}$ pool their shares, which are respectively $8, 10$, and 11. Writing the polynomial $a(x)$ as

$$a(x) = a_0 + a_1 x + a_2 x^2,$$

and computing $a(1)$, $a(3)$, and $a(5)$, the following three linear equations in $\mathbb{Z}_{17}$ are obtained:

$$\begin{aligned} a_0 + a_1 + a_2 &= 8 \\ a_0 + 3a_1 + 9a_2 &= 10 \\ a_0 + 5a_1 + 8a_2 &= 11. \end{aligned}$$

This system has a unique solution in $\mathbb{Z}_{17}$: $a_0 = 13$, $a_1 = 10$, and $a_2 = 2$. The key is therefore $K = a_0 = 13$. □

Clearly, it is important that the system of t linear equations has a unique solution, as in Example 11.7. There are various ways to show that this is always the case. Perhaps the nicest way to address this question is to appeal to the Lagrange interpolation formula for polynomials, which was presented in Theorem 11.3. This theorem states that the desired polynomial $a(x)$ of degree at most $t-1$ is unique, and it provides an explicit formula that can be used to compute $a(x)$. The formula for $a(x)$ is as follows:

$$a(x) = \sum_{j=1}^{t} \left(y_{i_j} \prod_{1 \leq k \leq t, k \neq j} \frac{x - x_{i_k}}{x_{i_j} - x_{i_k}} \right) \bmod p.$$

A group B of t participants can compute $a(x)$ by using the interpolation formula. But a simplification is possible, because the participants in B do not need to know the whole polynomial $a(x)$. It is sufficient for them to deduce the constant term $K = a(0)$. Hence, they can compute the following expression, which is obtained by substituting $x = 0$ into the Lagrange interpolation formula:

$$K = \sum_{j=1}^{t} \left(y_{i_j} \prod_{1 \leq k \leq t, k \neq j} \frac{x_{i_k}}{x_{i_k} - x_{i_j}} \right) \bmod p.$$

Suppose we define

$$b_j = \prod_{1 \leq k \leq t, k \neq j} \frac{x_{i_k}}{x_{i_k} - x_{i_j}} \bmod p,$$

$1 \leq j \leq t$. (Note that the b_j's can be precomputed, if desired, and their values are not secret.) Then we have

$$K = \sum_{j=1}^{t} b_j y_{i_j} \bmod p.$$

Hence, the key is a linear combination (modulo p) of the t shares.

To illustrate this approach, let's recompute the key from Example 11.7.

Example 11.7 (Cont.) The participants $\{P_1, P_3, P_5\}$ can compute b_1, b_2, and b_3 according to the formula given above. For example, they would obtain

$$\begin{aligned} b_1 &= \frac{x_3 x_5}{(x_3 - x_1)(x_5 - x_1)} \bmod 17 \\ &= 3 \times 5 \times (2)^{-1} \times (4)^{-1} \bmod 17 \\ &= 3 \times 5 \times 9 \times 13 \bmod 17 \\ &= 4. \end{aligned}$$

Similarly, it can be computed that $b_2 = 3$ and $b_3 = 11$. Then, given shares $8, 10$, and 11 (respectively), they would obtain

$$K = 4 \times 8 + 3 \times 10 + 11 \times 11 \bmod 17 = 13,$$

as before. □

What happens if a subset B of $t-1$ participants attempt to compute K? Suppose they hypothesize a value $y_0 \in \mathbb{Z}_p$ for the key K. In the *Shamir Threshold Scheme*, the key is $K = a_0 = a(0)$. Recall that the $t-1$ shares held by B are obtained by evaluating the polynomial $a(x)$ at $t-1$ elements of $\mathbb{Z}_p$. Now, applying Theorem 11.3 again, there is a unique polynomial $a_{y_0}(x)$ such that

$$y_{i_j} = a_{y_0}(x_{i_j}),$$

$1 \leq j \leq t-1$, and such that

$$y_0 = a_{y_0}(0).$$

That is, there is a polynomial $a_{y_0}(x)$ that is consistent with the $t-1$ shares known to B and which also has y_0 as the key. Since this is true for any possible value $y_0 \in \mathbb{Z}_p$, it follows that no value of the key can be ruled out, and thus a group of $t-1$ participants can obtain no information about the key.

For example, suppose that P_1 and P_3 try to compute K, given shares as in Example 11.7. Thus P_1 has the share 8 and P_3 has the share 10. For any possible value y_0 of the key, there is a unique polynomial $a_{y_0}(x)$ that takes on the value 8 at $x = 1$, the value 10 at $x = 3$, and the value y_0 at $x = 0$. Using the interpolation formula, this polynomial is seen to be

$$a_{y_0}(x) = 6y_0(x-1)(x-3) + 13x(x-3) + 13x(x-1) \bmod 17.$$

The subset $\{P_1, P_3\}$ has no way of knowing which of these polynomials is the correct one, and hence they have no information about the value of K.

11.5.2 A Simplified (t,t)-threshold Scheme

The last topic of this section is a simplified construction for threshold schemes in the special case $w = t$. This construction will work for any key set $\mathcal{K} = \mathbb{Z}_m$, and it has $\mathcal{S} = \mathbb{Z}_m$. (For this scheme, it is not required that m be prime, and it is not necessary that $m \geq w + 1$.) If D wants to share the key $K \in \mathbb{Z}_m$, he carries out the steps in Cryptosystem 11.2.

Cryptosystem 11.2: *Simplified (t,t)-Threshold Scheme*

1. D secretly chooses (independently at random) $t-1$ elements of $\mathbb{Z}_m$, $y_1, \dots, y_{t-1}$.
2. D computes
$$y_t = K - \sum_{i=1}^{t-1} y_i \bmod m.$$
3. For $1 \leq i \leq t$, D gives the share y_i to P_i.

Observe that the t participants can compute K by the formula

$$K = \sum_{i=1}^{t} y_i \bmod m.$$

Can $t-1$ participants compute K? Clearly, the first $t-1$ participants cannot do so, since they receive $t-1$ independent random numbers as their shares. Consider the $t-1$ participants in the set $\mathcal{P}\backslash\{P_j\}$, where $1 \leq j \leq t-1$. These $t-1$ participants possess the shares

$$y_1, \dots, y_{j-1}, y_{j+1}, \dots, y_{t-1}$$

and

$$K - \sum_{i=1}^{t-1} y_i.$$

By summing their shares, they can compute $K - y_j$. However, they do not know the random value y_j, and hence they have no information as to the value of K. Consequently, we have a (t,t)-threshold scheme.

Example 11.8 Suppose that $m = 10$ and $t = 4$ in Cryptosystem 11.2. Suppose also that the shares for the four participants are $y_1 = 7$, $y_2 = 2$, $y_3 = 4$, and $y_4 = 2$. The key is therefore

$$K = 7 + 2 + 4 + 2 \bmod 10 = 5.$$

Suppose that the first three participants try to determine K. They know that $y_1 + y_2 + y_3 \bmod 10 = 3$, but they do not know the value of y_4. There is a one-to-one correspondence between the ten possible values of y_4 and the ten possible values of the key K:

$$\begin{array}{ccc} y_4 = 0 & \Leftrightarrow & K = 3 \\ y_4 = 1 & \Leftrightarrow & K = 4 \\ \vdots & \vdots & \vdots \\ y_4 = 9 & \Leftrightarrow & K = 2. \end{array}$$

□

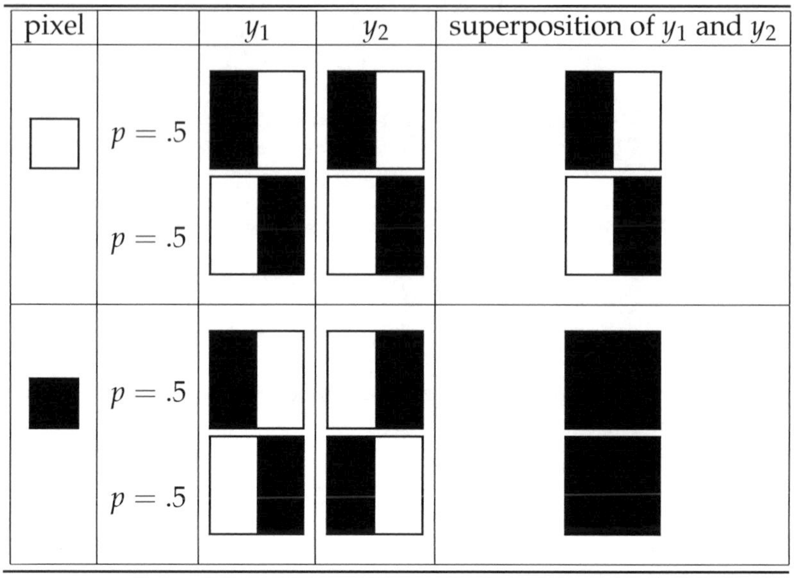
pixel
y_1
y_2
superposition of y_1 and y_2
$p = .5$
$p = .5$
$p = .5$
$p = .5$

FIGURE 11.5: A 2-out-of-2 visual threshold scheme

11.5.3 Visual Threshold Schemes

So far, we have considered the secret in a threshold scheme to be an element of a finite group (or finite field). Naor and Shamir suggested that the secret might be a rectangular image, $\mathcal{I}$, which could be composed of black and white pixels. They considered a threshold scheme scenario where shares for this "secret" are also images consisting of black and white pixels, and reconstruction of the secret corresponds to superimposing a subset of shares. If the shares are printed on transparencies, then the reconstruction is accomplished "visually" by simply stacking the shares. Thus, the term ***visual threshold scheme*** is used to describe a scheme of this kind.

The fact that reconstruction is done by the human visual system means that there is no need to trust a possibly malicious computer to perform the reconstruction correctly. This is an interesting feature of these kinds of schemes.

In a (t, w)-visual threshold scheme, there would be w transparencies. If any t of them are superimposed, then the secret image $\mathcal{I}$ should be recognizable. However, no subset of $t - 1$ (or fewer) shares should reveal any information about $\mathcal{I}$. The difference between a visual threshold scheme and a "traditional" threshold scheme only involves the reconstruction of the secret. The security condition is the same in the two types of schemes.

Initially, it might appear to be impossible to construct a visual threshold scheme that satisfies the security requirements. The reason is as follows: Suppose that a particular pixel P on a share y_i is black. Whenever any subset set of shares (including y_i) is superimposed, the result must be black. This means that, in the secret image $\mathcal{I}$, the reconstructed pixel P must also be black. Therefore, we may

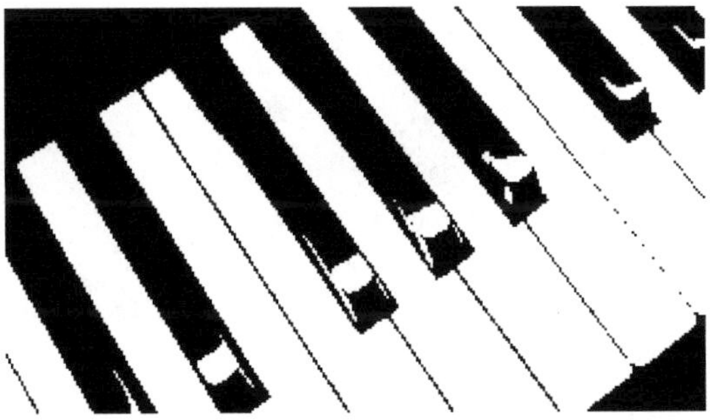

FIGURE 11.6: The original image

obtain some information about the pixel P in the secret image $\mathcal{I}$ by examining one of the shares. Of course this violates the security requirement of the scheme.

Naor and Shamir found an elegant way to avoid this difficulty. We will now describe their 1994 construction of a $(2,2)$-visual threshold scheme.

Figure 11.5 illustrates the Naor-Shamir scheme, by specifying an algorithm for encoding a single pixel in an image. This algorithm would be applied for every pixel P in the image $\mathcal{I}$ in order to construct the two shares.

The basic idea is that a pixel P is replaced by a 2×2 grid of four "subpixels" in each of the two shares. If the original pixel P is white, then a random coin flip is used to choose one of the first two rows of Figure 11.5. Similarly, if the original pixel P is black, then a random coin flip is used to choose one of the last two rows of Figure 11.5. The pixel P is split into two shares, as determined by the chosen row in Figure 11.5.

Because each pixel is expanded into a 2×2 grid of four subpixels, this means that the share is four times larger than the original image (twice as wide and twice as high). For this reason, we say that the ***share expansion*** of the scheme is equal to four.

This method of splitting each pixel into four subpixels in each of the two shares enables the desired security condition to be realized. Suppose we look at a pixel P in the share y_1. The two left subpixels in P are black and the other two are white, or vice versa. Moreover, each of these two possibilities (black/white and white/black) is equally likely to occur, independent of whether the corresponding pixel in the secret image $\mathcal{I}$ is black or white. Thus the share y_1 provides no information as to whether the pixel P is black or white. An identical argument applies to the share y_2. Furthermore, assuming that all the pixels in $\mathcal{I}$ are split into shares using independent random coin flips, no information can be obtained by looking at any group of pixels in a single share.

We also need to consider what happens when we superimpose the two shares y_1 and y_2 (in this analysis, we refer to the last column of Figure 11.5).

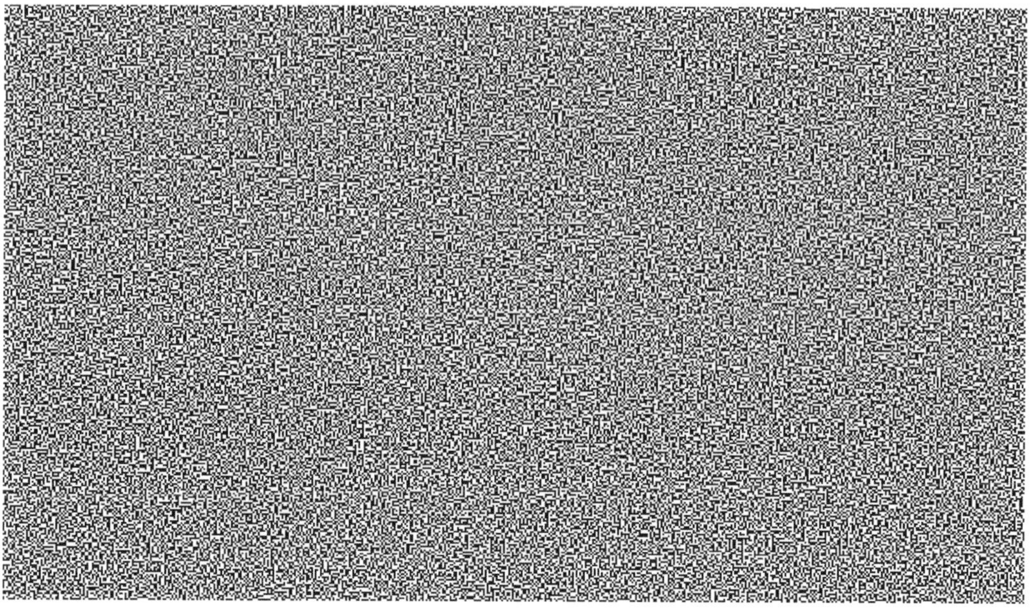

Share y_1

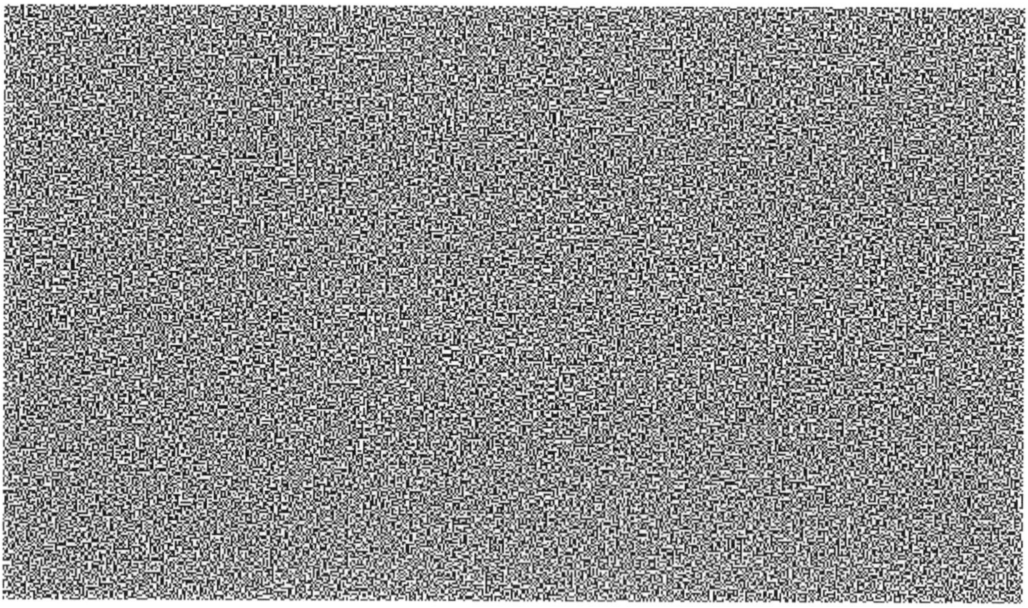

Share y_2

FIGURE 11.7: The two shares

Consider a single pixel P in the image $\mathcal{I}$. If P is black, then we obtain four black subpixels when we superimpose the two shares, so a black pixel is reconstructed correctly. On the other hand, if P is white, then we get two black subpixels and two white subpixels when we superimpose the two shares. Therefore, it may appear to be gray (i.e., halfway between black and white).

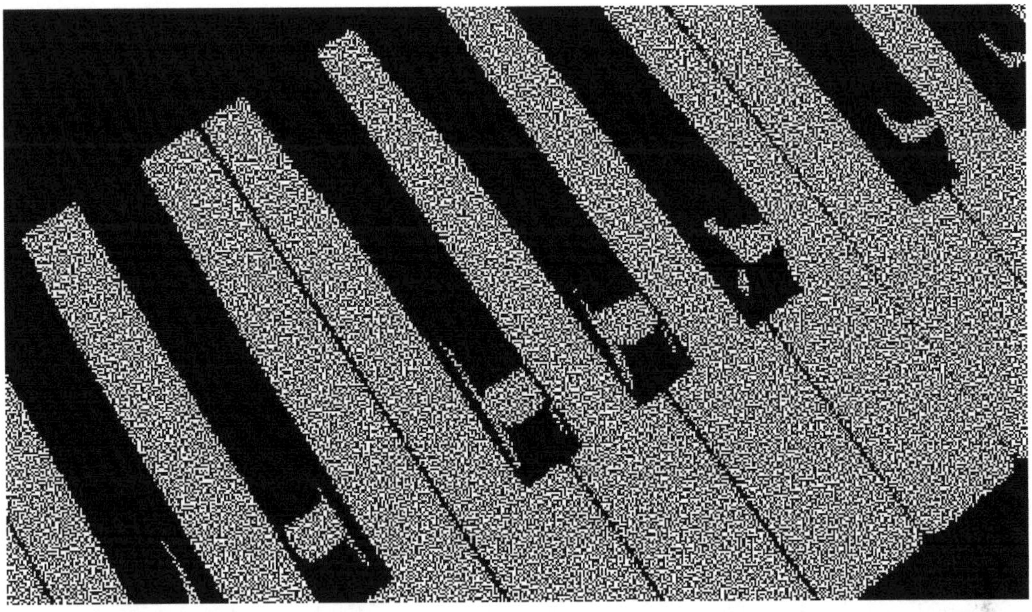

FIGURE 11.8: Superposition of shares y_1 and y_2 (reconstructed image)

Thus, we might say that the reconstructed pixel (which consists of two subpixels) has a ***gray level*** of 1 if P is black, and a gray level of $1/2$ if P is white. The resulting constructed image will suffer a 50% loss of contrast when compared to the original image $\mathcal{I}$, but $\mathcal{I}$ should still be recognizable in the reconstructed image.

In Figures 11.6, 11.7, and 11.8, we present an example to show how an entire image can be encrypted into two shares and then reconstructed. The image used in Figure 11.6, a piano keyboard, remains recognizable after reconstruction (see Figure 11.8), despite the 50% loss of contrast that occurs. Of course, more "complicated" images may be difficult to recognize when the two shares are superimposed. This is partly due to the 50% loss of contrast, and partly due to the two following non-mathematical reasons:

- Transparencies are floppy and may be hard to align precisely, and they slide around easily.
- When the transparencies are created, say by using a printer, the heat produced in the printing process can distort the plastic in the transparencies, making them even more difficult to align correctly.

Generally speaking, it is best to use rather "simple" images. Experimentation is useful to determine which images are suitable for sharing using this algorithm.

11.6 Notes and References

For a comprehensive book on the key establishment problem, see Boyd and Mathuria [44].

The *Blom Key Predistribution Scheme* was presented in [36]. For a generalization of this scheme, see Blundo *et al.* [38]. The *Lee-Stinson Linear KPS* appears in [119, 120].

The *Needham-Schroeder SKDS* is from [154] and the Denning-Sacco attack is from [68]. For information on *Kerberos*, see Kohl and Neuman [114] and Kohl, Neuman, and T'so [115].

The *Bellare-Rogaway SKDS* was described in [20]. Secure session key distribution schemes using public-key cryptography are discussed in Blake-Wilson and Menezes [33].

The *Logical Key Hierarchy* is (independently) due to Wallner, Harder, and Agee [197] and Wong and Lam [203].

Threshold schemes were invented independently by Blakley [35] and Shamir [175]. Visual cryptography was first proposed by Naor and Shamir [145].

Exercises

11.1 Suppose that $p = 150001$ and $\alpha = 7$ in the *Diffie-Hellman Key Predistribution Scheme*. (It can be verified that α is a generator of $\mathbb{Z}_p{}^*$.) Suppose that the private keys of U, V, and W are $a_U = 101459$, $a_V = 123967$, and $a_W = 99544$.

(a) Compute the public keys of U, V, and W.

(b) Show the computations performed by U to obtain $K_{U,V}$ and $K_{U,W}$.

(c) Verify that V computes the same key $K_{U,V}$ as U does.

(d) Explain why $\mathbb{Z}_{150001}{}^*$ is a very poor choice of a setting for the *Diffie-Hellman Key Predistribution Scheme* (notwithstanding the fact that p is too small for the scheme to be secure).

HINT Consider the factorization of $p - 1$.

11.2 Suppose the *Blom KPS* with $k = 2$ is implemented for a set of five users, U, V, W, X, and Y. Suppose that $p = 97$, $r_U = 14$, $r_V = 38$, $r_W = 92$, $r_X = 69$ and $r_Y = 70$. The secret g polynomials are as follows:

$$g_U(x) = 15 + 15x + 2x^2 \qquad g_V(x) = 95 + 77x + 83x^2$$
$$g_W(x) = 88 + 32x + 18x^2 \qquad g_X(x) = 62 + 91x + 59x^2$$
$$g_Y(x) = 10 + 82x + 52x^2.$$

(a) Compute the keys for all $\binom{5}{2} = 10$ pairs of users.

(b) Verify that $K_{U,V} = K_{V,U}$.

11.3 Suppose that the *Blom KPS* is implemented with security parameter k. Suppose that a coalition of k users, say $W_1, \ldots, W_k$, pool their secret information. Additionally, assume that a key $K_{U,V}$ is exposed, where U and V are two other users.

(a) Describe how the coalition can determine the polynomial $g_U(x)$ by polynomial interpolation, using known values of $g_U(x)$ at $k+1$ points.

(b) Having computed $g_U(x)$, describe how the coalition can compute the bivariate polynomial $f(x,y)$ by bivariate polynomial interpolation.

(c) Illustrate the preceding two steps, by determining the polynomial $f(x,y)$ in the sample implementation of the *Blom KPS* where $k = 2$, $p = 34877$, and $r_i = i$ $(1 \leq i \leq 4)$, supposing that

$$\begin{aligned} g_1(x) &= 13952 + 21199x + 19701x^2, \\ g_2(x) &= 25505 + 24549x + 15346x^2, \quad \text{and} \\ K_{3,4} &= 9211. \end{aligned}$$

11.4 Consider the *Lee-Stinson Linear KPS* with parameters p and k.

(a) Suppose $K_{i,j}$ and $K_{i',j'}$ are two keys in the scheme, where $i \neq i'$. Prove that there is a unique node in the scheme that contains both of these keys.

(b) Suppose that $U_{a,b}$ and $U_{a',b'}$ are two nodes that do not have a common key. Prove that there are exactly $(k-1)^2$ nodes that have a common key with $U_{a,b}$ and a (different) common key with $U_{a',b'}$.

11.5 We describe a secret-key based three-party session key distribution scheme in Protocol 11.8. In this scheme, K_{Alice} is a secret key shared by Alice and the *TA*, and K_{Bob} is a secret key shared by Bob and the *TA*.

(a) State all consistency checks that should be performed by Alice, Bob, and the *TA* during a session of the protocol.

(b) The protocol is vulnerable to an attack if the *TA* does not perform the necessary consistency checks you described in part (a). Suppose that Oscar replaces $ID(Bob)$ by $ID(Oscar)$, and he also replaces y_B by

$$y_O = e_{K_{Oscar}}(ID(Alice) \parallel ID(Bob) \parallel r'_B)$$

in step 2, where r'_B is random. Describe the possible consequences of this attack if the *TA* does not carry out its consistency checks properly.

(c) In this protocol, encryption is being done to ensure both confidentiality and data integrity. Indicate which pieces of data require encryption for the purposes of confidentiality, and which ones only need to be authenticated. Rewrite the protocol, using MACs for authentication in the appropriate places.

Protocol 11.8: SESSION KEY DISTRIBUTION SCHEME

1. Alice chooses a random number, r_A. Alice sends $ID(Alice)$, $ID(Bob)$, and
$$y_A = e_{K_{Alice}}(ID(Alice) \parallel ID(Bob) \parallel r_A)$$
to Bob.

2. Bob chooses a random number, r_B. Bob sends $ID(Alice)$, $ID(Bob)$, y_A and
$$y_B = e_{K_{Bob}}(ID(Alice) \parallel ID(Bob) \parallel r_B)$$
to the *TA*.

3. The *TA* decrypts y_A using the key K_{Alice} and it decrypts y_B using the key K_{Bob}, thus obtaining r_A and r_B. It chooses a random session key, K, and computes
$$z_A = e_{K_{Alice}}(r_A \parallel K)$$
and
$$z_B = e_{K_{Bob}}(r_B \parallel K).$$
z_A is sent to Alice and z_B is sent to Bob.

4. Alice decrypts z_A using the key K_{Alice}, obtaining K; and Bob decrypts z_B using the key K_{Bob}, obtaining K.

11.6 We describe a public-key protocol, in which Alice chooses a random session key and transmits it to Bob in encrypted form, in Protocol 11.9 (this is another example of key transport). In this scheme, K_{Bob} is Bob's public encryption key. Alice and Bob also have private signing keys and public verification keys for a signature scheme.

(a) Determine if the above protocol is a secure mutual identification scheme. If it is, then analyze an active adversary's probability of successfully deceiving Alice or Bob, given suitable assumptions on the security of the signature scheme. If it is not, then demonstrate an attack on the scheme.

(b) What type of key authentication or confirmation is provided by this protocol (from Alice to Bob, and from Bob to Alice)? Justify your answer briefly.

11.7 Suppose we want to simultaneously revoke r users, say $U_{i_1}, \dots, U_{i_r}$, in the *Logical Key Hierarchy*. Assuming that the tree depth is equal to d and the

Protocol 11.9: PUBLIC-KEY KEY TRANSPORT SCHEME

1. Bob chooses a random challenge, r_1. He sends r_1 and $\mathbf{Cert}(Bob)$ to Alice.

2. Alice verifies Bob's public encryption key, K_{Bob}, on the certificate $\mathbf{Cert}(Bob)$. Then Alice chooses a random session key, K, and computes

$$z = e_{K_{Bob}}(K).$$

She also computes

$$y_1 = \mathbf{sig}_{Alice}(r_1 \parallel z \parallel ID(Bob))$$

and sends $\mathbf{Cert}(Alice)$, z and y_1 to Bob.

3. Bob verifies Alice's public verification key, $\mathbf{ver}_{Alice}$, on the certificate $\mathbf{Cert}(Alice)$. Then he verifies that

$$\mathbf{ver}_{Alice}(r_1 \parallel z \parallel ID(Bob), y_1) = true.$$

If this is not the case, then Bob "rejects." Otherwise, Bob decrypts z obtaining the session key K, and "accepts." Finally, Bob computes

$$y_2 = \mathbf{sig}_{Bob}(z \parallel ID(Alice))$$

and sends y_2 to Alice.

4. Alice verifies Bob's public verification key, $\mathbf{ver}_{Bob}$, on the certificate $\mathbf{Cert}(Bob)$. Then she checks that

$$\mathbf{ver}_{Bob}(z \parallel ID(Alice), y_2) = true.$$

If so, then Alice "accepts"; otherwise, Alice "rejects."

tree nodes are labeled as described in Section 11.4, we can assume that $2^d \leq U_{i_1} < \cdots < U_{i_r} \leq 2^{d+1} - 1$.

(a) Informally describe an algorithm that can be used to determine which keys in the tree need to be updated.

(b) Describe the broadcast that is used to update the keys. Which keys are used to encrypt the new, updated keys?

(c) Illustrate your algorithm by describing the updated keys and the broadcast if users 18, 23, and 29 are to be revoked in a tree with depth $d = 4$ (this tree is depicted in Figure 9.1). How much smaller is the broadcast in this case, as compared to the three broadcasts that would be required to revoke these three users one at a time in the basic *Logical Key Hierarchy*?

11.8 Write a computer program to compute the key for the *Shamir* (t, w)-*Threshold Scheme* implemented in $\mathbb{Z}_p$. That is, given t public x-coordinates, $x_1, x_2, \ldots, x_t$, and t y-coordinates $y_1, \ldots, y_t$, compute the resulting key using the Lagrange interpolation formula.

(a) Test your program if $p = 31847$, $t = 5$, and $w = 10$, with the following shares:

$$\begin{aligned} x_1 &= 413 & y_1 &= 25439 \\ x_2 &= 432 & y_2 &= 14847 \\ x_3 &= 451 & y_3 &= 24780 \\ x_4 &= 470 & y_4 &= 5910 \\ x_5 &= 489 & y_5 &= 12734 \\ x_6 &= 508 & y_1 &= 12492 \\ x_7 &= 527 & y_2 &= 12555 \\ x_8 &= 546 & y_3 &= 28578 \\ x_9 &= 565 & y_4 &= 20806 \\ x_{10} &= 584 & y_5 &= 21462 \end{aligned}$$

Verify that the same key is computed by using several different subsets of five shares.

(b) Having determined the key, compute the share that would be given to a participant with x-coordinate equal to 10000. (Note that this can be done without computing the whole secret polynomial $a(x)$.)

11.9 (a) Suppose that the following are the nine shares in a $(5, 9)$-*Shamir Threshold Scheme* implemented in $\mathbb{Z}_{94875355691}$:

i	x_i	y_i
1	11	537048626
2	22	89894377870
3	33	65321160237
4	44	18374404957
5	55	24564576435
6	66	87371334299
7	77	60461341922
8	88	10096524973
9	99	81367619987

Exactly one of these shares is defective (i.e., incorrect). Your task is to determine which share is defective, and then figure out its correct value, as well as the value of the secret.

The "primitive operations" in your algorithm are polynomial interpolations and polynomial evaluations. Try to minimize the number of polynomial interpolations you perform.

HINT The question can be answered using at most three polynomial interpolations.

(b) Suppose that a (t, w)-*Shamir Threshold Scheme* has exactly one defective share, and suppose that $w - t \geq 2$. Describe how it is possible to determine which share is defective using at most $\lceil \frac{w}{w-t} \rceil$ polynomial interpolations. Why is this problem impossible to solve if $w - t = 1$?

(c) Suppose that a (t, w)-*Shamir Threshold Scheme* has exactly τ defective shares, and suppose that $w \geq (\tau + 1)t$. Describe how it is possible to determine which shares are defective using at most $\tau + 1$ polynomial interpolations.

11.10 Devise a $(2, 3)$-visual threshold scheme with pixel expansion equal to 9. Each pixel in each of three shares is replaced by a 3×3 grid of subpixels. The gray level of a reconstructed black pixel should be equal to $2/3$ and the gray level of a reconstructed white pixel should be equal to $1/3$. Finally, no 3×3 grid of subpixels (from a single share) should give any information as to whether the reconstructed pixel will be white or black.

Chapter 12

Key Agreement Schemes

In this chapter, we turn our attention to key agreement schemes (or KAS), in which two users can establish a new session key via an interactive protocol that does not require the active participation of a *TA*.

12.1 Introduction

This chapter is a companion to the previous chapter, where we discussed key predistribution schemes and session key distribution schemes. Both of these kinds of key distribution require a trusted authority (*TA*) to select keys and distribute them to network users. In this chapter, we focus on key agreement schemes (KAS), in which two users can establish a new session key via an interactive protocol that does not require the active participation of a *TA*. Note that we are mainly discussing key agreement schemes in the public-key setting.

Throughout this chapter, we will use the same terminology and notation as we did in Chapter 11. The reader should review the introductory material pertaining to key agreement that was presented in Section 11.1 before proceeding further.

The rest of this chapter is organized as follows. Section 12.1.1 gives a brief overview of *Transport Layer Security*. Section 12.2 introduces the *Diffie-Hellman Key Agreement Scheme* and some variations, and it also discusses various security proofs for these types of schemes. Section 12.3 provides a short treatment of key derivation functions. In Section 12.4, we examine the *MTI Key Agreement Scheme*. Deniability and key updating are discussed in Sections 12.5 and 12.6, respectively. Finally, conference key agreement is the topic of Section 12.7.

12.1.1 Transport Layer Security (TLS)

In practice, one of the most commonly used key agreement protocols is *Transport Layer Security* (*TLS*). We discuss this as our first example. A TLS session can be used, for example, to facilitate online purchases from a company's web page using a web browser. Suppose a ***client*** (Alice) wants to purchase something from a ***server*** (Bob, Inc.). In order to do this, the client and server must establish a session key using an appropriate key agreement mechanism. There are various methods supported by *TLS*. We describe one of them here, which is basically a form of key

client		server
	I'm Alice $\longrightarrow$	
	$\longleftarrow$ I'm Bob, Inc.	
	$\longleftarrow$ $PK, \mathbf{sig}_{CA}(PK)$	
verify PK		
generate MS	$y = e_{PK}(MS)$ $\longrightarrow$	
		$MS = d_{PK}(y)$
$K_1, K_2 = \mathbf{KDF}(MS)$		$K_1, K_2 = \mathbf{KDF}(MS)$

FIGURE 12.1: Setting up a TLS Session

transport. The main steps in setting up a TLS session are summarized in Figure 12.1.

In more detail, here is what takes place: First, Alice and Bob, Inc. introduce themselves. This is called a "hello," and no cryptographic tools are used in this step. At this time, Alice and Bob, Inc. also agree on the specific cryptographic algorithms they will use in the rest of the protocol.

Next, Bob Inc. authenticates himself to Alice; he sends her a certificate containing a copy of his public key, *PK*, signed by a trusted certification authority, *CA*. Alice verifies the *CA*'s signature on *PK* using the *CA*'s public verification key (which would have been bundled with the web browser software running on Alice's computer).

Now Alice and Bob, Inc. proceed to determine two common secret keys. Alice generates a random master secret, *MS*, using an appropriate pseudorandom number generator. She encrypts *MS* using Bob, Inc.'s public key and sends the resulting ciphertext to Bob, Inc., who then decrypts the ciphertext, obtaining *MS*.

Now Alice and Bob Inc. independently generate the same two keys K_1 and K_2 from *MS*. This step will use a prespecified ***key derivation function***, denoted by **KDF**. The function **KDF** is usually based on a hash function; key derivation functions are discussed in more detail in Section 12.3. Because it is only one party, namely Alice, who determines the resulting keys, this is an example of key transport.

Finally, now that Alice and Bob, Inc. have both derived the same two secret keys, they use these keys to authenticate and encrypt the messages they send to each other. The key K_1 would be used to authenticate data using a message authentication code, and the key K_2 would be used to encrypt and decrypt data using a secret key cryptosystem. Therefore, the *TLS* protocol enables secure communication between Alice and Bob, Inc.

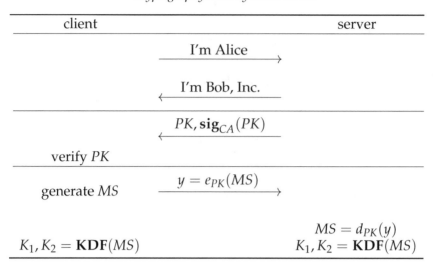

Protocol 12.1: DIFFIE-HELLMAN KAS

The public domain parameters consist of a group $(G, \cdot)$ and an element $\alpha \in G$ having order n.

1. U chooses a_U at random, where $0 \leq a_U \leq n-1$. Then she computes
$$b_U = \alpha^{a_U}$$
and sends b_U to V.
2. V chooses a_V at random, where $0 \leq a_V \leq n-1$. Then he computes
$$b_V = \alpha^{a_V}$$
and sends b_V to U.
3. U computes
$$K = (b_V)^{a_U}$$
and V computes
$$K = (b_U)^{a_V}.$$

It is interesting to note that only the server (Bob, Inc.) is required to supply a certificate during a *TLS* Session.[1] The client (Alice) may not even have a public key (or a certificate). This is a common state of affairs at present in electronic commerce: companies setting up web pages for business purposes require certificates, but users do not need a certificate in order to make an online purchase. From the company's point of view, the important point is not that Alice is really who she claims to be. It is more important that Alice's credit card number, which is supplied as part of the ensuing financial transaction, is valid and remains uncompromised. The credit card number and any personal information supplied by Alice will be encrypted (and authenticated, via a MAC) using the keys that are created in the TLS session.

12.2 Diffie-Hellman Key Agreement

The first and best-known key agreement scheme is the *Diffie-Hellman KAS*. This was actually the very first published realization of public key cryptography, which occurred in 1976. The *Diffie-Hellman KAS* is presented as Protocol 12.1.

[1]However, we note that *TLS* provides optional support for authentication of the client by the server if the client has a certificate.

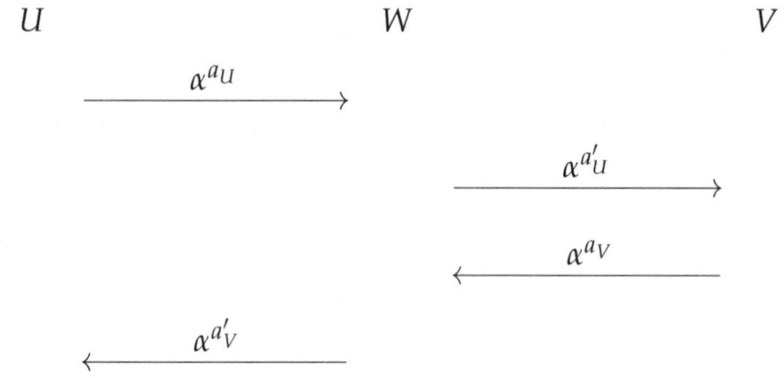

FIGURE 12.2: Intruder-in-the-middle attack

Protocol 12.1 is very similar to *Diffie-Hellman Key Predistribution* (Protocol 11.1), which was described in the previous chapter. The difference is that the exponents a_U and a_V of users U and V (respectively) are chosen anew each time the scheme is run, instead of being fixed. Also, there are no long-lived keys in this scheme.

At the end of a session of the *Diffie-Hellman KAS*, U and V have computed the same key,

$$K = \alpha^{a_U a_V} = \mathbf{CDH}(\alpha, b_U, b_V).$$

(Here, as usual, **CDH** refers to the **Computational Diffie-Hellman** problem, which was defined in Section 7.7.3 The notation $\mathbf{CDH}(\alpha, b_U, b_V)$ refers to the desired output for the instance (α, b_U, b_V).) Further, assuming that the **Decision Diffie-Hellman** problem is intractable, a passive adversary cannot compute any information about K.

It is well-known that the *Diffie-Hellman KAS* has a serious weakness in the presence of an active adversary. The *Diffie-Hellman KAS* is supposed to work like this:

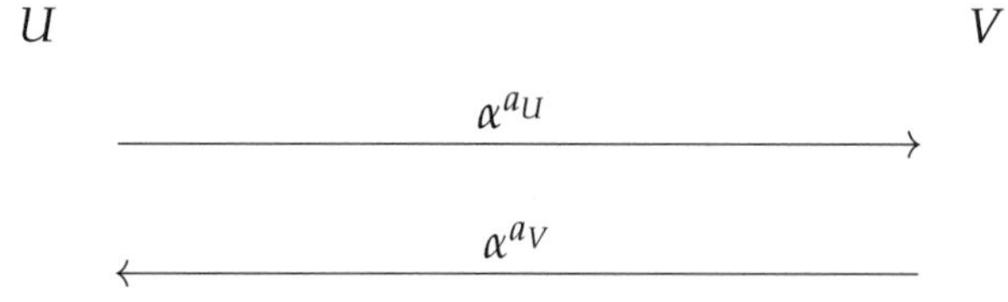

Unfortunately, the scheme is vulnerable to an active adversary who uses an ***intruder-in-the-middle*** attack.[2] The intruder-in-the-middle attack on the *Diffie-Hellman KAS* works in the following way. W will intercept messages between U and V and substitute his own messages, as indicated in Figure 12.2.

At the end of the session, U has actually established the secret key $\alpha^{a_U a'_V}$ with W, and V has established the secret key $\alpha^{a'_U a_V}$ with W. When U tries to encrypt a

[2]There is an episode of the popular 1950s television comedy *The Lucy Show* in which Vivian Vance is having dinner in a restaurant with a date, and Lucille Ball is hiding under the table. Vivian and her date decide to hold hands under the table. Lucy, trying to avoid detection, holds hands with each of them and they think they are holding hands with each other.

message to send to V, W will be able to decrypt it but V will not be able to do so. (A similar situation holds if V sends a message to U.)

Clearly, it is essential for U and V to make sure that they are exchanging messages (and keys!) with each other and not with W. Before exchanging keys, U and V might carry out a separate protocol to establish each other's identity, for example by using a secure mutual identification scheme. But this offers no protection against an intruder-in-the-middle attack if W simply remains inactive until after U and V have proved their identities to each other. A more promising approach is to design a key agreement scheme that authenticates the participants' identities at the same time as the key is being established. A KAS of this type will be called an ***authenticated key agreement scheme***.

Informally, we define an authenticated key agreement scheme to be a key agreement scheme that satisfies the following properties:

mutual identification

The scheme is a secure mutual identification scheme, as defined in Section 10.3.1: no honest participant in a session of the scheme will "accept" after any flow in which an adversary is active.

key agreement

If there is no active adversary, then both participants will compute the same new session key K. Moreover, no information about the value of K can be computed by the (passive) adversary.

12.2.1 The Station-to-station Key Agreement Scheme

In this section, we describe an authenticated key agreement scheme that is a modification of the *Diffie-Hellman KAS*. The scheme makes use of certificates which, as usual, are signed by a *TA*. Each user U will have a signature scheme with a verification algorithm $\mathbf{ver}_U$ and a signing algorithm $\mathbf{sig}_U$. The *TA* also has a signature scheme with a public verification algorithm $\mathbf{ver}_{TA}$. Each user U has a certificate

$$\mathbf{Cert}(U) = (ID(U), \mathbf{ver}_U, \mathbf{sig}_{TA}(ID(U), \mathbf{ver}_U)),$$

where $ID(U)$ is certain identification information for U. (These certificates are the same as the ones described in Section 8.6.)

The authenticated key agreement scheme known as the *Station-to-station KAS* (or *STS*, for short) is due to Diffie, Van Oorschot, and Wiener. Protocol 12.2 is a slight simplification. The basic idea of Protocol 12.2 is to combine the *Diffie-Hellman KAS* with a secure mutual identification scheme, where the exponentiated values b_U and b_V function as the random challenges in the identification scheme. If we follow this recipe, using Protocol 10.10 as the underlying identification scheme, then the result is Protocol 12.2.

Roughly speaking, signing the random challenges provides mutual authentication. Furthermore, these challenges, being computed according to the *Diffie-Hellman KAS*, allow both U and V to compute the same key, $K = \mathbf{CDH}(\alpha, b_U, b_V)$.

Protocol 12.2: SIMPLIFIED STATION-TO-STATION KAS

The public domain parameters consist of a group $(G, \cdot)$ and an element $\alpha \in G$ having order n.

1. U chooses a random number a_U, $0 \leq a_U \leq n-1$. Then she computes
$$b_U = \alpha^{a_U}$$
and she sends **Cert**(U) and b_U to V.

2. V chooses a random number a_V, $0 \leq a_V \leq n-1$. Then he computes
$$\begin{aligned} b_V &= \alpha^{a_V} \\ K &= (b_U)^{a_V}, \quad \text{and} \\ y_V &= \mathbf{sig}_V(ID(U) \parallel b_V \parallel b_U). \end{aligned}$$
Then V sends **Cert**(V), b_V and y_V to U.

3. U verifies y_V using $\mathbf{ver}_V$. If the signature y_V is not valid, then she "rejects" and quits. Otherwise, she "accepts," she computes
$$\begin{aligned} K &= (b_V)^{a_U}, \quad \text{and} \\ y_U &= \mathbf{sig}_U(ID(V) \parallel b_U \parallel b_V), \end{aligned}$$
and she sends y_U to V.

4. V verifies y_U using $\mathbf{ver}_U$. If the signature y_U is not valid, then he "rejects"; otherwise, he "accepts."

12.2.2 Security of STS

In this section, we discuss the security properties of the simplified *STS* scheme. For future reference, the information exchanged in a session of the scheme (excluding certificates) is illustrated as follows:

$$\begin{array}{ccc} U & & V \\ & \xrightarrow{\alpha^{a_U}} & \\ & \xleftarrow{\alpha^{a_V}, \mathbf{sig}_V(ID(U) \parallel \alpha^{a_V} \parallel \alpha^{a_U})} & \\ & \xrightarrow{\mathbf{sig}_U(ID(V) \parallel \alpha^{a_U} \parallel \alpha^{a_V})} & \end{array}$$

First, let's see how the use of signatures protects against the intruder-in-the-middle attack mentioned earlier. Suppose, as before, that W intercepts α^{a_U} and replaces it

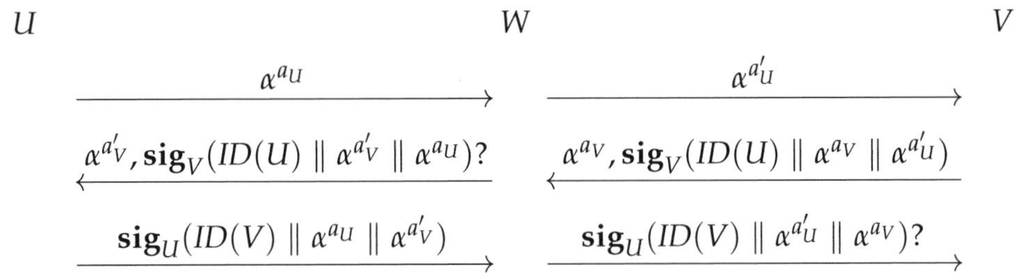

FIGURE 12.3: Thwarted intruder-in-the-middle attack on STS

with $\alpha^{a'_U}$. W then receives α^{a_V} and

$$\mathbf{sig}_V(ID(U) \parallel \alpha^{a_V} \parallel \alpha^{a'_U})$$

from V. He would like to replace α^{a_V} with $\alpha^{a'_V}$, as before. However, this means that he must also replace the signature by

$$\mathbf{sig}_V(ID(U) \parallel \alpha^{a'_V} \parallel \alpha^{a_U}).$$

Unfortunately for W, he cannot compute V's signature on the string $ID(U) \parallel \alpha^{a'_V} \parallel \alpha^{a_U}$ because he doesn't know V's signing algorithm $\mathbf{sig}_V$. Similarly, W is unable to replace $\mathbf{sig}_U(ID(V) \parallel \alpha^{a_U} \parallel \alpha^{a'_V})$ by $\mathbf{sig}_U(ID(V) \parallel \alpha^{a'_U} \parallel \alpha^{a_V})$ because he does not know U's signing algorithm.

This situation is illustrated in Figure 12.3, in which the question marks indicate signatures that the adversary is unable to compute. It is the judicious use of signatures that provides for mutual identification of U and V. This in turn thwarts the intruder-in-the-middle attack.

Of course, we want to be convinced that the scheme is secure against all possible attacks, not just one particular attack. However, from the way that the scheme is designed, we can use previous results to provide a general proof of security of *STS*. In doing so, we need to say more precisely what assurances are provided regarding knowledge of the session key.

First, we claim that *STS* is a secure mutual identification scheme. This fact can be proven using the methods described in Chapter 10. So, if an adversary is active, he will be detected by the honest participants in the session.

On the other hand, if the adversary is passive, then the session will terminate with both parties "accepting" (provided they behave honestly). That is, U and V successfully identify themselves to each other, and they both compute the key K as in the *Diffie-Hellman KAS*. The adversary cannot compute any information about the key K, assuming the intractability of the **Decision Diffie-Hellman** problem. In summary, an active adversary will be detected, and an inactive adversary is thwarted due to the intractability of the **Decision Diffie-Hellman** problem (exactly as was the case in the *Diffie-Hellman KAS*).

Now, using the properties discussed above, let us see what we can infer about the *STS* scheme if U or V "accepts." First, suppose that U "accepts." Because *STS* is a secure mutual identification scheme, U can be confident that she has really been communicating with V (her "intended peer") and that the adversary was inactive before the last flow. Assuming that V is honest and that he has executed the scheme according to its specifications, U can be confident that V can compute the value of K, and that no one other than V can compute the value of K.

Let us consider in a bit more detail why U should believe that V can compute K. The reason for this belief is that U has received V's signature on the values α^{a_U} and α^{a_V}, so it is reasonable for U to infer that V knows these two values. Now, assuming that V executed the scheme according to its specifications, U can infer that V knows the value of a_V. V is able to compute the value of K, provided that he knows the values of α^{a_U} and a_V. Of course, there is no guarantee to U that V has actually computed K at the time when V "accepts."

The analysis from the point of view of V is very similar. If V "accepts," then he is confident that he has really been communicating with U (his intended peer) and that the key K can be computed by U and by no one else. However, there is an asymmetry in the assurances provided to U and to V. When V "accepts," he can be confident that U has already "accepted" (provided that U is honest). However, when U "accepts," she has no way of knowing if V will subsequently "accept," because she does not know if V will actually receive the message being sent to him in the last flow of the session (for example, an adversary might intercept or corrupt this last flow, causing V to "reject"). A similar situation occurred in the setting of mutual identification schemes and was discussed in Section 10.2.2.

It is useful to define a few variations of properties relating to the users' knowledge of the computed session key, K. Suppose that V is the intended peer of U in a key agreement scheme. Here are three "levels" of assurance regarding key agreement that could be provided to U (or to V):

implicit key authentication
: We say that a key agreement scheme provides implicit key authentication to U if U is assured that no one other than V can compute K (in particular, the adversary should not be able to compute K).

implicit key confirmation
: We say that a key agreement scheme provides implicit key confirmation if U is assured that V can compute K (assuming that V executed the scheme according to its specifications), and no one other than V can compute K.

explicit key confirmation
: We say that a key agreement scheme provides explicit key confirmation if U is assured that V has computed K, and no one other than V can compute K.

We have presented two variations on the idea of key confirmation. The notion of key confirmation discussed in Chapter 11 (in connection with session key distribution schemes) was the "explicit" version. In general, explicit key confirmation is provided by using the newly constructed session key to encrypt a known value (or random challenge). *Kerberos* and *Needham-Schroeder* both attempt to provide explicit key confirmation by exactly this method.

The *STS* scheme does not make immediate use of the new session key, so we don't have explicit key confirmation. However, because both parties sign the exchanged exponentials, we achieve the slightly weaker property of implicit key confirmation. (Furthermore, as we mentioned in Chapter 11, it is always possible to augment any key agreement or key distribution scheme so it achieves explicit key confirmation, if so desired.)

Finally, note that the *Bellare-Rogaway Session Key Distribution Scheme* provides implicit key authentication; there is no attempt in that scheme to provide either party with any assurance that their intended peer has received (or can compute) the session key.

Summarizing the discussion in this section, we have established the following theorem.

THEOREM 12.1 *The Station-to-station KAS is an authenticated key agreement scheme that provides implicit key confirmation to both parties, assuming that the* ***Decision Diffie-Hellman*** *problem is intractable.*

12.2.3 Known Session Key Attacks

The security result proven in the last section basically considers one session of *STS* in isolation. However, in a realistic scenario involving a network with many users, there could be many sessions of *STS* taking place, involving many different users. In order to make a convincing argument that *STS* is secure, we need to consider the possible influence that different sessions might have on each other.

Therefore, we investigate security under a known session key attack (this attack model was defined in Section 11.1). In this scenario, the adversary, say Oscar, observes several sessions of a key agreement scheme, say $\mathcal{S}_1, \mathcal{S}_2, \dots, \mathcal{S}_t$. These sessions may be sessions involving other network users, or they may include Oscar himself as one of the participants. We will assume for convenience that all sessions use the same group and the same generator α.

As part of the attack model, Oscar is allowed to request that the session keys for the sessions $\mathcal{S}_1, \mathcal{S}_2, \dots, \mathcal{S}_t$ be revealed to him. Oscar's goal is to determine a session key (or information about a session key) for some other ***target session***, say $\mathcal{S}$, in which Oscar is not a participant. Furthermore, we do not require that the session $\mathcal{S}$ takes place after all the other sessions $\mathcal{S}_1, \mathcal{S}_2, \dots, \mathcal{S}_t$ have completed. In particular, we allow parallel session attacks (similar to those considered in the context of identification schemes).

In this section, we study the security of the *STS Key Agreement Scheme* against known session key attacks. First, suppose Oscar observes a session $\mathcal{S}$ between two users U and V. The information transmitted in this session (excluding signatures and certificates) consists of the two values $b_{\mathcal{S},U}$ and $b_{\mathcal{S},V}$. (We are including the name of the session, $\mathcal{S}$, as a subscript to make it clear that these values are associated with a particular session.) Oscar hopes ultimately to be able to determine some information about the value of the key $K_{\mathcal{S}}$ computed by U and V in the session $\mathcal{S}$. Observe that computing the key $K_{\mathcal{S}}$ is the same as solving the **Computational Diffie-Hellman** problem for the instance $(\alpha, b_{\mathcal{S},U}, b_{\mathcal{S},V})$, i.e., computing $K_{\mathcal{S}} = \textbf{CDH}(\alpha, b_{\mathcal{S},U}, b_{\mathcal{S},V})$.

Once Oscar has the pair $(b_{\mathcal{S},U}, b_{\mathcal{S},V})$, he is free to engage in various other sessions in an attempt to find out some information about $K_{\mathcal{S}}$. However, we only allow Oscar to request a key for a session $\mathcal{S}'$ from a user in the session $\mathcal{S}'$ who "accepts." Therefore Oscar cannot be active in a session and then request a session key from a user who does not "accept," because *STS* is a secure identification scheme.

However, Oscar can take part in a session $\mathcal{S}'$ as one of the participants, possibly not following the rules of *STS*. In particular, Oscar might transmit a value $b_{\mathcal{S}',Oscar}$ to his peer in the session $\mathcal{S}'$, without knowing the corresponding value $a_{\mathcal{S}',Oscar}$ such that $b_{\mathcal{S}',Oscar} = \alpha^{a_{\mathcal{S}',Oscar}}$. In accordance with the known session key attack model, Oscar would be allowed to request that the value of key $K_{\mathcal{S}'}$ be revealed to him. (If Oscar followed the "rules" of *STS*, then he would be able to compute $K_{\mathcal{S}'}$ himself. However, we are considering a situation where Oscar cannot compute $K_{\mathcal{S}'}$, but where we allow Oscar to be informed of its value, anyway.)

Suppose that Oscar takes part in such a session $\mathcal{S}'$ with a peer W. Then Oscar chooses a value $b_{\mathcal{S}',Oscar}$ in any way that he wishes. For example, Oscar might require W to initiate the session $\mathcal{S}'$ and then choose $b_{\mathcal{S}',Oscar}$ to be some complicated function of $b_{\mathcal{S}',W}$. However, we do assume that W chooses a random value $b_{\mathcal{S}',W}$ in the subgroup generated by α, by first choosing $a_{\mathcal{S}',W}$ uniformly at random and then computing $b_{\mathcal{S}',W} = \alpha^{a_{\mathcal{S}',W}}$.

After the session completes, Oscar requests and is given the value $K_{\mathcal{S}'} = \textbf{CDH}(b_{\mathcal{S}',Oscar}, b_{\mathcal{S}',W}) = (b_{\mathcal{S}',Oscar})^{a_{\mathcal{S}',W}}$. Oscar can then record the outcome of the session $\mathcal{S}'$ and the value of the session key $K_{\mathcal{S}'}$ in the form of a triple of values

$$(b_{\mathcal{S}',Oscar}, b_{\mathcal{S}',W}, \textbf{CDH}(b_{\mathcal{S}',Oscar}, b_{\mathcal{S}',W})).$$

After a number of such sessions, Oscar accumulates a list of triples (i.e., a ***transcript***) $\mathcal{T}$, where each triple $T \in \mathcal{T}$ has the form given above. We assume that Oscar has some polynomial-time algorithm A such that $\mathrm{A}(\mathcal{T}, (b_{\mathcal{S},U}, b_{\mathcal{S},V}))$ computes some partial information about the key $K_{\mathcal{S}}$ when $\mathcal{T}$ is constructed by the method described above.

We will argue that the hypothesized algorithm A cannot exist, assuming the intractability of the **DDH** problem. The way that we establish the non-existence of A is to show that it is possible to replace the transcript $\mathcal{T}$ by a simulated transcript $\mathcal{T}_{sim}$, which can be created by Oscar without taking part in any sessions and without requesting that any session keys be revealed to him.

We now show how Oscar can efficiently construct a simulated transcript $\mathcal{T}_{sim}$. Let's consider a typical triple on $\mathcal{T}$, which has the form

$$T = (b_1, b_2, b_3 = \textbf{CDH}(b_1, b_2)).$$

As mentioned above, b_1 is chosen by Oscar, using whatever method he desires (i.e., we allow the possibility that b_1 depends in some way on b_2), and b_2 is chosen randomly by Oscar's peer. Then $\mathbf{CDH}(b_1, b_2)$ is revealed to Oscar. Consider the following method of constructing a simulated triple, T_{sim}:

1. Oscar chooses a random value a_2 and computes $b_2 = \alpha^{a_2}$,
2. Oscar chooses b_1 as before,
3. Oscar computes $b_3 = (b_1)^{a_2}$ (observe that $b_3 = \mathbf{CDH}(b_1, b_2)$), and
4. Oscar defines $T_{sim} = (b_1, b_2, b_3)$.

Basically, the simulation replaces a random choice of b_2 made by Oscar's peer with a random choice of b_2 made by Oscar. Nothing else changes. However, when Oscar chooses b_2 as described above, he can compute the value of b_3 himself.

We claim that a triple T is indistinguishable from a simulated triple T_{sim}. More precisely, it holds that

$$\mathbf{Pr}[T = (b_1, b_2, b_3)] = \mathbf{Pr}[T_{sim} = (b_1, b_2, b_3)]$$

for all triples of the form $(b_1, b_2, b_3 = \mathbf{CDH}(b_1, b_2))$. In fact, this is almost trivial to confirm, because b_1 is chosen exactly the same way in both T and T_{sim}, b_2 is chosen uniformly at random in both T and T_{sim}, and $b_3 = \mathbf{CDH}(b_1, b_2)$ in both T and T_{sim}.

This simulation of triples can be extended to simulate transcripts. The simulated transcript $\mathcal{T}_{sim}$ is built up triple by triple, in the same way as $\mathcal{T}$, except that each triple $T \in \mathcal{T}$ is replaced by a simulated triple T_{sim}. The resulting simulated transcripts are indistinguishable from real transcripts.

Because of this indistinguishability property, it follows immediately that A behaves exactly the same when given a transcript $\mathcal{T}$ as it does when it is given a simulated transcript $\mathcal{T}_{sim}$. That is, the outputs $\mathrm{A}(\mathcal{T}_{sim}, (b_{\mathcal{S},U}, b_{\mathcal{S},V}))$ have exactly the same probability distribution as outputs $\mathrm{A}(\mathcal{T}, (b_{\mathcal{S},U}, b_{\mathcal{S},V}))$. This means that, whatever Oscar can do using a known session key attack, he can also do using a completely passive attack in which no sessions (other than $\mathcal{S}$) take place. But such an attack is not possible, given that **DDH** is intractable. This contradiction completes our proof, and therefore we have the following theorem.

THEOREM 12.2 *The Station-to-station key agreement scheme is an authenticated key agreement scheme that is secure against known session key attacks and which provides implicit key confirmation to both parties, assuming that the* ***Decision Diffie-Hellman*** *problem is intractable.*

12.3 Key Derivation Functions

Key agreement schemes enable two parties to establish a common shared secret. This shared secret is not usually used in unmodified form as a secret key.

For example, if we are using some version of the *Diffie-Hellman KAS*, the shared secret would be $\alpha^{a_U a_V}$, which might be an element of $\mathbb{Z}_p{}^*$, where p is a 2048-bit prime. The desired secret key could be an *AES* key, which is a bit string of length 128.

It is therefore common practice to employ a ***key derivation function***, denoted **KDF**, to derive one or more shared keys from the shared secret. The function **KDF** will take as input a shared secret and output a bit string of sufficient length to construct one or more keys of specified lengths. Roughly speaking, in order for a key derivation function to be considered secure, it should not be possible for an adversary to distinguish the output of the function **KDF** from a truly random string of the same length, assuming that the adversary does not have any information about the input to the function **KDF**.

In this section, we describe one common method of realizing key derivation functions that is based on a cryptographic hash function. The particular technique we present is called **KDM**; it is an example of ***one-step key derivation***, which is one of several methods that are recommended by NIST.

See Algorithm 12.1 for the detailed description of **KDM**. The inputs to Algorithm 12.1 consist of the following:

1. Z, the shared secret, which could be obtained from a suitable key agreement scheme, for example.
2. L, which is the length of the bit string to be output by the function **KDM**.
3. *FixedInput*, which is additional input to the function **KDM**. *FixedInput* could incorporate information such as the identities of the two parties who are establishing a secret key, some kind of string identifying the particular session, and/or the information exchanged in the key agreement scheme in order to establish the shared secret.

The function H is a cryptographic hash function that outputs a bit string (the message digest) of length *H_outputlen*. The output of Algorithm 12.1 is *DerivedKeyingMaterial*, which is a bit string of length L.

The basic idea of Algorithm 12.1 is to compute the hash function H a sufficient number of times to obtain the desired number of keying bits. Each time H is computed, the input to H is changed by incrementing the *counter*.

For example, suppose that H is instantiated with *SHA-224*, so $H\_outputlen = 224$, and suppose that we desire $L = 600$ key bits. Then $reps = 3$ and *Result* is 672 bits in length. The first 600 bits of *Result* is the *DerivedKeyingMaterial*.

12.4 MTI Key Agreement Schemes

Matsumoto, Takashima, and Imai have constructed several interesting key agreement schemes by modifying the *Diffie-Hellman KAS*. These schemes, which

Algorithm 12.1: KDM($Z, L, FixedInput$)

external *hash*
$reps \leftarrow \lceil L/H\_outputlen \rceil$
$counter \leftarrow 0$
comment: *counter* is a four-byte unsigned integer

$Result \leftarrow \epsilon$
comment: ϵ denotes an empty string

for $i \leftarrow 1$ **to** *reps*
do {
  $counter \leftarrow counter + 1$
  $Result \leftarrow Result \parallel H(counter \parallel Z \parallel FixedInput)$
  comment: "$\parallel$" denotes concatenation of bit strings
}
$DerivedKeyingMaterial \leftarrow$ the leftmost L bits of *Result*
return (*DerivedKeyingMaterial*)

we call *MTI* schemes, do not require that U and V compute any signatures. They are termed ***two-flow key agreement schemes***, because there are only two separate transmissions of information performed in each session of the scheme (one from U to V and one from V to U). In contrast, the *STS KAS* is a three-pass scheme.

We present one of the *MTI* key agreement schemes, namely, the *MTI/A0 KAS*, as Protocol 12.3.

We present a toy example to illustrate the workings of this scheme.

Example 12.1 Suppose $p = 27803$, $n = p - 1$ and $\alpha = 5$. The public domain parameters for the scheme consist of the group $(\mathbb{Z}_p{}^*, \cdot)$ and α. Here p is prime and α is a generator of $(\mathbb{Z}_p{}^*, \cdot)$, so the order of α is equal to n.

Assume U chooses secret exponent $a_U = 21131$; then she will compute

$$b_U = 5^{21131} \bmod 27803 = 21420,$$

which is placed on her certificate. As well, assume V chooses secret exponent $a_V = 17555$. Then he will compute

$$b_V = 5^{17555} \bmod 27803 = 17100,$$

which is placed on his certificate.

Now suppose that U chooses $r_U = 169$; then she will send the value

$$s_U = 5^{169} \bmod 27803 = 6268$$

to V. Suppose that V chooses $r_V = 23456$; then he will send the value

$$s_V = 5^{23456} \bmod 27803 = 26759$$

Protocol 12.3: MTI/A0 KAS

The public domain parameters consist of a group $(G, \cdot)$ and an element $\alpha \in G$ having order n.
Each user T has a long-term private key a_T, where $0 \leq a_T \leq n-1$, and a corresponding long-term public key

$$b_T = \alpha^{a_T}.$$

The value b_T is included in T's certificate and is signed by the TA.

1. U chooses r_U at random, $0 \leq r_U \leq n-1$, and computes

$$s_U = \alpha^{r_U}.$$

   Then U sends **Cert**(U) and s_U to V.

2. V chooses r_V at random, $0 \leq r_V \leq n-1$, and computes

$$s_V = \alpha^{r_V}.$$

   Then V sends **Cert**(V) and s_V to U.

   Finally, V computes the session key

$$K = {s_U}^{a_V} {b_U}^{r_V},$$

   where he obtains the value b_U from **Cert**(U).

3. U computes the session key

$$K = {s_V}^{a_U} {b_V}^{r_U},$$

   where she obtains the value b_V from **Cert**(V).

At the end of the session, U and V have both computed the same session key

$$K = \alpha^{r_U a_V + r_V a_U}.$$

to U.

Now U can compute the key

$$\begin{aligned} K_{U,V} &= {s_V}^{a_U} {b_V}^{r_U} \bmod p \\ &= 26759^{21131} 17100^{169} \bmod 27803 \\ &= 21600, \end{aligned}$$

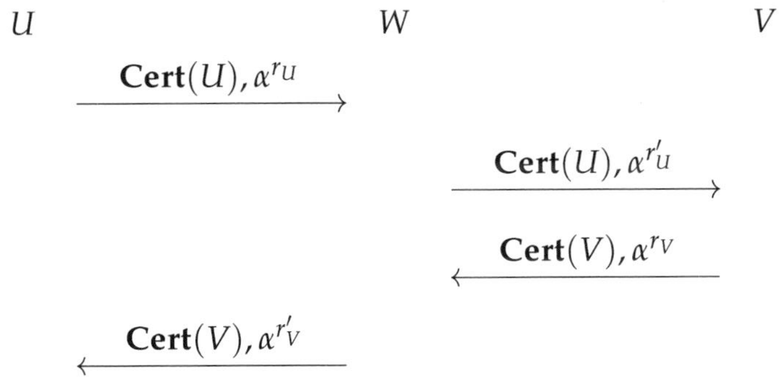

FIGURE 12.4: Unsuccessful intruder-in-the-middle attack on MTI/A0

and V can compute the (same) key

$$\begin{aligned} K_{U,V} &= s_U{}^{a_V} b_U{}^{r_V} \bmod p \\ &= 6268^{17555} 21420^{23456} \bmod 27803 \\ &= 21600. \end{aligned}$$

□

For future reference, the information transmitted during a session of the scheme is depicted as follows:

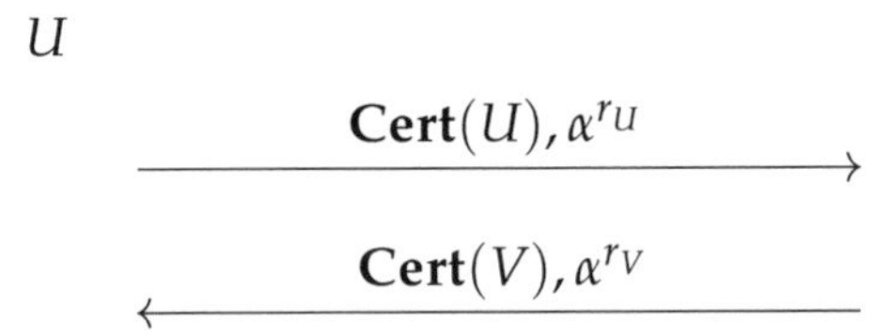
U
V
$\mathbf{Cert}(U), \alpha^{r_U}$
$\mathbf{Cert}(V), \alpha^{r_V}$

Let's now examine the security of the scheme. It is not too difficult to show that the security of the *MTI/A0* key agreement scheme against a passive adversary is exactly the same as in the *Diffie-Hellman* key agreement scheme—see the Exercises. As with many schemes, proving security in the presence of an active adversary is problematic. We will not attempt to prove anything in this regard, and we limit ourselves to some informal arguments.

Here is one threat we might consider: Without the use of signatures during the scheme, it might appear that there is no protection against an intruder-in-the-middle attack. Indeed, it is possible that W might alter the values that U and V send to each other. In Figure 12.4, we depict one typical scenario that might arise, which is analogous to the original intruder-in-the-middle attack on the *Diffie-Hellman KAS*.

In this situation, U and V will compute different keys: U will compute

$$K = \alpha^{r_U a_V + r'_V a_U},$$

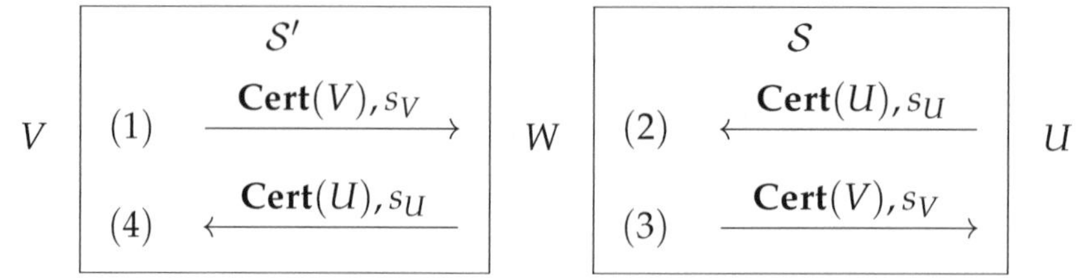

FIGURE 12.5: Known session key attack on MTI/A0

while V will compute

$$K = \alpha^{r'_U a_V + r_V a_U}.$$

However, neither of the computations of keys by U or V can be carried out by W, since they require knowledge of the secret exponents a_U and a_V, respectively. So even though U and V have computed different keys (which will of course be useless to them), neither of these keys can be computed by W, nor can he obtain any information about these keys (assuming the intractability of the **DDH** problem).

If this were the only possible attack on the scheme, then we would be able to say that the scheme provides implicit key authentication. This is because, even in the presence of this attack, both U and V are assured that the other is the only user in the network that could compute the key that they have computed. However, we will show in the next section that there are additional attacks which can be carried out by an adversary in the known session key attack model.

12.4.1 Known Session Key Attacks on MTI/A0

We begin by presenting a parallel session known session key attack on *MTI/A0*. This attack is a known session key attack utilizing a parallel session, hence the awkward terminology. The adversary, W, is an active participant in two sessions: W pretends to be V in a session $\mathcal{S}$ with U; and W pretends to be U in a parallel session $\mathcal{S}'$ with V. The actions taken by W are illustrated in Figure 12.5.

The flows in the two sessions are labeled in the order in which they occur. (1) and (2) represent the initial flows in the sessions $\mathcal{S}'$ and $\mathcal{S}$, respectively. Then the information in flow (1) is copied to flow (3), and the information in flow (2) is copied to flow (4) by W. Since the two sessions are being executed in parallel, we have a parallel session attack.

After the two sessions have completed, W requests the key K for session $\mathcal{S}'$, which he is allowed to do in a known session key attack. Of course, K is also the key for session $\mathcal{S}$, so W achieves his goal of computing the key for a session in which he is an active adversary and for which he has not requested the session key. This represents a successful attack in the known session key attack model.

The parallel session attack can be carried out because the key is a symmetric function of the inputs provided by the two parties:

$$K((r_U, a_U), (r_V, a_V)) = K((r_V, a_V), (r_U, a_U)).$$

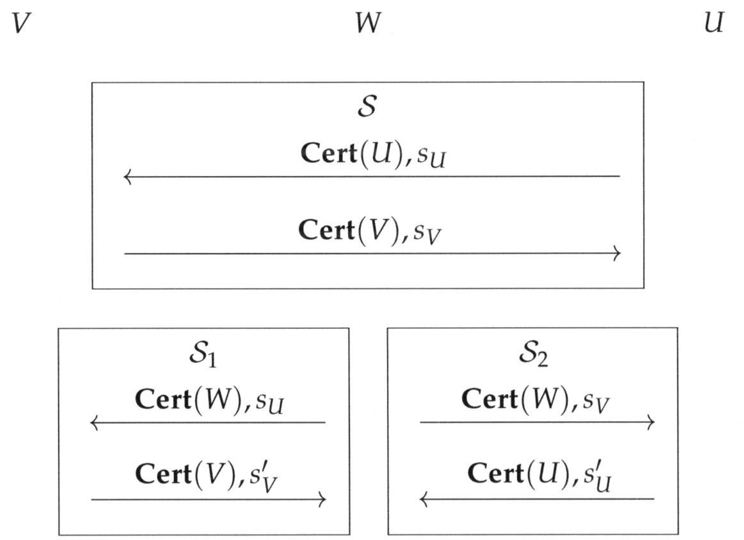

FIGURE 12.6: Burmester triangle attack on MTI/A0

To thwart the attack, we should eliminate this symmetry property. This could be done, for example, by using a key derivation function, say **KDF**. Suppose that the actual session key K was defined to be

$$K = \mathbf{KDF}(\alpha^{r_U a_V} \parallel \alpha^{r_V a_U})$$

U (the initiator of the session) would compute

$$K = \mathbf{KDF}({b_V}^{r_U} \parallel {s_V}^{a_U})$$

while V (the responder of the session) would compute

$$K = \mathbf{KDF}({s_U}^{a_V} \parallel {b_U}^{r_V}).$$

With this modified method of constructing a session key, the previous attack no longer works. This is because the two sessions $\mathcal{S}$ and $\mathcal{S}'$ now have different keys: the key for session $\mathcal{S}$ is

$$K_{\mathcal{S}} = \mathbf{KDF}(\alpha^{r_U a_V} \parallel \alpha^{r_V a_U}),$$

while the key for session $\mathcal{S}'$ is

$$K_{\mathcal{S}'} = \mathbf{KDF}(\alpha^{r_V a_U} \parallel \alpha^{r_U a_V}).$$

If **KDF** is a secure key derivation function, then there will be no way for W to compute $K_{\mathcal{S}}$ given $K_{\mathcal{S}'}$, or to compute $K_{\mathcal{S}'}$ given $K_{\mathcal{S}}$.

There is another known session key attack on *MTI/A0* which is called the ***Burmester triangle attack***. This attack is depicted in Figure 12.6.

We describe the triangle attack in more detail now. First, W observes a session

$\mathcal{S}$ between U and V. Then W participates in two additional sessions $\mathcal{S}_1$ and $\mathcal{S}_2$ with V and U, respectively. In these two sessions, W transmits values s_U and s_V that are copied from $\mathcal{S}$. (Of course, W does not know the exponents r_U and r_V corresponding to s_U and s_V, respectively.) Then, after the sessions $\mathcal{S}_1$ and $\mathcal{S}_2$ have concluded, W requests the keys for these two sessions, which is permitted in a known session key attack.

The session keys K, K_1, and K_2 for the sessions $\mathcal{S}$, $\mathcal{S}_1$, and $\mathcal{S}_2$ (respectively) are as follows:

$$\begin{aligned} K &= \alpha^{r_U a_V + r_V a_U} \\ K_1 &= \alpha^{r_U a_V + r'_V a_W} \\ K_2 &= \alpha^{r'_U a_W + r_V a_U}. \end{aligned}$$

Given K_1 and K_2, W is able to compute K as follows:

$$K = \frac{K_1 K_2}{(s'_V s'_U)^{a_W}}.$$

Therefore this is a successful known session key attack.

The triangle attack can also be defeated through the use of a key derivation function, as described above. It is conjectured that this modified version of *MTI/A0* is secure against known session key attacks.

12.5 Deniable Key Agreement Schemes

The concept of ***deniability*** provides an interesting counterpoint to the idea of non-repudiation, which is a central requirement of signature schemes. Recall that non-repudiation (in the context of a signature scheme) means that someone who has signed a message cannot later plausibly deny having done so. This is useful in any context where a signature should be considered to be a binding commitment, such as signing a contract.

On the other hand, there are situations where Alice and Bob might wish to engage in a private conversation, but neither of them desires that any third party should be able to prove that they had a particular conversation, even if the keys used to encrypt that conversation are leaked at some later time. In other words, the conversation is secure, but it affords plausible deniability to the participants in the event of a future key compromise.

Informally, a key agreement scheme is said to be ***deniable*** if this property is achieved when the resulting session keys are used to encrypt a conversation between the parties executing the key agreement protocol. To be more precise, we consider the following scenario.

1. An adversary gains access to the private keys belonging to V (this adversary could be V himself, or some third party.

Protocol 12.4: X3DH KAS

The public domain parameters consist of a group $(G, \cdot)$, an element $\alpha \in G$ having order n, and a key derivation function denoted by **KDF**.
Each user T has a long-term private key a_T, where $0 \leq a_T \leq n-1$, and a corresponding long-term public key

$$b_T = \alpha^{a_T}.$$

The value b_T is included in T's certificate and is signed by the TA.

1. U chooses r_U at random, $0 \leq r_U \leq n-1$, and computes

$$s_U = \alpha^{r_U}.$$

   Then U sends $\mathbf{Cert}(U)$ and s_U to V.

2. V chooses r_V at random, $0 \leq r_V \leq n-1$, and computes

$$s_V = \alpha^{r_V}.$$

   Then V sends $\mathbf{Cert}(V)$ and s_V to U.

   Finally, V computes the session key

$$K = \mathbf{KDF}({s_U}^{r_V} \parallel {s_U}^{a_V} \parallel {b_U}^{r_V}),$$

   where he obtains the value b_U from $\mathbf{Cert}(U)$.

3. U computes the session key

$$K = \mathbf{KDF}({s_V}^{r_U} \parallel {b_V}^{r_U} \parallel {s_V}^{a_U}),$$

   where she obtains the value b_V from $\mathbf{Cert}(V)$.

At the end of the session, U and V have both computed the same session key

$$K = \mathbf{KDF}(\alpha^{r_U r_V} \parallel \alpha^{r_U a_V} \parallel \alpha^{r_V a_U}).$$

2. The adversary produces a purported transcript of a session of a key agreement scheme involving U and V.

3. The goal is to determine if the transcript constitutes evidence that the session between U and V actually took place. If so, this would implicate U.

For some key agreement schemes, it turns out that it is possible to simulate (or forge) a transcript that is identical to a real transcript. For such a scheme, there is no way to determine if the session in question actually occurred. So the adversary is thus unable to implicate U and such a scheme would therefore be deniable.

To illustrate this concept of deniability, it is useful to contrast the basic Diffie-Hellman (Protocol 12.1), which is deniable, with *STS* (Protocol 12.2), which is not deniable.

First we look at basic *Diffie-Hellman*. Suppose for the purpose of discussion that U and V use Protocol 12.1 to derive a session key. Then, at some later time, V wishes to implicate U. V can store and reveal all the information he sent or received from U, along with his own private keys. In terms of the key agreement protocol, this information (i.e., the transcript) would consist of the following information:

- V's private key, a_V, and
- the public keys $b_U = \alpha^{a_U}$ and $b_V = \alpha^{a_V}$.

From this transcript, the key $K = \alpha^{a_U a_V}$ can be computed using the formula $K = {b_U}^{a_V}$. However, there is no convincing evidence that it was U who shared this key with V, because V could have simply created the public key b_U himself. The entire transcript could be forged by the adversary, and so we would say that basic *Diffie-Hellman* is deniable.

On the other hand, *STS* is not deniable. The reason for this is that both U and V sign the public keys they exchange during the protocol. Now the transcript would consist of

- V's private key, a_V,
- the public keys $b_U = \alpha^{a_U}$ and $b_V = \alpha^{a_V}$, and
- the signatures $\mathbf{sig}_U(ID(V) \parallel b_U \parallel b_V)$ and $\mathbf{sig}_V(ID(U) \parallel b_V \parallel b_U)$.

This transcript provides convincing evidence that the associated key $K = \alpha^{a_U a_V} = {b_U}^{a_V} = {b_V}^{a_U}$ was created in a session involving U and V. This because the public keys b_U and b_V, along with $ID(V)$, were signed by U. Therefore, V can implicate U, and U cannot plausibly deny that she took part in the given session of the key agreement protocol with V.

Thus, if deniability is a desired property of the key agreement scheme, then *STS* does not provide a satisfactory solution. On the other hand, basic *Diffie-Hellman*, while deniable, is susceptible to intruder-in-the-middle attacks. So the interesting question is how to design deniable key agreement schemes that are secure against intruder-in-the-middle attacks. We now present a recent method, known as *X3DH*, which is incorporated into the *Signal* messaging protocol.[3] *X3DH* is quite similar to *MTI* in some respects, but it uses three Diffie-Hellman keys instead of two. The *X3DH* key agreement scheme is presented as Protocol 12.4.

---

[3] *Signal* is a messaging protocol that has achieved widespread use since its development by Open Whisper Systems in 2013, notably in applications such as *WhatsApp*.

We do not discuss the security properties of *X3DH* in detail. However, we mention that *X3DH* is deniable and it provides perfect forward secrecy (see the Exercises). We also observe that a basic intruder-in-the middle attack does not succeed because the adversary cannot modify the long-term public keys without detection (since they are retrieved from a certificate). The adversary can modify the public keys s_U and s_V, changing them to s'_U and s'_V, respectively. However, in this situation, he would not be able to compute the resulting modified keys defined by the protocol. U would compute the key

$$\mathbf{KDF}(\alpha^{r_U r'_V} \parallel \alpha^{r_U a_V} \parallel \alpha^{r'_V a_U}).$$

However, the adversary cannot compute this key because he does not know the value of $\alpha^{r_U a_V}$. V would compute the key

$$\mathbf{KDF}(\alpha^{r'_U r_V} \parallel \alpha^{r'_U a_V} \parallel \alpha^{r_V a_U}).$$

The adversary does not know the value of $\alpha^{r_V a_U}$, so he cannot compute this key either.

12.6 Key Updating

Key updating schemes provide methods of updating keys on a regular basis. Ideally, the compromise of a key should not affect the security of previously-used keys (this is the "perfect forward secrecy" property, as defined in Section 11.1), nor should it allow the adversary to determine keys that are established in the future. One obvious way to approach this problem would be to execute a *Diffie-Hellman KAS* every time a message is sent. Each key is used only once and then deleted, and "new" keys have no dependence on old keys. Of course, *Diffie-Hellman* requires "expensive" operations such as exponentiations in finite fields, so we might seek less costly alternatives.

We already described the *Logical Key Hierarchy*, which is a type of key updating for dynamic networks, in Section 11.4. In Section 11.4, the reason for updating keys was to allow users to join or leave the network without impacting the security of the other network users. On the other hand, in this section, we have a pair of users who wish to communicate over a long period of time, and they wish to update their keys periodically.

We will now describe in simplified form some of the key updating techniques that are used in the *Signal* protocol. One of the goals of *Signal* is to provide ***end-to-end encryption***, which ensures that only the communicating parties can decrypt the encrypted communications. No third party, including the service provider, should have the technological capability to decrypt messages.

One design element incorporated into *Signal* is sometimes termed a ***Diffie-Hellman ratchet***. This manages to reduce the amount of key computation (as compared to setting up a new Diffie-Hellman key for every message sent) by about

25%. The idea is as follows: Every time U sends a message to V (or vice versa), the sender chooses new public and private keys (e.g., a_U and $b_U = \alpha^{a_U}$ in the case of user U) and sends the new public key along with a message that is encrypted under the old Diffie-Hellman key. The next Diffie-Hellman key to be used by the recipient is computed from the new public key along with the recipient's old private key. See Protocol 12.5 for the details of this protocol.

In practice, an authenticated version of *Diffie-Hellman* might be preferred. We are just describing the updating (i.e., ratcheting) process using basic *Diffie-Hellman* for simplicity.

If U and V used a *Diffie-Hellman KAS* to compute a new key every time a message is sent, they would each have to perform two exponentiations per message sent. Here, each party performs three exponentiations to compute two successive keys (after the initial key K_{00} is computed using two exponentiations by both parties). This is how the 25% speedup is achieved.

Observe that an adversary who manages to access a user's private key will only be able to use it to compute two successive keys.

The second major type of key updating or key ratcheting that is incorporated into *Signal* makes use of a key derivation function denoted by **KDF**. The function **KDF** has two inputs and two outputs. The two inputs are

1. a constant value C, and
2. a ***KDF key***, say K_i,

and the two outputs are

1. a "new" KDF key, say K_{i+1}, and
2. an ***output key***, denoted by OK_{i+1}.

We denote this process by the notation

$$\mathbf{KDF}(C, K_i) = (K_{i+1}, OK_{i+1}).$$

KDF is used to iteratively construct a ***KDF chain***. This requires an initial KDF key K_0. Then a sequence of output keys is produced as follows:

$$\begin{aligned}
\mathbf{KDF}(C, K_0) &= (K_1, OK_1) \\
\mathbf{KDF}(C, K_1) &= (K_2, OK_2) \\
\mathbf{KDF}(C, K_2) &= (K_3, OK_3),
\end{aligned}$$

etc. The output keys $OK_1, OK_2, \dots$ are used to encrypt and decrypt messages.

A KDF chain is faster than a public key ratchet because it is based on a fast hash function. However, the security properties are weaker. An adversary who compromises a KDF key K_i (and who knows the value of the constant C) can compute all subsequent output keys, beginning with OK_{i+1}. (However, assuming that the function **KDF** is one-way, the adversary cannot compute any previous output

Protocol 12.5: DIFFIE-HELLMAN RATCHET

The public domain parameters consist of a group $(G, \cdot)$ and an element $\alpha \in G$ having order n.

1. U chooses a private key a_0 and computes a corresponding public key α^{a_0}. She sends α^{a_0} to V.

2. V chooses a private key b_0 and computes a corresponding public key α^{b_0}. He also computes the Diffie-Hellman key

$$K_{00} = (\alpha^{a_0})^{b_0}$$

and he sends α^{b_0} to U along with a message encrypted with K_{00}, say $y_{00} = e_{K_{00}}(x_{00})$.

3. U receives α^{b_0} and y_{00}. She computes the Diffie-Hellman key

$$K_{00} = (\alpha^{b_0})^{a_0}$$

and then she uses K_{00} to decrypt y_{00}. U then chooses a new private key a_1 and she computes the corresponding public key α^{a_1}. She sends α^{a_1} to V. Finally, U computes the Diffie-Hellman key

$$K_{10} = (\alpha^{b_0})^{a_1}$$

and she uses K_{10} to encrypt a message $y_{10} = e_{K_{10}}(x_{10})$. The value y_{10} is sent to V.

4. V receives α^{a_1} and y_{10}. He computes the Diffie-Hellman key

$$K_{10} = (\alpha^{a_1})^{b_0}$$

and then he uses K_{10} to decrypt y_{10}. V then chooses a new private key b_1 and computes the corresponding public key α^{b_1}. He sends α^{b_1} to U. Finally, V computes the Diffie-Hellman key

$$K_{11} = (\alpha^{a_1})^{b_1}$$

and he uses K_{11} to encrypt a message $y_{11} = e_{K_{11}}(x_{11})$. The value y_{11} is sent to U.

5. The preceding two steps are repeated as often as desired.

keys.) So a particular KDF chain should not be used for an extended period of time.

The *Signal* protocol uses both public-key ratchets and KDF chains. The combination of these two techniques is called the *double ratchet*. We do not go into the details. However, roughly speaking, the keys created in the public-key ratchet are not used to encrypt messages. They are instead used to initiate KDF chains. At any point in time, there are two active KDF chains that are maintained by U and V. U has a ***sending chain*** whose output keys are used to encrypt messages that U sends to V, and a ***receiving chain*** whose output keys are used to decrypt messages that U receives from V. V also has a sending chain and a receiving chain. The sending chain for V is identical to the receiving chain for U and the receiving chain for V is identical to the sending chain for U. Whenever the public-key ratcheting scheme is applied, it is used to derive two new initial KDF keys, one for each of these two KDF chains.

12.7 Conference Key Agreement Schemes

A ***conference key agreement scheme*** (or, ***CKAS***) is a key agreement scheme in which a subset of two or more users in a network can construct a common secret key (i.e., a group key). In this section, we discuss (without proof) two conference key agreement schemes. The first CKAS we present was described in 1994 by Burmester and Desmedt. We also present the 1996 CKAS due to Steiner, Tsudik, and Waidner.

Both of these schemes are modifications of the *Diffie-Hellman KAS* in which m users, say $U_0, \ldots, U_{m-1}$, compute a common secret key. The schemes are set in a subgroup of a finite group in which the **Decision Diffie-Hellman** problem is intractable.

The *Burmester-Desmedt CKAS* is presented as Protocol 12.6. It is not hard to verify that all the participants in a session of this CKAS will compute the same key, Z, provided that the participants behave correctly and there is no active adversary who changes any of the transmitted messages. Suppose we define

$$Y_i = b_i{}^{a_{i+1}} = \alpha^{a_i a_{i+1}}$$

for all i (where all subscripts are to be reduced modulo m). Then

$$X_i = \left(\frac{b_{i+1}}{b_{i-1}}\right)^{a_i} = \left(\frac{\alpha^{a_{i+1}}}{\alpha^{a_{i-1}}}\right)^{a_i} = \frac{\alpha^{a_{i+1}a_i}}{\alpha^{a_{i-1}a_i}} = \frac{Y_i}{Y_{i-1}}$$

for all i. Then the following equations confirm that the key computation works

Protocol 12.6: BURMESTER-DESMEDT CONFERENCE KAS

The public domain parameters consist of a group $(G, \cdot)$ and an element $\alpha \in G$ having order n.
Note: all subscripts are to be reduced modulo m in this scheme, where m is the number of participants in the scheme.

1. For $0 \leq i \leq m-1$, U_i chooses a random number a_i, where $0 \leq a_i \leq n-1$. Then he computes
$$b_i = \alpha^{a_i}$$
and he sends b_i to U_{i+1} and U_{i-1}.
2. For $0 \leq i \leq m-1$, U_i computes
$$X_i = (b_{i+1}/b_{i-1})^{a_i}.$$
Then U_i broadcasts X_i to the $m-1$ other users.
3. For $0 \leq i \leq m-1$, U_i computes
$$Z = {b_{i-1}}^{a_i m}\, {X_i}^{m-1}\, {X_{i+1}}^{m-2} \cdots {X_{i-2}}^{1}.$$
Then
$$Z = \alpha^{a_0 a_1 + a_1 a_2 + \cdots + a_{m-1} a_0}$$
is the secret conference key which is computed by $U_0, \ldots, U_{m-1}$.

correctly:

$$\begin{aligned}
& {b_{i-1}}^{a_i m}\, {X_i}^{m-1}\, {X_{i+1}}^{m-2} \cdots {X_{i-2}}^{1} \\
&= {Y_{i-1}}^{m} \left(\frac{Y_i}{Y_{i-1}}\right)^{m-1} \left(\frac{Y_{i+1}}{Y_i}\right)^{m-2} \cdots \left(\frac{Y_{i-2}}{Y_{i-3}}\right)^{1} \\
&= Y_{i-1}\, Y_i \cdots Y_{i-2} \\
&= \alpha^{a_{i-1} a_i + a_i a_{i+1} + \cdots + a_{i-2} a_{i-1}} \\
&= Z.
\end{aligned}$$

Protocol 12.6 takes place in two stages. In the first stage, every participant sends a message to his or her two neighbors, where we view the m participants as being arranged in a ring of size m. In the second stage, each participant broadcasts one piece of information to everyone else. All the transmissions in each of the two stages can be done in parallel.

Protocol 12.7: STEINER-TSUDIK-WAIDNER CONFERENCE KAS

The public domain parameters consist of a group $(G, \cdot)$ and an element $\alpha \in G$ having order n.

Stage 1.

U_0 chooses a random number a_0, computes α^{a_0}, and sends $\mathcal{L}_0 = (\alpha^{a_0})$ to U_1.

For $i = 1, \ldots, m-2$, U_i receives the list $\mathcal{L}_{i-1}$ from U_{i-1}. Then U_i chooses a random number a_i and computes $\alpha^{a_0 a_1 \ldots a_i} = (\alpha^{a_0 a_1 \ldots a_{i-1}})^{a_i}$. Then he sends the list $\mathcal{L}_i = \mathcal{L}_{i-1} \parallel \alpha^{a_0 a_1 \ldots a_i}$ to U_{i+1}.

U_{m-1} receives the list $\mathcal{L}_{m-2}$ from U_{m-2}. Then he chooses a random number a_{m-1} and computes $\alpha^{a_0 a_1 \ldots a_{m-1}} = (\alpha^{a_0 a_1 \ldots a_{m-2}})^{a_{m-1}}$. Then he constructs the list $\mathcal{L}_{m-1} = \mathcal{L}_{m-2} \parallel \alpha^{a_0 a_1 \ldots a_{m-1}}$.

Stage 2.

U_{m-1} extracts the conference key $Z = \alpha^{a_0 a_1 \ldots a_{m-1}}$ from $\mathcal{L}_{m-1}$. For every other element $y \in \mathcal{L}_{m-1}$, U_{m-1} computes the value $y^{a_{m-1}}$. Then U_{m-1} constructs the list of $m-1$ values

$$\mathcal{M}_{m-1} = (\alpha^{a_{m-1}}, \alpha^{a_0 a_{m-1}}, \alpha^{a_0 a_1 a_{m-1}}, \ldots, \alpha^{a_0 a_1 \ldots a_{m-3} a_{m-1}})$$

and he sends $\mathcal{M}_{m-1}$ to U_{m-2}.

For $i = m-2, \ldots, 1$, U_i receives the list $\mathcal{M}_{i+1}$ from U_{i+1}. He computes the conference key $Z = (\alpha^{a_0 \ldots a_{i-1} a_{i+1} \ldots a_{m-1}})^{a_i}$ from the last element in $\mathcal{M}_{i+1}$. For every other element $y \in \mathcal{L}_{i+1}$, U_i computes the value y^{a_i}. Then U_i constructs the list of i values

$$\mathcal{M}_i = (\alpha^{a_i \ldots a_{m-1}}, \alpha^{a_0 a_i \ldots a_{m-1}}, \alpha^{a_0 a_1 a_i \ldots a_{m-1}}, \ldots, \alpha^{a_0 a_1 \ldots a_{i-2} a_i \ldots a_{m-1}})$$

and he sends $\mathcal{M}_i$ to U_{i-1}.

U_0 receives the list $\mathcal{M}_1$ from U_1. He computes the conference key $Z = (\alpha^{a_1 \ldots a_{m-1}})^{a_0}$ from the (only) element in $\mathcal{M}_1$.

Overall, each participant transmits two pieces of information and each participant receives $m+1$ pieces of information during one session of the scheme. This is quite efficient, but it requires the existence of a broadcast channel.

Steiner, Tsudik, and Waidner suggested a CKAS that is more "sequential" in nature, but which does not require a broadcast channel. Their scheme is presented as Protocol 12.7.

Stage 1	Stage 2	
U_3	U_3	$Z = (\alpha^{a_0a_1a_2})^{a_3}$
$\uparrow$	$\downarrow$	
$(\alpha^{a_0}, \alpha^{a_0a_1}, \alpha^{a_0a_1a_2})$	$(\alpha^{a_3}, \alpha^{a_0a_3}, \alpha^{a_0a_1a_3})$	
$\uparrow$	$\downarrow$	
U_2	U_2	$Z = (\alpha^{a_0a_1a_3})^{a_2}$
$\uparrow$	$\downarrow$	
$(\alpha^{a_0}, \alpha^{a_0a_1})$	$(\alpha^{a_2a_3}, \alpha^{a_0a_2a_3})$	
$\uparrow$	$\downarrow$	
U_1	U_1	$Z = (\alpha^{a_0a_2a_3})^{a_1}$
$\uparrow$	$\downarrow$	
(α^{a_0})	$(\alpha^{a_1a_2a_3})$	
$\uparrow$	$\downarrow$	
U_0	U_0	$Z = (\alpha^{a_1a_2a_3})^{a_0}$

FIGURE 12.7: Information transmitted in the Steiner, Tsudik, and Waidner CKAS with four participants

A session of Protocol 12.7 takes place in two stages. In the first stage, information is transmitted sequentially from U_0 to U_1, from U_1 to $U_2, \dots,$ and finally from U_{m-2} to U_{m-1}. For $i \geq 1$, each user U_i receives a list of values from U_{i-1}, computes one new value, and appends it to the list. By the end of the first stage, a list of m values is held by U_{m-1}.

Then stage 2 begins. In stage 2, information is transmitted in the opposite order to stage 1. Each participant in turn computes the session key from the last element in the current list and then modifies the remaining elements in the list. At the end of this stage, every participant has computed the same session key, $Z = \alpha^{a_0a_1 \dots a_{m-1}}$. See Figure 12.7 for a diagram illustrating the information transmitted in a session of Protocol 12.7 in which there are four participants.

It is not difficult to count the number of messages transmitted and received by each participant in Protocol 12.7. For example, for $0 \leq i \leq m-2$, it is easily seen that U_i transmits $2i + 1$ messages, while U_{m-1} transmits $m - 1$ messages. The total number of messages transmitted is $m^2 - m$.

Neither Protocol 12.6 nor Protocol 12.7 provides any kind of authentication. Security against active adversaries would require additional information to be transmitted such as signatures, certificates, etc. It is also not immediately obvious that these schemes are secure even against a passive adversary (as usual, under the assumption that the **Decision Diffie-Hellman** problem is intractable). However, it has in fact been proven that the *Steiner-Tsudik-Waidner Conference KAS* is secure in this setting.

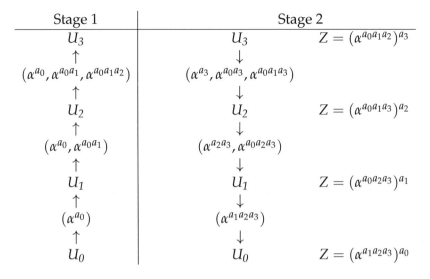

12.8 Notes and References

Diffie and Hellman presented their key agreement scheme in [71]. The idea of key exchange was discovered independently by Merkle [135]. The *Station-to-station KAS* is due to Diffie, van Oorschot, and Wiener [72]. The variation of *STS* that we present in Protocol 12.2 is essentially the same as "Protocol SIG-DH" from [54]. For an overview of key agreement schemes based on the **Diffie-Hellman** problems, see Blake-Wilson and Menezes [34].

Key derivation functions currently recommended by NIST can be found in [9].

The schemes of Matsumoto, Takashima, and Imai can be found in [129]. The triangle attack is from Burmester [51].

A detailed description of the *double ratchet* technique is presented in [162]. The *X3DH KAS* is from [126]. For a rigorous security analysis of the *Signal* protocol, see Cohn-Gordon *et al.* [60].

The *Burmester-Desmedt Conference KAS* is described in [52] and the *Steiner-Tsudik-Waidner Conference KAS* is from [190].

Boyd and Mathuria [44] is a book that contains a great deal of information on the topics covered in this chapter.

Exercises

12.1 Suppose that U and V take part in a session of the *Diffie-Hellman KAS* with $p = 27001$ and $\alpha = 101$. Suppose that U chooses $a_U = 21768$ and V chooses $a_V = 9898$. Show the computations performed by both U and V, and determine the key that they will compute.

12.2 Consider the modification of the *STS KAS* that is presented as Protocol 12.8. In this modification of the protocol, the signatures omit the intended receiver. Show how this renders the protocol insecure, by describing an intruder-in-the-middle attack. Discuss the consequences of this attack, in terms of key authentication properties and how they are violated. (This attack is known as an ***unknown key-share attack***.)

12.3 Discuss whether the property of perfect forward secrecy (which was defined in Section 11.1) is achieved in the *STS KAS*, assuming that the secret signing keys of one or more users are revealed.

12.4 Suppose that U and V carry out the *MTI/A0 KAS* with $p = 30113$ and $\alpha = 52$. Suppose that U has $a_U = 8642$ and he chooses $r_U = 28654$; and V has $a_V = 24673$ and she chooses $r_V = 12385$. Show the computations performed by both U and V, and determine the key that they will compute.

Protocol 12.8: MODIFIED STATION-TO-STATION KAS

The public domain parameters consist of a group $(G, \cdot)$ and an element $\alpha \in G$ having order n.

1. U chooses a random number a_U, $0 \leq a_U \leq n-1$. Then she computes

$$b_U = \alpha^{a_U}$$

and she sends $\mathbf{Cert}(U)$ and b_U to V.

2. V chooses a random number a_V, $0 \leq a_V \leq n-1$. Then he computes

$$\begin{aligned} b_V &= \alpha^{a_V} \\ K &= (b_U)^{a_V}, \quad \text{and} \\ y_V &= \mathbf{sig}_V(b_V \parallel b_U). \end{aligned}$$

Then V sends $\mathbf{Cert}(V)$, b_V and y_V to U.

3. U verifies y_V using $\mathbf{ver}_V$. If the signature y_V is not valid, then she "rejects" and quits. Otherwise, she "accepts," she computes

$$\begin{aligned} K &= (b_V)^{a_U}, \quad \text{and} \\ y_U &= \mathbf{sig}_U(b_U \parallel b_V), \end{aligned}$$

and she sends y_U to V.

4. V verifies y_U using $\mathbf{ver}_U$. If the signature y_U is not valid, then he "rejects"; otherwise, he "accepts."

12.5 Discuss whether the property of perfect forward secrecy is achieved in *MTI/A0* for a session key that was established between U and V in the following two cases:

(a) one LL-key a_U is revealed.

(b) both LL-keys a_U and a_V are revealed.

12.6 If a passive adversary tries to compute the key K constructed by U and V by using the *MTI/A0 KAS*, then he is faced with an instance of what we might term the **MTI** problem, which we present as Problem 12.1.

Prove that any algorithm that can be used to solve the **MTI** problem can be used to solve the **Computational Diffie-Hellman** problem, and vice versa. (i.e., give Turing reductions between these two problems).

12.7 Analyze the deniability properties of *X3DH*. Specifically, show how an ad-

Problem 12.1: MTI

Instance: $I = (p, \alpha, \beta, \gamma, \delta, \epsilon)$, where p is prime, $\alpha \in \mathbb{Z}_p{}^*$ is a primitive element, and $\beta, \gamma, \delta, \epsilon \in \mathbb{Z}_p{}^*$.
Question: Compute $\beta^{\log_\alpha \gamma} \delta^{\log_\alpha \epsilon} \bmod p$.

versary with access to V's private keys can forge a transcript that appears to correspond to a session between U and V. Carefully describe how the associated transcript is created and how the associated key is computed.

12.8 Show that *X3DH* provides perfect forward secrecy. That is, assume that an adversary records all the information transmitted in a particular session between U and V and later obtains the long-term private keys belonging to U and V. Despite learning all this information, the adversary should be unable to compute the session key for this specific session.

12.9 The purpose of this question is to perform the required computations in a session of the *Burmester-Desmedt Conference KAS*. Suppose we take $p = 128047$, $\alpha = 8$, and $n = 21341$. (It can be verified that p is prime and the order of α in $\mathbb{Z}_p{}^*$ is equal to n.) Suppose there are $m = 4$ participants, and they choose secret values $a_0 = 4499$, $a_1 = 9854$, $a_2 = 19887$, and $a_3 = 10002$.

(a) Compute the values b_0, b_1, b_2, and b_3.

(b) Compute the values X_0, X_1, X_2, and X_3.

(c) Show the computations performed by U_0, U_1, U_2, and U_3 to construct the conference key Z.

12.10 Show all the computations performed in a session of the *Steiner-Tsudik-Waidner Conference KAS* involving four participants. Use the same values of p, α, n, a_0, a_1, a_2, and a_3 as in the previous exercise.

Chapter 13

Miscellaneous Topics

In this chapter, we introduce a selection of further topics of interest to cryptographers, including Paillier encryption; identity-based encryption; copyright protection utilizing tracing techniques; and blockchain technology.

13.1 Identity-based Cryptography

The use of public-key cryptography requires a mechanism to authenticate public keys. This is commonly done using certificates, which were introduced in Section 8.6. However, there are many potential problems with certificates, including the fact that distributing and authenticating certificates can be cumbersome. Shamir suggested in 1984 that the use of certificates could be eliminated by an identity-based approach.

The basic idea of identity-based cryptography is that the public key for a user U is obtained by applying a public hash function h to the user's identity string, $ID(U)$. The corresponding private key is generated by a central trusted authority (denoted by *TA*). This private key is then transmitted to the user U, using a secure channel, after that user proves his or her identity to the *TA*. In identity-based cryptography, the *TA* issues a private key rather than a certificate. The resulting public key and private key will be used in an encryption scheme, signature scheme, or other cryptographic scheme. Identity-based cryptography also requires some fixed, public system parameters (including a certain "master key") that will be used by everyone.

One significant advantage of identity-based cryptography is that it removes the need for certificates. However, we require a convenient and reliable method of associating an identity string (an email address, for example) with a person. Further, it is still necessary to put a great deal of trust in the *TA* when using identity-based cryptography. In fact, in this setting, the *TA* knows the values of all private keys, which need not be the case if certificates are used to authenticate public keys.

Designing an identity-based cryptosystem is not an easy exercise. Unfortunately, there does not seem to be an obvious or straightforward way to turn an arbitrary public-key cryptosystem into an identity-based cryptosystem. To illustrate, suppose that we tried to transform the *RSA Cryptosystem* into an identity-based cryptosystem in a naive way. We might envisage a situation where the *TA* chooses the RSA modulus $n = pq$ to be used as the public master key. The factors p and q would be the master private key.

The public RSA key of a user U is an encryption exponent and a private key is a decryption exponent. However, once U has a public key and corresponding private key, then he or she can easily factor n (we showed how to do this in Section 6.7.2). Once U knows the private master key, he can impersonate the *TA* and issue private keys to anyone else, as well as compute anyone else's private key. So this method of creating an identity-based cryptosystem fails utterly.

As can be seen from the above example, identity-based cryptography necessitates devising a system where a user's public and private key cannot be used to determine the private master key of the *TA*.

Here is a detailed description of the required operations in an identity-based (public-key) encryption scheme.

master key generation

The *TA* generates a ***master public key*** M^{pub} and a corresponding ***master private key*** M^{priv}. The ***master key*** is $M = (M^{pub}, M^{priv})$. A public hash function h is used to derive a user's public key from their ID string. The master key and the hash function comprise the ***system parameters***.

user key generation

When a user U identifies himself to the *TA*, the *TA* uses a function **extract** to compute U's private key K_U^{priv}, as follows:

$$K_U^{priv} = \mathbf{extract}(M, K_U^{pub})$$

where U's public key is

$$K_U^{pub} = h(ID(U)).$$

User U's key is $K_U = (K_U^{pub}, K_U^{priv})$.

encryption

U's public key K_U^{pub} defines a public encryption rule e_{K_U} that can be used (by anyone) to encrypt messages sent to U.

decryption

U's private key K_U^{priv} defines a private decryption rule d_{K_U} that U will use to decrypt messages he receives.

13.1.1 The Cocks Identity-based Cryptosystem

In this section, we discuss the *Cocks Identity-based Cryptosystem*, which is presented as Cryptosystem 13.1. Cryptosystem 13.1 depends on certain properties of Jacobi symbols. The cryptosystem is based on arithmetic in $\mathbb{Z}_n$, where $n = pq$ and p and q are distinct primes, each congruent to 3 modulo 4. Let $\mathbf{QR}(n)$ denote the set of quadratic residues modulo n:

$$\mathbf{QR}(n) = \left\{ x \in \mathbb{Z}_n : \left(\frac{x}{p}\right) = \left(\frac{x}{q}\right) = 1 \right\}.$$

As well, $\widetilde{\mathbf{QR}}(n)$ denotes the set of ***pseudo-squares modulo*** n:

$$\widetilde{\mathbf{QR}}(n) = \left\{ x \in \mathbb{Z}_n : \left(\frac{x}{p}\right) = \left(\frac{x}{q}\right) = -1 \right\}.$$

The security of the *Cocks Identity-based Cryptosystem*, which will be discussed later, is related to the difficulty of the **Composite Quadratic Residues** problem in $\mathbb{Z}_n$, which we present as Problem 13.1.

Problem 13.1: Composite Quadratic Residues

Instance: A positive integer n that is the product of two unknown distinct odd primes p and q, and an integer $x \in \mathbb{Z}_n{}^*$ such that the Jacobi symbol $(\frac{x}{n}) = 1$.
Question: Is $x \in \mathbf{QR}(n)$?

We note that we already defined a computational version of this problem as Problem 10.1. In the decision problem we defined above, however, we only require a yes/no answer.

The **Composite Quadratic Residues** problem requires us to distinguish quadratic residues modulo n from pseudo-squares modulo n. This can be no more difficult than factoring n. For, if the factorization $n = pq$ can be computed, then it is a simple matter to calculate $(\frac{x}{p})$, say. Given that $(\frac{x}{n}) = 1$, it follows from the multiplicative property of Jacobi symbols that x is a quadratic residue modulo n if and only if $(\frac{x}{p}) = 1$.

We note that, in Chapter 6, we defined the **Quadratic Residues** problem modulo a prime and showed that it is easy to solve; here we have a composite modulus.

There does not seem to be any way to solve the **Composite Quadratic Residues** problem efficiently if the factorization of n is not known. So it is commonly conjectured that this problem is intractable if it is infeasible to factor n.

Several aspects of Cryptosystem 13.1 require explanation. First, we have stated that the hash function h produces outputs that are always elements in $\mathbf{QR}(n) \cup \widetilde{\mathbf{QR}}(n)$. This is equivalent to saying that $0 < h(x) < n$ and the Jacobi symbol $(\frac{h(x)}{n}) = 1$ for all $x \in \{0,1\}^*$. In practice, one might compute $(\frac{h(x)}{n})$. If it is equal to -1, then we would multiply $h(x)$ by some fixed value $a \in \mathbb{Z}_n$ having Jacobi symbol equal to -1. This value a could be predetermined and made public. In any event, we will assume that some method has been specified so that $h(x) \in \mathbf{QR}(n) \cup \widetilde{\mathbf{QR}}(n)$ for all relevant values of x.

The generation of a user's private key is basically a matter of extracting a square root modulo n, as was done in the decryption operation of the *Rabin Cryptosystem*. This computation can be carried out by the *TA* because the *TA* knows the factorization of n.

Cryptosystem 13.1: *Cocks Identity-based Cryptosystem*

Let p, q be two distinct primes such that $p \equiv q \equiv 3 \bmod 4$, and define $n = pq$.

System parameters: The master key $M = (M^{pub}, M^{priv})$, where

$$M^{pub} = n$$

and

$$M^{priv} = (p, q).$$

As well, $h : \{0,1\}^* \rightarrow \mathbb{Z}_n$ is a public hash function with the property that $h(x) \in \mathbf{QR}(n) \cup \widetilde{\mathbf{QR}}(n)$ for all $x \in \{0,1\}^*$.

User key generation: For a user U, the key $K_U = (K_U^{pub}, K_U^{priv})$, where

$$K_U^{pub} = h(ID(U))$$

and

$$(K_U^{priv})^2 = \begin{cases} K_U^{pub} & \text{if } K_U^{pub} \in \mathbf{QR}(n) \\ -K_U^{pub} & \text{if } K_U^{pub} \in \widetilde{\mathbf{QR}}(n). \end{cases}$$

Encryption: A plaintext is an element in the set $\{1, -1\}$. To encrypt a plaintext element $x \in \{1, -1\}$, the following steps are performed:

1. Choose two random values $t_1, t_2 \in \mathbb{Z}_n$ such that the Jacobi symbols $\left(\frac{t_1}{n}\right) = \left(\frac{t_2}{n}\right) = x$.
2. Compute

$$y_1 = t_1 + K_U^{pub} (t_1)^{-1} \bmod n$$

and

$$y_2 = t_2 - K_U^{pub} (t_2)^{-1} \bmod n.$$

3. The ciphertext $y = (y_1, y_2)$.

Decryption: A ciphertext $y = (y_1, y_2)$ is decrypted as follows:

1. If $(K_U^{priv})^2 = K_U^{pub}$, then define $s = y_1$; otherwise, define $s = y_2$.
2. Compute the Jacobi symbol

$$x = \left(\frac{s + 2K_U^{priv}}{n}\right).$$

3. The decrypted plaintext is x.

Note that the *TA* will only compute square roots of numbers having a special, predetermined form, namely, $h(ID(U))$ or $-h(ID(U))$, for a user U, say. This is important, because a square root oracle can be used to factor n, as we showed in Section 6.8.1. This attack cannot be carried out in the context of the *Cocks Identity-based Cryptosystem* because U cannot use the *TA* as an oracle to extract square roots of arbitrary elements of $\mathbb{Z}_n$.

Suppose a user V wishes to encrypt a plaintext $x = \pm 1$ to send to U. This requires V to generate two random elements $t_1, t_2 \in \mathbb{Z}_n$, both having Jacobi symbols equal to x. This would be done by choosing random elements of $\mathbb{Z}_n$ and computing their Jacobi symbols, until elements with the desired Jacobi symbols are obtained. (Recall that the computation of Jacobi symbols modulo n can be performed efficiently without knowing the factorization of n.) If V wishes to encrypt a long string of plaintext elements, then each element must be encrypted independently, using different random values of t.

In order to decrypt a ciphertext y, only one of the two values y_1 and y_2 is required. So U chooses the appropriate one and the other one can be discarded. The reason why both y_1 and y_2 are transmitted to U is that V does not know whether U's private key is a square root of K_U^{pub} or a square root of $-K_U^{pub}$.

Now, let's show that the decryption operation works correctly, i.e., any encryption of x can successfully be decrypted, given the relevant private key. Suppose that U receives a ciphertext (y_1, y_2). Suppose further that $(K_U^{priv})^2 = K_U^{pub}$; then we will show that

$$\left(\frac{y_1 + 2K_U^{priv}}{n}\right) = x.$$

(If $(K_U^{priv})^2 = -K_U^{pub}$, then the decryption process is modified, but the proof of correctness is similar.) The following sequence of equations follows from basic properties of Jacobi symbols:

$$\begin{aligned}
\left(\frac{y_1 + 2K_U^{priv}}{n}\right) &= \left(\frac{t_1 + K_U^{pub}\,(t_1)^{-1} + 2K_U^{priv}}{n}\right)\\
&= \left(\frac{t_1 + 2K_U^{priv} + (K_U^{priv})^2\,(t_1)^{-1}}{n}\right)\\
&= \left(\frac{t_1(1 + 2K_U^{priv}\,(t_1)^{-1} + (K_U^{priv})^2\,(t_1)^{-2})}{n}\right)\\
&= \left(\frac{t_1}{n}\right)\left(\frac{1 + 2K_U^{priv}\,(t_1)^{-1} + (K_U^{priv})^2\,(t_1)^{-2}}{n}\right)\\
&= \left(\frac{t_1}{n}\right)\left(\frac{(1 + K_U^{priv}\,(t_1)^{-1})^2}{n}\right)\\
&= \left(\frac{t_1}{n}\right)\left(\frac{1 + K_U^{priv}\,(t_1)^{-1}}{n}\right)^2
\end{aligned}$$

$$= \left(\frac{t_1}{n}\right)$$
$$= x.$$

In the derivation of the penultimate line, we are using the fact that

$$\left(\frac{1 + K_U^{priv} (t_1)^{-1}}{n}\right) = \pm 1,$$

which can be proven easily (see the Exercises).

Next, let's consider the security of the scheme. We will prove that a decryption oracle for Cryptosystem 13.1 can be used to solve the **Composite Quadratic Residues** problem in $\mathbb{Z}_n$. Thus the cryptosystem is provably secure provided that the **Composite Quadratic Residues** problem is intractable.

First, we begin with an important technical lemma.

LEMMA 13.1 *Suppose that* $x = \pm 1$ *and* $\left(\frac{t}{n}\right) = x$*, where* x *and* t *are unknown. If* $(K_U^{priv})^2 \equiv K_U^{pub} \pmod{n}$*, then the value*

$$t - K_U^{pub}\, t^{-1} \bmod n$$

provides no information about x*. Similarly, if* $(K_U^{priv})^2 \equiv -K_U^{pub} \pmod{n}$*, then the value*

$$t + K_U^{pub}\, t^{-1} \bmod n$$

provides no information about x*.*

PROOF Suppose that

$$(K_U^{priv})^2 \equiv K_U^{pub} \pmod{n}$$

and denote

$$y = t - K_U^{pub}\, t^{-1} \bmod n.$$

Then

$$t^2 - ty - K_U^{pub} \equiv 0 \pmod{n},$$

so

$$t^2 - ty - K_U^{pub} \equiv 0 \pmod{p}$$

and

$$t^2 - ty - K_U^{pub} \equiv 0 \pmod{q}.$$

The first congruence has two solutions modulo p, and the product of these two solutions is congruent to $-K_U^{pub}$ modulo p. If the two solutions are r_1 and r_2, then we have

$$\left(\frac{r_2}{p}\right) = \left(\frac{-r_1\, K_U^{pub}}{p}\right) = \left(\frac{-r_1\, (K_U^{priv})^2}{p}\right) = \left(\frac{-r_1}{p}\right) = -\left(\frac{r_1}{p}\right).$$

Algorithm 13.1: COCKS-ORACLE-RESIDUE-TESTING(n, a)

comment: $\left(\frac{a}{n}\right) = 1$

external COCKS-DECRYPT
choose $x \in \{1, -1\}$ randomly
choose a random $t \in \mathbb{Z}_n$ such that $\left(\frac{t}{n}\right) = x$
$y_1 \leftarrow t + a\,t^{-1} \bmod n$
choose a random $y_2 \in {\mathbb{Z}_n}^*$
$y \leftarrow (y_1, y_2)$
$x' \leftarrow$ COCKS-DECRYPT(n, a, y)
if $x' = x$
  then return ("$a \in \mathbf{QR}(n)$")
  else return ("$a \in \widetilde{\mathbf{QR}}(n)$")

Note that the last equality holds because $p \equiv 3 \bmod 4$ and hence $\left(\frac{-1}{p}\right) = -1$.

A similar property holds for the two solutions of the second congruence. If these two solutions are s_1 and s_2, then

$$\left(\frac{s_2}{q}\right) = -\left(\frac{s_1}{q}\right).$$

Now, the congruence modulo n has four solutions for t. It is easy to see that two of the solutions have Jacobi symbol $\left(\frac{t}{n}\right) = 1$ and two of them have $\left(\frac{t}{n}\right) = -1$. Therefore it is not possible to compute any information about the Jacobi symbol $\left(\frac{t}{n}\right)$.

The second part of this lemma is proven in a similar manner. ∎

Suppose that COCKS-DECRYPT is a decryption oracle for the *Cocks Identity-based Cryptosystem*. That is, COCKS-DECRYPT(K_U^{pub}, n, y) correctly outputs the value of x whenever y is a valid encryption of x. We will show how to use the algorithm COCKS-DECRYPT to determine if K_U^{pub} is a quadratic residue or a pseudo-square modulo n. This algorithm is presented as Algorithm 13.1.

We will analyze Algorithm 13.1 informally. First, let's discuss the operations it performs. The input a is an element of $\mathbf{QR}(n) \cup \widetilde{\mathbf{QR}}(n)$. We treat a as a public key for the *Cocks Cryptosystem* and encrypt a random plaintext x. However, we only compute y_1 according to the encryption rule; y_2 is just a random element of ${\mathbb{Z}_n}^*$. Then we give the pair (y_1, y_2) to the decryption oracle COCKS-DECRYPT. The oracle outputs a decryption x'. Then Algorithm 13.1 reports that a is a quadratic residue modulo n if and only if $x = x'$.

Suppose that $a \in \mathbf{QR}(n)$. Then Lemma 13.1 says that, even if y_2 were computed according to the encryption rule, it would provide no information about x. So COCKS-DECRYPT can correctly compute x from y_1 alone. In this case, Algorithm 13.1 correctly states that a is a quadratic residue.

On the other hand, suppose that $a \in \widetilde{\mathbf{QR}}(n)$. Then Lemma 13.1 says that y_1 provides no information about x. Clearly y_2 provides no information about x because y_2 is random. Therefore, the value x' that is returned by COCKS-DECRYPT is going to be equal to x exactly half the time, because x is random and y is independent of x. Therefore the output of Algorithm 13.1 will be correct with probability 1/2.

The situation is analogous to that of a biased Monte Carlo algorithm. If $x \neq x'$, then we can be sure that $a \in \widetilde{\mathbf{QR}}(n)$. On the other hand, if $x = x'$, we cannot say with certainty that $a \in \mathbf{QR}(n)$; it may be only that COCKS-DECRYPT guessed the value of x correctly. So we should run Algorithm 13.1 several times on the same input. If it always reports that $a \in \mathbf{QR}(n)$, then we have some confidence that this conclusion is correct. The analysis of the probability of correctness of this approach is the same as was done in Section 6.4.

The above discussion assumed that COCKS-DECRYPT always outputs the correct answer if it is given a correctly formed ciphertext. A more complicated analysis shows that we can obtain a (possibly) unbiased Monte Carlo algorithm for **Composite Quadratic Residues** with error probability as small as desired, provided that COCKS-DECRYPT has an error probability less than 1/2.

13.1.2 The Boneh-Franklin Identity-based Cryptosystem

One drawback to the *Cocks Identity-based Cryptosystem* is the fact that it only encrypts one bit at a time. In this section we discuss the *Boneh-Franklin Identity-based Cryptosystem*, presented as Cryptosystem 13.2, which is better suited for encrypting larger amounts of plaintext.

The *Boneh-Franklin Identity-based Cryptosystem* is an example of ***pairing-based cryptography***, in which a pairing on an elliptic curve is used in the construction of various types of cryptographic schemes (we refer the reader to Definition 7.5 for the definition of a pairing.) The security of the *Boneh-Franklin Identity-based Cryptosystem* relies on the hardness of the **Bilinear Diffie-Hellman** problem, which is presented as Problem 13.2.

Problem 13.2: Bilinear Diffie-Hellman

Instance: Additive groups $(G_1, +)$ and $(G_2, +)$ of prime order q and a multiplicative group $(G_3, \cdot)$ of order q, along with a pairing $e_q : G_1 \times G_2 \rightarrow G_3$, an element $P \in G_1$, an element $Q \in G_2$ having order q, and elements $aQ, bQ \in G_2$ for some $a, b \in {\mathbb{Z}_q}^*$.

Question: Find the unique element $W \in G_3$, such that

$$W = e_q(P, Q)^{ab}.$$

It is possible to use a suitably-chosen elliptic curve to construct groups G_1 and G_2 and a pairing for which the **Bilinear Diffie-Hellman** problem is believed to be difficult. It is interesting to observe that, if we can solve the **CDH** problem (Prob-

Cryptosystem 13.2: *Boneh-Franklin Identity-based Cryptosystem*

Let q be a prime, and let G_1 and G_2 be groups of order q, along with a pairing $e_q : G_1 \times G_2 \to G_3$. Let P be a generator of G_2, and let n be a positive integer.

System parameters: The master key $M = (M^{pub}, M^{priv})$, where

$$M^{priv} = s$$

and

$$M^{pub} = sP.$$

As well, $h_1 : \{0,1\}^* \to G_1 \setminus \{0\}$ and $h_2 : G_2 \to \{0,1\}^n$ are public hash functions.

User key generation: For a user U, the key $K_U = (K_U^{pub}, K_U^{priv})$, where

$$K_U^{pub} = h_1(ID(U))$$

and

$$K_U^{priv} = sK_U^{pub}.$$

Encryption: A plaintext is an element in the set $\{0,1\}^n$. To encrypt a plaintext element $x \in \{0,1\}^n$, the following steps are performed:

1. Choose a random value $r \in \mathbb{Z}_q{}^*$.
2. Compute

   $$y_1 = rP$$

   and

   $$y_2 = x \oplus h_2(e_q(K_U^{pub}, M^{pub})^r).$$

3. The ciphertext $y = (y_1, y_2)$.

Decryption: A ciphertext $y = (y_1, y_2)$ is decrypted as follows:

1. Compute

   $$x = y_2 \oplus h_2(e_q(K_U^{priv}, y_1)).$$

2. The decrypted plaintext is x.

lem 7.3) in either G_2 or G_3, then we can solve the **Bilinear Diffie-Hellman** problem. We illustrate by assuming we can solve the **CDH** problem in G_2. Thus, given Q, aQ, and bQ, we can compute abQ. Now we evaluate the pairing $e_q(P, abQ)$. Since $e_q(P, abQ) = e_q(P, Q)^{ab}$, we have solved this instance of the **Bilinear Diffie-**

Hellman problem. For the solution when we can solve **CDH** in G_3, see the Exercises.

Cryptosystem 13.2 is a description of the basic version of the *Boneh-Franklin Identity-based Cryptosystem,* which is secure against chosen plaintext attacks (Boneh and Franklin also showed that this basic scheme can be extended to give a scheme that provides chosen ciphertext security.) It uses a cyclic group G_1 of prime order q, constructed from a subgroup of points on an elliptic curve, which is equipped with a pairing e_q. The master private key is an element $s \in \mathbb{Z}_q$, and the master public key is sP, where P is a publicly known generator of G_2. It requires a hash function h_1 that maps user identities onto points in G_1, and a second hash function h_2 that maps elements in the range of e_q onto binary strings of length n. Encryption is carried out by computing the pairing of a user's public key with a random point in G_1, and applying h_2 to the result, in order to create a mask that is added (mod 2) to the message.

Now we show that the decryption operation is successful in recovering the plaintext. We begin by observing that

$$\begin{aligned} e_q(K_U^{priv}, y_1) &= e_q(sK_U^{pub}, rP) \\ &= e_q(K_U^{pub}, P)^{sr} \\ &= e_q(K_U^{pub}, sP)^r \\ &= e_q(K_U^{pub}, M^{pub})^r. \end{aligned}$$

Therefore we have that

$$\begin{aligned} y_2 \oplus h_2(e_q(K_U^{priv}, y_1)) &= x \oplus h_2(e_q(K_U^{pub}, M^{pub})^r) \oplus h_2(e_q(K_U^{pub}, M^{pub})^r) \\ &= x, \end{aligned}$$

as required.

Boneh and Franklin proved that Cryptosystem 13.2 is secure against chosen plaintext attacks in a model where the attacker is allowed to learn private keys corresponding to identities other than the one they have chosen to attack. Here we will consider the more restricted case where we fix an identity U and we do not allow the attacker to query private keys for other identities. We show that an algorithm DISTINGUISH that solves the problem of **Ciphertext Distinguishability** for two plaintexts x_1 and x_2 in this model can be used to obtain an algorithm BDH-SOLVER that solves the **Bilinear Diffie-Hellman** problem. The algorithm BDH-SOLVER is presented as Algorithm 13.2.

We suppose that DISTINGUISH can distinguish between the encryptions of two plaintexts x_1 and x_2 with probability $1/2 + \epsilon$, and that DISTINGUISH makes at most q_{h_2} calls to the oracle h_2. The algorithm BDH-SOLVER uses the inputs P, aQ, and bQ to construct the public keys and the ciphertext that are input to DISTINGUISH,

Algorithm 13.2: BDH-SOLVER(P, Q, aQ, bQ, q_{h_2})

external e_q
global $RList, h2List$
procedure SIM$h_2(r)$
  $i \leftarrow 1$
  $found \leftarrow$ **false**
  while $i \leq q_{h_2}$ **and not** $found$
    do { **if** $RList[i] = r$
      then $found \leftarrow$ **true**
      else $i \leftarrow i + 1$ }
  if $found$
    then return $(h2List[i])$
    else let h be chosen at random
  $RList[i] \leftarrow r$
  $h2List[i] \leftarrow h$
  return (h)

main
  choose $j_0 \in \{1, \ldots, q_{h_2}\}$ at random
  $M^{pub} \leftarrow aQ$
  $K_U^{pub} \leftarrow P$
  $y_1 \leftarrow bQ$
  choose y_2 at random
  insert the code for DISTINGUISH$(x_1, x_2, M^{pub}, K_U^{pub}, (y_1, y_2))$ here
  return $(RList[j_0])$

according to the following formulas:

$$\begin{aligned} M^{pub} &= aQ, \\ K_U^{pub} &= P, \\ y_1 &= bQ, \\ y_2 &= \text{random}. \end{aligned}$$

The ordered pair (y_1, y_2) is therefore an encryption of x_i if and only if

$$y_2 = x_i \oplus h_2(e_q(P, aQ)^b) = x_i \oplus h_2(e_q(P, Q)^{ab}).$$

Since h_2 is a random oracle, there is no way to distinguish an encryption of x_1 from an encryption of x_2 without querying h_2 with the input value $e_q(P, Q)^{ab}$. However, $e_q(P, Q)^{ab}$ is the desired solution to the given instance of the **Bilinear Diffie-Hellman** problem. Therefore, roughly speaking, an algorithm DISTINGUISH that succeeds with a reasonable probability must have queried h_2 with an input value that is the solution to the given instance of the **Bilinear Diffie-Hellman** problem.

Algorithm BDH-SOLVER replaces the random oracle h_2 by the function SIM$h_2(r)$. This function returns a random value in response to each request, while checking to ensure that repeated requests are answered consistently. As long as DISTINGUISH does not make a query on the value $e_q(P,Q)^{ab}$, then SIM$h_2(r)$ is a perfect simulation of a random oracle, and so DISTINGUISH will behave as though it is interacting with a random oracle.

If DISTINGUISH makes an h_2 query on the value $e_q(P,Q)^{ab}$, then SIM$h_2(r)$ replies with a random response that is not, in general, consistent with (y_1, y_2) being a valid encryption of either x_1 or x_2. Hence, at this point, it no longer provides a perfect simulation of a random oracle. So we cannot say anything about the output of DISTINGUISH after this query takes place. However, the success of BDH-SOLVER depends only on whether DISTINGUISH queries the value $e_q(P,Q)^{ab}$, and the fact that the random oracle simulation is perfect until the time when that query is made implies that we can bound this success probability by analyzing the likelihood of DISTINGUISH querying this target value when it interacts with a true random oracle.

Let β denote the probability that DISTINGUISH makes an h_2 query on the "target value" $e_q(P,Q)^{ab}$ when interacting with a true random oracle. If DISTINGUISH does not query this value, then $h_2(e_q(P,Q)^{ab})$, and hence the correct decryption of (y_1, y_2), is independent of all the other values that it sees. Therefore, it learns no information about the true plaintext. This implies that it has success probability $1/2$ in this case. Therefore we have that

$$\begin{aligned}
&\frac{1}{2}+\epsilon \\
&\quad = \mathbf{Pr}[\text{DISTINGUISH succeeds}] \\
&\quad = (1-\beta)\frac{1}{2} + \beta\,\mathbf{Pr}[\text{DISTINGUISH succeeds after querying the target value}] \\
&\quad \leq \frac{1}{2}(1-\beta)+\beta \\
&\quad \leq \frac{1}{2}+\frac{1}{2}\beta,
\end{aligned}$$

which implies $\beta \geq 2\epsilon$. Hence, we deduce that DISTINGUISH queries the target value with probability at least 2ϵ. There are at most q_{h_2} hash queries that are made by DISTINGUISH. We have no idea which of these queries is actually the target value, so BDH-SOLVER randomly chooses one of these q_{h_2} hash queries. This random guess is defined to be its solution to the **Bilinear Diffie-Hellman** problem. Thus, it succeeds in computing the correct value of $e_q(P,Q)^{ab}$ with probability at least $2\epsilon/q_{h_2}$.

13.2 The Paillier Cryptosystem

The *Paillier Cryptosystem* is an RSA-like cryptosystem that has an interesting homomorphic property. We noted in Chapter 6 that

$$e_K(x_1)e_K(x_2) = e_K(x_1x_2)$$

for any two RSA plaintexts x_1 and x_2. That is, the product of two RSA ciphertexts is the encryption of the product of the two corresponding RSA plaintexts. In the *Paillier Cryptosystem*, the product of two ciphertexts is the encryption of the sum of the two corresponding plaintexts.

The term "homomorphic property" derives from the notion of a ***group homomorphism***. Suppose G is an abelian group with group operation "$\cdot$" and H is an abelian group with group operation "$\star$". A homomorphism from $(G, \cdot)$ to $(H, \star)$ is a mapping $f : G \rightarrow H$ that satisfies the condition

$$f(x_1) \star f(x_2) = f(x_1 \cdot x_2)$$

for all $x_1, x_2 \in G$. The *RSA* encryption operation is a homomorphism from $(\mathbb{Z}_n, \cdot)$ to $(\mathbb{Z}_n, \cdot)$, whereas the *Paillier* encryption operation is a homomorphism from $(\mathbb{Z}_n, +)$ to $(\mathbb{Z}_{n^2}, \cdot)$.

One of the potential applications of homomorphic encryption is ***computing on encrypted data***. Suppose we have non-negative integer values $x_1, \dots, x_k$. Perhaps these values are sensitive, and hence they are encrypted and stored as ciphertexts $y_1, \dots, y_k$. Now suppose we want to compute the sum $x_1 + \cdots + x_k$. We could do this by first decrypting the k ciphertexts and then computing the sum of the k plaintexts. However, if encryption is performed using the *Paillier Cryptosystem*, there is an alternative. Namely, we could multiply the k ciphertexts together and then decrypt the result. This allows the same sum to be computed using only one decryption operation rather than k decryption operations.

The *Paillier Cryptosystem*, which is described in Cryptosystem 13.3, involves computations in $\mathbb{Z}_{n^2}$, where n is the product of two distinct odd primes p and q, as in *RSA*. As was the case in *RSA*, we have $\phi(n) = (p-1)(q-1)$.

Example 13.1 Suppose $p = 541$ and $q = 613$; then $n = 331633$, $n^2 = 109980446689$, $g = 331634$, $\phi(n) = 330480$, and $\phi(n)^{-1} \bmod n = 120803$.

Suppose we want to encrypt the plaintext $x = 239588$ and we choose the random value $r = 230550$. Then the ciphertext is

$$y = 331634^{239588} 230550^{331633} \bmod 109980446689 = 3599380886.$$

To decrypt the ciphertext, we compute

$$\begin{aligned} x &= \left(\frac{3599380886^{330480} \bmod 109980446689 - 1}{331633} \times 120803 \right) \bmod 331633 \\ &= 239588. \end{aligned}$$

□

Cryptosystem 13.3: *Paillier Cryptosystem*

Let $n = pq$, where p and q are distinct odd primes, so $\phi(n) = (p-1)(q-1)$. Suppose that

$$\gcd(n, \phi(n)) = 1.$$

This gcd condition will hold provided that $p \nmid (q-1)$ and $q \nmid (p-1)$.
Let $\mathcal{P} = \mathbb{Z}_n$, $\mathcal{C} = {\mathbb{Z}_{n^2}}^*$, define $g = n+1$, and let

$$\mathcal{K} = \{(n, g, p, q)\}.$$

The values n and g comprise the public key, and the values p and q form the private key. The value $\phi(n) = (p-1)(q-1)$ can be computed from the private key.
For $K = (n, g, p, q)$, the encryption operation is

$$e_K(x, r) = g^x r^n \bmod n^2$$

where $x \in \mathbb{Z}_n$ is the plaintext and $r \in {\mathbb{Z}_n}^*$ is random, and

$$d_K(y) = \left(\frac{(y^{\phi(n)} \bmod n^2) - 1}{n}\right) \times (\phi(n)^{-1} \bmod n) \bmod n,$$

where $y \in {\mathbb{Z}_{n^2}}^*$ is the ciphertext.

We make a few preliminary observations. First, the encryption operation is randomized in the *Paillier Cryptosystem*. A ciphertext ends up being twice as long as a plaintext, since a plaintext is an element of $\mathbb{Z}_n$ whereas a ciphertext is an element of $\mathbb{Z}_{n^2}$. The encryption operation is basically two exponentiations modulo n^2 and decryption requires one exponentiation modulo n^2. The value $\phi(n)^{-1} \bmod n$ that is used in the decryption operation can be precomputed. Finally, the division operation during decryption is integer division (without remainder).

Perhaps the trickiest aspect of the *Paillier Cryptosystem* is proving that applying the decryption operation to a ciphertext results in the original plaintext. We will prove this shortly, but we first establish the simpler result that this cryptosystem satisfies a homomorphic property. From the encryption operation, as described in Cryptosystem 13.3, it is easy to see that

$$e_K(x_1, r_1) e_K(x_2, r_2) = g^{x_1 + x_2} (r_1 r_2)^n \bmod n^2 = e_K(x_1 + x_2, r_1 r_2),$$

where addition and multiplication of the x_i's and r_i's is done modulo n. Thus, the product of two ciphertexts is the encryption of the sum of the two corresponding plaintexts. From this, it follows easily that

$$e_K(x, r)^c = e_K(cx, r^c)$$

for any positive integer c.

Let's now verify that decryption of a Paillier ciphertext always yields the correct plaintext. We will make use of the following two lemmas.

LEMMA 13.2 *For any integers $n \geq 2$ and $t \geq 1$, it holds that*

$$(n+1)^t \equiv 1 + tn \pmod{n^2}.$$

PROOF If $t = 1$, the result is obvious. If $t \geq 2$ and we expand $(n+1)^t$ using the binomial theorem[1], we obtain

$$(n+1)^t = 1 + tn + \text{terms divisible by } n^2.$$

∎

LEMMA 13.3 *Suppose $n = pq$, where p and q are distinct primes. Then, for any $r \in \mathbb{Z}_{n^2}{}^*$, it holds that*

$$r^{n\phi(n)} \equiv 1 \pmod{n^2}.$$

PROOF We have $r^{n\phi(n)} = r^{p(p-1)q(q-1)}$. Since $n = pq$, the group $\mathbb{Z}_{n^2}{}^*$ has order $p(p-1)q(q-1)$ from Theorem 2.2. The desired result then follows immediately from Lagrange's theorem (Theorem 6.4). ∎

Now suppose that $y = e_K(x, r) = g^x r^n \bmod n^2$. The first step of the decryption process is to compute

$$z = y^{\phi(n)} \bmod n^2.$$

Clearly, we have

$$z = g^{x\phi(n)} r^{n\phi(n)} \bmod n^2 = g^{x\phi(n)} \bmod n^2,$$

from Lemma 13.3. Thus we have eliminated the dependence on r.

Now, since $z = g^{x\phi(n)} \bmod n^2$ and $g = n+1$, Lemma 13.3 tells us that

$$z \equiv 1 + x\phi(n)n \pmod{n^2}.$$

From this congruence, it is easy to verify that $n \mid (z-1)$ and

$$x\phi(n) \equiv \frac{z-1}{n} \pmod{n^2}.$$

Using the fact that $\gcd(n, \phi(n)) = 1$, we know that $\phi(n)^{-1} \bmod n$ exists, and hence

$$x = \left(\frac{z-1}{n} \pmod{n^2}\right) \times (\phi(n)^{-1} \bmod n) \bmod n.$$

---

[1] The ***binomial theorem*** gives a formula for expressing $(x+y)^t$ as a polynomial in x and y, namely $(x+y)^t = \sum_{i=0}^{t} \binom{t}{i} x^i y^{t-i}$.

The security of the *Paillier Cryptosystem* depends on the intractability of the so-called n**th residue** problem.

Problem 13.3: n**th residue**

Instance: A positive integer $n = pq$, where p and q are distinct odd primes, and an integer $y \in \mathbb{Z}_{n^2}{}^*$.
Question: Does there exist an integer $z \in \mathbb{Z}_{n^2}{}^*$ such that $y = z^n \bmod n^2$? In other words, is y an n*th residue* modulo n^2?

It is believed that the n**th residue** problem is intractable for integers n that are the product of two large primes p and q. The security proof for the *Paillier Cryptosystem* involves a reduction, showing that any algorithm that can decrypt Paillier ciphertexts can be used to solve the n**th residue** problem. This is in fact almost immediate, once we have proven the following theorem.

THEOREM 13.4 *Suppose $n = pq$, where p and q are distinct odd primes, and let $y \in \mathbb{Z}_{n^2}{}^*$. Then y is an nth residue modulo n^2 if and only if $d_K(y) = 0$, where d_K is the decryption function of the associated* Paillier Cryptosystem.

PROOF If y is an encryption of 0, then $y = r^n \bmod n^2$ for some r and hence y is an nth residue modulo n^2.

To prove the converse, we assume that $y = t^n \bmod n^2$ for some $t \in \mathbb{Z}_{n^2}{}^*$. When we decrypt y, we first compute

$$y^{\phi(n)} \bmod n^2 = t^{n\phi(n)} \bmod n^2.$$

But $t^{n\phi(n)} \bmod n^2 = 1$ from Lemma 13.3. It then follows immediately that $d_K(y) = 0$. ∎

The reduction is now easy. Suppose that we have a decryption oracle for the *Paillier Cryptosystem*, denoted by d_K. Given any $y \in \mathbb{Z}_{n^2}{}^*$, we compute $d_K(y)$. Then we report that y is an nth residue modulo n^2 if and only if $d_K(y) = 0$.

13.3 Copyright Protection

Protection against copyright violation is an important, but very difficult, challenge in the Internet age. Digital content can easily be copied and transmitted over computer networks. Content may be encrypted before it is transmitted; however, all content must eventually be decrypted before it will be intelligible to an end user. After content is decrypted, it can potentially be copied.

Hardware-based solutions, such as tamper-resistant hardware, for example, can provide a limited amount of protection. Other approaches include algorithms (and coding methods) that enable ***tracing***. This allows content to be traced to its rightful owner, which discourages people from unauthorized copying of digital data. In this section, we describe some types of "codes" that can be used for tracing.

Before continuing further, it is useful to distinguish some different types of copyright violation. There are many potential threats. Here are two threats that we introduce as typical examples.

illegal content redistribution
As mentioned above, encrypted content is invariably decrypted once it gets to its authorized destination. Decrypted content can then be copied and transmitted to others, for example in an illegal ***pirate broadcast***.

illegal key redistribution
Assuming that content is encrypted, there must be a mechanism for the content to be decrypted by an end user. The keys used to decrypt the content may be copied and distributed to other users. Alternatively, these keys may be combined to create a new ***pirate decoder***, which can subsequently be used to decrypt encrypted content illegally.

13.3.1 Fingerprinting

We first address the problem of illegal content redistribution. Suppose that every copy of some digital data, D, contains a unique ***fingerprint***, F. For example, there might be 1 megabyte of binary data, and a fingerprint might consist of 100 "special" bits "hidden" in the data in a manner that is hard to detect. (Sometimes the process of embedding hidden identifying data is called ***watermarking***.)

In this scenario, the vendor can maintain a database that keeps track of all the different fingerprints, as well as the rightful owners of the corresponding copies of the data D. Then any exact copy of the data can be traced back to its owner. Unfortunately, there are some serious flaws with this approach. For example, if the fingerprint is easily recognized, then it can be modified or destroyed, thus making the data impossible to trace. A second threat is that coalitions may be able to recognize fingerprints or parts of fingerprints—even if individual users cannot do so—and then create a new copy of the data with the fingerprint destroyed.

Here is a more precise mathematical model that will facilitate studying this problem. For concreteness, suppose that each copy of the data consists of L bits of ***content***, say C, and an ℓ-bit ***fingerprint***, F; hence, the data has the form $D = (C, F)$. All the data is represented over some fixed alphabet. For example, binary data uses the alphabet $\{0, 1\}$. We will assume that all copies of the data have the same content but different fingerprints, so we have $D_1 = (C, F_1)$, $D_2 = (C, F_2)$, etc. Furthermore, we will assume that the fingerprint bits[2] always occur in the same (secret) positions in all copies of the data; e.g., bits $b_{i_1}, \dots, b_{i_\ell}$ are fingerprint bits.

[2]We will use the term "fingerprint bits" to denote the positions in which the fingerprints occur. The term "bits" suggests that the data has a binary form, but we will use this term even if the data is defined over a non-binary alphabet.

FIGURE 13.1: The marking assumption

Fingerprinting problems are usually studied assuming that a certain ***marking assumption*** holds. This assumption is stated as follows:

> *Given some number of copies of the data, say $D_1, D_2, \ldots, D_w$, the only bits that can be identified as fingerprint bits by a coalition are those bits b such that $D_i[b] \neq D_j[b]$ for some i, j.*

In other words, we are assuming that the fingerprints are hidden well enough that no particular bit can be identified as a fingerprint bit by a coalition of bad guys unless the coalition possesses two copies of the data in which the bit in question takes on different values.

The diagram in Figure 13.1 illustrates the idea behind the marking assumption. This diagram contains two grids made up of black and white "pixels." It can be verified that there are exactly three "pixels" in which the two grids differ. According to the marking assumption, only these three pixels can be recognized as fingerprint bits.

Let's consider the kinds of attacks that a coalition can carry out, assuming that the marking assumption holds. A bit of thought shows that the marking assumption implies that the actual content is irrelevant, and the problem reduces to studying combinatorial properties of the set of fingerprints. As described above, given w copies of the data, some bits can be identified as fingerprint bits. Then a new "pirate" copy of the data can be constructed, by setting values of these identified fingerprint bits from one of the copies of the data in an arbitrary fashion. The resulting data is $D' = (C, F')$, where F' is a newly created ***hybrid fingerprint***. The fundamental question is whether a hybrid fingerprint can be "traced" if the fingerprints are constructed in a suitable way. The notion of hybrid fingerprints is defined precisely in Definition 13.1.

For example, suppose that

$$\mathcal{C}_0 = \{(1,1,2),(2,3,2)\}.$$

In the descendant code, the first co-ordinate can be 1 or 2, the second co-ordinate can be 1 or 3, and the last co-ordinate must be 2. Therefore, it is easy to see that

$$\mathbf{desc}(\{(1,1,2),(2,3,2)\}) = \{(1,1,2),(2,3,2),(1,3,2),(2,1,2)\}.$$

Definition 13.1: An (ℓ, n, q)-***code*** is a subset $\mathcal{C} \subseteq Q^\ell$ such that $|Q| = q$ and $|\mathcal{C}| = n$. That is, we have n ***codewords***, each of which is an ℓ-tuple of elements from the alphabet Q. A codeword is the same thing as a fingerprint.

Let $\mathcal{C}_0 \subseteq \mathcal{C}$ (i.e., $\mathcal{C}_0$ is a subset of codewords). Define $\mathbf{desc}(\mathcal{C}_0)$ to consist of all ℓ-tuples $\mathbf{f} = (f_1, \ldots, f_\ell)$ such that, for all $1 \leq i \leq \ell$, there exists a codeword $\mathbf{c} = (c_1, \ldots, c_\ell) \in \mathcal{C}_0$ such that $f_i = c_i$. The set $\mathbf{desc}(\mathcal{C}_0)$ consists of all the hybrid fingerprints that can be constructed from the fingerprints in $\mathcal{C}_0$; it is called the ***descendant code*** of $\mathcal{C}_0$.

Finally, for any $\mathbf{c} \in \mathcal{C}_0$ and for any $\mathbf{f} \in \mathbf{desc}(\mathcal{C}_0)$, we say that $\mathbf{c}$ is a ***parent*** of $\mathbf{f}$ in the code $\mathbf{desc}(\mathcal{C}_0)$.

In this example, the descendant code consists of the two original codewords and two new hybrid fingerprints.

For an integer $w \geq 2$, the w-***descendant code*** of $\mathcal{C}$, denoted $\mathbf{desc}_w(\mathcal{C})$, consists of the following set of ℓ-tuples:

$$\mathbf{desc}_w(\mathcal{C}) = \bigcup_{\mathcal{C}_0 \subseteq \mathcal{C}, |\mathcal{C}_0| \leq w} \mathbf{desc}(\mathcal{C}_0).$$

The w-descendant code consists of all hybrid fingerprints that could be produced by a coalition of size at most w.

13.3.2 Identifiable Parent Property

We now turn to the "inverse" process, namely trying to determine the coalition that constructed a hybrid fingerprint. Suppose that $\mathbf{f} \in \mathbf{desc}_w(\mathcal{C})$. We define the set of ***suspect coalitions*** for $\mathbf{f}$ as follows:

$$\mathbf{susp}_w(\mathbf{f}) = \{\mathcal{C}_0 \subseteq \mathcal{C} : |\mathcal{C}_0| \leq w, \mathbf{f} \in \mathbf{desc}(\mathcal{C}_0)\}.$$

The set $\mathbf{susp}_w(\mathbf{f})$ consists of all the coalitions of size at most w that could have produced the hybrid fingerprint $\mathbf{f}$ by following the process described above. Ideally, $\mathbf{susp}_w(\mathbf{f})$ would consist of one and only one set. In this case, we would have some evidence that this subset in fact created the hybrid fingerprint (of course, we can never rule out the possibility that some other coalition, necessarily of size exceeding w, is in fact the guilty subset).

Even if $\mathbf{susp}_w(\mathbf{f})$ consists of more than one set, we still may be able to extract some useful information by looking at the sets in $\mathbf{susp}_w(\mathbf{f})$. For example, suppose that there exists a codeword $\mathbf{c} \in \mathcal{C}$ such that $\mathbf{c} \in \mathcal{C}_0$ for all $\mathcal{C}_0 \in \mathbf{susp}_w(\mathbf{f})$. Any such codeword can be identified as guilty (under the assumption that the coalition has size at most w), even if we are not able to identify the complete guilty subset.

The above-mentioned property can be stated in an equivalent form as follows:

$$\bigcap_{\mathcal{C}_0 \in \mathbf{susp}_w(\mathbf{f})} \mathcal{C}_0 \neq \emptyset. \tag{13.1}$$

We say that $\mathcal{C}$ is a ***w-identifiable parent property code*** (or ***w-IPP code***) provided that (13.1) is satisfied for all $\mathbf{f} \in \mathbf{desc}_w(\mathcal{C})$. Further, in a w-IPP code, if

$$\mathbf{c} \in \bigcap_{\mathcal{C}_0 \in \mathbf{susp}_w(\mathbf{f})} \mathcal{C}_0,$$

then $\mathbf{c}$ is called an ***identifiable parent*** of $\mathbf{f}$.

Example 13.2 We present a $(3,6,3)$ code, and consider coalitions of size at most two:

$$\begin{array}{lll} \mathbf{c}_1 = (0,1,1), & \mathbf{c}_2 = (1,0,1), & \mathbf{c}_3 = (1,1,0), \\ \mathbf{c}_4 = (2,0,2), & \mathbf{c}_5 = (1,0,2), & \mathbf{c}_6 = (2,1,0). \end{array}$$

Consider the hybrid fingerprint $\mathbf{f}_1 = (1,1,1)$. It is not difficult to verify that

$$\mathbf{susp}_2(\mathbf{f}_1) = \{\{1,2\},\{1,3\},\{2,3\},\{1,5\},\{2,6\}\}.$$

This hybrid fingerprint $\mathbf{f}_1$ violates property (13.1), so the code is not a 2-IPP code.

On the other hand, consider $\mathbf{f}_2 = (0,1,2)$. Here it can be seen that

$$\mathbf{susp}_2(\mathbf{f}_2) = \{\{1,4\},\{1,5\}\}.$$

Observe that property (13.1) is satisfied for the hybrid fingerprint $\mathbf{f}_2$. Hence, $\mathbf{c}_1$ is an identifiable parent of $\mathbf{f}_2$ (under the assumption that a coalition of size at most two created $\mathbf{f}_2$), because

$$\{1,4\} \cap \{1,5\} = \{1\}.$$

□

Example 13.3 We present a $(3,7,5)$ 2-IPP code:

$$\begin{array}{llll} \mathbf{c}_1 = (0,0,0), & \mathbf{c}_2 = (0,1,1), & \mathbf{c}_3 = (0,2,2), & \mathbf{c}_4 = (1,0,3), \\ \mathbf{c}_5 = (2,0,4), & \mathbf{c}_6 = (3,3,0), & \mathbf{c}_7 = (4,4,0). & \end{array}$$

We show that the property (13.1) holds for all relevant hybrid fingerprints $\mathbf{f}$. Suppose that $\mathbf{f} = (f_1, f_2, f_3)$ is a hybrid fingerprint created by a coalition of size two. If any co-ordinate of $\mathbf{f}$ is non-zero, then at least one parent of $\mathbf{f}$ can be identified, as indicated in the following exhaustive list of possibilities:

$$\begin{array}{llll} f_1 = 1 \Rightarrow \mathbf{c}_4; & f_1 = 2 \Rightarrow \mathbf{c}_5; & f_1 = 3 \Rightarrow \mathbf{c}_6; & f_1 = 4 \Rightarrow \mathbf{c}_7 \\ f_2 = 1 \Rightarrow \mathbf{c}_2; & f_2 = 2 \Rightarrow \mathbf{c}_3; & f_2 = 3 \Rightarrow \mathbf{c}_6; & f_2 = 4 \Rightarrow \mathbf{c}_7 \\ f_3 = 1 \Rightarrow \mathbf{c}_2; & f_3 = 2 \Rightarrow \mathbf{c}_3; & f_3 = 3 \Rightarrow \mathbf{c}_4; & f_3 = 4 \Rightarrow \mathbf{c}_5. \end{array}$$

Finally, if $\mathbf{f} = (0,0,0)$, then $\mathbf{c}_1$ is an identifiable parent. □

13.3.3 2-IPP Codes

In general, it is not an easy task to do any of the following:

1. construct a w-IPP code;
2. verify whether a given code is a w-IPP code; or
3. find an efficient algorithm to identify a parent, given an ℓ-tuple in the w-descendant code of a w-IPP code.

In reference to the third task, it is of particular interest to design w-IPP codes for which efficient parent-identifying algorithms can be constructed.

In this section, we will pursue these questions in the easiest case, $w = 2$. We will provide a nice characterization of 2-IPP codes that involves two kinds of hash families. We first introduce "perfect hash families" in Definition 13.2. A related concept, that of "separating hash families," is defined in Definition 13.3.

Definition 13.2: An (n, m, w)-***perfect hash family*** is a set of functions, say $\mathcal{F}$, such that $|X| = n$, $|Y| = m$, $f : X \rightarrow Y$ for each $f \in \mathcal{F}$, and for any $X_1 \subseteq X$ such that $|X_1| = w$, there exists at least one $f \in \mathcal{F}$ such that $f|_{X_1}$ is one-to-one.[3] When $|\mathcal{F}| = N$, an (n, m, w)-perfect hash family will be denoted by $\mathrm{PHF}(N; n, m, w)$.

A $\mathrm{PHF}(N; n, m, w)$ can be depicted as an $n \times N$ array with entries from Y, having the property that in any w rows there exists at least one column such that the w entries in the given w rows are distinct. Here the columns of the array are labeled by the functions in $\mathcal{F}$, the rows are labeled by the elements in X, and the entry in row x and column f of the array is $f(x)$.

Perfect hash families have been widely studied in the context of information retrieval algorithms. However, as we shall see, perfect hash families have close connections to w-IPP codes.

The following example serves to illustrate the two previous definitions.

Example 13.4 Consider the following seven by three array:

0	0	0
0	1	1
0	2	2
1	0	3
2	0	4
3	3	0
4	4	0

[3]The notation $f|_{X_1}$ denotes the restriction of the function f to the subset X_1 of the domain. The requirement that $f|_{X_1}$ is one-to-one means that $f(x) \neq f(x')$ for all $x, x' \in X_1$ such that $x \neq x'$.

Definition 13.3: An $(n, m, \{w_1, w_2\})$-***separating hash family*** is a set of functions, say $\mathcal{F}$, such that $|X| = n$, $|Y| = m$, $f : X \rightarrow Y$ for each $f \in \mathcal{F}$, and for any $X_1, X_2 \subseteq X$ such that $|X_1| = w_1$, $|X_2| = w_2$ and $X_1 \cap X_2 = \emptyset$, there exists at least one $f \in \mathcal{F}$ such that

$$\{f(x) : x \in X_1\} \cap \{f(x) : x \in X_2\} = \emptyset.$$

The notation SHF$(N; n, m, \{w_1, w_2\})$ will be used to denote an $(n, m, \{w_1, w_2\})$-separating hash family with $|\mathcal{F}| = N$.

An SHF$(N; n, m, \{w_1, w_2\})$ can be depicted as an $n \times N$ array with entries from the set Y, having the property that in any w_1 rows and any w_2 disjoint rows there exists at least one column such that the entries in the given w_1 rows are distinct from the entries in the given w_2 rows.

It can be verified that the above array is simultaneously a PHF$(3; 7, 5, 3)$ and an SHF$(3; 7, 5, \{2, 2\})$.

We note, however, that the array is not a PHF$(3; 7, 5, 4)$. Consider rows 1, 2, 4, and 6. None of the three columns contain distinct entries in all four of the given rows. □

We will now derive an efficient algorithm to determine if a given (ℓ, n, q) code, say C, is a 2-IPP code. Suppose the codewords are written in the form of an n by ℓ array, say $A(C)$, and suppose that $A(C)$ is not a PHF$(\ell; n, q, 3)$. Then there exist three rows, r_1, r_2, r_3 of A that violate the perfect hash family property. For every column c, let f_c be an element that is repeated (i.e., it occurs in at least two of the three given rows r_1, r_2, r_3 in column c). Now, define $\mathbf{f} = (f_1, \ldots, f_\ell)$. Clearly

$$\{r_1, r_2\}, \{r_1, r_3\}, \{r_2, r_3\} \in \mathbf{susp}_2(\mathbf{f}).$$

Therefore, C is not a 2-IPP code, because the intersection of these three 2-subsets is the empty set.

Next, suppose that $A(C)$ is not an SHF$(\ell; n, q, \{2, 2\})$. Then there exist two sets of two rows of $A(C)$, say $\{r_1, r_2\}$ and $\{r_3, r_4\}$, that violate the separating hash family property. For every column c, let f_c be an element that occurs in column c in one of rows r_1 and r_2, and again in column c in one of rows r_3 and r_4. Define $\mathbf{f} = (f_1, \ldots, f_\ell)$. Clearly,

$$\{r_1, r_2\}, \{r_3, r_4\} \in \mathbf{susp}_2(\mathbf{f}).$$

Therefore, C is not a 2-IPP code, because the intersection of these two 2-subsets is the empty set.

From the above discussion, we see that a necessary condition for C to be a 2-IPP code is that $A(C)$ is simultaneously a PHF$(\ell; n, q, 3)$ and an SHF$(\ell; n, q, \{2, 2\})$.

The converse is also true (see the Exercises), and therefore we have the following theorem.

THEOREM 13.5 *An (ℓ, n, q) code $\mathcal{C}$ is a 2-IPP code if and only if $A(\mathcal{C})$ is simultaneously a PHF$(\ell; n, q, 3)$ and an SHF$(\ell; n, q, \{2,2\})$.*

As a corollary, an $(\ell, n, 2)$ code cannot be a 2-IPP code. For $n \geq 3$, it follows from Theorem 13.5 that an (ℓ, n, q) code, $\mathcal{C}$, can be tested to determine if it is a 2-IPP code in polynomial time as a function of n.

Now we consider identification of parents in a 2-IPP code. Suppose that $\mathcal{C}$ is a 2-IPP code and $\mathbf{f} \in \mathbf{desc}_2(\mathcal{C}) \backslash \mathcal{C}$. Thus $\mathbf{f}$ is not a codeword, and there is at least one subset of two codewords for which $\mathbf{f}$ is in the descendant subcode. The fact that $\mathcal{C}$ is a 2-IPP code severely constrains the possible structure of $\mathbf{susp}_2(\mathbf{f})$. It can be shown that one of two possible scenarios must hold:

1. either $\mathbf{susp}_2(\mathbf{f})$ consists of a single set of two codewords, or
2. $\mathbf{susp}_2(\mathbf{f})$ consists of a two or more sets of two codewords, all of which contain a fixed codeword. For example,

$$\mathbf{susp}_2(\mathbf{f}) = \{\{\mathbf{c}_1, \mathbf{c}_2\}, \{\mathbf{c}_1, \mathbf{c}_3\}, \{\mathbf{c}_1, \mathbf{c}_4\}\}$$

   would fall into this case.

In the first case, we can identify both parents of $\mathbf{f}$. In the second case, we can identify one parent (namely, $\mathbf{c}_1$, in the example provided).

In a 2-IPP code, we only consider suspect coalitions of size two. Given $\mathbf{f}$, we can examine all the $\binom{n}{2}$ subsets of two codewords. For each 2-subset $\{\mathbf{c}, \mathbf{d}\}$, we can check to see if $\mathbf{f} \in \mathbf{desc}(\{\mathbf{c}, \mathbf{d}\})$. This will yield an algorithm having complexity $\Theta(n^2)$, which will identify a parent in an arbitrary 2-IPP code.

There are many constructions for 2-IPP codes. We present a simple and efficient construction for certain 2-IPP codes with $\ell = 3$, which is due to Hollmann, van Lint, Linnartz, and Tolhuizen. Suppose that $r \geq 2$ is an integer, let $q = r^2 + 2r$, and define

$$\begin{aligned}
S &= \{1, \ldots, r\} \quad (|S| = r) \\
M &= \{r+1, \ldots, 2r\} \quad (|M| = r) \\
L &= \{2r+1, \ldots, q\} \quad (|L| = r^2) \\
\mathcal{C}_1 &= \{(s_1, s_2, rs_1 + s_2 + r) : s_1, s_2 \in S\} \subseteq S \times S \times L \\
\mathcal{C}_2 &= \{(m, sr + m, s) : m \in M, s \in S\} \subseteq M \times L \times S \\
\mathcal{C}_3 &= \{(rm_1 + m_2 - r^2, m_1, m_2) : m_1, m_2 \in M\} \subseteq L \times M \times M.
\end{aligned}$$

Example 13.5 We construct a $(3, 27, 15)$ 2-IPP code by following the recipe given above. We have $r = 3$, $S = \{1,2,3\}$, $M = \{4,5,6\}$, and $L = \{7,\ldots,15\}$. C_1, C_2, and C_3 each consist of nine codewords, as indicated here:

$$\begin{array}{lll}
\mathbf{c}_1 = (1,1,7), & \mathbf{c}_2 = (1,2,8), & \mathbf{c}_3 = (1,3,9), \\
\mathbf{c}_4 = (2,1,10), & \mathbf{c}_5 = (2,2,11), & \mathbf{c}_6 = (2,3,12), \\
\mathbf{c}_7 = (3,1,13), & \mathbf{c}_8 = (3,2,14), & \mathbf{c}_9 = (3,3,15), \\ \hline
\mathbf{c}_{10} = (4,7,1), & \mathbf{c}_{11} = (5,8,1), & \mathbf{c}_{12} = (6,9,1), \\
\mathbf{c}_{13} = (4,10,2), & \mathbf{c}_{14} = (5,11,2), & \mathbf{c}_{15} = (6,12,2), \\
\mathbf{c}_{16} = (4,13,3), & \mathbf{c}_{17} = (5,14,3), & \mathbf{c}_{18} = (6,15,3), \\ \hline
\mathbf{c}_{19} = (7,4,4), & \mathbf{c}_{20} = (8,4,5), & \mathbf{c}_{21} = (9,4,6), \\
\mathbf{c}_{22} = (10,5,4), & \mathbf{c}_{23} = (11,5,5), & \mathbf{c}_{24} = (12,5,6), \\
\mathbf{c}_{25} = (13,6,4), & \mathbf{c}_{26} = (14,6,5), & \mathbf{c}_{27} = (15,6,6).
\end{array}$$

We claim that $C_1 \cup C_2 \cup C_3$ is a 2-IPP code with $n = 3r^2$. Furthermore, this code has an $O(1)$ time algorithm to find an identifiable parent. Actually, we show how to find an identifiable parent (which will prove implicitly that the code is a 2-IPP code). The main steps in a parent-identifying algorithm are as follows:

1. If $\mathbf{f} = (f_1, f_2, f_3)$ has a co-ordinate in L, then a parent is easily identified.

   For example, suppose that $f_2 = 13$. Then $3s + m = 13$, where $s \in \{1,2,3\}$ and $m \in \{4,5,6\}$. Hence $s = 3$ and $m = 4$, and therefore $(4,13,3)$ is an identifiable parent.

2. If $\mathbf{f}$ has no co-ordinate in L, then it is possible to compute $i \neq j$ such that the two parents of $\mathbf{f}$ are in C_i and C_j. The parent that contributed two co-ordinates to $\mathbf{f}$ can then be identified.

   For example, suppose that $\mathbf{f} = (1,3,2)$. The parents of $\mathbf{f}$ must be from C_1 and C_2. The parent from C_1 contributes f_1 and f_2, and hence $(1,3,9)$ is an identifiable parent.

□

The reasoning used in the above example to identify a parent will work for any code in this family. The complexity of the resulting parent-identification algorithm is independent of n (i.e., it has complexity $O(1)$).

Summarizing the results of this section, we have the following theorem.

THEOREM 13.6 *For all integers $r \geq 2$ there exists a $(3, 3r^2, r^2 + 2r)$-code that is a 2-IPP code. Furthermore, this code has a parent-identifying algorithm having complexity $O(1)$.*

13.3.4 Tracing Illegally Redistributed Keys

Suppose that every user in a network is given a ***decoder box*** that allows encrypted broadcasts to be decrypted. We might refer to such a scheme as a ***broadcast encryption scheme***. In general, every decoder box contains a different collection of keys. Suppose that a coalition of w malicious users creates a pirate decoder by combining keys from their decoder boxes in a suitable way. A pirate decoder will be able to decrypt broadcasts and hence it could be sold on the black market.

The set of keys in each decoder box can be thought of as a codeword in a certain code, and the keys in a pirate decoder can be thought of as a codeword in the w-descendant code. If the code is traceable (e.g., if it satisfies the w-IPP property), then a pirate decoder can be traced back to at least one member of the coalition that created it. Thus, if a pirate decoder is confiscated, then at least one of the guilty parties can be determined.

Let us briefly discuss how this broadcast encryption scheme works. First, the *TA* chooses ℓ sets of keys, denoted $\mathcal{K}_1, \ldots, \mathcal{K}_\ell$, where each $\mathcal{K}_i$ consists of q keys chosen from $\mathbb{Z}_m$, for some fixed m. For $1 \leq i \leq \ell$, let $\mathcal{K}_i = \{k_{i,j} : 1 \leq j \leq q\}$. A decoder box contains ℓ keys, one from each set $\mathcal{K}_i$.

The secret key $K \in \mathbb{Z}_m$ (which is used to encrypt the broadcast content, $\mathcal{M}$) is split into ℓ shares using an (ℓ, ℓ) threshold scheme (we use the threshold scheme described in Section 11.5.2). The shares are denoted $s_1, \ldots, s_\ell$, where

$$s_1 + \cdots + s_\ell \equiv K \pmod{m}.$$

Then K is used to encrypt $\mathcal{M}$, and for $1 \leq i \leq \ell$, every $k_{i,j}$ is used to encrypt s_i (so each share is encrypted under q different keys). The entire broadcast consists of the following information:

$$y = e_K(M) \quad \text{and} \quad (e_{k_{i,j}}(s_i) : 1 \leq i \leq \ell, 1 \leq j \leq q).$$

After receiving the broadcast, a user U who possesses a decoder box can perform the following operations:

1. The user U can decrypt all ℓ shares of K, because they have one key from each of the ℓ sets $\mathcal{K}_1, \mathcal{K}_2, \ldots, \mathcal{K}_\ell$.

2. The user U can then reconstruct K from the ℓ decrypted shares.

3. The user U can then use the key K to decrypt y, thus obtaining the content $\mathcal{M}$.

Each decoder box corresponds to a codeword $\mathbf{c} \in Q^\ell$, where $Q = \{1, \ldots, q\}$, in an obvious way:

keys in decoder box	codeword
$\{k_{1,j_1}, k_{2,j_2}, \ldots, k_{\ell,j_\ell}\}$	$(j_1, j_2, \ldots, j_\ell)$.

Denote by $\mathcal{C}$ the set of codewords corresponding to all the decoder boxes in the scheme. The keys in a pirate decoder form a codeword in the w-descendant code $\mathbf{desc}_w(\mathcal{C})$.

There is a special class of w-IPP codes that have very efficient tracing algorithms. These tracing algorithms are based on the idea of "nearest neighbor decoding" that is used in error-correcting codes. This concept was introduced in Section

9.3 for linear codes, but exactly the same method can be employed for an arbitrary (i.e., linear or nonlinear) code.

First, we recall a couple of definitions. As before, $\mathbf{dist}(\mathbf{c}, \mathbf{d})$ denotes the Hamming distance between two vectors $\mathbf{c}, \mathbf{d} \in Q^\ell$. That is,

$$\mathbf{dist}(\mathbf{c}, \mathbf{d}) = |\{i : \mathbf{c}_i \neq \mathbf{d}_i\}|.$$

Then, for a vector $\mathbf{f} \in Q^\ell$, a nearest neighbor to $\mathbf{f}$ is any codeword $\mathbf{c} \in \mathcal{C}$ such that $\mathbf{dist}(\mathbf{f}, \mathbf{c})$ is as small as possible. A nearest neighbor to $\mathbf{f}$ is denoted by $\mathbf{nn}(\mathbf{f})$.

The code $\mathcal{C}$ is said to be a *w-TA code* if the following property holds for all $\mathbf{f} \in \mathbf{desc}_w(\mathcal{C})$:

$$\mathbf{nn}(\mathbf{f}) \in \bigcap_{\mathcal{C}_0 \in \mathbf{susp}_w(\mathbf{f})} \mathcal{C}_0. \tag{13.2}$$

In other words, a w-TA code is a w-IPP code in which nearest neighbor decoding always yields an identifiable parent.

Here is a small example to illustrate.

Example 13.6 We present a certain $(5, 16, 4)$ code:

$$\begin{array}{ll}
\mathbf{c}_1 = (1,1,1,1,1) & \mathbf{c}_2 = (1,2,2,2,2) \\
\mathbf{c}_3 = (1,3,3,3,3) & \mathbf{c}_4 = (1,4,4,4,4) \\
\mathbf{c}_5 = (2,1,2,3,4) & \mathbf{c}_6 = (2,2,1,4,3) \\
\mathbf{c}_7 = (2,3,4,1,2) & \mathbf{c}_8 = (2,4,3,2,1) \\
\mathbf{c}_9 = (3,1,4,2,3) & \mathbf{c}_{10} = (3,2,3,1,4) \\
\mathbf{c}_{11} = (3,3,2,4,1) & \mathbf{c}_{12} = (3,4,1,3,2) \\
\mathbf{c}_{13} = (4,1,3,4,2) & \mathbf{c}_{14} = (4,2,4,3,1) \\
\mathbf{c}_{15} = (4,3,1,2,4) & \mathbf{c}_{16} = (4,4,2,1,3).
\end{array}$$

It can be proven that this code is a 2-TA code. Therefore nearest neighbor decoding can be used to identify parents.

Consider the vector $\mathbf{f} = (2, 3, 2, 4, 4)$. This is a vector in the 2-descendant code. If we compute the distance from $\mathbf{f}$ to all the codewords, then we see that

$$\mathbf{dist}(\mathbf{f}, \mathbf{c}_5) = \mathbf{dist}(\mathbf{f}, \mathbf{c}_{11}) = 2$$

and

$$\mathbf{dist}(\mathbf{f}, \mathbf{c}_i) \geq 3$$

for all $i \neq 5, 11$. Hence $\mathbf{c}_5$ and $\mathbf{c}_{11}$ are both identifiable parents of $\mathbf{f}$. □

One sufficient condition for a code to be a w-TA code is for it to have a large minimum distance between distinct codewords. Therefore, as we did in Section 9.3 with linear codes, we define

$$\mathbf{dist}(\mathcal{C}) = \min\{\mathbf{dist}(\mathbf{c}, \mathbf{d}) : \mathbf{c}, \mathbf{d} \in \mathcal{C}, \mathbf{c} \neq \mathbf{d}\}.$$

The next theorem provides a useful, easily tested, condition relating to TA codes.

THEOREM 13.7 *Suppose that $\mathcal{C}$ is an (ℓ, n, q)-code in which*

$$\mathbf{dist}(\mathcal{C}) > \ell\left(1 - \frac{1}{w^2}\right).$$

Then $\mathcal{C}$ is a w-TA code.

PROOF We will use the following notation. Denote $d = \mathbf{dist}(\mathcal{C})$. For any vectors, $\mathbf{c}$, $\mathbf{d}$, define

$$\mathbf{match}(\mathbf{c}, \mathbf{d}) = \ell - \mathbf{dist}(\mathbf{c}, \mathbf{d}).$$

Now, suppose that $\mathbf{c} = \mathbf{nn}(\mathbf{f})$ and suppose that $\mathcal{C}_0 \in \mathbf{susp}_w(\mathbf{f})$. We need to prove that $\mathbf{c} \in \mathcal{C}_0$.

First, because $\mathbf{f} \in \mathbf{desc}(\mathcal{C}_0)$, it follows that

$$\sum_{\mathbf{c}' \in \mathcal{C}_0} \mathbf{match}(\mathbf{f}, \mathbf{c}') \geq \ell.$$

Then, because $|\mathcal{C}_0| \leq w$, it follows that there exists a codeword $\mathbf{c}' \in \mathcal{C}_0$ such that

$$\mathbf{match}(\mathbf{f}, \mathbf{c}') \geq \frac{\ell}{w}.$$

Hence,

$$\mathbf{match}(\mathbf{f}, \mathbf{c}) \geq \frac{\ell}{w},$$

because $\mathbf{c}$ is the nearest neighbor to $\mathbf{f}$.

Next, let $\mathbf{b} \in \mathcal{C} \backslash \mathcal{C}_0$. Because $\mathbf{f} \in \mathbf{desc}(\mathcal{C}_0)$, we have that

$$\begin{aligned} \mathbf{match}(\mathbf{f}, \mathbf{b}) &\leq \sum_{\mathbf{c}' \in \mathcal{C}_0} \mathbf{match}(\mathbf{c}', \mathbf{b}) \\ &\leq w(\ell - d). \end{aligned}$$

Now, notice that $d > \ell(1 - 1/w^2)$ is equivalent to

$$w(\ell - d) < \frac{\ell}{w}.$$

Therefore, it follows that $\mathbf{match}(\mathbf{f}, \mathbf{b}) < \mathbf{match}(\mathbf{f}, \mathbf{c})$ for all codewords $\mathbf{b} \notin \mathcal{C}_0$. Hence, $\mathbf{c} \in \mathcal{C}_0$, and we have proven that the code is a w-TA code. ∎

We close this section by describing an easy construction for certain w-TA codes.[4] Suppose q is prime and $t < q$. Define the set $\mathcal{P}(q, t)$ to consist of all polynomials $a(x) \in \mathbb{Z}_q[x]$ having degree at most $t - 1$. For a positive integer $\ell < q$, define

$$\mathcal{C}(q, \ell, t) = \{(a(0), a(1), \ldots, a(\ell - 1)) : a(x) \in \mathcal{P}(q, t)\}.$$

We claim that $\mathcal{C} = \mathcal{C}(q, \ell, t)$ is an (ℓ, q^t, q) code such that $\mathbf{dist}(\mathcal{C}) = \ell - t + 1$.

[4]The codes we describe are in fact linear codes that are known as ***Reed-Solomon codes***.

This is easy to see, because any two distinct polynomials of degree not exceeding $t-1$ can agree on at most $t-1$ points (recall that a polynomial of degree not exceeding $t-1$ is completely determined by its values at t points by means of the Lagrange interpolation formula, which we presented as Theorem 11.3).

Suppose we define

$$t = \left\lceil \frac{\ell}{w^2} \right\rceil;$$

then

$$t < \frac{\ell}{w^2} + 1.$$

Therefore,

$$\mathbf{dist}(\mathcal{C}) > \ell \left(1 - \frac{1}{w^2}\right).$$

Hence, by Theorem 13.7, we have a w-TA code with $n = q^{\left\lceil \frac{\ell}{w^2} \right\rceil}$.

Summarizing, we obtain the following.

THEOREM 13.8 *Suppose q is prime, $\ell \leq q$ and $w \geq 2$ is an integer. Then there is an $\left(\ell, q^{\left\lceil \frac{\ell}{w^2} \right\rceil}, q\right)$-code that is a w-TA code.*

13.4 Bitcoin and Blockchain Technology

In this section, we discuss several of the technical aspects of blockchain technology, which is used in cryptocurrencies such as BITCOIN. The main cryptographic tools used in these applications are signatures and hash functions, including, specifically, *Merkle trees,* which were introduced in Section 9.5.3.

BITCOIN was invented by Satoshi Nakamoto[5], who published a document entitled *Bitcoin: A Peer-to-Peer Electronic Cash System* in October, 2008. The objective of cryptocurrencies such as BITCOIN is to support financial transactions without the requirement of a central "bank." The underlying idea is that there is a distributed ***public ledger*** of transactions that is maintained and verified voluntarily by millions of people on the internet. This public ledger is called the ***blockchain***.

Conceptually, a ***transaction*** is a transfer of a fixed amount of digital cash (which we will call ***bitcoin***) from one account to another. The role of an "account" is played by a ***bitcoin address***. A bitcoin address is just a random-looking sequence of numbers; it is in fact a message digest obtained by hashing a public (verification) key for a signature scheme. So, for example, a transaction could be thought of as a message M of the form

transfer X bitcoins from address A_1 to address A_2.

---

[5]It is interesting to note that Satoshi Nakamoto is a pseudonym and the true identity of the inventor of BITCOIN is not known at the time of writing this book.

header :	$h(header(B_i)), nonce, V(1)$
transactions :	$T_1, T_2, \ldots, T_m$
Merkle tree :	$V(2), V(3), \ldots$

FIGURE 13.2: Structure of block B_{i+1}

The message M would be transmitted along with

1. a signature y on M that is created using the private signing key $\mathbf{sig}_{K_1}$ corresponding to the address A_1, and
2. the public key $\mathbf{ver}_{K_1}$ corresponding to the address A_1.

This would allow any interested party (the owner of address A_2, for example) to verify that this transfer of funds was authorized by the owner of address A_1. To do this, one would check that

1. A_1 is the hash of $\mathbf{ver}_{K_1}$, and
2. $\mathbf{ver}_{K_1}(M, y) = \text{true}$.

Now, for this transaction to be "valid," it must also be the case that there are at least X bitcoins currently associated with the address A_1. In more technical parlance, the ***unspent transaction output*** associated with bitcoin address A_1 should be at least X. This fact could be verified by examining the public blockchain.

A transaction also may contain a ***transaction fee,*** which we consider later in this discussion.

Since bitcoin addresses are message digests corresponding to signature verification keys, there is no intrinsic requirement that a bitcoin address should be easily linked to any individual. Thus, some level of anonymity is supported (more precisely, a sequence of transactions involving the same bitcoin address would provide ***pseudonymity***). However, in practice, many people publicize bitcoin addresses for which they are the owners, which would certainly not be compatible with a desire for anonymity.

The blockchain can be thought of a sequence of ***blocks,*** which contain all the bitcoin transactions since the technology was first implemented. Each block will contain a large number of transactions (perhaps 2000–3000 or thereabouts), and every transaction should appear in a unique block. The blocks will be in a chronological order, which is maintained by linking each new block to the previous block in the blockchain. This is accomplished by including the hash of the header of a block B_i in the header of the next block, B_{i+1}. The very first block was the ***genesis block,*** which created the initial bitcoins out of "thin air."

More precisely, the structure of a typical block is depicted in Figure 13.2. In additional detail, the information that will be contained in a block B_{i+1} is enumerated as follows:

Protocol 13.1: MINING A NEW BLOCK IN BITCOIN

1. New transactions are broadcast, so any node in the BITCOIN network can accumulate a list of new transactions.

2. Once a sufficient number of transactions are accumulated, any node can attempt to create a new block B_{i+1}, which contains these transactions along with a valid proof-of-work.

3. One of the transactions in B_{i+1} will be a transaction that credits the address of the block creator with a certain amount of bitcoins (determined by a public formula), along with any transaction fees in the other transactions in B_{i+1}.

4. After a new block B_{i+1} is created, it is also broadcast.

3. Other nodes will accept the new block if and only if it contains valid transactions as well as a valid proof-of-work (this can be verified by examining B_{i+1} and previous blocks in the blockchain). (By "accepting" a block B_{i+1}, we mean that the next block B_{i+2} in the blockchain will link back to B_{i+1}.)

1. a ***block header***, denoted $header(B_{i+1})$, which consists of
   (a) a hash of the block header of the previous block, i.e., $h(header(B_i))$
   (b) a ***nonce*** (more about this later!)
   (c) the root node $V(1)$ of the Merkle tree formed by hashing the transactions $T_1, T_2, \ldots, T_m$, and
2. a list of new transactions $T_1, T_2, \ldots, T_m$ and the remaining nodes in the Merkle tree.

It is reasonable to ask who will create new blocks, and why. Actually, anyone can attempt to create a new block, but it takes considerable effort to do so. On the other hand, there is a financial incentive for creating a new block that adheres to a certain requirement, called "proof-of-work," which we will discuss a bit later. There are in fact several interesting technical aspects to the creation of new blocks, but we first discuss the process to be followed. This process, which is termed ***mining***, comprises the steps listed in Protocol 13.1.

There are three important aspects about blockchains that we still need to explore:

1. What is a proof-of-work?

2. How do we deal with ***forking***, where two nodes create a new block roughly simultaneously?

3. How do we prevent ***double spending*** (where a dishonest node tries to spend more bitcoins than are associated with a particular address, perhaps by broadcasting two transactions spending the current balance)?

The idea of ***proof-of-work*** is that it takes a considerable amount of computational output to create a new block. The "work" that is targeted is that of creating the block header so its hash value has a certain form. More specifically, suppose we consider the hash of $header(B_{i+1})$:

$$h(header(B_{i+1})) = h(h(header(B_i)) \parallel nonce \parallel V(1)).$$

The requirement is that this hash value should have a particular form. For example, suppose we stipulate that $h(header(B_{i+1}))$ should begin with S zeros (when considered as a binary string). If we think of the output of a hash function as being a random string, then this would occur with probability $1/2^S$. If we pick random values for *nonce*, then we would expect that, on average, 2^S random choices of *nonce* would be tested before we encounter an output of the specified form (note that the remaining inputs to the hash function are not altered). By choosing a suitable value for S (perhaps $S \approx 20$), we can require the node that is mining the new block to do a considerable amount of work (at least on average). In practice, as of 2018, new blocks are created approximately every fifteen minutes.

One issue that we need to consider is what happens if two new blocks are created simultaneously (or, at least, at close to the same time). As mentioned above, this phenomenon is called forking. When forking occurs, we have two blocks that probably contain many of the same transactions, which is a problem because the blockchain should only contain one copy of each transaction. If forking is not dealt with, then the result could be two separate branches extending the blockchain, which would undoubtedly lead to many difficulties and ambiguities. So it is essential, when a fork occurs, that one block is deemed to be the "winner" and the other one should be ignored.

The block that was constructed first would be considered to be the winner. But this could lead to ambiguous situations when it is not completely obvious which of two blocks was constructed first. However, in practice, such situations normally resolve themselves quickly, by using the convention that the longest branch wins when forking occurs. When it is obvious that one branch is longer than the other, all blocks in the shorter branch are deemed invalid and all transactions in these blocks are considered not to have been validated (unless they have been validated in the longer branch).

Here is a typical scenario. Suppose that two blocks B_{i+1} and B'_{i+1} create a fork in the blockchain. If B_{i+2} is then created (as a successor to B_{i+1}), then the fork consisting of B_{i+1} and B_{i+2} is longer than the fork consisting of B'_{i+1}. In recognition of this fact, the next block to be created would likely be B_{i+3}, extending this fork further. See Figure 13.3.

The actual rule that is used to designate a transaction as being confirmed is that it should be contained in a block B_i in the blockchain, and there are at least five additional blocks following B_i in the blockchain. If blocks are created at the

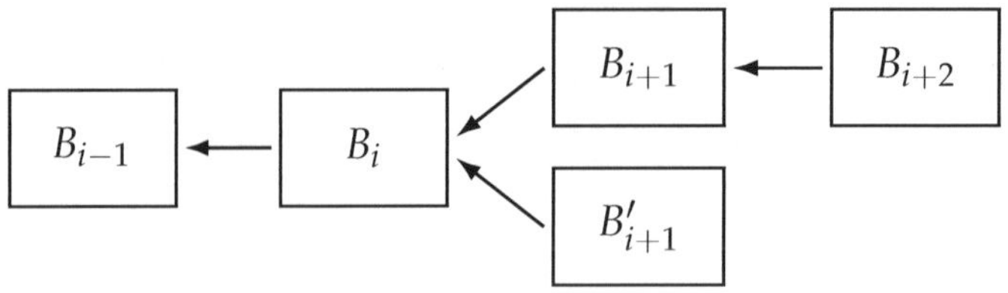

FIGURE 13.3: A Fork in the Blockchain

rate of one new block every 15 minutes, then it would take about 90 minutes for a given transaction to be confirmed.

It should be noted that there may be situations where it takes some time to reach a consensus after an instance of forking has arisen. One such occurrence took place in March 2013, when it took roughly six hours to reach consensus. The consensus was achieved only after the two forks had reached 24 blocks in length and there was widespread agreement to abandon one of the two forks!

Being able to maintain an unambiguous blockchain by reconciling any forking that occurs also has the benefit of preventing double spending (as mentioned above, double spending refers to the situation where someone tries to transfer some amount of bitcoins to two different addresses, hoping that both of these transactions will be accepted). However, as soon as one of the two transactions is validated, the other one should not be validated. The only way that both transactions could be validated (at least temporarily) would be if these two transactions ended up in different forks. However, given that one of the two forks will eventually be deleted, the transaction in the deleted fork will ultimately not be validated.

There is one other aspect of the *Bitcoin* design that we want to mention, namely, the presence of Merkle trees in blocks. The advantage of using Merkle trees is that it makes validation of a previously accepted transaction more efficient. The idea is to verify the root of the Merkle tree containing a particular transaction in the same way that a particular public key is validated when a Merkle tree is used in the context of signature schemes (as described in Section 9.5.3).

13.5 Notes and References

The concept of identity-based cryptography was introduced by Shamir [176] in 1984. The *Cocks Identity-based Cryptosystem,* which we presented in Section 13.1.1, was published in 2001 (see [59]). Research into identity-based cryptosystems exploded after the publication of the system proposed by Boneh and Franklin [42] in 2002, which was the first really practical system.

Paillier's public-key cryptosystem is from [161]. There has been much recent interest in ***fully homomorphic encryption***, in which multiplication and/or addition of plaintexts correspond to the corresponding operations on plaintexts. The breakthrough paper [86] by Gentry in 2009 gave the first potentially practical solution to this more general problem. It is an example of lattice-based cryptography. Since 2009, there has been a considerable amount of additional research into refining and implementing these techniques.

Boneh and Shaw [43] introduced the model used for fingerprinting in a cryptographic context; IPP codes were defined by Hollmann, van Lint, Linnartz, and Tolhuizen [97]. Much of Section 13.3.3 is based on [97]. Chor, Fiat, Naor, and Pinkas introduced traitor tracing for broadcast encryption schemes; see [57]. Theorem 13.7 is from [57] and Theorem 13.8 was proven in [189].

MacWilliams and Sloane [125] is a standard reference for coding theory; for a recent textbook, see Huffman and Pless [98].

Bitcoin was first described in the white paper [144]. For a readable tutorial introduction, we recommend [155].

Exercises

13.1 In the *Cocks Identity-based Cryptosystem*, verify that

$$\left(\frac{1 + K_U^{priv}\,(t_1)^{-1}}{n}\right) = \pm 1.$$

13.2 Suppose the *Cocks Identity-based Cryptosystem* is implemented with master public key $n = 16402692653$, and suppose that a user U has public key $K_U^{pub} = 9305496225$.

(a) Let $t_1 = 3975333024$ and $t_2 = 4892498575$. Verify that $\left(\frac{t_1}{n}\right) = \left(\frac{t_2}{n}\right) = -1$.

(b) Encrypt the plaintext $x = -1$ using the "random" values t_1 and t_2, obtaining the ciphertext (y_1, y_2).

(c) Given that $K_U^{priv} = 96465$, verify that the decryption of (y_1, y_2) is equal to x.

13.3 Suppose you are given an instance of the **BDH** problem, specifically, consisting of the following:

- additive groups $(G_1, +)$ and $(G_2, +)$ of prime order q and a multiplicative group $(G_3, \cdot)$ of order q,
- a pairing $e_q : G_1 \times G_2 \to G_3$,
- an element $P \in G_1$,
- an element $Q \in G_2$ having order q, and

- elements $aQ, bQ \in G_2$ for some $a, b \in \mathbb{Z}_q{}^*$.

Show that, if you can solve the **CDH** problem in G_3, then you can solve the given instance of the **BDH** problem.

13.4 The purpose of this question is to perform some computations using the *Paillier Cryptosystem*. Suppose $p = 1041817$ and $q = 716809$.

(a) Suppose $x_1 = 726095811532$, $r_1 = 270134931749$, $x_2 = 450864083576$, and $r_2 = 378141346340$. Compute $y_1 = e_K(x_1, r_1)$ and $y_2 = e_K(x_2, r_2)$.

(b) Let $y_3 = y_1 y_2 \bmod n^2$. Compute $x_3 = d_K(y_3)$ using the decryption algorithm for the *Paillier Cryptosystem*.

(c) Verify that $x_3 \equiv x_1 + x_2 \pmod{n}$.

13.5 Suppose that $n = pq$, where p and q are the values from the previous exercise. Determine if 229805443172001836784448 is an nth residue modulo n^2.

13.6 Prove the "if" part of Theorem 13.5; i.e., that an (ℓ, n, q) code $\mathcal{C}$ is a 2-IPP code if $A(\mathcal{C})$ is simultaneously a PHF$(\ell; n, q, 3)$ and an SHF$(\ell; n, q, \{2, 2\})$.

13.7 (a) Consider the $(3, 3r^2, r^2 + 2r)$ 2-IPP code $\mathcal{C}$ that was described in Section 13.3.3. Give a complete description of a $O(1)$ time algorithm TRACE, which takes as input a triple $\mathbf{f} = (f_1, f_2, f_3)$ and attempts to determine an identifiable parent of $\mathbf{f}$. If $\mathbf{f} \in \mathcal{C}$, then TRACE$(\mathbf{f}) = \mathbf{f}$; if $\mathbf{f} \in \mathbf{desc}_2(\mathcal{C}) \backslash \mathcal{C}$, then TRACE$(\mathbf{f})$ should find an identifiable parent of $\mathbf{f}$; and TRACE$(\mathbf{f})$ should return the output "fail," if $\mathbf{f} \notin \mathbf{desc}_2(\mathcal{C})$.

(b) Illustrate the execution of your algorithm in the case $r = 10$ $(q = 120)$ for the following triples: $(13, 11, 17)$; $(44, 9, 14)$; $(18, 108, 9)$.

13.8 We describe a $(4, r^3, r^2)$ 2-IPP code due to Hollman, van Lint, Linnartz, and Tolhuizen. The alphabet is $Q = \mathbb{Z}_r \times \mathbb{Z}_r$. The code $\mathcal{C} \subseteq Q^4$ consists of the following set of r^3 4-tuples:

$$\{((a, b), (a, c), (b, c), (a + b \bmod r, c)) : a, b, c \in \mathbb{Z}_r\}.$$

(a) Give a complete description of a $O(1)$ time algorithm TRACE, which takes as input a 4-tuple $\mathbf{f} = ((\alpha_1, \alpha_2), (\beta_1, \beta_2), (\gamma_1, \gamma_2), (\delta_1, \delta_2))$ and attempts to determine an identifiable parent of $\mathbf{f}$. The output of TRACE should be as follows:

- if $\mathbf{f} \in \mathcal{C}$, then TRACE$(\mathbf{f}) = \mathbf{f}$;
- if $\mathbf{f} \in \mathbf{desc}_2(\mathcal{C}) \backslash \mathcal{C}$, then TRACE$(\mathbf{f})$ should find one identifiable parent of $\mathbf{f}$; and
- TRACE$(\mathbf{f})$ should return the output "fail," if $\mathbf{f} \notin \mathbf{desc}_2(\mathcal{C})$.

In order for the algorithm to be an $O(1)$ time algorithm, there should be no linear searches, for example. You can assume that an arithmetic operation can be done in $O(1)$ time, however.

HINT In designing the algorithm, you will need to consider several cases. Many of the cases (and resulting subcases) are quite similar, however. You could initially divide the problem into the following four cases:

- $\alpha_1 \neq \beta_1$
- $\alpha_2 \neq \gamma_1$
- $\beta_2 \neq \gamma_2$
- $\alpha_1 = \beta_1, \alpha_2 = \gamma_1$, and $\beta_2 = \gamma_2$.

(b) Illustrate the execution of your algorithm in detail in the case $r = 100$ for each of the following 4-tuples **f**:

$$((37,71),(37,96),(71,96),(12,96))$$
$$((25,16),(83,54),(16,54),(41,54))$$
$$((19,11),(19,12),(11,15),(30,12))$$
$$((32,40),(32,50),(50,40),(82,30))$$

13.9 Consider the 3-TA code constructed by applying Theorem 13.8 with $\ell = 19$ and $q = 101$. This code is a $(19, 101^3, 101)$-code.

(a) Write a computer program to construct the 101^3 codewords in this code.

(b) Given the vector

$$\mathbf{f} = (14,66,46,56,13,31,50,30,77,32,0,93,48,37,16,66,24,42,9)$$

in the 3-descendant code, compute a parent of **f** using nearest neighbor decoding.

13.10 There are many online programs to compute *SHA-1* message digests. Typically, the input will be given in ascii form and the output will be a sequence of 40 hexadecimal characters $(0,\ldots,9,A,B,C,D,F)$. By computing the *SHA-1* message digests of the strings $0,1,2,3\ldots$, determine the smallest positive integer x whose *SHA-1* message digest starts with the hexadecimal digit 0. Then determine the smallest positive integer x whose corresponding *SHA-1* message digest starts with hexadecimal digits 00.

Appendix A

Number Theory and Algebraic Concepts for Cryptography

A.1 Modular Arithmetic

Definition A.1.1 (congruences) Suppose a and b are integers, and m is a positive integer. Then we write $a \equiv b \pmod{m}$ if m divides $b - a$. The phrase $a \equiv b \pmod{m}$ is called a ***congruence***, and it is read as "a is ***congruent*** to b modulo m." The integer m is called the ***modulus***.

Definition A.1.2 (modular reduction) Suppose we divide a and b by m, obtaining integer quotients and remainders, where the remainders are between 0 and $m - 1$. That is, $a = q_1 m + r_1$ and $b = q_2 m + r_2$, where $0 \le r_1 \le m - 1$ and $0 \le r_2 \le m - 1$. Then it is not difficult to see that $a \equiv b \pmod{m}$ if and only if $r_1 = r_2$. We will use the notation $a \bmod m$ (without parentheses) to denote the remainder when a is divided by m, i.e., the value r_1 above. Thus $a \equiv b \pmod{m}$ if and only if $a \bmod m = b \bmod m$. If we replace a by $a \bmod m$, we say that a is ***reduced modulo*** m. This process is called ***modular reduction***.

Example A.1.3 To compute 101 mod 7, we write $101 = 7 \times 14 + 3$. Since $0 \le 3 \le 6$, it follows that $101 \bmod 7 = 3$. As another example, suppose we want to compute $(-101) \bmod 7$. In this case, we write $-101 = 7 \times (-15) + 4$. Since $0 \le 4 \le 6$, it follows that $(-101) \bmod 7 = 4$.

Remark A.1.4 Some computer programming languages define $a \bmod m$ to be the remainder in the range $-m + 1, \ldots, m - 1$ having the same sign as a. For example, $(-101) \bmod 7$ would be -3, rather than 4 as we defined it above. But for our purposes, it is much more convenient to define $a \bmod m$ always to be non-negative.

Definition A.1.5 (arithmetic modulo m) We now define arithmetic modulo m: $\mathbb{Z}_m$ is the set $\{0, \ldots, m - 1\}$, equipped with two operations, $+$ (addition) and $\cdot$ (multiplication). Addition and multiplication in $\mathbb{Z}_m$ work exactly like real addition and multiplication, except that the results are reduced modulo m.

Example A.1.6 Suppose we want to compute $11 + 13$ in $\mathbb{Z}_{16}$. As integers, we have $11 + 13 = 24$. Then we reduce 24 modulo 16 as described above: $24 = 1 \times 16 + 8$, so $24 \bmod 16 = 8$, and hence $11 + 13 = 8$ in $\mathbb{Z}_{16}$.

Remark A.1.7 Suppose $0 \leq a, b < n$. Then $0 \leq a + b < 2n$. When we compute $a + b$ in $\mathbb{Z}_n$, the result is the integer $a + b$ if $a + b < n$, or $a + b - n$ if $a + b \geq n$.

Example A.1.8 Suppose we want to compute 11×13 in $\mathbb{Z}_{16}$. As integers, we have $11 \times 13 = 143$. Then we reduce 143 modulo 16 as described above: $143 = 8 \times 16 + 15$, so $143 \bmod 16 = 15$, and hence $11 \times 13 = 15$ in $\mathbb{Z}_{16}$.

A.2 Groups

Definition A.2.1 (group) A ***group*** is a pair $G = (X, \star)$, where X is a set and $\star$ is a binary operation defined on X, that satisfies the following properties:

- The operation $\star$ is ***associative***, i.e., $(a \star b) \star c = a \star (b \star c)$ for any $a, b, c \in X$.
- There is an element $id \in X$ called the ***identity***, such that $a \star id = id \star a = a$ for any $a \in X$.
- For every $a \in X$, there exists an element $b \in X$ called the ***inverse*** of a, such that $a \star b = b \star a = id$.

Definition A.2.2 A group $G = (X, \star)$ is ***abelian*** if the the operation $\star$ is ***commutative***, i.e., $a \star b = b \star a$ for any $a, b \in X$.

Definition A.2.3 A group $G = (X, \star)$ is a ***finite group*** if X is a finite set.

Definition A.2.4 The ***order*** of a finite group $G = (X, \star)$, denoted $\mathsf{ord}(G)$, is equal to $|X|$.

Remark A.2.5 For notational convenience, most group operations are written as multiplication or addition. If the group operation is multiplication, then the identity is usually denoted by 1 and the inverse of a by a^{-1}. If the group operation is addition, then the identity is usually denoted by 0 and the inverse of a by $-a$.

Remark A.2.6 If $-a = b$ in some additive group, then $-b = a$. A similar property holds for multiplicative groups.

Example A.2.7 (the additive group $\mathbb{Z}_n$) Let $n \geq 2$ be an integer. Then $(\mathbb{Z}_n, +)$ is a finite abelian group of order n, where $+$ denotes addition modulo n. The identity element is 0, and the inverse of a, usually denoted $-a$, is $(-a) \bmod n$.

Remark A.2.8 Suppose $1 \leq a < n$. Then $(-a) \bmod n = n - a$.

Example A.2.9 The additive inverses of the elements in $(\mathbb{Z}_{10}, +)$ are as follows: $-0 = 0$, $-1 = 9$, $-2 = 8$, $-3 = 7$, $-4 = 6$, $-5 = 5$, $-6 = 4$, $-7 = 3$, $-8 = 2$, and $-9 = 1$.

Example A.2.10 (the multiplicative group $\mathbb{Z}_p{}^*$) Let $p \geq 2$ be a prime. Define $\mathbb{Z}_p{}^* = \mathbb{Z}_p \backslash \{0\}$. Then $(\mathbb{Z}_p{}^*, \cdot)$ is a finite abelian group of order $p - 1$, where $\cdot$ denotes multiplication modulo p. The identity element is 1, and the inverse of a, usually denoted a^{-1}, can be computed efficiently using the EXTENDED EUCLIDEAN ALGORITHM (see Theorem A.2.69).

Example A.2.11 The multiplicative inverses of the elements in $(\mathbb{Z}_{11}{}^*, \cdot)$ are as follows: $1^{-1} = 1$, $2^{-1} = 6$, $3^{-1} = 4$, $4^{-1} = 3$, $5^{-1} = 9$ $6^{-1} = 2$, $7^{-1} = 8$, $8^{-1} = 7$, $9^{-1} = 5$, and $10^{-1} = 10$.

Definition A.2.12 For an integer $n \geq 2$, $\phi(n)$ denotes the number of positive integers less than n that are relatively prime to n. The function $\phi(n)$ is known as the ***Euler totient function***.

Theorem A.2.13 *$\phi(n)$ can be computed from the following formula: suppose that n has prime power factorization*

$$n = \prod_{i=1}^{\ell} p_i{}^{e_i}$$

(i.e., the p_i's are distinct primes and $e_i \geq 1$ for $1 \leq i \leq \ell$). Then

$$\phi(n) = \prod_{i=1}^{\ell} p_i{}^{e_i - 1}(p_i - 1) = \prod_{i=1}^{\ell} \left(p_i{}^{e_i} - p_i{}^{e_i - 1}\right).$$

Corollary A.2.14 *Here are some special cases of Theorem A.2.13.*

A.1 If p is prime, then $\phi(p) = p - 1$.

A.2 If p is prime, then $\phi(p^e) = p^e - p^{e-1}$.

A.3 If $p_1, \ldots, p_\ell$ are distinct primes, then

$$\phi\left(\prod_{i=1}^{\ell} p_i\right) = \prod_{i=1}^{\ell}(p_i - 1).$$

Example A.2.15 (the multiplicative group $\mathbb{Z}_n{}^*$) This example generalizes Example A.2.7. Let $n \geq 2$ be an integer. Define

$$\mathbb{Z}_n{}^* = \mathbb{Z}_n \backslash \{d \in \mathbb{Z}_n : \gcd(d, n) > 1\}.$$

Then $(\mathbb{Z}_n{}^*, \cdot)$ is a finite abelian group where $\cdot$ denotes multiplication modulo n. The identity element is 1, and the inverse of a, usually denoted a^{-1}, can be computed efficiently using the EXTENDED EUCLIDEAN ALGORITHM (see Theorem A.2.69). The order of $(\mathbb{Z}_n{}^*, \cdot)$ is equal to $\phi(n)$.

Remark A.2.16 To verify that $a^{-1} \bmod n = b$ in $\mathbb{Z}_n{}^*$, it is sufficient to check that $ab - 1$ is divisible by n.

Example A.2.17 The order of $(\mathbb{Z}_{20}{}^*, \cdot)$ is equal to $\phi(20) = (2^2 - 2)(5-1) = 8$. The elements in $(\mathbb{Z}_{20}{}^*, \cdot)$ are $1, 3, 7, 9, 11, 13, 17$, and 19. The multiplicative inverses are as follows: $1^{-1} = 1, 3^{-1} = 7, 7^{-1} = 3, 9^{-1} = 9, 11^{-1} = 11, 13^{-1} = 17, 17^{-1} = 13$, and $19^{-1} = 19$.

Example A.2.18 The *RSA Cryptosystem* is constructed using the group $\mathbb{Z}_n{}^*$, where $n = pq$ and p and q are distinct odd primes. For such an integer n, the order of $(\mathbb{Z}_n{}^*, \cdot)$ is equal to $(p-1)(q-1)$.

Example A.2.19 (matrices with non-zero determinant) Let $n \geq 2$. The set of $n \times n$ matrices with entries from $\mathbb{Z}_p$ (where p is prime) having non-zero determinant is a multiplicative group. The identity is the $n \times n$ matrix with 1s on the diagonal and 0s elsewhere. This is a non-abelian group, since matrix multiplication is not commutative.

Example A.2.20 (elliptic curves) Let $p > 3$ be prime. An ***elliptic curve*** is the set of solutions $(x, y) \in \mathbb{Z}_p \times \mathbb{Z}_p$ to the congruence $y^2 \equiv x^3 + ax + b \pmod{p}$, where $a, b \in \mathbb{Z}_p$ are constants such that $4a^3 + 27b^2 \not\equiv 0 \pmod{p}$, together with a special point $\mathcal{O}$ called the ***point at infinity***. Suppose we denote the set of points on the elliptic curve by $\mathcal{E}$. It is possible to define an addition operation on $\mathcal{E}$ so that $(\mathcal{E}, +)$ is an abelian group. Addition is defined as follows (where all arithmetic operations are performed in $\mathbb{Z}_p$): Suppose $P = (x_1, y_1)$ and $Q = (x_2, y_2)$ are points on $\mathcal{E}$. If $x_2 = x_1$ and $y_2 = -y_1$, then $P + Q = \mathcal{O}$; otherwise $P + Q = (x_3, y_3)$, where

$$x_3 = \lambda^2 - x_1 - x_2 \quad \text{and} \quad y_3 = \lambda(x_1 - x_3) - y_1,$$

and

$$\lambda = \begin{cases} (y_2 - y_1)(x_2 - x_1)^{-1}, & \text{if } P \neq Q \\ (3x_1{}^2 + a)(2y_1)^{-1}, & \text{if } P = Q. \end{cases}$$

Finally, define $P + \mathcal{O} = \mathcal{O} + P = P$ for all $P \in \mathcal{E}$.

A.2.1 Orders of Group Elements

Definition A.2.21 (orders of group elements) For a finite group $(X, \star)$, define the ***order*** of an element $a \in X$, denoted $\mathbf{ord}(a)$, to be the smallest positive integer m such that

$$\underbrace{a \star a \star \cdots \star a}_{m} = id.$$

If the group operation is written multiplicatively, then

$$\underbrace{a \star a \star \cdots \star a}_{m}$$

is written as an exponentiation, a^m. If the group operation is written additively, then the same expression is written as a multiplication, ma. The identity element is defined to have order 1. Any nonidentity element has order greater than 1.

Example A.2.22 In the group $(\mathbb{Z}_n, +)$, the element 1 has order n.

Theorem A.2.23 *For a finite group $(X, \star)$, the order of any $a \in X$ divides the order of the group, i.e.,* $\mathbf{ord}(a) | \mathbf{ord}(G)$.

Corollary A.2.24 *In a group of prime order p, every nonidentity element has order p.*

Theorem A.2.25 *For a finite group $(X, \cdot)$ and for any $a \in X$, the order of $b = a^i$ is*

$$\mathbf{ord}(b) = \frac{\mathbf{ord}(a)}{\gcd(\mathbf{ord}(a), i)}.$$

(Here, for concreteness, we assume that the group operation is written multiplicatively.)

Example A.2.26 If $\mathbf{ord}(a) = 100$ and $b = a^{35}$, then

$$\mathbf{ord}(b) = \frac{100}{\gcd(100, 35)} = \frac{100}{5} = 20.$$

Corollary A.2.27 *In the group $(\mathbb{Z}_n, +)$, the element b has order $n/\gcd(n, b)$.*

Theorem A.2.28 *If $\mathbf{ord}(a) = i$, then $a^{-1} = a^{i-1}$. More generally, $a^i = a^j$ if and only if* $i \equiv j \pmod{\mathbf{ord}(a)}$.

A.2.2 Cyclic Groups and Primitive Elements

Definition A.2.29 (cyclic group) A finite abelian group $(X, \star)$ is a ***cyclic group*** if there exists an element $a \in X$ having order equal to $|X|$. Such an element is called a ***generator*** of the group.

Example A.2.30 Let $n \geq 2$ be an integer. Then $(\mathbb{Z}_n, +)$ is a cyclic group, and 1 is a generator. Further, an element $a \in \mathbb{Z}_n$ is a generator of $(\mathbb{Z}_n, +)$ if and only if $\gcd(a, n) = 1$. The number of generators of $(\mathbb{Z}_n, +)$ is $\phi(n)$.

Example A.2.31 Let $p \geq 2$ be a prime. Then $({\mathbb{Z}_p}^*, \cdot)$ is a cyclic group of order $p - 1$, and a generator of this group is called a ***primitive element modulo*** p.

Theorem A.2.32 $({\mathbb{Z}_n}^*, \cdot)$ *is a cyclic group (of order $\phi(n)$) if and only if $n = 2, 4, p^k$ or $2p^k$, where p is an odd prime and k is a positive integer.*

Theorem A.2.33 *$\alpha \in {\mathbb{Z}_p}^*$ is a primitive element if and only if*

$$\alpha^{(p-1)/q} \not\equiv 1 \pmod{p}$$

for all primes q such that $q | (p - 1)$.

Remark A.2.34 Using Theorem A.2.33, it is simple to test whether a given element $\alpha \in {\mathbb{Z}_p}^*$ is a primitive element (where p is an odd prime) provided that the factorization of $p - 1$ is known.

Example A.2.35 Suppose $p = 13$ and $\alpha = 2$. The factorization of 12 into prime powers is $12 = 2^2 3^1$. Therefore, to verify that 2 is a primitive element modulo 13, it is sufficient to check that $2^6 \not\equiv 1 \pmod{13}$ and $2^4 \not\equiv 1 \pmod{13}$. This is much faster than checking all 12 powers of α.

Theorem A.2.36 *The number of primitive elements in* $(\mathbb{Z}_p{}^*, \cdot)$ *is* $\phi(p-1) = \phi(\phi(p))$.

Example A.2.37 The number of primitive elements in $(\mathbb{Z}_{73}{}^*, \cdot)$ is $\phi(72)$. Since $72 = 2^3 \times 3^2$, there are $(2^3 - 2^2)(3^2 - 3) = 24$ primitive elements in $(\mathbb{Z}_{73}{}^*, \cdot)$.

A.2.3 Subgroups and Cosets

Definition A.2.38 (subgroup) Suppose $G = (X, \star)$ is a finite group and $Y \subseteq X$. We say that $H = (Y, \star)$ is a ***subgroup*** of G if H is also a (finite) group.

Theorem A.2.39 *Suppose* $G = (X, \star)$ *is a finite group and* $Y \subseteq X$. *Then* $H = (Y, \star)$ *is a subgroup of* G *if and only if it is* ***closed****, i.e., if* $h_1 \star h_2 \in H$ *for all* $h_1, h_2 \in H$.

Definition A.2.40 (coset) Suppose $H = (Y, \star)$ is a subgroup of the group $G = (X, \star)$. For any $a \in X$, define the ***right coset*** Ya as follows:

$$Ya = \{y \star a : y \in Y\}.$$

Also, define the ***left coset*** aY as follows:

$$aY = \{a \star y : y \in Y\}.$$

Theorem A.2.41 *Suppose* $H = (Y, \star)$ *is a subgroup of* $G = (X, \star)$. *Then,* $|Ya| = |Y|$ *for all* a. *Furthermore, two right cosets* Ya *and* Ya' *(or two left cosets* aY *and* $a'Y$*) are either identical or disjoint.*

Corollary A.2.42 *A group* X *can be partitioned into right (or left) cosets of any subgroup* Y.

Theorem A.2.43 *Suppose* $H = (Y, \star)$ *is a subgroup of* $G = (X, \star)$*, and* $a, b \in X$. *Then* $Ya = Yb$ *if and only if* $ab^{-1} \in Y$.

Theorem A.2.44 (Lagrange's theorem) *Suppose* $H = (Y, \star)$ *is a subgroup of the finite group* $G = (X, \star)$. *Then* $\mathbf{ord}(H)$ *divides* $\mathbf{ord}(G)$.

Definition A.2.45 Suppose that $G = (X, \star)$ is a finite group and $y \in X$. Define $\langle a \rangle = \{a^i : i \geq 0\}$.

Remark A.2.46 It is easy to see that $(\langle a \rangle, \star)$ is a cyclic subgroup of $(X, \star)$ and $\mathbf{ord}(\langle a \rangle) = \mathbf{ord}(a)$. We say that $(\langle a \rangle, \star)$ is the ***subgroup generated by*** a. Lagrange's theorem therefore shows that $\mathbf{ord}(a) | \mathbf{ord}(G)$, as stated previously in Theorem A.2.23.

Example A.2.47 Consider the group $G = (\mathbb{Z}_{19}{}^*, \cdot)$, which is a cyclic group of order 18. It can be verified that 2 is a primitive element in G. The element $2^3 = 8$ generates a subgroup H of order $18/3 = 6$, where

$$H = \{1, 8, 7, 18, 11, 12\}.$$

There are two additional cosets of H, namely

$$2H = \{2, 16, 14, 17, 3, 5\}$$

and

$$4H = \{4, 13, 9, 15, 6, 10\}.$$

Example A.2.48 Suppose that $G = (X, \star)$ is a finite group of order n and a is a generator of G. Suppose that m is a divisor of n. Then there is a unique subgroup of G having order m, namely, $\langle a^{n/m} \rangle$.

Example A.2.49 $G = (\mathbb{Z}_{90}, +)$ has subgroups of orders $1, 2, 3, 5, 6, 9, 10, 15, 18, 30, 45$, and 90.

A.2.4 Group Isomorphisms and Homomorphisms

Definition A.2.50 An ***isomorphism*** from a group $G = (X, \star)$ to a group $H = (Y, *)$ is a bijection $\varphi : X \to Y$ such that $\varphi(a \star a') = \varphi(a) * \varphi(a')$ for all $a, a' \in X$.

Theorem A.2.51 *If $\varphi : X \to Y$ is an isomorphism from $G = (X, \star)$ to $H = (Y, *)$, then G and H have the same order. Furthermore, for any $x \in X$, $\mathbf{ord}(x) = \mathbf{ord}(\varphi(x))$.*

Theorem A.2.52 *Any two cyclic groups of the same order n are isomorphic.*

Corollary A.2.53 *If $G = (X, \star)$ is any finite group, and $a \in X$, then $(\langle a \rangle, \star)$ is isomorphic to $(\mathbb{Z}_{\mathbf{ord}(a)}, +)$.*

Example A.2.54 $G = (\mathbb{Z}_{12}, +)$ is isomorphic to $(\mathbb{Z}_{13}{}^*, \cdot)$. An isomorphism is given by the mapping $\varphi : \mathbb{Z}_{12} \to \mathbb{Z}_{13}{}^*$ defined by $\varphi(i) = \alpha^i \bmod 13$, where α is a primitive element in $(\mathbb{Z}_{13}{}^*, \cdot)$.

Definition A.2.55 A ***homomorphism*** from a group $G = (X, \star)$ to a group $H = (Y, *)$ is a mapping $\varphi : X \to Y$ such that $\varphi(a \star a') = \varphi(a) * \varphi(a')$ for all $a, a' \in X$.

Remark A.2.56 A homomorphism φ from a group $G = (X, \star)$ to a group $H = (Y, *)$ is an isomorphism if and only if it is a bijection from X to Y.

Example A.2.57 Suppose $1 \leq m < n$. Then the mapping $\varphi : \mathbb{Z}_n \to \mathbb{Z}_n$ defined by the rule $\varphi(a) = am \bmod n$ is a homomorphism from $(\mathbb{Z}_n, +)$ to $(\mathbb{Z}_n, +)$. If $\gcd(n, m) = 1$, then this mapping is an isomorphism from $(\mathbb{Z}_n, +)$ to $(\mathbb{Z}_n, +)$.

A.2.5 Quadratic Residues

Definition A.2.58 (quadratic residue) Suppose p is an odd prime and a is an integer. Then a is defined to be a ***quadratic residue*** modulo p if $a \not\equiv 0 \pmod{p}$ and the congruence $y^2 \equiv a \pmod{p}$ has a solution $y \in \mathbb{Z}_p$. If $a \not\equiv 0 \pmod{p}$ and a is not a quadratic residue modulo p, then a is defined to be a ***quadratic non-residue*** modulo p.

Remark A.2.59 If a is a quadratic residue modulo an odd prime p, then a has exactly two square roots modulo p. Furthermore, these two square roots sum to 0 modulo p.

Example A.2.60 3 is a quadratic residue modulo 23. The two square roots of 3 modulo 23 are 7 and 16. Note that $7 + 16 = 23$.

Definition A.2.61 (Legendre symbol) Suppose p is an odd prime. For any integer a, define the ***Legendre symbol*** $\left(\frac{a}{p}\right)$ as follows:

$$\left(\frac{a}{p}\right) = \begin{cases} 0 & \text{if } a \equiv 0 \pmod{p} \\ 1 & \text{if } a \text{ is a quadratic residue modulo } p \\ -1 & \text{if } a \text{ is a quadratic non-residue modulo } p. \end{cases}$$

Theorem A.2.62 *Suppose p is an odd prime. Then*

$$\left(\frac{a}{p}\right) = a^{(p-1)/2} \bmod p.$$

Remark A.2.63 Suppose p is an odd prime. Then the mapping $a \mapsto \left(\frac{a}{p}\right)$ is a homomorphism from $(\mathbb{Z}_p{}^*, \cdot)$ to $(\{1, -1\}, \cdot)$.

Remark A.2.64 The product of two quadratic residues modulo p is again a quadratic residue modulo p. The product of two quadratic nonresidues modulo p is a quadratic residue modulo p. The product of a quadratic residue and a quadratic nonresidue modulo p is a quadratic nonresidue modulo p.

Theorem A.2.65 *Suppose $p \equiv 3 \pmod{4}$ is prime and suppose y is a quadratic residue modulo p. Then the two square roots of y modulo p are $\pm y^{(p+1)/4} \bmod p$.*

Example A.2.66 Suppose we take $p = 23$ and $y = 3$. Then $\left(\frac{3}{23}\right) = 3^{11} \bmod 23 = 1$. Hence, 3 is a quadratic residue modulo 23. The two square roots of 3 modulo 23 are $\pm 3^6 \bmod 23$, i.e., 7 and 16.

A.2.6 Euclidean Algorithm

Algorithm A.2.67 (EUCLIDEAN ALGORITHM) The EUCLIDEAN ALGORITHM computes the greatest common divisor of two positive integers, say a and b. The algorithm sets r_0 to be a and r_1 to be b, and performs the following sequence of divisions:

$$\begin{array}{rcll} r_0 & = & q_1 r_1 + r_2, & 0 < r_2 < r_1 \\ r_1 & = & q_2 r_2 + r_3, & 0 < r_3 < r_2 \\ \vdots & \vdots & \vdots & \vdots \\ r_{m-2} & = & q_{m-1} r_{m-1} + r_m, & 0 < r_m < r_{m-1} \\ r_{m-1} & = & q_m r_m. & \end{array}$$

The algorithm terminates when a division yields a remainder of 0. The last nonzero remainder, r_m, is the greatest common divisor of a and b.

Example A.2.68 We compute the greatest common divisor of 34 and 99. The EUCLIDEAN ALGORITHM proceeds as follows:

$$\begin{array}{rcl} 99 & = & 2 \times 34 + 31 \\ 34 & = & 1 \times 31 + 3 \\ 31 & = & 10 \times 3 + 1 \\ 3 & = & 3 \times 1 + 0. \end{array}$$

Hence, $\gcd(34, 99) = 1$.

Algorithm A.2.69 (EXTENDED EUCLIDEAN ALGORITHM) Given two integers a and b, the EXTENDED EUCLIDEAN ALGORITHM computes integers s and t such that $as + bt = \gcd(a, b)$. Algorithm 6.2 is a detailed description of this algorithm.

Example A.2.70 The EXTENDED EUCLIDEAN ALGORITHM can be used to express 1 as a combination of 99 and 34: $11 \times 99 - 32 \times 34 = 1$.

Theorem A.2.71 (multiplicative inverses in $\mathbb{Z}_n$) *Let $n \geq 2$. A multiplicative inverse $a^{-1} \bmod n$ exists if and only if $\gcd(a, n) = 1$. In this case, given inputs a and n, the* EXTENDED EUCLIDEAN ALGORITHM *will compute integers s and t such that $as + nt = 1$. Then $a^{-1} \equiv s \pmod{n}$.*

Example A.2.72 We noted in the previous example that $11 \times 99 - 32 \times 34 = 1$. Therefore, $34^{-1} \bmod 99 = -32 \bmod 99 = 67$.

Theorem A.2.73 (linear congruences mod n) *Suppose $\gcd(a, n) = 1$. Then the linear congruence $ax \equiv c \pmod{n}$ has a unique solution modulo n, given by the formula $x = a^{-1} c \bmod n$.*

Example A.2.74 Suppose we wish to solve the linear congruence $34x \equiv 25 \pmod{99}$. We have already computed $34^{-1} \bmod 99 = 67$. Therefore the solution to the linear congruence is $x = 67 \times 25 \bmod 99 = 91$.

Theorem A.2.75 (linear congruences mod n) *Suppose* $\gcd(a, n) = d > 1$. *If* $c \not\equiv 0 \pmod{d}$, *then the linear congruence* $ax \equiv c \pmod{n}$ *has no solutions. If* $c \equiv 0 \pmod{d}$, *then the linear congruence* $ax \equiv c \pmod{n}$ *is equivalent to linear congruence* $a'x \equiv c' \pmod{n'}$, *where* $a' = a/d$, $c' = c/d$ *and* $n' = n/d$. *This congruence has a unique solution modulo* n' *by Theorem A.2.73, say* $x \equiv x_0 \pmod{n'}$. *The original congruence has* d *solutions modulo* n, *namely,* $x = x_0 + in' \bmod n$, *for* $0 \leq i \leq d-1$.

Example A.2.76 Consider the linear congruence $22x \equiv 55 \pmod{99}$. It is easy to compute $\gcd(22, 99) = 11$. Since $11 | 55$, the original congruence is equivalent to $2x \equiv 5 \pmod{9}$. The solution to this "reduced" congruence is $x \equiv 7 \pmod{9}$. The original congruence has the solutions $x \equiv 7, 16, 25, 34, 43, 52, 61, 70, 79, 88, 97 \pmod{99}$.

A.2.7 Direct Products

Definition A.2.77 (direct product) Suppose that $G = (X, \star)$ and $G' = (X', *)$ are groups. The ***direct product*** $G \times G'$ is the group defined as follows: $G \times G' = (X \times X', \circ)$, where

$$(a, a') \circ (b, b') = (a \star b, a' * b')$$

for all $a, b \in X$ and all $a', b' \in X'$.

Theorem A.2.78 *Suppose* $\gcd(m, n) = 1$. *Then* $(\mathbb{Z}_m, +) \times (\mathbb{Z}_n, +)$ *is isomorphic to* $(\mathbb{Z}_{mn}, +)$ *and* $(\mathbb{Z}_m{}^*, \cdot) \times (\mathbb{Z}_n{}^*, \cdot)$ *is isomorphic to* $(\mathbb{Z}_{mn}{}^*, \cdot)$.

Remark A.2.79 Suppose $(a, a') \in G \times G'$. If the order of a is equal to d and the order of a' is equal to d', then the order of (a, a') is equal to the least common multiple of d and d'.

Example A.2.80 $(\mathbb{Z}_n, +) \times (\mathbb{Z}_n, +)$ is not isomorphic to $(\mathbb{Z}_{n^2}, +)$. One way to see this is to observe that every element of $(\mathbb{Z}_n, +) \times (\mathbb{Z}_n, +)$ has order dividing n, whereas $(\mathbb{Z}_{n^2}, +)$ contains an element of order n^2 (namely, 1).

Remark A.2.81 Definition A.2.77 can be extended in the obvious way to define a direct product of more than two groups.

Theorem A.2.82 (Fundamental theorem of abelian groups) *Every finite abelian group is isomorphic to a direct product of cyclic groups of prime power order.*

Example A.2.83 The factorization of 36 into prime powers is $36 = 2^2 3^2$. There are precisely four nonisomorphic groups of order 36, namely, $\mathbb{Z}_4 \times \mathbb{Z}_9$, $\mathbb{Z}_2 \times \mathbb{Z}_2 \times \mathbb{Z}_9$, $\mathbb{Z}_4 \times \mathbb{Z}_3 \times \mathbb{Z}_3$, and $\mathbb{Z}_2 \times \mathbb{Z}_2 \times \mathbb{Z}_3 \times \mathbb{Z}_3$.

A.3 Rings

Definition A.3.1 (ring) A ***ring*** is a triple $R = (X, \cdot, +)$, where X is a finite set and $\cdot$ and $+$ are a binary operations defined on X, that satisfies the following properties:

- $(X, +)$ is an abelian group with identity 0.
- Multiplication is ***associative***, i.e., for any $a, b, c \in X$, $(ab)c = a(bc)$.
- The ***distributive property*** is satisfied, i.e., for any $a, b, c \in X$, $(a + b)c = (ac) + (bc)$ and $a(b + c) = (ab) + (ac)$.

Definition A.3.2 A ring $R = (X, \cdot, +)$ is a ***finite ring*** if X is a finite set.

Definition A.3.3 A ring $R = (X, \cdot, +)$ is a ***ring with identity*** if X contains a multiplicative identity, denoted by 1.

Definition A.3.4 A ring $R = (X, \cdot, +)$ is a ***commutative ring*** if multiplication is commutative.

Example A.3.5 Some familiar examples of commutative rings include the integers, $\mathbb{Z}$; the real numbers, $\mathbb{R}$; and the complex numbers, $\mathbb{C}$. These are all infinite rings.

Example A.3.6 $(\mathbb{Z}_m, \cdot, +)$ is a finite ring for any $m \geq 2$.

Example A.3.7 (matrices) Let $n \geq 2$. The set of $n \times n$ matrices with entries from $\mathbb{Z}_p$ is a ring, but not a commutative ring.

Example A.3.8 (ring of polynomials) Suppose $(A, \cdot, +)$ is a field (see Section A.4 for the definition) and x is an indeterminate. Let $A[x]$ denote the set of all polynomials with coefficients from A. Then $(A[x], \cdot, +)$ is a commutative ring with identity 1.

Example A.3.9 $\mathbb{Z}_2[x]$ denotes the ring of polynomials with coefficients from $\mathbb{Z}_2$. We add and multiply polynomials in the usual way, but we reduce the coefficients modulo 2. For example, if $a(x) = x^2 + 1$ and $b(x) = x^2 + x + 1$, we would have $a(x) + b(x) = x$ and $a(x)b(x) = x^4 + x^3 + x + 1$.

Algorithm A.3.10 (EUCLIDEAN ALGORITHM FOR POLYNOMIALS) The greatest common divisor of two polynomials $a(x)$ and $b(x)$ can be computed using the EUCLIDEAN ALGORITHM FOR POLYNOMIALS. It is a straightforward modification of the EUCLIDEAN ALGORITHM for integers. The algorithm sets $r_0(x)$ to be $a(x)$ and $r_1(x)$ to be $b(x)$, and performs the following sequence of divisions:

$$\begin{array}{rcll} r_0(x) & = & q_1(x)r_1(x) + r_2(x), & 0 < \mathsf{deg}(r_2) < \mathsf{deg}(r_1) \\ r_1(x) & = & q_2(x)r_2(x) + r_3(x), & 0 < \mathsf{deg}(r_3) < \mathsf{deg}(r_2) \\ \vdots & \vdots & \vdots & \vdots \\ r_{m-2}(x) & = & q_{m-1}(x)r_{m-1}(x) + r_m(x), & 0 < \mathsf{deg}(r_m) < \mathsf{deg}(r_{m-1}) \\ r_{m-1}(x) & = & q_m(x)r_m(x). & \end{array}$$

The algorithm terminates when a division yields a remainder of 0. The last nonzero remainder, $r_m(x)$, is the greatest common divisor of $a(x)$ and $b(x)$.

Example A.3.11 We compute the greatest common divisor of x^4+x+1 and x^3+x in $\mathbb{Z}_2[x]$. The EUCLIDEAN ALGORITHM FOR POLYNOMIALS proceeds as follows:

$$\begin{aligned} x^4+x+1 &= x(x^3+x)+x^2+x+1 \\ x^3+x &= (x+1)(x^2+x+1)+x+1 \\ x^2+x+1 &= x(x+1)+1 \\ x &= x(1)+0. \end{aligned}$$

Hence, 1 is the greatest common divisor of x^4+x+1 and x^3+x in $\mathbb{Z}_2[x]$.

A.3.1 The Chinese Remainder Theorem

Definition A.3.12 (direct product) Suppose that $R = (X, \cdot, +)$ and $R' = (X', \cdot, +)$ are rings. The ***direct product*** $R \times R'$ is the ring defined as follows: $R \times R' = (X \times X', \cdot, +)$, where

$$(a, a') \cdot (b, b') = (a \cdot b, a' \cdot b')$$

and

$$(a, a') + (b, b') = (a + b, a' + b')$$

for all $a, b \in X$ and all $a', b' \in X'$.

Remark A.3.13 Definition A.3.12 can be extended in the obvious way to define a direct product of more than two rings.

Definition A.3.14 An ***isomorphism*** from a ring $R = (X, \cdot, +)$ to a ring $S = (Y, \cdot, +)$ is a bijection $\varphi : X \rightarrow Y$ such that $\varphi(a \cdot a') = \varphi(a) \cdot \varphi(a')$ for all $a, a' \in X$ and $\varphi(a + a') = \varphi(a) + \varphi(a')$ for all $a, a' \in X$.

Theorem A.3.15 *Suppose $M = m_1 \times m_2 \times \cdots \times m_r$, where $\gcd(m_i, m_j) = 1$ for all $i \neq j$. Then the ring $(\mathbb{Z}_M, \cdot, +)$ is isomorphic to the ring $(\mathbb{Z}_{m_1} \times \cdots \times \mathbb{Z}_{m_r}, \cdot, +)$. (This theorem generalizes Theorem A.2.78.)*

Remark A.3.16 Define $\chi : \mathbb{Z}_M \rightarrow \mathbb{Z}_{m_1} \times \cdots \times \mathbb{Z}_{m_r}$, as follows:

$$\chi(a) = (a \bmod m_1, \ldots, a \bmod m_r).$$

Then χ can be shown to be an isomorphism of the two rings $(\mathbb{Z}_M, \cdot, +)$ and $(\mathbb{Z}_{m_1} \times \cdots \times \mathbb{Z}_{m_r}, \cdot, +)$.

Remark A.3.17 For $1 \leq i \leq r$, define $M_i = M/m_i$ and $y_i = M_i^{-1} \bmod m_i$. Then the inverse function $\chi^{-1} : \mathbb{Z}_{m_1} \times \cdots \times \mathbb{Z}_{m_r} \rightarrow \mathbb{Z}_M$ is

$$\chi^{-1}(a_1, \ldots, a_r) = \sum_{i=1}^{r} a_i M_i y_i \bmod M.$$

Example A.3.18 Suppose $r = 3$, $m_1 = 7$, $m_2 = 11$, and $m_3 = 13$. Then $M = 1001$. We compute $M_1 = 143$, $M_2 = 91$, and $M_3 = 77$, and then $y_1 = 5$, $y_2 = 4$, and $y_3 = 12$. Then the function $\chi^{-1} : \mathbb{Z}_7 \times \mathbb{Z}_{11} \times \mathbb{Z}_{13} \rightarrow \mathbb{Z}_{1001}$ is the following:

$$\chi^{-1}(a_1, a_2, a_3) = (715a_1 + 364a_2 + 924a_3) \bmod 1001.$$

Remark A.3.19 The fact that the function χ^{-1} constitutes an isomorphism is an important result that is commonly known as the Chinese Remainder Theorem.

Theorem A.3.20 (Chinese remainder theorem) *Suppose $m_1, \ldots, m_r$ are pairwise relatively prime positive integers, and suppose $a_1, \ldots, a_r$ are integers. Then the system of r congruences $x \equiv a_i \pmod{m_i}$ ($1 \leq i \leq r$) has a unique solution modulo $M = m_1 \times \cdots \times m_r$, which is given by $x = \chi^{-1}(a_1, \ldots, a_m)$.*

Example A.3.21 *Using the formula developed in Example A.3.18, the system of congruences*

$$\begin{aligned} x &\equiv 3 \pmod{7} \\ x &\equiv 6 \pmod{11} \\ x &\equiv 5 \pmod{13} \end{aligned}$$

has the solution

$$\begin{aligned} x &\equiv 715 \times 3 + 364 \times 6 + 924 \times 5 \pmod{1001} \\ &\equiv 8949 \pmod{1001} \\ &\equiv 941 \pmod{1001}. \end{aligned}$$

A.3.2 Ideals and Quotient Rings

Definition A.3.22 (ideal) Suppose $R = (X, \cdot, +)$ is a commutative ring. An ***ideal*** is a subset $I \subseteq X$ that satisfies the following properties:

- $(I, +)$ is an abelian group, and
- $ab \in I$ whenever $a \in X$ and $b \in I$.

Definition A.3.23 (principal ideal) Suppose $R = (X, \cdot, +)$ is a commutative ring and let $c \in X$. The ***principal ideal*** generated by c, which is denoted by (c), is the subset defined as follows:

$$(c) = \{ac : a \in X\}.$$

It is easy to see that a principal ideal is always an ideal.

Definition A.3.24 (quotient ring) Suppose $R = (X, \cdot, +)$ is a commutative ring and $I = (c)$ is a principal ideal. The ***quotient ring*** R/I is constructed as follows. $R/I = (Y, \cdot, +)$, where Y consists of the (additive) cosets of I in $(X, +)$. The sum of two cosets $I + a$ and $I + b$ is defined to be $I + (a + b)$, for any $a, b \in X$, and the product of the two cosets $I + a$ and $I + b$ is defined to be $I + ab$.

Example A.3.25 The quotient ring $\mathbb{Z}_2[x]/(x^3+1)$ is obtained from the ring $\mathbb{Z}_2[x]$ by equating x^3+1 and 0. Since coefficients are in $\mathbb{Z}_2$, this is the same thing as saying that $x^3 = 1$. Then $x^4 = x$, $x^5 = x^2$, etc. In general, computations in $\mathbb{Z}_2[x]/(x^3+1)$ are the same as in $\mathbb{Z}_2[x]$, except that all exponents are reduced modulo 3. There are eight polynomials in $\mathbb{Z}_2[x]/(x^3+1)$, namely, $0, 1, x, x+1, x^2, x^2+1, x^2+x$, and x^2+x+1.

Example A.3.26 The quotient ring $\mathbb{Z}_3[x]/(x^2+1)$ is obtained from the ring $\mathbb{Z}_3[x]$ by equating x^2+1 and 0. Since coefficients are in $\mathbb{Z}_3$, this is the same thing as saying that $x^2 = 2$. We would compute $(x+1)(2x+1)$ as follows:

$$\begin{aligned}(x+1)(2x+1) &= 2x^2+3x+1\\ &= 2x^2+1\\ &= 1+1\\ &= 2.\end{aligned}$$

There are nine polynomials in $\mathbb{Z}_3[x]/(x^2+1)$, namely, $0, 1, 2, x, x+1, x+2, 2x, 2x+1$, and $2x+2$.

Definition A.3.27 (principal ring) Suppose $R = (X, \cdot, +)$ is a commutative ring. We say that R is a ***principal ring*** if every ideal is a principal ideal.

Example A.3.28 One example of a principal ring is $(\mathbb{Z}, \cdot, +)$.

Example A.3.29 Since $(\mathbb{Z}, \cdot, +)$ is a principal ring, it follows that any ideal I in this ring consists of all the multiples (positive and negative) of a positive integer c, i.e., $I = (c)$. The quotient ring $\mathbb{Z}/(c)$ is simply $\mathbb{Z}_c$.

A.4 Fields

Definition A.4.1 (field) A ring $R = (X, \cdot, +)$ is a ***field*** if it is a commutative ring with identity such that every non-zero element has a multiplicative inverse (i.e., $(R\backslash\{0\}, \cdot)$ is an abelian group).

Example A.4.2 $(\mathbb{Z}_n, \cdot, +)$ is a finite field if and only if n is prime.

Remark A.4.3 As mentioned earlier, multiplicative inverses modulo a prime p can be computed using the EXTENDED EUCLIDEAN ALGORITHM.

Remark A.4.4 Suppose n can be factored as $n = n_1 n_2$. Then the product $n_1 n_2 = 0$ in the ring $\mathbb{Z}_n$. It follows that neither n_1 nor n_2 is invertible in $(\mathbb{Z}_n, \cdot)$. To see this, suppose $rn_1 = 1$. Then $rn_1 n_2 = 1 \times n_2 = n_2 \neq 0$ (where all computations are modulo n). However, $rn_1 n_2 = r \times 0 = 0$, so we have a contradiction.

Remark A.4.5 The direct product of two fields is not a field.

Definition A.4.6 (irreducible polynomial) Suppose A is a field. A polynomial $f(x) \in A[x]$ is ***irreducible*** if $f(x)$ cannot be written as a product of two polynomials $f_1(x)f_2(x)$, where $f_1(x)$ and $f_2(x)$ both have positive degree.

Example A.4.7 In the ring $\mathbb{Z}_2[x]$, we have that $x^2+1=(x+1)(x+1)$, so x^2+1 is reducible. Because $x^2+x=x(x+1)$, this polynomial is also reducible. However, x^2+x+1 is irreducible.

Example A.4.8 Suppose that A is any finite field and suppose n is a positive integer. Then there is at least one irreducible polynomial of degree n in $(A[x], \cdot, +)$.

Theorem A.4.9 *There exists a finite field of order q if and only if $q = p^k$ where p is prime and $k \geq 1$ is an integer.*

Definition A.4.10 A finite field of order $q = p^k$ (where p is prime) is said to have ***characteristic*** p.

Theorem A.4.11 *Suppose p is prime and $k \geq 2$. A finite field of order p^k can be constructed as follows. Let $f(x) \in \mathbb{Z}_p[x]$ be an irreducible polynomial of degree k. Then the quotient ring $\mathbb{Z}_p[x]/(f(x))$ is a finite field of order p^k.*

Example A.4.12 Since x^2+x+1 is irreducible in $\mathbb{Z}_2[x]$, it follows that $\mathbb{Z}_2[x]/(x^2+x+1)$ is a finite field of order four. The polynomials in $\mathbb{Z}_2[x]/(x^2+x+1)$ are $0, 1, x$ and $x+1$. The multiplicative inverse of x is $x+1$, since $x(x+1) = x^2+x = (x+1)+x = 1$.

Remark A.4.13 Multiplicative inverses in a finite field $\mathbb{Z}_p[x]/(f(x))$ can be computed using the EXTENDED EUCLIDEAN ALGORITHM FOR POLYNOMIALS. The idea is to modify the EUCLIDEAN ALGORITHM FOR POLYNOMIALS by computing a sequence of polynomials $r_i(x), q_i(x), s_i(x)$, and $t_i(x)$, analogous to modifying the EUCLIDEAN ALGORITHM to obtain the EXTENDED EUCLIDEAN ALGORITHM.

Example A.4.14 The EXTENDED EUCLIDEAN ALGORITHM FOR POLYNOMIALS can be used to express 1 as a combination of x^4+x+1 and x^3+x in $\mathbb{Z}_2[x]$:

$$(x^2+x+1)(x^4+x+1)+(x^3+x^2)(x^3+x)=1.$$

Example A.4.15 We observe that x^4+x+1 is an irreducible polynomial in $\mathbb{Z}_2[x]$. We noted in the previous example that $(x^2+x+1)(x^4+x+1)+(x^3+x^2)(x^3+x)=1$. Therefore, the inverse of x^3+x in $\mathbb{Z}_2[x]/(x^4+x+1)$ is x^3+x^2.

Remark A.4.16 For any polynomial $f(x) \in \mathbb{Z}_p[x]$ having degree k, the additive group $(\mathbb{Z}_p[x]/(f(x)), +)$ is isomorphic to $(\mathbb{Z}_p)^k$.

Theorem A.4.17 *All finite fields of a given order n are isomorphic.*

Remark A.4.18 We denote the unique (up to isomorphism) finite field of order q by $\mathbb{F}_q$.

Example A.4.19 The field $\mathbb{F}_8$ can be constructed as either $\mathbb{Z}_2[x]/(x^3+x+1)$ or $\mathbb{Z}_2[x]/(x^3+x^2+1)$, since both x^3+x+1 and x^3+x^2+1 are irreducible polynomials in $\mathbb{Z}_2[x]$. However, the two constructions yield isomorphic fields.

Theorem A.4.20 *The multiplicative group* $(\mathbb{F}_q\backslash\{0\},\cdot)$ *is cyclic.*

Definition A.4.21 A generator of $(\mathbb{F}_q\backslash\{0\},\cdot)$ is called a ***primitive element in*** $\mathbb{F}_q$.

Example A.4.22 The polynomial x is a primitive element of $\mathbb{F}_8 = \mathbb{Z}_2[x]/(x^3+x+1)$. The powers of x are as follows:

$$\begin{aligned}
x^0 &= 1\\
x^1 &= x\\
x^2 &= x^2\\
x^3 &= x+1\\
x^4 &= x^2+x\\
x^5 &= x^2+x+1\\
x^6 &= x^2+1.
\end{aligned}$$

Appendix B

Pseudorandom Bit Generation for Cryptography

B.1 Bit Generators

Remark B.1.1 There are many situations in cryptography where it is important to be able to generate random numbers, random bitstrings, etc. For example, cryptographic keys are normally generated uniformly at random from a specified keyspace, and many encryption schemes and signature schemes require random numbers to be generated during their execution. Generating random numbers nondeterministically by means of coin tosses or other physical processes is time-consuming and expensive. In practice, it is more common to use a ***pseudo-random bit generator***.

Definition B.1.2 (bit generator) Let k, ℓ be positive integers such that $\ell \geq k+1$. A ***(k, ℓ)-bit generator*** is a function $f : (\mathbb{Z}_2)^k \rightarrow (\mathbb{Z}_2)^\ell$ that can be computed in polynomial time (as a function of k). The input $s_0 \in (\mathbb{Z}_2)^k$ is called the ***seed***, and the output $f(s_0) \in (\mathbb{Z}_2)^\ell$ is called the ***generated bitstring***. It will always be required that ℓ is a polynomial function of k.

Remark B.1.3 The bit generator f is deterministic, so the bitstring $f(s_0)$ is dependent only on the seed.

Example B.1.4 A linear feedback shift register (LFSR), as described in Section 2.1.7, can be thought of as a bit generator. Given a k-bit seed, an LFSR of degree k can be used to produce as many as $2^k - k - 1$ further bits before repeating.

Example B.1.5 More generally, any keystream generator for a synchronous stream cipher is an example of a bit generator. Examples include the ***combination generator***, the ***filter generator***, and the ***shrinking generator***. (These are discussed in Section 4.8.)

Remark B.1.6 Roughly speaking, a bit generator takes a short random seed and expands it to a long string of random-looking bits. On the other hand, a hash function takes a (possibly) long string of input bits (which may or may not be random) and shrinks it to a short random-looking output. Hash functions are therefore often used as key derivation functions, which were discussed in Section 12.3. A typical setting is one where Alice and Bob have agreed on a 2048-bit shared secret value, say using the *Diffie-Hellman KAS*. Alice and Bob want to obtain a 128-bit *AES* key, so the 2048-bit shared secret value could be the input to an appropriate key derivation function. The 128-bit output would comprise the *AES* key.

Definition B.1.7 (linear congruential generator) Suppose M is a positive integer and suppose $1 \leq a, b \leq M-1$. Define $k = \lceil \log_2 M \rceil$ and let ℓ be chosen such that $k+1 \leq \ell \leq M-1$. The seed is an integer s_0, where $0 \leq s_0 \leq M-1$ (observe that the binary representation of a seed is a bitstring of length not exceeding k). Now, define

$$s_i = (as_{i-1} + b) \bmod M$$

for $1 \leq i \leq \ell$, and then define

$$f(s_0) = (z_1, z_2, \ldots, z_\ell),$$

where

$$z_i = s_i \bmod 2,$$

$1 \leq i \leq \ell$. Then f is a (k, ℓ)*-linear congruential generator.*

Example B.1.8 Suppose we construct a $(5, 10)$-bit generator by taking $M = 31$, $a = 3$, and $b = 5$ in the *linear congruential generator*. Suppose we consider the mapping $s \mapsto 3s + 5 \bmod 31$. Then $13 \mapsto 13$, and the other 30 residues are permuted in a cycle of length 30, namely

$$0, 5, 20, 3, 14, 16, 22, 9, 1, 8, 29, 30, 2, 11, 7, 26,$$
$$21, 6, 23, 12, 10, 4, 17, 25, 18, 28, 27, 24, 15, 19.$$

If the seed s_0 is anything other than 13, then the seed specifies a starting point in this cycle, and the next 10 elements, reduced modulo 2, form the sequence $s_1, s_2, \ldots$.

The 31 possible bitstrings produced by this generator are shown in Table B.1. For example, the sequence constructed from the seed 0 is obtained by taking the ten integers following 0 in the above list, namely, $5, 20, 3, 14, 16, 22, 9, 1, 8, 29$, and reducing them modulo 2.

Definition B.1.9 (RSA generator) Suppose p, q are two $(k/2)$-bit primes, and define $n = pq$. Suppose b is chosen such that $\gcd(b, \phi(n)) = 1$. As always, n and b are public while p and q are secret. A seed s_0 is any element of $\mathbb{Z}_n{}^*$, so s_0 has k bits. For $i \geq 1$, define

$$s_{i+1} = s_i{}^b \bmod n,$$

and then define

$$f(s_0) = (z_1, z_2, \ldots, z_\ell),$$

where

$$z_i = s_i \bmod 2,$$

$1 \leq i \leq \ell$. Then f is a (k, ℓ)*-RSA generator.*

TABLE B.1: Bitstrings produced by the *linear congruential generator*

s_0	sequence	s_0	sequence
0	1010001101	16	0110100110
1	0100110101	17	1001011010
2	1101010001	18	0101101010
3	0001101001	19	0101000110
4	1100101101	20	1000110100
5	0100011010	21	0100011001
6	1000110010	22	1101001101
7	0101000110	23	0001100101
8	1001101010	24	1101010001
9	1010011010	25	0010110101
10	0110010110	26	1010001100
11	1010100011	27	0110101000
12	0011001011	28	1011010100
13	1111111111	29	0011010100
14	0011010011	30	0110101000
15	1010100011		

TABLE B.2: Bitstrings produced by the *RSA generator*

i	s_i	z_i	i	s_i	z_i	i	s_i	z_i
0	75634		1	31483	1	2	31238	0
3	51968	0	4	39796	0	5	28716	0
6	14089	1	7	5923	1	8	44891	1
9	62284	0	10	11889	1	11	43467	1
12	71215	1	13	10401	1	14	77444	0
15	56794	0	16	78147	1	17	72137	1
18	89592	0	19	29022	0	20	13356	0

Example B.1.10 We now give an example of the *RSA generator*. Suppose $n = 91261 = 263 \times 347$, $b = 1547$, and $s_0 = 75634$. The first 20 bits produced by the *RSA generator* are shown in Table B.2.

The bitstring resulting from this seed is

$$10000111011110011000.$$

Definition B.1.11 (Blum-Blum-Shub (BBS) generator) Let p, q be two $(k/2)$-bit primes such that $p \equiv q \equiv 3 \bmod 4$, and define $n = pq$. Let $\mathbf{QR}(n)$ denote the set of quadratic residues modulo n. A seed s_0 is any element of $\mathbf{QR}(n)$. For $0 \leq i \leq \ell - 1$, define

$$s_{i+1} = s_i^2 \bmod n,$$

TABLE B.3: Bitstrings produced by the *BBS generator*

i	s_i	z_i	i	s_i	z_i	i	s_i	z_i
0	20749		1	143135	1	2	177671	1
3	97048	0	4	89992	0	5	174051	1
6	80649	1	7	45663	1	8	69442	0
9	186894	0	10	177046	0	11	137922	0
12	123175	1	13	8630	0	14	114386	0
15	14863	1	16	133015	1	17	106065	1
18	45870	0	19	137171	1	20	48060	0

and then define

$$f(s_0) = (z_1, z_2, \ldots, z_\ell),$$

where

$$z_i = s_i \bmod 2,$$

$1 \leq i \leq \ell$.

Remark B.1.12 One way to choose an appropriate seed for the *BBS generator* is to select an element $s_{-1} \in \mathbb{Z}_n{}^*$ and compute $s_0 = (s_{-1})^2 \bmod n$. This ensures that $s_0 \in \mathbf{QR}(n)$.

Remark B.1.13 In the *BBS generator*, given a seed $s_0 \in \mathbf{QR}(n)$, we compute the sequence $s_1, s_2, \ldots, s_\ell$ by successive squarings modulo n, and then we reduce each s_i modulo 2 to obtain z_i. It follows that

$$z_i = \left(s_0{}^{2^i} \bmod n\right) \bmod 2,$$

$1 \leq i \leq \ell$.

Example B.1.14 Here is an example of the *BBS generator*. Suppose $n = 192649 = 383 \times 503$ and $s_0 = 101355^2 \bmod n = 20749$. The first 20 bits produced by the *BBS generator* are shown in Table B.3. Hence the bitstring resulting from this seed is

$$11001110000100111010.$$

Remark B.1.15 There are three "Deterministic Random Bit Generators" that were recommended by NIST in June 2015. They are denoted by *Hash_DRBG* (based on a hash function), *HMAC_DRBG* (based on *HMAC*), and *CTR_DRBG* (based on block cipher encryption in counter mode). We describe the basic methodology (with some simplifications) of these three generators now.

Definition B.1.16 *Hash_DRBG* uses a hash function h and a seed s_0. Then the generator outputs

$$h(s_0) \parallel h(s_0+1) \parallel h(s_0+2) \cdots .$$

Definition B.1.17 *HMAC_DRBG* is based on *HMAC* (section 5.5.1), which uses a key K. Let s_0 be the seed. Define $s_{i+1} = HMAC_K(s_i)$ for $i = 0, 1, \ldots$. Then the generator outputs

$$s_1 \parallel s_2 \parallel s_3 \cdots .$$

Definition B.1.18 *CTR_DRBG* uses an encryption function e_K for a secret-key cryptosystem such as *AES*, as well as a seed s_0. Then the generator outputs

$$e_K(s_0) \parallel e_K(s_0 + 1) \parallel e_K(s_0 + 2) \cdots .$$

Remark B.1.19 *Dual_EC_DRBG* was a generator that was recommended for use by NIST in 2006, along with the other three generators mentioned in Remark B.1.15. Unlike the other three generators, *Dual_EC_DRBG* is not based on a symmetric key primitive; rather, it is based on arithmetic in certain elliptic curves.

Definition B.1.20 *Dual_EC_DRBG* has system parameters consisting of an elliptic curve E defined by an equation $y^2 = x^3 + ax + b$ over $\mathbb{F}_p$, and two points on E denoted by P and Q. The length of p is typically 192 bits or 256 bits.

For a point $R = (x, y) \in E$, let $\mathbf{X}(R) = x$, so $\mathbf{X}(R)$ denotes the x-co-ordinate of a point on the elliptic curve. An element $x \in \mathbb{F}_p$ will be represented as an n-bit binary string, where $n = \log_2 p$ (so $n = 192$ or 256). For a positive integer $m \leq n$, let $\mathbf{Trunc}(x, m)$ denote the $n - m$ low-order bits of x. *Dual_EC_DRBG* uses the value $m = 16$.

The generator will output a sequence $r_1, r_2, \ldots$, where $r_i \in \{0,1\}^{n-m}$, for $i = 1, 2, \ldots$. The *internal state* of the generator at time i is denoted by s_i. The initial value s_0 is the seed. The r_i and s_i values are computed using the following relations:

$$\begin{aligned} s_i &= \mathbf{X}(s_{i-1}P) \\ r_i &= \mathbf{Trunc}(\mathbf{X}(s_iQ), m), \end{aligned}$$

for $i \geq 1$.

Remark B.1.21 *Dual_EC_DRBG* can be proven to be secure, depending on certain computational assumptions. However, *Dual_EC_DRBG* was very controversial due to the involvement of the NSA in its design, as well as the fact that the system parameters of the generator could easily be configured to contain a trapdoor. The trapdoor would allow someone who does not know the value of the seed to successfully predict future bits, after observing a relatively small number of initial bits produced by the generator. The trapdoor in *Dual_EC_DRBG* is the discrete logarithm a such that $P = aQ$. It turns out that knowledge of the value of a compromises the security of the generator. Note that it is straightforward for the entity who sets up the generator to do it in such a way that they know the value of the trapdoor a.

Remark B.1.22 The Snowden leaks in 2013 included the revelation of the NSA *Bullrun* program, whose purpose was "to covertly introduce weaknesses into the encryption standards followed by hardware and software developers around the world." NIST finally removed *Dual_EC_DRBG* as a recommended bit generator in April 2014, probably due to the increased controversy that resulted from the Snowden leaks.

Example B.1.23 Suppose we are given a value r_i that is output by the DRBG. We can compute a list of 2^{16} "candidates" for $\mathbf{X}(s_iQ)$ simply by enumerating all combinations of the 16 missing bits. For each candidate, determine if it is the x-coordinate of a point on E. We will then get a list of at most 2^{17} "candidate points." One of these candidate points is s_iQ.

In order to compute r_{i+1}, we need to know s_{i+1}, which is computed as $s_{i+1} = \mathbf{X}(s_iP)$. However, because $P = aQ$, we have

$$s_iP = as_iQ, \tag{B.1}$$

where, as noted above, s_iQ is one of the candidate points. Thus we have

$$\begin{aligned} r_{i+1} &= \mathbf{Trunc}(\mathbf{X}(s_{i+1}Q)) \\ &= \mathbf{Trunc}(\mathbf{X}(\mathbf{X}(s_iP)Q)) \\ &= \mathbf{Trunc}(\mathbf{X}(\mathbf{X}(as_iQ)Q). \end{aligned} \tag{B.2}$$

The value s_iQ in equation (B.2) is one of the candidate points.

Now, suppose we are also given r_{i+1}. With high probability, this allows us to identify which candidate point is the correct one. Thus we can determine the correct value of s_iQ and s_iP from (B.1). Since $s_{i+1} = \mathbf{X}(s_iP)$, we can compute the internal state at time $i+1$ and hence we can compute all subsequent values (and internal states) of the generator.

B.2 Security of Pseudorandom Bit Generators

Remark B.2.1 Intuitively, a pseudorandom bit generator is cryptographically secure if a string of ℓ bits produced by the generator appears to be "random." That is, it should be impossible in an amount of time that is polynomial in k (equivalently, polynomial in ℓ) to distinguish a string of ℓ bits produced by a PRBG from a string of ℓ truly random bits.

Remark B.2.2 The *linear congruential generator* is not cryptographically secure. Keystream generators for asynchronous stream ciphers are intended to be secure, but most schemes used in practice do not have proofs of security. The *RSA generator* and *BBS generator* are cryptographically secure, provided that certain computational assumptions hold. However, they are not often used in practice because their efficiency is much lower than alternative designs.

Remark B.2.3 The notion of a string of bits appearing to be "random" is formalized using the concept of indistinguishability, which is defined now.

Definition B.2.4 (distinguishability of probability distributions) Suppose p_0 and p_1 are two probability distributions on the set $(\mathbb{Z}_2)^\ell$ of all bitstrings of length ℓ. For $j = 0, 1$ and $z^\ell \in (\mathbb{Z}_2)^\ell$, the quantity $p_j(z^\ell)$ denotes the probability of the string z^ℓ occurring in the distribution p_j.

Let $\mathbf{dst} : (\mathbb{Z}_2)^\ell \rightarrow \{0, 1\}$ be a function and let $\epsilon > 0$. For $j = 0, 1$, define

$$E_{\mathbf{dst}}(p_j) = \sum_{\{z^\ell \in (\mathbb{Z}_2)^\ell \,:\, \mathbf{dst}(z^\ell) = 1\}} p_j(z^\ell).$$

The quantity $E_{\mathbf{dst}}(p_j)$ represents the average (i.e., expected) value of the output of **dst** over the probability distribution p_j, for $j = 0, 1$.

We say that **dst** is an ϵ-***distinguisher*** of p_0 and p_1 provided that

$$|E_{\mathbf{dst}}(p_0) - E_{\mathbf{dst}}(p_1)| \geq \epsilon,$$

and we say that p_0 and p_1 are ϵ-***distinguishable*** if there exists an ϵ-distinguisher of p_0 and p_1.

A distinguisher, say **dst**, is a ***polynomial-time distinguisher*** provided that $\mathbf{dst}(z^\ell)$ can be computed in polynomial time as a function of ℓ.

Remark B.2.5 The intuition behind the definition of a distinguisher is as follows. The function (or algorithm) **dst** tries to decide if a given bitstring z^ℓ of length ℓ is more likely to have arisen from probability distribution p_0 or from probability distribution p_1. The output $\mathbf{dst}(z^\ell)$ represents the distinguisher's guess as to which of these two probability distributions is more likely to have produced z^ℓ. Then **dst** is an ϵ-distinguisher provided that the values of these two expectations are at least ϵ apart.

Remark B.2.6 We next describe one particular method that can potentially be used to distinguish a random string from a nonrandom string.

Definition B.2.7 A ***next bit predictor*** is defined as follows. Let f be a (k, ℓ)-bit generator. Suppose $1 \leq i \leq \ell - 1$, and we have a function $\mathbf{nbp} : (\mathbb{Z}_2)^{i-1} \rightarrow \mathbb{Z}_2$, which takes as input an $(i-1)$-tuple $z^{i-1} = (z_1, \ldots, z_{i-1})$. This $(i-1)$-tuple represents the first $i-1$ bits produced by f (given an unknown, random, k-bit seed). Then the function **nbp** attempts to predict the next bit produced by f, namely, z_i. We say that the function **nbp** is an ϵ-***i-th bit predictor*** if **nbp** can predict the i-th bit of the generated bitstring (given the first $i-1$ bits) with probability at least $1/2 + \epsilon$, where $\epsilon > 0$.

Remark B.2.8 Example B.1.23 basically describes a next bit predictor for *Dual_EC_DRBG*.

Remark B.2.9 The reason for the expression $1/2 + \epsilon$ in this definition is that any "reasonable" predicting algorithm will predict any bit of a truly random bitstring with probability $1/2$. If a generated bitstring is not truly random, then it may be possible to predict a given bit with higher probability. (Note that it is unnecessary to consider functions that predict a given bit with probability less than $1/2$, because in this case a function that replaces every prediction z by $1 - z$ will predict the bit with probability greater than $1/2$.)

Remark B.2.10 One of the main results in the theory of pseudorandom bit generators, due to Yao, is that a next bit predictor is a ***universal test***. Roughly speaking, a bit generator is "secure" if and only if there does not exist any polynomial-time ϵ-i-th bit predictor for the generator, except perhaps for very small values of ϵ.

Remark B.2.11 The security of the *Blum-Blum-Shub generator* can be proven, assuming the intractability of the **Composite Quadratic Residues** problem, which was defined as Problem 13.1.

B.3 Notes and References

A thorough treatment of pseudorandom bit generators can be found in the book by Luby [124]. Knuth [110] discusses random number generation (mostly in a non-cryptographic context) in considerable detail.

Properties of the *RSA Generator* are studied in Alexi, Chor, Goldreich, and Schnorr [3]. The *BBS Generator* is described by Blum, Blum, and Shub in [37].

The four generators originally recommended by NIST are found in the 2012 publication [6]; however, note that earlier versions of this publication date back to 2006. This publication was superceded by [7] in 2015, in which *Dual_EC_DRBG* was withdrawn.

The *Dual_EC_DRBG* trapdoor was first pointed out by Shumow and Ferguson at the CRYPTO 2007 Rump Session [180]. For a nice discussion of the *Dual_EC_DRBG* trapdoor, including some of the political context, see Hales [91].

A security proof for *Dual_EC_DRBG*, assuming certain values of the system parameters, has been given by Brown and Gjøsteen [49].

The basic theory of secure pseudorandom bit generators is due to Yao [206], who proved the universality of the next bit test.

Bibliography

[1] CARLISLE ADAMS AND STEVE LLOYD. *Understanding PKI: Concepts, Standards, and Deployment Considerations, Second Edition*. Addison Wesley, 2003.

[2] LEONARD ADLEMAN. A subexponential algorithm for the discrete logarithm problem with applications to cryptography. In *20th Annual Symposium on Foundations of Computer Science*, pages 55–60. IEEE, 1979.

[3] WERNER ALEXI, BENNY CHOR, ODED GOLDREICH, AND CLAUS SCHNORR. RSA and Rabin functions: certain parts are as hard as the whole. *SIAM Journal on Computing*, **17** (1988), 194–209.

[4] JEE HEA AN, YEVGENIY DODIS, AND TAL RABIN. On the security of joint signature and encryption. *Lecture Notes in Computer Science*, **2332** (2002), 83–107. (EUROCRYPT 2002.)

[5] RAZVAN BARBULESCU, PIERRICK GAUDRY, ANTOINE JOUX, AND EMMANUEL THOMÉ. A heuristic quasi-polynomial algorithm for discrete logarithm in finite fields of small characteristic. *Lecture Notes in Computer Science*, **8441** (2014), 1–16. (EUROCRYPT 2014.)

[6] ELAINE BARKER AND JOHN KELSEY. *Recommendation for random number generation using deterministic random bit generators*. National Institute of Standards and Technology (NIST) Special Publication 800-90A, 2012.

[7] ELAINE BARKER AND JOHN KELSEY. *Recommendation for random number generation using deterministic random bit generators*. National Institute of Standards and Technology (NIST) Special Publication 800-90A, Revision 1, 2015.

[8] ELAINE BARKER AND ALLEN ROGINSKY. *Transitions: Recommendation for Transitioning the Use of Cryptographic Algorithms and Key Lengths*. National Institute of Standards and Technology (NIST) Special Publication 800-131A, revision 1, 2015.

[9] ELAINE BARKER, LILY CHEN, AND RICH DAVIS. *Recommendation for Key-Derivation Methods in Key-Establishment Schemes*. Draft National Institute of Standards and Technology (NIST) Special Publication 800-56C, revision 1, August 2017.

[10] FRIEDRICH BAUER. *Decrypted Secrets: Methods and Maxims of Cryptology, Second Edition*. Springer, 2000.

[11] PIERRE BEAUCHEMIN AND GILLES BRASSARD. A generalization of Hellman's extension to Shannon's approach to cryptography. *Journal of Cryptology*, **1** (1988), 129–131.

[12] PIERRE BEAUCHEMIN, GILLES BRASSARD, CLAUDE CRÉPEAU, CLAUDE GOUTIER, AND CARL POMERANCE. The generation of random numbers that are probably prime. *Journal of Cryptology*, **1** (1988), 53–64.

[13] HENRY BEKER AND FRED PIPER. *Cipher Systems: The Protection of Communications*. John Wiley and Sons, 1983.

[14] MIHIR BELLARE, SHAFI GOLDWASSER, AND DANIELE MICCIANCIO. "Pseudo-random" number generation within cryptographic algorithms: the DSS case. *Lecture Notes in Computer Science*, **1294** (1997), 277–292. (CRYPTO '97.)

[15] MIHIR BELLARE, JOE KILIAN, AND PHILLIP ROGAWAY. The security of the cipher block chaining message authentication code. *Journal of Computer and System Sciences*, **61** (2000), 362–399.

[16] MIHIR BELLARE AND CHANATHIP NAMPREMPRE. Authenticated encryption: relations among notions and analysis of the generic composition paradigm. *Lecture Notes in Computer Science* **1976**, (2000), 531–545. (ASIACRYPT 2000.)

[17] MIHIR BELLARE AND ADRIAN PALACIO. GQ and Schnorr identification schemes: proofs of security against impersonation under active and concurrent attacks. *Lecture Notes in Computer Science*, **2442** (2002), 162–177. (CRYPTO 2002.)

[18] MIHIR BELLARE AND PHILLIP ROGAWAY. Entity authentication and key distribution. *Lecture Notes in Computer Science*, **773** (1994), 232–249. (CRYPTO '93.)

[19] MIHIR BELLARE AND PHILLIP ROGAWAY. Optimal asymmetric encryption. *Lecture Notes in Computer Science*, **950** (1995), 92–111. (EUROCRYPT '94.)

[20] MIHIR BELLARE AND PHILLIP ROGAWAY. Provably secure session key distribution: the three party case. In *27th Annual ACM Symposium on Theory of Computing*, pages 57–66. ACM Press, 1995.

[21] MIHIR BELLARE AND PHILLIP ROGAWAY. The exact security of digital signatures: how to sign with RSA and Rabin. *Lecture Notes in Computer Science*, **1070** (1996), 399–416. (EUROCRYPT '96.)

[22] MIHIR BELLARE AND PHILLIP ROGAWAY. Random oracles are practical: a paradigm for designing efficient protocols. In *First ACM Conference on Computer and Communications Security*, pages 62–73. ACM Press, 1993.

[23] DANIEL BERNSTEIN, TANJA LANGE, AND CHRISTIANE PETERS. Attacking and defending the McEliece cryptosystem. *Lecture Notes in Computer Science,* **5299** (2008), 31–46. (PQCrypto 2008.)

[24] DANIEL BERNSTEIN, JOHANNES BUCHMANN, AND ERIK DAHMEN, EDS. *Post-quantum Cryptography.* Springer, 2009.

[25] DANIEL BERNSTEIN AND TANJA LANGE. Post-quantum cryptography. *Nature* **549** (2017), 188–194.

[26] GUIDO BERTONI, JOAN DAEMEN, MICHAËL PEETERS, AND GILLES VAN ASSCHE. Sponge functions. Ecrypt Hash Workshop 2007, May 2007. Available from `https://keccak.team/files/SpongeFunctions.pdf`.

[27] GUIDO BERTONI, JOAN DAEMEN, MICHAËL PEETERS, AND GILLES VAN ASSCHE. The Keccak SHA-3 submission, January 2011. Available from `https://keccak.team/files/Keccak-submission-3.pdf`.

[28] ALBRECHT BEUTELSPACHER. *Cryptology.* Mathematical Association of America, 1994.

[29] ELI BIHAM AND ADI SHAMIR. Differential cryptanalysis of DES-like cryptosystems. *Journal of Cryptology,* **4** (1991), 3–72.

[30] ALEX BIRYUKOV, DANIEL DINU, AND DMITRY KHOVRATOVICH. Argon2: new generation of memory-hard functions for password hashing and other applications. In *IEEE European Symposium on Security and Privacy,* pages 292–302. IEEE, 2016.

[31] ALEX BIRYUKOV, ORR DUNKELMAN, NATHAN KELLER, DMITRY KHOVRATOVICH, AND ADI SHAMIR. Key recovery attacks of practical complexity on AES-256 variants with up to 10 rounds. *Lecture Notes in Computer Science,* **6110** (2010), 299–319. (EUROCRYPT 2010.)

[32] JOHN BLACK, SHAI HALEVI, HUGO KRAWCZYK, TED KROVETZ, AND PHILLIP ROGAWAY. UMAC: fast and secure message authentication. *Lecture Notes in Computer Science,* **1666** (1999), 216–233. (CRYPTO '99.)

[33] SIMON BLAKE-WILSON AND ALFRED MENEZES. Entity authentication and authenticated key transport protocols employing asymmetric techniques. *Lecture Notes in Computer Science,* **1361** (1998), 137–158. (Fifth International Workshop on Security Protocols.)

[34] SIMON BLAKE-WILSON AND ALFRED MENEZES., Authenticated Diffie-Hellman key agreement protocols. *Lecture Notes in Computer Science,* **1556** (1999), 339–361. (Selected Areas in Cryptography '98.)

[35] G. R. (BOB) BLAKLEY. Safeguarding cryptographic keys. *Federal Information Processing Standard Conference Proceedings,* **48** (1979), 313–317.

[36] R. BLOM. An optimal class of symmetric key generation schemes. *Lecture Notes in Computer Science*, **209** (1985), 335–338. (EUROCRYPT '84.)

[37] LENORE BLUM, MANUEL BLUM, AND MICHAEL SHUB. A simple unpredictable random number generator. *SIAM Journal on Computing*, **15** (1986), 364–383.

[38] CARLO BLUNDO, ALFREDO DE SANTIS, AMIR HERZBERG, SHAY KUTTEN, UGO VACCARO, AND MOTI YUNG. Perfectly-secure key distribution for dynamic conferences. *Lecture Notes in Computer Science*, **740** (1993), 471–486. (CRYPTO '92.)

[39] ANDREY BOGDANOV, DMITRY KHOVRATOVICH, AND CHRISTIAN RECHBERGER. Biclique cryptanalysis of the full AES. *Lecture Notes in Computer Science*, **7073** (2011), 344–371. (ASIACRYPT 2011.)

[40] DAN BONEH. The decision Diffie-Hellman problem. *Lecture Notes in Computer Science*, **1423** (1998), 48–63. (Proceedings of the Third Algorithmic Number Theory Symposium.)

[41] DAN BONEH AND GLENN DURFEE. Cryptanalysis of RSA with private key d less than $N^{0.292}$. *IEEE Transactions on Information Theory*, **46** (2000), 1339–1349.

[42] DAN BONEH AND MATTHEW FRANKLIN. Identity-based encryption from the Weil pairing. *Lecture Notes in Computer Science*, **2139** (2001), 213–229. (CRYPTO 2001.)

[43] DAN BONEH AND JAMES SHAW. Collusion-secure fingerprinting for digital data. *IEEE Transactions on Information Theory*, **44** (1998), 1897–1905.

[44] COLIN BOYD AND ANISH MATHURIA. *Protocols for Authentication and Key Establishment*. Springer, 2003.

[45] GILLES BRASSARD AND PAUL BRATLEY. *Fundamentals of Algorithmics*. Prentice Hall, 1995.

[46] RICHARD BRENT. An improved Monte Carlo factorization method. *BIT*, **20** (1980), 176–184.

[47] DAVID BRESSOUD AND STAN WAGON. *A Course in Computational Number Theory*. Wiley, 2008.

[48] JON BRODKIN. Kim Dotcom claims he invented two-factor authentication but he wasn't first. *Ars Technica*, May 23, 2013. `https://arstechnica.com/information-technology/2013/05/kim-dotcom-claims-he-invented-two-factor-authentication-but-he-wasnt-first/`

[49] DANIEL R. L. BROWN AND KRISTIAN GJØSTEEN. A security analysis of the NIST SP 800-90 elliptic curve random number generator. *Lecture Notes in Computer Science*, **4622** (2007), 466–481. (CRYPTO 2007.)

[50] JOHANNES BUCHMANN, ERIK DAHMEN, AND ANDREAS HÜLSING. XMSS — a practical forward secure signature scheme based on minimal security assumptions. *Lecture Notes in Computer Science*, **7071** (2011), 117–129. (PQCrypto 2011.)

[51] MIKE BURMESTER. On the risk of opening distributed keys. *Lecture Notes in Computer Science*, **839** (1994), 308–317 (CRYPTO '94.)

[52] MIKE BURMESTER AND YVO DESMEDT. A secure and efficient conference key distribution system. *Lecture Notes in Computer Science*, **950** (1994), 275–286 (EUROCRYPT '94.)

[53] DAVID BURTON. *Elementary Number Theory, 7th Edition.* McGraw-Hill, 2010.

[54] R. CANETTI AND H. KRAWCZYK. Analysis of key-exchange protocols and their use for building secure channels. *Lecture Notes in Computer Science*, **2045** (2001), 453–474 (EUROCRYPT 2001.)

[55] J. LAWRENCE CARTER AND MARK WEGMAN. Universal classes of hash functions. *Journal of Computer and System Sciences*, **18** (1979), 143–154.

[56] MARK CHATEAUNEUF, ALAN LING, AND DOUGLAS STINSON. Slope packings and coverings, and generic algorithms for the discrete logarithm problem. *Journal of Combinatorial Designs*, **11** (2003), 36–50.

[57] BENNY CHOR, AMOS FIAT, MONI NAOR, AND BENNY PINKAS. Tracing traitors. *IEEE Transactions on Information Theory*, **46** (2000), 893–910.

[58] CARLOS CID, SEAN MURPHY, AND MATTHEW ROBSHAW. *Algebraic Aspects of the Advanced Encryption Standard.* Springer, 2006.

[59] CLIFFORD COCKS. An identity based encryption scheme based on quadratic residues. *Lecture Notes in Computer Science*, **2260** (2001), 360–363. (Eighth IMA International Conference on Cryptography and Coding.)

[60] KATRIEL COHN-GORDON, CAS CREMERS, BENJAMIN DOWLING, LUKE GARRATT, AND DOUGLAS STEBILA. A formal security analysis of the Signal messaging protocol. Cryptology ePrint Archive: Report 2016/1013. `https://eprint.iacr.org/2016/1013.pdf`

[61] DON COPPERSMITH. Fast evaluation of logarithms in fields of characteristic two *IEEE Transactions on Information Theory* **30** (1984), 587–594.

[62] NICOLAS T. COURTOIS. Fast algebraic attacks on stream ciphers with linear feedback. *Lecture Notes in Computer Science*, **2729** (2003), 176–194. (CRYPTO 2003.)

[63] JOAN DAEMEN AND VINCENT RIJMEN. *The Design of Rijndael: AES — The Advanced Encryption Standard*. Springer, 2002.

[64] IVAN DAMGÅRD. A design principle for hash functions. *Lecture Notes in Computer Science*, **435** (1990), 416–427. (CRYPTO '89.)

[65] CHRISTOPHE DE CANNIÈRE. Trivium: a stream cipher construction inspired by block cipher design principles. *Lecture Notes in Computer Science*, **4176** (2006), 171–186. (International Conference on Information Security, ISC 2006.)

[66] CHRISTOPHE DE CANNIÈRE AND BART PRENEEL. Trivium: a stream cipher construction inspired by block cipher design principles. *eSTREAM submitted papers*, available from `http://www.ecrypt.eu.org/stream/papersdir/2006/021.pdf`.

[67] JOHN DELAURENTIS. A further weakness in the common modulus protocol for the RSA cryptosystem. *Cryptologia*, **8** (1984), 253–259.

[68] DOROTHY DENNING AND GIOVANNI SACCO. Timestamps in key distribution protocols. *Communications of the ACM*, **24** (1981), 533–536.

[69] WHITFIELD DIFFIE. The first ten years of public-key cryptography. In *Contemporary Cryptology, The Science of Information Integrity*, pages 135–175. IEEE Press, 1992.

[70] WHITFIELD DIFFIE AND MARTIN HELLMAN. Multiuser cryptographic techniques. *Federal Information Processing Standard Conference Proceedings*, **45** (1976), 109–112.

[71] WHITFIELD DIFFIE AND MARTIN HELLMAN. New directions in cryptography. *IEEE Transactions on Information Theory*, **22** (1976), 644–654.

[72] WHITFIELD DIFFIE, PAUL VAN OORSCHOT, AND MICHAEL WIENER. Authentication and authenticated key exchanges. *Designs, Codes and Cryptography*, **2** (1992), 107–125.

[73] JINTAI DING AND DIETER SCHMIDT. Rainbow, a new multivariable polynomial signature scheme. *Lecture Notes in Computer Science*, **3531** (2005), 164–175. (ACNS 2005.)

[74] JINTAI DING, XIANG XIE, AND XIAODONG LIN. A simple provably secure key exchange scheme based on the learning with errors problem. Cryptology ePrint Archive: Report 2012/688. `https://eprint.iacr.org/2012/688.pdf`

[75] CHRIS DODS, NIGEL SMART, AND MARTIJN STAM. Hash based digital signature schemes. *Lecture Notes in Computer Science*, **3796** (2006), 96–116. (Cryptography and Coding 2005.)

[76] MORRIS DWORKIN. *Recommendation for Block Cipher Modes of Operation: Methods and Techniques*. National Institute of Standards and Technology (NIST) Special Publication 800-38A, 2001.

[77] MORRIS DWORKIN. *Recommendation for Block Cipher Modes of Operation: the CMAC Mode for Authentication*. National Institute of Standards and Technology (NIST) Special Publication 800-38B, 2005 (updated 2016).

[78] MORRIS DWORKIN. *Recommendation for Block Cipher Modes of Operation: the CCM Mode for Authentication and Confidentiality*. National Institute of Standards and Technology (NIST) Special Publication 800-38D, 2004.

[79] MORRIS DWORKIN. *Recommendation for Block Cipher Modes of Operation: Galois/Counter Mode (GCM) and GMAC*. National Institute of Standards and Technology (NIST) Special Publication 800-38D, 2007.

[80] TAHER ELGAMAL. A public key cryptosystem and a signature scheme based on discrete logarithms. *IEEE Transactions on Information Theory*, **31** (1985), 469–472.

[81] ANDREAS ENGE. Bilinear pairings on elliptic curves. ArXiv report 1301.5520v2, Feb. 15, 2014. `https://arxiv.org/abs/1301.5520v2`.

[82] URIEL FEIGE, AMOS FIAT, AND ADI SHAMIR. Zero-knowledge proofs of identity. *Journal of Cryptology*, **1** (1988), 77–94.

[83] AMOS FIAT AND ADI SHAMIR. How to prove yourself: practical solutions to identification and signature problems. *Lecture Notes in Computer Science*, **263** (1987), 186–194. (CRYPTO '86.)

[84] STEPHEN GALBRAITH. *Mathematics of Public Key Cryptography*. Cambridge University Press, 2012.

[85] STEPHEN GALBRAITH AND PIERRICK GAUDRY. Recent progress on the elliptic curve discrete logarithm problem. *Designs, Codes and Cryptography*, **78** (2016), 51–72.

[86] CRAIG GENTRY. Fully homomorphic encryption using ideal lattices. In *41st Annual Symposium on Theory of Computing*, pages 169–178. ACM, 2009.

[87] EDGAR N. GILBERT, F. JESSIE MACWILLIAMS, AND NEIL J. A. SLOANE. Codes which detect deception. *Bell Systems Technical Journal*, **53** (1974), 405–424.

[88] CHARLES GOLDIE AND RICHARD PINCH. *Communication Theory*. Cambridge University Press, 1991.

[89] SHAFI GOLDWASSER AND SILVIO MICALI. Probabilistic encryption. *Journal of Computer and Systems Science*, **28** (1984), 270–299.

[90] SHAFI GOLDWASSER, SILVIO MICALI, AND PO TONG. Why and how to establish a common code on a public network. In *23rd Annual Symposium on the Foundations of Computer Science*, pages 134–144. IEEE Press, 1982.

[91] THOMAS C. HALES. The NSA Back Door to NIST. *Notices of the AMS*, **61** (2014), 190–192.

[92] DARREL HANKERSON, ALFRED MENEZES, AND SCOTT VANSTONE. *Guide to Elliptic Curve Cryptography*. Springer, 2004.

[93] HOWARD M. HEYS. A tutorial on linear and differential cryptanalysis. *Cryptologia*, **26** (2002), 189–221.

[94] HOWARD M. HEYS AND STAFFORD E. TAVARES. Substitution-permutation networks resistant to differential and linear cryptanalysis. *Journal of Cryptology*, **9** (1996), 1–19.

[95] M. JASON HINEK. *Cryptanalysis of RSA and Its Variants*. Chapman and Hall/CRC, 2009.

[96] JEFFREY HOFFSTEIN, JILL PIPHER, AND JOSEPH SILVERMAN. *An Introduction to Mathematical Cryptography*. Springer, 2008.

[97] HENK HOLLMANN, JACK VAN LINT, JEAN-PAUL LINNARTZ, AND LUDO TOLHUIZEN. On codes with the identifiable parent property. *Journal of Combinatorial Theory A*, **82** (1998), 121–133.

[98] W. CARY HUFFMAN AND VERA PLESS. *Fundamentals of Error-Correcting Codes*. Cambridge University Press, 2003.

[99] TETSU IWATA AND KAORU KUROSAWA. OMAC: one-key CBC MAC. *Lecture Notes in Computer Science*, **2887** (2003), 129–153. (Fast Software Encryption 2003.)

[100] DON JOHNSON, ALFRED MENEZES, AND SCOTT VANSTONE. The elliptic curve digital signature algorithm (ECDSA). *International Journal on Information Security*, **1** (2001), 36–63.

[101] ANTOINE JOUX. A new index calculus algorithm with complexity $L(1/4 + o(1))$ in small characteristic. *Lecture Notes in Computer Science*, **8282** (2014), 355–379. (Selected Areas in Cryptography 2013.)

[102] ANTOINE JOUX, ANDREW ODLYZKO, AND CÉCILE PIERRO. The past, evolving present, and future of the discrete logarithm. In *Open Problems in Mathematics and Computational Science*, pages 5–36. Springer, 2014

[103] DAVID KAHN. *The Codebreakers: The Comprehensive History of Secret Communication from Ancient Times to the Internet*. Scribner, 1996.

[104] JONATHAN KATZ AND YEHUDA LINDELL. *Introduction to Modern Cryptography, Second Edition*. Chapman and Hall/CRC, 2014.

[105] AVIAD KIPNIS, JACQUES PATARIN, AND LOUIS GOUBIN. Unbalanced oil and vinegar signature schemes. *Lecture Notes in Computer Science*, **1592** (1999), 206–222. (EUROCRYPT '99.)

[106] RUDOLF KIPPENHAHN. *Code Breaking, A History and Exploration*. Overlook Press, 1999.

[107] ANDREAS KLEIN. *Stream Ciphers*. Springer, 2013.

[108] THORSTEN KLEINJUNG, KAZUMARO AOKI, JENS FRANKE, ARJEN LENSTRA, EMMANUEL THOMÉ, JOPPE BOS, PIERRICK GAUDRY, ALEXANDER KRUPPA, PETER MONTGOMERY, DAG ARNE OSVIK, HERMAN TE RIELE, ANDREY TIMOFEEV, AND PAUL ZIMMERMANN. Factorization of a 768-Bit RSA modulus. *Lecture Notes in Computer Science*, **6223** (2010), 333–350. (CRYPTO 2010.)

[109] LARS R. KNUDSEN AND MATTHEW ROBSHAW. *The Block Cipher Companion*. Springer, 2011.

[110] DONALD E. KNUTH. *The Art of Computer Programming, Volume 2, Seminumerical Algorithms, Second Edition*. Addison-Wesley, 1998.

[111] NEAL KOBLITZ. *A Course in Number Theory and Cryptography, Second Edition*. Springer, 1994.

[112] NEAL KOBLITZ. Elliptic curve cryptosystems. *Mathematics of Computation*, **48** (1987), 203–209.

[113] NEAL KOBLITZ AND ALFRED MENEZES. Another look at HMAC. *Journal of Mathematical Cryptology* **7** (2013), 225–251.

[114] JOHN T. KOHL AND B. CLIFFORD NEUMAN. *The Kerberos Network Authentication Service (V5)*. Network Working Group Request for Comments 1510, 1993.

[115] JOHN T. KOHL, B. CLIFFORD NEUMAN, AND THEODORE Y. T'SO. The evolution of the Kerberos authentication system. In *Distributed Open Systems* pages 78–94. IEEE Computer Society Press, 1994.

[116] LOREN M. KOHNFELDER. *Towards a practical public-key cryptosystem*. Bachelor's Thesis, MIT, 1978.

[117] KAORU KUROSAWA, TOSHIYA ITO, AND MASASCHI TAKEUCHI. Public key cryptosystem using a reciprocal number with the same intractability as factoring a large number. *Cryptologia*, **12** (1988), 225–233.

[118] DAVID LAY, STEVEN LAY, AND JUDI MCDONALD. *Linear Algebra and Its Applications, 5th Edition.* Pearson, 2015.

[119] JOOYOUNG LEE AND DOUGLAS R. STINSON. A combinatorial approach to key predistribution for distributed sensor networks. *IEEE Wireless Communications and Networking Conference (WCNC 2005)*, vol. 2, pp. 1200–1205.

[120] JOOYOUNG LEE AND DOUGLAS R. STINSON. On the construction of practical key predistribution schemes for distributed sensor networks using combinatorial designs. *ACM Transactions on Information and System Security* **11-2** (2008), article No. 1, 35 pp.

[121] ARJEN LENSTRA AND HENDRIK LENSTRA, JR. (EDS.) *The Development of the Number Field Sieve. Lecture Notes in Mathematics*, vol. 1554. Springer, 1993.

[122] RUDOLF LIDL AND HARALD NIEDERREITER. *Finite Fields, Second Edition.* Cambridge University Press, 1997.

[123] JAN C. A. VAN DER LUBBE. *Basic Methods of Cryptography*. Cambridge, 1998.

[124] MICHAEL LUBY. *Pseudorandomness and Cryptographic Applications*. Princeton University Press, 1996.

[125] F. JESSIE MACWILLIAMS AND NEIL J. A. SLOANE. *The Theory of Error-correcting Codes*, North-Holland, 1977.

[126] MOXIE MARLINSPIKE AND TREVOR PERRIN (EDITOR). The X3DH key agreement protocol. Open Whisper Systems, November 4, 2016. `https://signal.org/docs/specifications/x3dh/`

[127] KEITH M. MARTIN. *Everyday Cryptography: Fundamental Principles and Applications, Second Edition.* Oxford University Press, 2017.

[128] MITSURU MATSUI. Linear cryptanalysis method for DES cipher. *Lecture Notes in Computer Science*, **765** (1994), 386–397. (EUROCRYPT '93.)

[129] TSUTOMU MATSUMOTO, YOUICHI TAKASHIMA, AND HIDEKI IMAI. On seeking smart public-key distribution systems. *Transactions of the IECE (Japan)*, **69** (1986), 99–106.

[130] TSUTOMU MATSUMOTO AND HIDEKI IMAI. Public quadratic polynomial-tuples for efficient signature-verification and message-encryption. *Lecture Notes in Computer Science*, **330** (1988), 419–453. (EUROCRYPT '88.)

[131] ROBERT MCELIECE. A public-key cryptosystem based on algebraic coding theory. DSN Progress Report **42-44** (1978), 114–116.

[132] WILLI MEIER AND OTHMAR STAFFELBACH. Fast correlation attacks on certain stream ciphers. *Journal of Cryptology* **1** (1989) 159–176.

[133] ALFRED. J. MENEZES, TATSUAKI OKAMOTO, AND SCOTT A. VANSTONE. Reducing elliptic curve logarithms to logarithms in a finite field. *IEEE Transactions on Information Theory*, **39** (1993), 1639–1646.

[134] ALFRED J. MENEZES, PAUL C. VAN OORSCHOT, AND SCOTT A. VANSTONE. *Handbook of Applied Cryptography*. CRC Press, 1996.

[135] RALPH MERKLE. Secure communications over insecure channels. *Communications of the ACM*, **21** (1978), 294–299.

[136] RALPH MERKLE. A certified digital signature. *Lecture Notes in Computer Science*, **435** (1990), 218–238. (CRYPTO '89.)

[137] RALPH MERKLE. One way hash functions and DES. *Lecture Notes in Computer Science*, **435** (1990), 428–446. (CRYPTO '89.)

[138] GARY MILLER. Riemann's hypothesis and tests for primality. *Journal of Computer and Systems Science*, **13** (1976), 300–317.

[139] VICTOR MILLER. Uses of elliptic curves in cryptography. *Lecture Notes in Computer Science*, **218** (1986), 417–426. (CRYPTO '85.)

[140] CHRIS MITCHELL, FRED PIPER, AND PETER WILD. Digital signatures. In *Contemporary Cryptology, The Science of Information Integrity*, pages 325–378. IEEE Press, 1992.

[141] JUDY MOORE. Protocol failures in cryptosystems. In *Contemporary Cryptology, The Science of Information Integrity*, pages 541–558. IEEE Press, 1992.

[142] MICHELE MOSCA. Cybersecurity in an era with quantum computers: will we be ready? IACR ePrint Archive, report # 2015/1075. `https://eprint.iacr.org/2015/1075.pdf`

[143] GARY MULLEN AND DANIEL PANARIO, EDS. *Handbook of Finite Fields*. Chapman and Hall/CRC, 2013.

[144] SATOSHI NAKAMOTO. Bitcoin: a peer-to-peer electronic cash system. White paper, October 31, 2008. `https://bitcoin.org/bitcoin.pdf`

[145] MONI NAOR AND ADI SHAMIR. Visual cryptography. *Lecture Notes in Computer Science*, **950** (1995), 1–12. (EUROCRYPT '94.)

[146] NATIONAL INSTITUTE OF STANDARDS AND TECHNOLOGY. *Data Encryption Standard (DES)*. Federal Information Processing Standard (FIPS) Publication 46-3, October 1999. (Withdrawn on May 19, 2005.)

[147] NATIONAL INSTITUTE OF STANDARDS AND TECHNOLOGY. *Digital Signature Standard*. Federal Information Processing Standard (FIPS) Publication 186-4, July 2013.

[148] NATIONAL INSTITUTE OF STANDARDS AND TECHNOLOGY. *Entity Authentication Using Public Key Cryptography*. Federal Information Processing Standard (FIPS) Publication 196, February 1997. (Withdrawn on October 19, 2015.)

[149] NATIONAL INSTITUTE OF STANDARDS AND TECHNOLOGY. *Advanced Encryption Standard*. Federal Information Processing Standard (FIPS) Publication 197, 2001.

[150] NATIONAL INSTITUTE OF STANDARDS AND TECHNOLOGY. *The Keyed-Hash Message Authentication Code (HMAC)*. Federal Information Processing Standard (FIPS) Publication 198-1, 2008.

[151] NATIONAL INSTITUTE OF STANDARDS AND TECHNOLOGY. *Secure Hash Standard (SHS)*. Federal Information Processing Standard (FIPS) Publication 180-4, 2015.

[152] NATIONAL INSTITUTE OF STANDARDS AND TECHNOLOGY. *SHA-3 Standard: Permutation-Based Hash and Extendable-Output Functions*. Federal Information Processing Standard (FIPS) Publication 202, 2015.

[153] VASILII NECHAEV. On the complexity of a deterministic algorithm for a discrete logarithm. *Math. Zametki*, **55** (1994), 91–101.

[154] ROGER NEEDHAM AND MICHAEL SCHROEDER. Using encryption for authentication in large networks of computers. *Communications of the ACM*, **21** (1978), 993–999.

[155] MICHAEL NIELSEN. How the Bitcoin protocol actually works. December 6, 2013. `http://www.michaelnielsen.org/ddi/how-the-bitcoin-protocol-actually-works/`

[156] PHONG NGUYEN AND IGOR SHPARLINSKI. The insecurity of the digital signature algorithm with partially known nonces. *Journal of Cryptology*, **15** (2002), 151–176.

[157] CHRISTOF PAAR AND JAN PELZL. *Understanding Cryptography: A Textbook for Students and Practitioners*. Springer, 2010.

[158] JACQUES PATARIN. Hidden fields equations (HFE) and isomorphisms of polynomials (IP): two new families of asymmetric algorithms. *Lecture Notes in Computer Science*, **1070** (1996), 33–48. (EUROCRYPT '96.)

[159] JACQUES PATARIN. The Oil and Vinegar signature scheme. Presented at the *Dagstuhl Workshop on Cryptography*, September 1997.

[160] RENÉ PERALTA. Simultaneous security of bits in the discrete log. *Lecture Notes in Computer Science*, **219** (1986), 62–72. (EUROCRYPT '85.)

[161] PASCAL PAILLIER. Public-Key cryptosystems based on composite degree residuosity classes. *Lecture Notes in Computer Science*, **1592** (1999), 223–238. (EUROCRYPT '99.)

[162] TREVOR PERRIN (EDITOR) AND MOXIE MARLINSPIKE. The double ratchet algorithm. Open Whisper Systems, November 20, 2016. `https://signal.org/docs/specifications/doubleratchet/`

[163] STEPHEN POHLIG AND MARTIN HELLMAN. An improved algorithm for computing logarithms over GF(p) and its cryptographic significance. *IEEE Transactions on Information Theory*, **24** (1978), 106–110.

[164] DAVID POINTCHEVAL AND JACQUES STERN. Security arguments for signature schemes and blind signatures. *Journal of Cryptology*, **13** (2000), 361–396.

[165] JOHN POLLARD. Monte Carlo methods for index computation (mod p). *Mathematics of Computation*, **32** (1978), 918–924.

[166] BART PRENEEL. The state of cryptographic hash functions. *Lecture Notes in Computer Science*, **1561** (1999), 158–182. (Lectures on Data Security.)

[167] MICHAEL RABIN. Digitized signatures and public-key functions as intractable as factorization. *MIT Laboratory for Computer Science Technical Report*, LCS/TR-212, 1979.

[168] MICHAEL RABIN. Probabilistic algorithms for testing primality. *Journal of Number Theory*, **12** (1980), 128–138.

[169] ODED REGEV. The learning with errors problem, In *25th IEEE Conference on Computational Complexity*, pages 191–204. IEEE, 2010.

[170] RONALD RIVEST. The MD4 message digest algorithm. *Lecture Notes in Computer Science*, **537** (1991), 303–311. (CRYPTO '90.)

[171] RONALD RIVEST. The MD5 message digest algorithm. Internet Network Working Group RFC 1321, April 1992.

[172] RONALD RIVEST, ADI SHAMIR, AND LEONARD ADLEMAN. A method for obtaining digital signatures and public key cryptosystems. *Communications of the ACM*, **21** (1978), 120–126.

[173] ARTO SALOMAA. *Public-Key Cryptography*. Springer, 1990.

[174] CLAUS SCHNORR. Efficient signature generation by smart cards. *Journal of Cryptology*, **4** (1991), 161–174.

[175] ADI SHAMIR. How to share a secret. *Communications of the ACM*, **22** (1979), 612–613.

[176] ADI SHAMIR. Identity-based cryptosystems and signature schemes. *Lecture Notes in Computer Science*, **196** (1985), 47–53. (CRYPTO '84.)

[177] CLAUDE E. SHANNON. A mathematical theory of communication. *Bell Systems Technical Journal*, **27** (1948), 379–423, 623–656.

[178] CLAUDE E. SHANNON. Communication theory of secrecy systems. *Bell Systems Technical Journal*, **28** (1949), 656–715.

[179] VICTOR SHOUP. Lower bounds for discrete logarithms and related problems. *Lecture Notes in Computer Science*, **1233** (1997), 256–266. (EUROCRYPT '97.)

[180] DAN SHUMOW AND NEILS FERGUSON. On the possibility of a back door in the NIST SP800-90 Dual Ec Prng. *CRYPTO 2007 Rump Session.* `http://rump2007.cr.yp.to/15-shumow.pdf`

[181] THOMAS SIEGENTHALER. Decrypting a class of stream ciphers using ciphertext only, *IEEE Transactions on Computers* **34** (1985), 81–85.

[182] GUSTAVUS J. SIMMONS. A survey of information authentication. In *Contemporary Cryptology, The Science of Information Integrity*, pages 379–419. IEEE Press, 1992.

[183] SIMON SINGH. *The Code Book: The Science Of Secrecy From Ancient Egypt To Quantum Cryptography.* Anchor Books, 2000.

[184] NIGEL SMART. The discrete logarithm problem on elliptic curves of trace one. *Journal of Cryptology*, **12** (1999), 193–196.

[185] NIGEL SMART. *Cryptography Made Simple.* Springer, 2015.

[186] CLAYTON D. SMITH. *Digital Signcryption.* Masters Thesis, Department of Combinatorics and Optimization, University of Waterloo, 2005.

[187] JEROME SOLINAS. Efficient arithmetic on Koblitz curves. *Designs, Codes and Cryptography*, **19** (2000), 195–249.

[188] ROBERT SOLOVAY AND VOLKER STRASSEN. A fast Monte Carlo test for primality. *SIAM Journal on Computing*, **6** (1977), 84–85.

[189] JESSICA STADDON, DOUGLAS R. STINSON, AND RUIZHONG WEI. Combinatorial properties of frameproof and traceability codes. *IEEE Transactions on Information Theory*, **47** (2001), 1042–1049.

[190] MICHAEL STEINER, GENE TSUDIK, AND MICHAEL WAIDNER. Diffie-Hellman key distribution extended to group communication. In *Proceedings of the 3rd ACM Conference on Computer and Communications Security*, pages 31–37. ACM Press, 1996.

[191] MARC STEVENS, ELIE BURSZTEIN, PIERRE KARPMAN, ANGE ALBERTINI, AND YARIK MARKOV. The first collision for full SHA-1. *Lecture Notes in Computer Science*, **10401** (2017), 570–596. (Crypto 2017, Part I.)

[192] DOUGLAS STINSON. Some observations on the theory of cryptographic hash functions. *Designs, Codes and Cryptography*, **38** (2006), 259–277.

[193] CHENGDONG TAO, ADAMA DIENE, SHAOHUA TANG, AND JINTAI DING. Simple matrix scheme for encryption. *Lecture Notes in Computer Science*, **7932** (2013), 231–242. (PQCrypto 2013.)

[194] EDLYN TESKE. On random walks for Pollard's rho method. *Mathematics of Computation*, **70** (2001), 809–825.

[195] SERGE VAUDENAY. Security flaws induced by CBC padding—Applications to SSL, IPSEC, WTLS .... *Lecture Notes in Computer Science*, **2332** (2002), 534–545. (EUROCRYPT 2002.)

[196] SERGE VAUDENAY. *A Classical Introduction to Cryptography: Applications for Communications Security*. Springer, 2005.

[197] DEBBY M. WALLNER, ERIC J. HARDER, AND RYAN C. AGEE. Key management for multicast: issues and architectures. *Internet Request for Comments* 2627, June, 1999.

[198] LAWRENCE WASHINGTON. *Elliptic Curves: Number Theory and Cryptography, Second Edition*. Chapman & Hall/CRC, 2008.

[199] MARK WEGMAN AND J. LAWRENCE CARTER. New hash functions and their use in authentication and set equality. *Journal of Computer and System Sciences*, **22** (1981), 265–279.

[200] DOMINIC WELSH. *Codes and Cryptography*. Oxford Science Publications, 1988.

[201] MICHAEL WIENER. Cryptanalysis of short RSA secret exponents. *IEEE Transactions on Information Theory*, **36** (1990), 553–558.

[202] HUGH WILLIAMS. A modification of the RSA public-key encryption procedure. *IEEE Transactions on Information Theory*, **26** (1980), 726–729.

[203] CHUNG KEI WONG AND SIMON S. LAM. Digital signatures for flows and multicasts. *IEEE/ACM Transactions on Networking*, **7** (1999), 502–513.

[204] TAO XIE, FANBAO LIU, AND DENGGUO FENG. Fast Collision Attack on MD5. IACR ePrint Archive, report # 2013/170.

[205] SONG YAN. *Cryptanalytic Attacks on RSA*. Springer, 2008.

[206] ANDREW YAO. Theory and applications of trapdoor functions. In *Proceedings of the 23rd Annual Symposium on the Foundations of Computer Science*, pages 80–91. IEEE Press, 1982.

[207] YULIANG ZHENG. Digital signcryption or how to achieve cost(signature & encryption) $\ll$ cost(signature) + cost(encryption). *Lecture Notes in Computer Science*, **1294** (1997), 165–179. (CRYPTO '97.)

[208] Discrete logarithm records. `https://en.wikipedia.org/wiki/Discrete_logarithm_records`

Index